AF725458

GRE

Math Bible

JEFF KOLBY

DERRICK VAUGHN

KUNDA VAMSIDHAR

NOVA PRESS

Additional educational titles from Nova Press (available at novapress.net):

ISBN 1–889057–49–5

NOVA PRESS
11659 Mayfield Ave., Suite 1
Los Angeles, CA 90049

Phone: 1-800-949-6175
E-mail: info@novapress.net
Website: www.novapress.net

ABOUT THIS BOOK

If you don't have a pencil in your hand, get one now! Don't just read this book—write on it, study it, scrutinize it! In short, for the next four weeks, this book should be a part of your life. When you have finished the book, it should be marked-up, dog-eared, tattered and torn.

Although the GRE is a difficult test, it is a *very* learnable test. This is not to say that the GRE is "beatable." There is no bag of tricks that will show you how to master it overnight. You probably have already realized this. Some books, nevertheless, offer "inside stuff" or "tricks" which they claim will enable you to beat the test. These include declaring that answer-choices B, C, or D are more likely to be correct than choices A or E. This tactic, like most of its type, does not work. It is offered to give the student the feeling that he or she is getting the scoop on the test.

The GRE cannot be "beaten." But it can be mastered—through hard work, analytical thought, and by training yourself to think like a test writer. Many of the exercises in this book are designed to prompt you to think like a test writer. For example, you will find "Duals." These are pairs of similar problems in which only one property is different. They illustrate the process of creating GRE questions.

The GRE math sections are not easy—nor is this book. To improve your GRE math score, you must be willing to work; if you study hard and master the techniques in this book, your score will improve—significantly.

This book will introduce you to numerous analytic techniques that will help you immensely, not only on the GRE but in graduate school as well. For this reason, studying for the GRE can be a rewarding and satisfying experience.

To insure that you perform at your expected level on the actual GRE, you need to develop a level of mathematical skill that is greater than what is tested on the GRE. Hence, about 10% of the math problems in this book (labeled "Very Hard") are harder than actual GRE math problems.

Although the quick-fix method is not offered in this book, about 15% of the material is dedicated to studying how the questions are constructed. Knowing how the problems are written and how the test writers think will give you useful insight into the problems and make them less mysterious. Moreover, familiarity with the GRE's structure will help reduce your anxiety. The more you know about this test, the less anxious you will be the day you take it.

CONTENTS

ORIENTATION

Format of the Math Sections

The math section consists of three types of questions: *Quantitative Comparisons, Standard Multiple Choice,* and *Graphs.* They are designed to test your ability to solve problems, not to test your mathematical knowledge.

The math section is 45 minutes long and contains 28 questions. The questions can appear in any order.

FORMAT
About 14 Quantitative Comparisons
About 9 Standard Multiple Choice
About 5 Graphs

Level of Difficulty

GRE math is very similar to SAT math, though surprisingly slightly easier. The mathematical skills tested are very basic: only first year high school algebra and geometry (no proofs). However, this does not mean that the math section is easy. The medium of basic mathematics is chosen so that everyone taking the test will be on a fairly even playing field. This way, students who majored in math, engineering, or science don't have an undue advantage over students who majored in humanities. Although the questions require only basic mathematics and **all** have **simple** solutions, it can require considerable ingenuity to find the simple solution. If you have taken a course in calculus or another advanced math topic, don't assume that you will find the math section easy. Other than increasing your mathematical maturity, little you learned in calculus will help on the GRE.

Quantitative comparisons are the most common math questions. This is good news since they are mostly intuitive and require little math. Further, they are the easiest math problems on which to improve since certain techniques—such as substitution—are very effective.

As mentioned above, every GRE math problem has a simple solution, but finding that simple solution may not be easy. The intent of the math section is to test how skilled you are at finding the simple solutions. The premise is that if you spend a lot of time working out long solutions you will not finish as much of the test as students who spot the short, simple solutions. So, if you find yourself performing long calculations or applying advanced mathematics—stop. You're heading in the wrong direction.

Experimental Section

The GRE is a standardized test. Each time it is offered, the test has, as close as possible, the same level of difficulty as every previous test. Maintaining this consistency is very difficult—hence the experimental section. The effectiveness of each question must be assessed before it can be used on the GRE. A problem that one person finds easy another person may find hard, and vice versa. The experimental section measures the relative difficulty of potential questions; if responses to a question do not perform to strict specifications, the question is rejected.

The experimental section can be a verbal section or a math section. You won't know which section is experimental. You will know which type of section it is, though, since there will be an extra one of that type.

Because the "bugs" have not been worked out of the experimental section—or, to put it more directly, because you are being used as a guinea pig to work out the "bugs"—this portion of the test is often more difficult and confusing than the other parts.

This brings up an ethical issue: How many students have run into the experimental section early in the test and have been confused and discouraged by it? Crestfallen by having done poorly on, say, the first—though experimental—section, they lose confidence and perform below their ability on the rest of the test. Some testing companies are becoming more enlightened in this regard and are administering experimental sections as separate practice tests. Unfortunately, ETS has yet to see the light.

Knowing that the experimental section can be disproportionately difficult, if you do poorly on a particular section you can take some solace in the hope that it may have been the experimental section. In other words, do not allow one difficult section to discourage your performance on the rest of the test.

Research Section

You may also see a research section. This section, if it appears, will be identified and will be last. The research section will not be scored and will not affect your score on other parts of the test.

The CAT & the Old Paper-&-Pencil Test

The computer based GRE uses the same type of questions as the old paper-&-pencil test. The only difference is the medium, that is the way the questions are presented.

There are advantages and disadvantages to the CAT. Probably the biggest advantages are that you can take the CAT just about any time and you can take it in a small room with just a few other people—instead of in a large auditorium with hundreds of other stressed people. One the other hand, you cannot return to previously answered questions, it is easier to misread a computer screen than it is to misread printed material, and it can be distracting looking back and forth from the computer screen to your scratch paper.

Pacing

Although time is limited on the GRE, working too quickly can damage your score. Many problems hinge on subtle points, and most require careful reading of the setup. Because undergraduate school puts such heavy reading loads on students, many will follow their academic conditioning and read the questions quickly, looking only for the gist of what the question is asking. Once they have found it, they mark their answer and move on, confident they have answered it correctly. Later, many are startled to discover that they missed questions because they either misread the problems or overlooked subtle points.

To do well in your undergraduate classes, you had to attempt to solve every, or nearly every, problem on a test. Not so with the GRE. In fact, if you try to solve every problem on the test, you will probably damage your score. For the vast majority of people, the key to performing well on the GRE is not the number of questions they solve, within reason, but the percentage they solve correctly.

On the GRE, the first question will be of medium difficulty. If you answer it correctly, the next question will be a little harder. If you answer it incorrectly, the next question will be a little easier. Because the CAT "adapts" to your performance, early questions are more important than later ones. In fact, by about the fifth or sixth question the test believes that it has a general measure of your score, say, 500–600. The rest of the test is determining whether your score should be, say, 550 or 560. Because of the importance of the first five questions to your score, you should read and solve these questions slowly and carefully. Allot nearly one-third of the time for each section to the first five questions. Then work progressively faster as you work toward the end of the section.

Scoring the GRE

The three major parts of the test are scored independently. You will receive a verbal score, a math score, and a writing score. The verbal and math scores range from 200 to 800. The writing score is on a scale from 0 to 6. In addition to the scaled score, you will be assigned a percentile ranking, which gives the percentage of students with scores below yours. The following table relates the scaled scores to the percentile ranking.

Scaled Score	Verbal	Math
800	99	99
700	97	80
600	84	58
500	59	35
400	26	15
300	5	3

The following table lists the average scaled scores. Notice how much higher the average score for math is than for verbal. Even though the math section intimidates most people, it is very learnable. The verbal section is also very learnable, but it takes more work to master it.

Average Scaled Score

Verbal	Math	Total
470	570	1040

Skipping and Guessing

On the test, you cannot skip questions; each question must be answered before moving to the next question. However, if you can eliminate even one of the answer-choices, guessing can be advantageous. Unfortunately, you cannot return to previously answered questions.

On the test, your first question will be of medium difficulty. If you answer it correctly, the next question will be a little harder. If you again answer it correctly, the next question will be harder still, and so on. If your GRE skills are strong and you are not making any mistakes, you should reach the medium-hard or hard problems by about the fifth problem. Although this is not very precise, it can be quite helpful. Once you have passed the fifth question, you should be alert to subtleties in any seemingly simple problems.

Often students become obsessed with a particular problem and waste time trying to solve it. To get a top score, learn to cut your losses and move on. The exception to this rule is the first five questions of each section. Because of the importance of the first five questions to your score, you should read and solve these questions slowly and carefully.

If you are running out of time, randomly guess on the remaining questions. This is unlikely to harm your score. In fact, if you do not obsess about particular questions (except for the first five), you probably will have plenty of time to solve a sufficient number of questions.

Because the total number of questions answered contributes to the calculation of your score, you should answer ALL the questions—even if this means guessing randomly before time runs out.

The Structure of this Book

Because it can be rather dull to spend a lot of time reviewing basic math before tackling full-fledged GRE problems, the first few chapters present techniques that don't require much foundational knowledge of mathematics. Then, in latter chapters, review is introduced as needed.

The problems in the exercises are ranked Easy, Medium, Hard, and Very Hard. This helps you to determine how well you are prepared for the test.

Directions and Reference Material

Be sure you understand the directions below so that you do not need to read or interpret them during the test.

Directions

Solve each problem and decide which one of the choices given is best. Fill in the corresponding circle on your answer sheet. You can use any available space for scratchwork.

Notes

1. All numbers used are real numbers.
2. Figures are intended to provide information useful in answering the questions. However, unless a note states that a figure is drawn to scale, you should not solve these problems by estimating sizes by sight or by measurement.
3. All figures lie in a plane unless otherwise indicated. Position of points, angles, regions, etc. can be assumed to be in the order shown; and angle measures can be assumed to be positive.

Note 1 indicates that complex numbers, $i = \sqrt{-1}$, do not appear on the test.

Note 2 indicates that figures are not drawn accurately. Hence, an angle that appears to be 90° may not be or an object that appears congruent to another object may not be.

Note 3 indicates that two-dimensional figures do not represent three-dimensional objects. That is, the drawing of a circle is not representing a sphere, and the drawing of a square is not representing a cube.

Reference Information

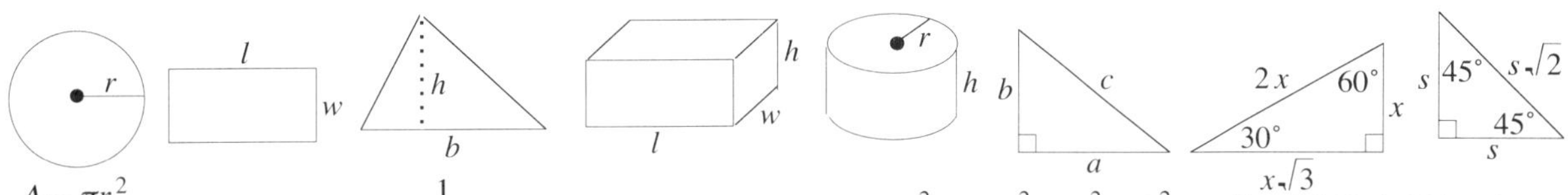

$A = \pi r^2$ $C = 2\pi r$ $A = lw$ $A = \frac{1}{2}bh$ $V = lwh$ $V = \pi r^2 h$ $c^2 = a^2 + b^2$ Special Right Triangles

The number of degrees of arc in a circle is 360.
The sum of the measures in degrees of the angles of a triangle is 180.

Although this reference material can be handy, be sure you know it well so that you do not waste time looking it up during the test.

Part One

MATH

Substitution

Substitution is a very useful technique for solving GRE math problems. It often reduces hard problems to routine ones. In the substitution method, we choose numbers that have the properties given in the problem and plug them into the answer-choices. A few examples will illustrate.

Example 1: If n is an even integer, which one of the following is an odd integer?

(A) n^2

(B) $\dfrac{n+1}{2}$

(C) $-2n-4$

(D) $2n^2-3$

(E) $\sqrt{n^2+2}$

We are told that n is an even integer. So, choose an even integer for n, say, 2 and substitute it into each answer-choice. Now, n^2 becomes $2^2 = 4$, which is not an odd integer. So eliminate (A). Next, $\dfrac{n+1}{2} = \dfrac{2+1}{2} = \dfrac{3}{2}$ is not an odd integer—eliminate (B). Next, $-2n-4=-2\cdot 2-4=-4-4=-8$ is not an odd integer—eliminate (C). Next, $2n^2-3=2(2)^2-3=2(4)-3=8-3=5$ is odd and hence the answer is possibly (D). Finally, $\sqrt{n^2+2}=\sqrt{2^2+2}=\sqrt{4+2}=\sqrt{6}$, which is not odd—eliminate (E). The answer is (D).

- When using the substitution method, be sure to check every answer-choice because the number you choose may work for more than one answer-choice. If this does occur, then choose another number and plug it in, and so on, until you have eliminated all but the answer. This may sound like a lot of computing, but the calculations can usually be done in a few seconds.

Example 2: If n is an integer, which of the following CANNOT be an integer?

(A) $\dfrac{n-2}{2}$

(B) $\sqrt{n}$

(C) $\dfrac{2}{n+1}$

(D) $\sqrt{n^2+3}$

(E) $\sqrt{\dfrac{1}{n^2+2}}$

Choose n to be 0. Then $\dfrac{n-2}{2}=\dfrac{0-2}{2}=\dfrac{-2}{2}=-1$, which is an integer. So eliminate (A). Next, $\sqrt{n}=\sqrt{0}=0$. Eliminate (B). Next, $\dfrac{2}{n+1}=\dfrac{2}{0+1}=\dfrac{2}{1}=2$. Eliminate (C). Next, $\sqrt{n^2+3}=\sqrt{0^2+3}=\sqrt{0+3}=\sqrt{3}$, which is *not* an integer—it *may* be our answer. However, $\sqrt{\dfrac{1}{n^2+2}}=\sqrt{\dfrac{1}{0^2+2}}=\sqrt{\dfrac{1}{0+2}}=\sqrt{\dfrac{1}{2}}$, which is *not* an integer as well. So, we choose another number, say, 1. Then $\sqrt{n^2+3}=\sqrt{1^2+3}=\sqrt{1+3}=\sqrt{4}=2$, which is an integer, eliminating (D). Thus, choice (E), $\sqrt{\dfrac{1}{n^2+2}}$, is the answer.

Example 3: If x, y, and z are positive integers such that $x < y < z$ and $x + y + z = 6$, then what is the value of z?

(A) 1
(B) 2
(C) 3
(D) 4
(E) 5

From the given inequality $x < y < z$, it is clear that the positive integers x, y, and z are different and are in the increasing order of size.

Assume $x > 1$. Then $y > 2$ and $z > 3$. Adding the inequalities yields $x + y + z > 6$. This contradicts the given equation $x + y + z = 6$. Hence, the assumption $x > 1$ is false. Since x is a positive integer, x must be 1.

Next, assume $y > 2$. Then $z > 3$ and $x + y + z = 1 + y + z > 1 + 2 + 3 = 6$, so $x + y + z > 6$. This contradicts the given equation $x + y + z = 6$. Hence, the assumption $y > 2$ is incorrect. Since we know y is a positive integer and greater than x (= 1), y must be 2.

Now, the substituting known values in equation $x + y + z = 6$ yields $1 + 2 + z = 6$, or $z = 3$. The answer is (C).

Method II (without substitution):
We have the inequality $x < y < z$ and the equation $x + y + z = 6$. Since x is a positive integer, $x \ge 1$. From the inequality $x < y < z$, we have two inequalities: $y > x$ and $z > y$. Applying the first inequality ($y > x$) to the inequality $x \ge 1$ yields $y \ge 2$ (since y is also a positive integer, given); and applying the second inequality ($z > y$) to the second inequality $y \ge 2$ yields $z \ge 3$ (since z is also a positive integer, given). Summing the inequalities $x \ge 1$, $y \ge 2$, and $z \ge 3$ yields $x + y + z \ge 6$. But we have $x + y + z = 6$, exactly. This happens only when $x = 1$, $y = 2$, and $z = 3$ (not when $x > 1$, $y > 2$, and $z > 3$). Hence, $z = 3$, and the answer is (C).

Problem Set A: Solve the following problems by using substitution.

➢ Easy

1. By how much is the greatest of five consecutive even integers greater than the smallest among them?

(A) 1
(B) 2
(C) 4
(D) 8
(E) 10

➢ Medium

2. Which one of the following could be an integer?

(A) Average of two consecutive integers.
(B) Average of three consecutive integers.
(C) Average of four consecutive integers.
(D) Average of six consecutive integers.
(E) Average of 6 and 9.

3. (The average of five consecutive integers starting from m) – (the average of six consecutive integers starting from m) =

   (A) –1/4
   (B) –1/2
   (C) 0
   (D) 1/2
   (E) 1/4

➢ Hard

4. The remainder when the positive integer m is divided by n is r. What is the remainder when $2m$ is divided by $2n$?

   (A) r
   (B) $2r$
   (C) $2n$
   (D) $m - nr$
   (E) $2(m - nr)$

5. If $1 < p < 3$, then which of the following could be true?

   (I) $p^2 < 2p$
   (II) $p^2 = 2p$
   (III) $p^2 > 2p$

   (A) I only
   (B) II only
   (C) III only
   (D) I and II only
   (E) I, II, and III

6. If $42.42 = k(14 + m/50)$, where k and m are positive integers and $m < 50$, then what is the value of $k + m$?

   (A) 6
   (B) 7
   (C) 8
   (D) 9
   (E) 10

7. If p and q are both positive integers such that $p/9 + q/10$ is also an integer, then which one of the following numbers could p equal?

   (A) 3
   (B) 4
   (C) 9
   (D) 11
   (E) 19

Answers and Solutions to Problem Set A

➢ Easy

1. Choose any 5 consecutive even integers—say—2, 4, 6, 8, 10. The largest in this group is 10, and the smallest is 2. Their difference is 10 – 2 = 8. The answer is (D).

➢ Medium

2. Choose any three consecutive integers, say, 1, 2, and 3. Forming their average yields $\frac{1+2+3}{3}=\frac{6}{3}=2$. Since 2 is an integer, the answer is (B).

Method II (without substitution):

Choice (A): Let a and $a + 1$ be the consecutive integers. The average of the two is $\frac{a+(a+1)}{2}=\frac{2a+1}{2}=a+\frac{1}{2}$, certainly not an integer since a is an integer. Reject.

Choice (B): Let a, $a + 1$, and $a + 2$ be the three consecutive integers. The average of the three numbers is $\frac{a+(a+1)+(a+2)}{3}=\frac{3a+3}{3}=a+1$, certainly an integer since a is an integer. Correct.

Choice (C): Let a, $a + 1$, $a + 2$, and $a + 3$ be the four consecutive integers. The average of the four numbers is $\frac{a+(a+1)+(a+2)+(a+3)}{4}=\frac{4a+6}{4}=a+\frac{3}{2}$, certainly not an integer since a is an integer. Reject.

Choice (D): Let a, $a + 1$, $a + 2$, $a + 3$, $a + 4$, and $a + 5$ be the six consecutive integers. The average of the six numbers is $\frac{a+(a+1)+(a+2)+(a+3)+(a+4)+(a+5)}{6}=\frac{6a+15}{6}=a+\frac{5}{2}$, certainly not an integer since a is an integer. Reject.

Choice (E): The average of 6 and 9 is $\frac{6+9}{2}=\frac{15}{2}=7.5$, not an integer. Reject.

The answer is (B).

3. Choose any five consecutive integers, say, –2, –1, 0, 1 and 2. (We chose these particular numbers to make the calculation as easy as possible. But any five consecutive integers will do. For example, 1, 2, 3, 4, and 5.) Forming the average yields (–1 + (–2) + 0 + 1 + 2)/5 = 0/5 = 0. Now, add 3 to the set to form 6 consecutive integers: –2, –1, 0, 1, 2, and 3. Forming the average yields

$\frac{-1+(-2)+0+1+2+3}{6}=$

$\frac{[-1+(-2)+0+1+2]+3}{6}=$

$\frac{[0]+3}{6}=$ since the average of –1 + (–2) + 0 + 1 + 2 is zero, their sum must be zero

3/6 =

1/2

(The average of five consecutive integers starting from m) – (The average of six consecutive integers starting from m) = (0) – (1/2) = –1/2.

The answer is (B).

Method II (without substitution):
The five consecutive integers starting from m are $m, m + 1$, $m + 2$, $m + 3$, and $m + 4$. The average of the five numbers equals

$$\frac{\text{the sum of the five numbers}}{5} =$$

$$\frac{m + (m+1) + (m+2) + (m+3) + (m+4)}{5} =$$

$$\frac{5m+10}{5} =$$

$$m + 2$$

The average of six consecutive integers starting from m are m, $m + 1$, $m + 2$, $m + 3$, $m + 4$, and $m + 5$. The average of the six numbers equals

$$\frac{\text{the sum of the six numbers}}{6} =$$

$$\frac{m + (m+1) + (m+2) + (m+3) + (m+4) + (m+5)}{6} =$$

$$\frac{6m+15}{6} =$$

$$m + 5/2 =$$

$$m + 2 + 1/2 =$$

$$(m + 2) + 1/2$$

(The average of five consecutive integers starting from m) – (The average of six consecutive integers starting from m) = $(m + 2) - [(m + 2) + 1/2] = -1/2$.

The answer is (B).

➢ Hard

4. As a particular case, suppose $m = 7$ and $n = 4$. Then $m/n = 7/4 = 1 + 3/4$. Here, the remainder r equals 3.

Now, $2m = 2 \cdot 7 = 14$ and $2n = 2 \cdot 4 = 8$. Hence, $2m/2n = 14/8 = 1 + 6/8$. Here, the remainder is 6. Now, let's choose the answer-choice that equals 6.

Choice (A): $r = 3 \neq 6$. Reject.
Choice (B): $2r = 2 \cdot 3 = 6$. Possible answer.
Choice (C): $2n = 2 \cdot 4 = 8 \neq 6$. Reject.
Choice (D): $m - nr = 7 - 4 \cdot 3 = -5 \neq 6$. Reject.
Choice (E): $2(m - nr) = 2(7 - 4 \cdot 3) = 2(-5) = -10 \neq 6$. Reject.

Hence, the answer is (B).

Method II (without substitution):
Since the remainder when m is divided by n is r, we can represent m as $m = kn + r$, where k is some integer. Now, $2m$ equals $2kn + 2r$. Hence, dividing $2m$ by $2n$ yields $2m/2n = (2kn + 2r)/2n = k + 2r/2n$. Since we are dividing by $2n$ (not by n), the remainder when divided by $2n$ is $2r$. The answer is (B).

5. If $p = 3/2$, then $p^2 = (3/2)^2 = 9/4 = 2.25$ and $2p = 2 \cdot 3/2 = 3$. Hence, $p^2 < 2p$, I is true, and clearly II ($p^2 = 2p$) and III ($p^2 > 2p$) are both false. This is true for all $1 < p < 2$.

If $p = 2$, then $p^2 = 2^2 = 4$ and $2p = 2 \cdot 2 = 4$. Hence, $p^2 = 2p$, II is true, and clearly I ($p^2 < 2p$) and III ($p^2 > 2p$) are both false.

If $p = 5/2$, then $p^2 = (5/2)^2 = 25/4 = 6.25$ and $2p = 2 \cdot 5/2 = 5$. Hence, $p^2 > 2p$, III is true, and clearly I ($p^2 < 2p$) and II($p^2 = 2p$) are both false. This is true for any $2 < p < 3$.

Hence, exactly one of the three choices I, II, and III is true simultaneously (for a given value of p). The answer is (E).

6. We are given that k is a positive integer and m is a positive integer less than 50. We are also given that $42.42 = k(14 + m/50)$.

Suppose $k = 1$. Then $k(14 + m/50) = 14 + m/50 = 42.42$. Solving for m yields $m = 50(42.42 - 14) = 50 \times 28.42$, which is not less than 50. Hence, $k \neq 1$.

Now, suppose $k = 2$. Then $k(14 + m/50) = 2(14 + m/50) = 42.42$, or $(14 + m/50) = 21.21$. Solving for m yields $m = 50(21.21 - 14) = 50 \times 7.21$, which is not less than 50. Hence, $k \neq 2$.

Now, suppose $k = 3$. Then $k(14 + m/50) = 3(14 + m/50) = 42.42$, or $(14 + m/50) = 14.14$. Solving for m yields $m = 50(14.14 - 14) = 50 \times 0.14 = 7$, which is less than 50. Hence, $k = 3$ and $m = 7$ and $k + m = 3 + 7 = 10$.

The answer is (E).

7. If p is not divisible by 9 and q is not divisible by 10, then $p/9$ results in a non-terminating decimal and $q/10$ results in a terminating decimal and the sum of the two would not result in an integer. [Because (a terminating decimal) + (a non-terminating decimal) is always a non-terminating decimal, and a non-terminating decimal is not an integer.]

Since we are given that the expression is an integer, p must be divisible by 9.

For example, if $p = 1$ and $q = 10$, the expression equals $1/9 + 10/10 = 1.11...$, not an integer.

If $p = 9$ and $q = 5$, the expression equals $9/9 + 5/10 = 1.5$, not an integer.

If $p = 9$ and $q = 10$, the expression equals $9/9 + 10/10 = 2$, an integer.

In short, p must be a positive integer divisible by 9. The answer is (C).

Substitution (Quantitative Comparisons): When substituting in quantitative comparison problems, don't rely on only positive whole numbers. You must also check negative numbers, fractions, 0, and 1 because they often give results different from those of positive whole numbers. Plug in the numbers 0, 1, 2, –2, and 1/2, in that order.

Example 1: Determine which of the two expressions below is larger, whether they are equal, or whether there is not enough information to decide. [The answer is (A) if Column A is larger, (B) if Column B is larger, (C) if the columns are equal, and (D) if there is not enough information to decide.]

Column A	$x \neq 0$	Column B
x		x^2

If $x = 2$, then $x^2 = 4$. In this case, Column B is larger. However, if x equals 1, then $x^2 = 1$. In this case, the two columns are equal. Hence, the answer is (D)—not enough information to decide.

Note! If, as above, you get a certain answer when a particular number is substituted and a different answer when another number is substituted (Double Case), then the answer is (D)—not enough information to decide.

Example 2: Let $\boxed{x}$ denote the greatest integer less than or equal to x. For example: $\boxed{5.5} = 5$ and $\boxed{3} = 3$. Now, which column below is larger?

Column A	$x \geq 0$	Column B
$\boxed{\sqrt{x}}$		x

If $x = 0$, then $\boxed{\sqrt{x}} = \boxed{\sqrt{0}} = \boxed{0} = 0$. In this case, Column A equals Column B. Now, if $x = 1$, then $\boxed{\sqrt{x}} = \boxed{\sqrt{1}} = 1$. In this case, the two columns are again equal. But if $x = 2$, then $\boxed{\sqrt{x}} = \boxed{\sqrt{2}} = 1$. Thus, in this case Column B is larger. This is a double case. Hence, the answer is (D)—not enough information to decide.

Problem Set B: Solve the following quantitative comparison problems by plugging in the numbers 0, 1, 2, –2, and 1/2 in that order—when possible.

➢ **Easy**

1.

Column A	$x > 0$	Column B
$x^2 + 2$		$x^3 - 2$

2.

Column A	$m > 0$	Column B
m^{10}		m^{100}

➢ **Medium**

3.

Column A	$x < 0$	Column B
$x^2 - x^5$		0

4.

Column A	$-1 < x < 0$	Column B
x		$1/x$

5.

Column A		Column B
ab^2		a^2b

6.

Column A	$y \neq 0$	Column B
x/y		xy

7.

Column A	$a < 0$	Column B
$1/a$		a

8.

Column A	$x = y \neq 0$	Column B
0		x/y

9. For all numbers x, $\boxed{x}$ denotes the value of x^3 rounded to the nearest multiple of ten.

Column A	Column B
$\boxed{x+1}$	$\boxed{x} + 1$

10.

Column A	$0 < x < 2$	Column B
x^2		$\sqrt{x}$

11.

Column A	$x > y > 0$	Column B
$x - y$		$\frac{x}{3} + \frac{y}{3}$

Note! In quantitative comparison problems, answer-choice (D), "not enough information," is as likely to be the answer as are choices (A), (B), or (C).

Answers and Solutions to Problem Set B

➢ Easy

1. Since $x > 0$, we need only look at $x = 1$, 2, and 1/2. If $x = 1$, then $x^2 + 2 = 3$ and $x^3 - 2 = -1$. In this case, Column A is larger. Next, if $x = 2$, then $x^2 + 2 = 6$ and $x^3 - 2 = 6$. In this case, the two columns are equal. This is a double case and therefore the answer is (D).

2. If $m = 1$, then $m^{10} = 1$ and $m^{100} = 1$. In this case, the two columns are equal. Next, if $m = 2$, then clearly m^{100} is greater than m^{10}. This is a double case, and the answer is (D).

➢ Medium

3. If $x = -1$, then $x^2 - x^5 = 2$ and Column A is larger. If $x = -2$, then $x^2 - x^5 = (-2)^2 - (-2)^5 = 4 + 32 =$ 36 and Column A is again larger. Finally, if $x = 1/2$, then $x^2 - x^5 = \frac{1}{4} + \frac{1}{32} = \frac{9}{32}$ and Column A is still larger. This covers the three types of negative numbers, so we can confidently conclude the answer is (A).

4. There is only one type of number between –1 and 0—negative fractions. So we need only choose one number, say, $x = -1/2$. Then $\frac{1}{x} = \frac{1}{-\frac{1}{2}} = -2$. Now, –1/2 is larger than –2 (since –1/2 is to the right of –2 on the number line). Hence, Column A is larger, and the answer is (A).

5. If $a = 0$, both columns equal zero. If $a = 1$ and $b = 2$, the two columns are unequal. This is a double case and the answer is (D).

6. If $x = y = 1$, then both columns equal 1. If $x = y = 2$, then $x/y = 1$ and $xy = 4$. In this case, the columns are unequal. The answer is (D).

7. If $a = -1$, both columns equal –1. If $a = -2$, the columns are unequal. The answer is (D).

8. If x and y are positive, then Column B is positive and hence larger than zero. If x and y are negative, then Column B is still positive since a negative divided by a negative yields a positive. This covers all possible signs for x and y. The answer is (B).

9. Suppose $x = 0$. Then $\boxed{x+1} = \boxed{0+1} = \boxed{1} = 0$,* and $\boxed{x} + 1 = \boxed{0} + 1 = 0 + 1 = 1$. In this case, Column B is larger. Next, suppose $x = 1$. Then $\boxed{x+1} = \boxed{1+1} = \boxed{2} = 10$, and $\boxed{x} + 1 = \boxed{1} + 1 = 0 + 1 = 1$. In this case, Column A is larger. The answer is (D).

10. If $x = 1$, then $x^2 = 1^2 = 1 = \sqrt{1} = \sqrt{x}$. In this case, the columns are equal. If $x = 1/2$, then $x^2 = \left(\frac{1}{2}\right)^2 =$ $\frac{1}{4} \neq \sqrt{\frac{1}{2}} = \sqrt{x}$. In this case, the columns are not equal and therefore the answer is (D).

11. If $x = 2$ and $y = 1$, then $x - y = 2 - 1 = 1 = \frac{3}{3} = \frac{2}{3} + \frac{1}{3} = \frac{x}{3} + \frac{y}{3}$. In this case, the columns are equal. If $x = 3$ and $y = 1$, then $x - y = 3 - 1 = 2 \neq \frac{3}{3} + \frac{1}{3} = \frac{4}{3} = \frac{x}{3} + \frac{y}{3}$. In this case, the columns are not equal and therefore the answer is (D).

---

* $\boxed{1} = 0$ because 0 is a multiple of 10: $0 = 0 \cdot 10$.

Substitution (Plugging In): Sometimes instead of making up numbers to substitute into the problem, we can use the actual answer-choices. This is called "Plugging In." It is a very effective technique, but not as common as Substitution.

Example 1: If $(a - b)(a + b) = 7 \times 13$, then which one of the following pairs could be the values of a and b, respectively?

(A) 7, 13
(B) 5, 15
(C) 3, 10
(D) −10, 3
(E) −3, −10

Substitute the values for a and b shown in the answer-choices into the expression $(a - b)(a + b)$:

Choice (A): $(7 - 13)(7 + 13) = -6 \times 20$

Choice (B): $(5 - 15)(5 + 15) = -10 \times 20$

Choice (C): $(3 - 10)(3 + 10) = -7 \times 13$

Choice (D): $(-10 - 3)(-10 + 3) = -13 \times (-7) = 7 \times 13$

Choice (E): $(-3 - (-10))(-3 + (-10)) = 7 \times (-13)$

Since only choice (D) equals the product 7×13, the answer is (D).

Example 2: If $a^3 + a^2 - a - 1 = 0$, then which one of the following could be the value of a?

(A) 0
(B) 1
(C) 2
(D) 3
(E) 4

Let's test which answer-choice satisfies the equation $a^3 + a^2 - a - 1 = 0$.

Choice (A): $a = 0$. $a^3 + a^2 - a - 1 = 0^3 + 0^2 - 0 - 1 = -1 \neq 0$. Reject.
Choice (B): $a = 1$. $a^3 + a^2 - a - 1 = 1^3 + 1^2 - 1 - 1 = 0$. Correct.
Choice (C): $a = 2$. $a^3 + a^2 - a - 1 = 2^3 + 2^2 - 2 - 1 = 9 \neq 0$. Reject.
Choice (D): $a = 3$. $a^3 + a^2 - a - 1 = 3^3 + 3^2 - 3 - 1 = 32 \neq 0$. Reject.
Choice (E): $a = 4$. $a^3 + a^2 - a - 1 = 4^3 + 4^2 - 4 - 1 = 75 \neq 0$. Reject.

The answer is (B).

Method II (This problem can also be solved by factoring.)

$$a^3 + a^2 - a - 1 = 0$$
$$a^2(a + 1) - (a + 1) = 0$$
$$(a + 1)(a^2 - 1) = 0$$
$$(a + 1)(a + 1)(a - 1) = 0$$
$$a + 1 = 0 \text{ or } a - 1 = 0$$

Hence, $a = 1$ or -1. The answer is (B).

Problem Set C:

Use the method of Plugging In to solve the following problems.

➤ Easy

1. If $(x - 3)(x + 2) = (x - 2)(x + 3)$, then $x =$

   (A) 3
   (B) –2
   (C) 0
   (D) 2
   (E) 3

2. Which one of the following is the solution of the system of equations given?

$$x + 2y = 7$$
$$x + y = 4$$

   (A) $x = 3, y = 2$
   (B) $x = 2, y = 3$
   (C) $x = 1, y = 3$
   (D) $x = 3, y = 1$
   (E) $x = 7, y = 1$

➤ Medium

3. If $x^2 + 4x + 3$ is odd, then which one of the following could be the value of x ?

   (A) 3
   (B) 5
   (C) 9
   (D) 13
   (E) 16

4. If $(2x + 1)^2 = 100$, then which one of the following COULD equal x ?

   (A) –11/2
   (B) –9/2
   (C) 11/2
   (D) 13/2
   (E) 17/2

➤ Hard

5. The number m yields a remainder p when divided by 14 and a remainder q when divided by 7. If $p = q + 7$, then which one of the following could be the value of m ?

   (A) 45
   (B) 53
   (C) 72
   (D) 85
   (E) 100

Answers and Solutions to Problem Set C

➤ Easy

1. If $x = 0$, then the equation $(x - 3)(x + 2) = (x - 2)(x + 3)$ becomes

$$(0 - 3)(0 + 2) = (0 - 2)(0 + 3)$$

$$(-3)(2) = (-2)(3)$$

$$-6 = -6$$

The answer is (C).

2. The given system of equations is $x + 2y - 7$ and $x + y - 4$. Now, just substitute each answer-choice into the two equations and see which one works (start checking with the easiest equation, $x + y = 4$):

Choice (A): $x = 3$, $y = 2$: Here, $x + y = 3 + 2 = 5 \neq 4$. Reject.

Choice (B): $x = 2$, $y = 3$: Here, $x + y = 2 + 3 = 5 \neq 4$. Reject.

Choice (C): $x = 1$, $y = 3$: Here, $x + y = 1 + 3 = 4 = 4$, and $x + 2y = 1 + 2(3) = 7$. Correct.

Choice (D): $x = 3$, $y = 1$: Here, $x + y = 3 + 1 = 4$, but $x + 2y = 3 + 2(1) = 5 \neq 7$. Reject.

Choice (E): $x = 7$, $y = 1$: Here, $x + y = 7 + 1 = 8 \neq 4$. Reject.

The answer is (C).

Method II (without substitution):
In the system of equations, subtracting the bottom equation from the top one yields $(x + 2y) - (x + y) = 7 - 4$, or $y = 3$. Substituting this result in the bottom equation yields $x + 3 = 4$. Solving the equation for x yields $x = 1$.

The answer is (C).

➤ Medium

3. Let's substitute the given choices for x in the expression $x^2 + 4x + 3$ and find out which one results in an odd number.

Choice (A): $x = 3$. $x^2 + 4x + 3 = 3^2 + 4(3) + 3 = 9 + 12 + 3 = 24$, an even number. Reject.
Choice (B): $x = 5$. $x^2 + 4x + 3 = 5^2 + 4(5) + 3 = 25 + 20 + 3 = 48$, an even number. Reject.
Choice (C): $x = 9$. $x^2 + 4x + 3 = 9^2 + 4(9) + 3 = 81 + 36 + 3 = 120$, an even number. Reject.
Choice (D): $x = 13$. $x^2 + 4x + 3 = 13^2 + 4(13) + 3 = 169 + 52 + 3 = 224$, an even number. Reject.
Choice (E): $x = 16$. $x^2 + 4x + 3 = 16^2 + 4(16) + 3 = 256 + 64 + 3 = 323$, an odd number. Correct.

The answer is (E).

Method II (without substitution):

$x^2 + 4x + 3$ = An Odd Number
$x^2 + 4x$ = An Odd Number – 3
$x^2 + 4x$ = An Even Number
$x(x + 4)$ = An Even Number. This happens only when x is even. If x is odd, $x(x + 4)$ is not even. Hence, x must be even. Since 16 is the only even answer-choice, the answer is (E).

4. Choice (A): $(2x + 1)^2 = \left(2\left[\frac{-11}{2}\right]+1\right)^2 = (-11+1)^2 = (-10)^2 = 100$. Since this value of x satisfies the equation, the answer is (A).

Method II (without substitution):
Square rooting both sides of the given equation $(2x + 1)^2 = 100$ yields two equations: $2x + 1 = 10$ and $2x + 1 = -10$. Solving the first equation for x yields $x = 9/2$, and solving the second equation for x yields $x = -11/2$. We have the second solution in choice (A), so the answer is (A).

➢ **Hard**

5. Select the choice that satisfies the equation $p = q + 7$.

Choice (A): Suppose $m = 45$. Then $m/14 = 45/14 = 3 + 3/14$. So, the remainder is $p = 3$. Also, $m/7 = 45/7 = 6 + 3/7$. So, the remainder is $q = 3$. Here, $p \neq q + 7$. So, reject the choice.

Choice (B): Suppose $m = 53$. Then $m/14 = 53/14 = 3 + 11/14$. So, the remainder is $p = 11$. Also, $m/7 = 53/7 = 7 + 4/7$. So, the remainder is $q = 4$. Here, $p = q + 7$. So, select the choice.

Choice (C): Suppose $m = 72$. Then $m/14 = 72/14 = 5 + 2/14$. So, the remainder is $p = 2$. Now, $m/7 = 72/7 = 10 + 2/7$. So, the remainder is $q = 2$. Here, $p \neq q + 7$. So, reject the choice.

Choice (D): Suppose $m = 85$. Then $m/14 = 85/14 = 6 + 1/14$. So, the remainder is $p = 1$. Now, $m/7 = 85/7 = 12 + 1/7$. So, the remainder is $q = 1$. Here, $p \neq q + 7$. So, reject the choice.

Choice (E): Suppose $m = 100$. Then $m/14 = 100/14 = 7 + 2/14$. So, the remainder is $p = 2$. Now, $m/7 = 100/7 = 14 + 2/7$. So, the remainder is $q = 2$. Here, $p \neq q + 7$. So, reject the choice.

Hence, the answer is (B).

Defined Functions

Defined functions are very common on the GRE, and at first most students struggle with them. Yet, once you get used to them, defined functions can be some of the easiest problems on the test. In this type of problem, you will be given a symbol and a property that defines the symbol. Some examples will illustrate.

Example 1: If $x * y$ represents the number of integers between x and y, then $(-2 * 8) + (2 * -8) =$

(A) 0
(B) 9
(C) 10
(D) 18
(E) 20

The integers between –2 and 8 are –1, 0, 1, 2, 3, 4, 5, 6, 7 (a total of 9). Hence, $-2 * 8 = 9$. The integers between –8 and 2 are: –7, –6, –5, –4, –3, –2, –1, 0, 1 (a total of 9). Hence, $2 * -8 = 9$. Therefore, $(-2 * 8) + (2 * -8) = 9 + 9 = 18$. The answer is (D).

Example 2: For any positive integer n, $n!$ denotes the product of all the integers from 1 through n. What is the value of $3!(7-2)!$?

(A) 2!
(B) 3!
(C) 5!
(D) 6!
(E) 10!

$3!(7-2)! = 3! \cdot 5!$

As defined, $3! = 3 \cdot 2 \cdot 1 = 6$ and $5! = 5 \cdot 4 \cdot 3 \cdot 2 \cdot 1$.

Hence, $3!(7-2)! = 3! \cdot 5! = 6(5 \cdot 4 \cdot 3 \cdot 2 \cdot 1) = 6 \cdot 5 \cdot 4 \cdot 3 \cdot 2 \cdot 1 = 6!$, as defined.

The answer is (D).

Example 3: A function @ is defined on positive integers as $@(a) = @(a-1) + 1$. If the value of $@(1)$ is 1, then $@(3)$ equals which one of the following?

(A) 0
(B) 1
(C) 2
(D) 3
(E) 4

The function @ is defined on positive integers by the rule $@(a) = @(a-1) + 1$.

Using the rule for $a = 2$ yields $@(2) = @(2-1) + 1 = @(1) + 1 = 1 + 1 = 2$. [Since $@(1) = 1$, given.]
Using the rule for $a = 3$ yields $@(3) = @(3-1) + 1 = @(2) + 1 = 2 + 1 = 3$. [Since $@(2) = 2$, derived.]

Hence, $@(3) = 3$, and the answer is (D).

You may be wondering how defined functions differ from the functions, $f(x)$, you studied in Intermediate Algebra and more advanced math courses. They *don't* differ. They are the same old concept you dealt with in your math classes. The function in Example 3 could just as easily be written as

$$f(a) = f(a - 1) + 1$$

The purpose of defined functions is to see how well you can adapt to unusual structures. Once you realize that defined functions are evaluated and manipulated just as regular functions, they become much less daunting.

Problem Set D:

➢ **Medium**

1.

Column A		Column B
	For any positive integer n, $n!$ denotes the product of all the integers from 1 through n.	
$1!(10 - 1)!$		$2!(10 - 2)!$

2. If $A*B$ is the greatest common factor of A and B, $A\$B$ is defined as the least common multiple of A and B, and $A \cap B$ is defined as equal to $(A*B)\ \$\ (A\$B)$, then what is the value of $12 \cap 15$?
   (A) 42
   (B) 45
   (C) 48
   (D) 52
   (E) 60

3.

Column A		Column B
	A function * is defined for all even positive integers n as the number of even factors of n other than n itself.	
*(48)		*(122)

➢ **Hard**

4.

Column A		Column B
	For any positive integer n, $\pi(n)$ represents the number of factors of n, inclusive of 1 and itself. a and b are prime numbers	
$\pi(a) + \pi(b)$		$\pi(a \times b)$

5. The function $\Delta(m)$ is defined for all positive integers m as the product of $m + 4$, $m + 5$, and $m + 6$. If n is a positive integer, then $\Delta(n)$ must be divisible by which one of the following numbers?
   (A) 4
   (B) 5
   (C) 6
   (D) 7
   (E) 11

6. Define x^* by the equation $x^* = \pi/x$. Then $((-\pi)^*)^* =$

(A) $-1/\pi$
(B) $-1/2$
(C) $-\pi$
(D) $1/\pi$
(E) π

7.

Column A		Column B
	$a * b = \frac{a}{b} - \frac{b}{a}$, $m > n > 0$	
$\frac{1}{m} * \frac{1}{n}$		$\frac{1}{n} * \frac{1}{m}$

8.

Column A	Column B
The number of multiples of 3 between 102 and 729, inclusive	$\frac{729 - 102}{3}$

Answers and Solutions to Problem Set D

➢ Medium

1. First, subtracting the numbers in the parentheses yields

Column A	Column B
1! · 9!	2! · 8!

As defined, 2! = 2 · 1 = 2 and 8! = 8 · 7 · 6 · 5 · 4 · 3 · 2 · 1. Hence, in Column B, we have

$$2! \cdot 8! = 2(8 \cdot 7 \cdot 6 \cdot 5 \cdot 4 \cdot 3 \cdot 2 \cdot 1)$$

Similarly, 1! equals the product of the integers from 1 through 1. Hence, 1! = 1. Also, 9! = 9 · 8 · 7 · 6 · 5 · 4 · 3 · 2 · 1. Hence, in Column A, we have

$$1! \cdot 9! = 1(9 \cdot 8 \cdot 7 \cdot 6 \cdot 5 \cdot 4 \cdot 3 \cdot 2 \cdot 1) = 9 \cdot 8 \cdot 7 \cdot 6 \cdot 5 \cdot 4 \cdot 3 \cdot 2 \cdot 1$$

Now, rewriting 9 · 8 · 7 · 6 · 5 · 4 · 3 · 2 · 1 as 9(8 · 7 · 6 · 5 · 4 · 3 · 2 · 1) yields

Column A	Column B
9(8 · 7 · 6 · 5 · 4 · 3 · 2 · 1)	2(8 · 7 · 6 · 5 · 4 · 3 · 2 · 1)

Finally, dividing both columns by 8 · 7 · 6 · 5 · 4 · 3 · 2 · 1 yields

Column A	Column B
9	2

The answer is (A).

2. According to the definitions given, 12∩15 equals (12*15) $ (12$15) = (GCF of 12 and 15) $ (LCM of 12 and 15) = 3$60 = LCM of 3 and 60 = 60. The answer is (E).

3. Prime factoring 48 yields 2 · 2 · 2 · 2 · 3. The even factors of 48 are

2
2 · 2 (= 4)
2 · 2 · 2 (= 8)
2 · 2 · 2 · 2 (= 16)
3 · 2 (= 6)
3 · 2 · 2 (= 12)
3 · 2 · 2 · 2 (= 24)
3 · 2 · 2 · 2 · 2 (= 48)

Not counting the last factor (48 itself), the total number of factors is 7. Hence, Column A equals 7.

Now, 122 = 2 · 61 (61 is a prime). So, 2 is the only even factor. Hence, *(122) = 1, which is less than 7 (= Column A). Hence, the answer is (A).

➢ Hard

4. The only factors of a prime number are 1 and itself. Hence, π(any prime number) = 2. So, $\pi(a) = 2$ and $\pi(b) = 2$, and therefore Column A equals $\pi(a) + \pi(b) = 2 + 2 = 4$.

Now, the factors of ab are 1, a, b, and ab itself. Hence, the total number of factors of ab is 4. In other words, $\pi(a \times b) = 4$. Hence, Column B also equals 4.

Since the columns are equal, the answer is (C).

5. By the given definition, $\Delta(n) = (n + 4)(n + 5)(n + 6)$, a product of three consecutive integers. There is exactly one multiple of 3 in every three consecutive positive integers. Also, at least one of the three numbers must be an even number. Hence, $\Delta(n)$ must be a multiple of both 2 and 3. Hence, $\Delta(n)$ must be a multiple of 6 (= 2 × 3), because 2 and 3 are primes. The answer is (C).

6. Working from the inner parentheses out, we get

$$((-\pi)^*)^* =$$
$$(\pi/(-\pi))^* =$$
$$(-1)^* =$$
$$\pi/(-1) =$$
$$-\pi$$

The answer is (C).

Method II:
We can rewrite this problem using ordinary function notation. Replacing the odd symbol x^* with $f(x)$ gives $f(x) = \pi / x$. Now, the expression $((-\pi)^*)^*$ becomes the ordinary composite function

$$f(f(-\pi)) =$$
$$f(\pi/(-\pi)) =$$
$$f(-1) =$$
$$\pi/(-1) =$$
$$-\pi$$

7. The function $a * b$ is defined to be $a/b - b/a$ for any numbers a and b. Applying this definition to the columns gives

Column A	Column B
$\frac{1}{m} * \frac{1}{n} =$	$\frac{1}{n} * \frac{1}{m} =$
$\frac{\frac{1}{m}}{\frac{1}{n}} - \frac{\frac{1}{n}}{\frac{1}{m}} =$	$\frac{\frac{1}{n}}{\frac{1}{m}} - \frac{\frac{1}{m}}{\frac{1}{n}} =$
$\frac{n}{m} - \frac{m}{n} =$	$\frac{m}{n} - \frac{n}{m} =$
$\frac{nn - mm}{mn} =$	$\frac{mm - nn}{mn} =$
$\frac{n^2 - m^2}{mn}$	$\frac{m^2 - n^2}{mn}$

The given inequality $m > n > 0$ indicates that both m and n are positive and therefore their product mn is positive. Multiplying both columns by mn to clear fractions yields

$n^2 - m^2$	$m^2 - n^2$

Adding n^2 and m^2 to both columns yields

$2n^2$	$2m^2$

Finally, dividing both columns by 2 yields

n^2	m^2

Since both m and n are positive and $m > n$, $m^2 > n^2$. Hence, Column B is greater than Column A, and the answer is (B).

8. The numbers 102 and 729 are themselves multiples of 3. Also, a multiple of 3 exists once in every three consecutive integers. Counting the multiples of 3 starting with 1 for 102, 2 (= 1 + (105 – 102)/3 = 1 + 1 = 2) for 105, 3 (= 1 + (108 – 102)/3 = 1 + 2 = 3) for 108, and so on, the count we get for 729 equals 1 + (729 – 102)/3 = 1 + Column B. Hence, Column A is greater than Column B by 1 unit. Hence, the answer is (A).

Math Notes

We'll discuss many of the concepts in this chapter in depth later. But for now, we need a brief review of these concepts for many of the problems that follow.

1. **To compare two fractions, cross-multiply. The larger product will be on the same side as the larger fraction.**

   Example: Given $\frac{5}{6}$ vs. $\frac{6}{7}$. Cross-multiplying gives $5 \cdot 7$ vs. $6 \cdot 6$, or 35 vs. 36. Now 36 is larger than 35, so $\frac{6}{7}$ is larger than $\frac{5}{6}$.

2. **Taking the square root of a fraction between 0 and 1 makes it larger.**

   Example: $\sqrt{\frac{1}{4}} = \frac{1}{2}$ and $\frac{1}{2}$ is greater than $\frac{1}{4}$.

   Caution: This is not true for fractions greater than 1. For example, $\sqrt{\frac{9}{4}} = \frac{3}{2}$. But $\frac{3}{2} < \frac{9}{4}$.

3. **Squaring a fraction between 0 and 1 makes it smaller.**

   Example: $\left(\frac{1}{2}\right)^2 = \frac{1}{4}$ and $\frac{1}{4}$ is less than $\frac{1}{2}$.

4. $\boldsymbol{ax^2 \neq (ax)^2}$. **In fact,** $\boldsymbol{a^2x^2 = (ax)^2}$.

   Example: $3 \cdot 2^2 = 3 \cdot 4 = 12$. But $(3 \cdot 2)^2 = 6^2 = 36$. This mistake is often seen in the following form: $-x^2 = (-x)^2$. To see more clearly why this is wrong, write $-x^2 = (-1)x^2$, which is negative. But $(-x)^2 = (-x)(-x) = x^2$, which is positive.

   Example: $-5^2 = (-1)5^2 = (-1)25 = -25$. But $(-5)^2 = (-5)(-5) = 5 \cdot 5 = 25$.

5. $\boldsymbol{\frac{1/a}{b} \neq \frac{1}{a/b}}$. **In fact,** $\boldsymbol{\frac{1/a}{b} = \frac{1}{ab}}$ **and** $\boldsymbol{\frac{1}{a/b} = \frac{b}{a}}$.

   Example: $\frac{1/2}{3} = \frac{1}{2} \cdot \frac{1}{3} = \frac{1}{6}$. But $\frac{1}{2/3} = 1 \cdot \frac{3}{2} = \frac{3}{2}$.

6. **$-(a + b) \neq -a + b$. In fact, $-(a + b) = -a - b$.**
   Example: $-(2 + 3) = -5$. But $-2 + 3 = 1$.
   Example: $-(2 + x) = -2 - x$.

7. **Memorize the following factoring formulas—they occur frequently on the GRE.**
   A. $x^2 - y^2 = (x + y)(x - y)$
   B. $x^2 \pm 2xy + y^2 = (x \pm y)^2$
   C. $a(b + c) = ab + ac$

8. Know these rules for radicals:

A. $\sqrt{x}\sqrt{y} = \sqrt{xy}$

B. $\sqrt{\dfrac{x}{y}} = \dfrac{\sqrt{x}}{\sqrt{y}}$

9. Pythagorean Theorem (For right triangles only):

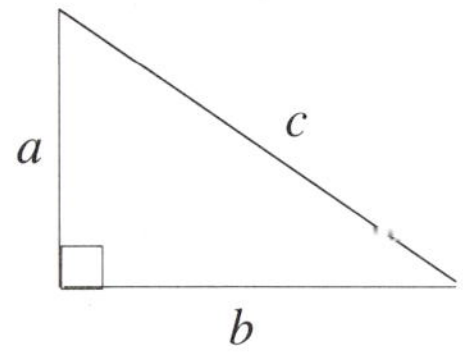

$$c^2 = a^2 + b^2$$

Example: What is the area of the triangle to the right?

(A) 6
(B) 7.5
(C) 8
(D) 11
(E) 15

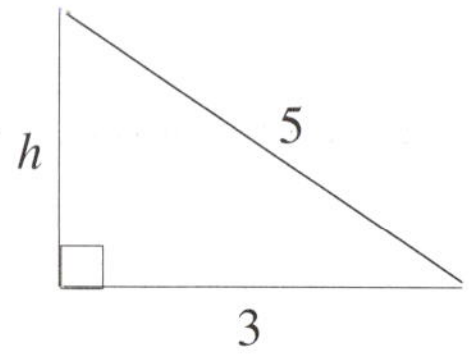

Since the triangle is a right triangle, the Pythagorean Theorem applies: $h^2 + 3^2 = 5^2$, where h is the height of the triangle. Solving for h yields $h = 4$. Hence, the area of the triangle is $\frac{1}{2}(base)(height) =$ $\frac{1}{2}(3)(4) = 6$. The answer is (A).

10. When parallel lines are cut by a transversal, three important angle relationships are formed:

Alternate interior angles are equal.

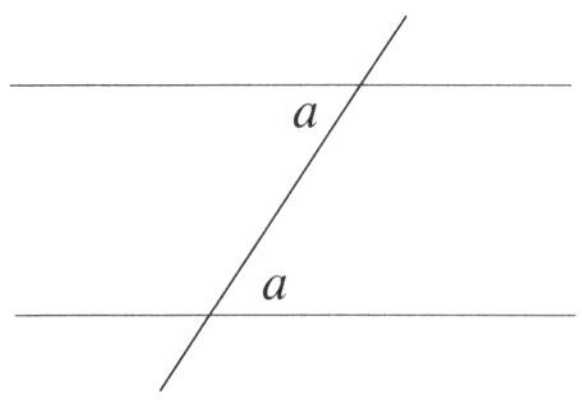

Corresponding angles are equal.

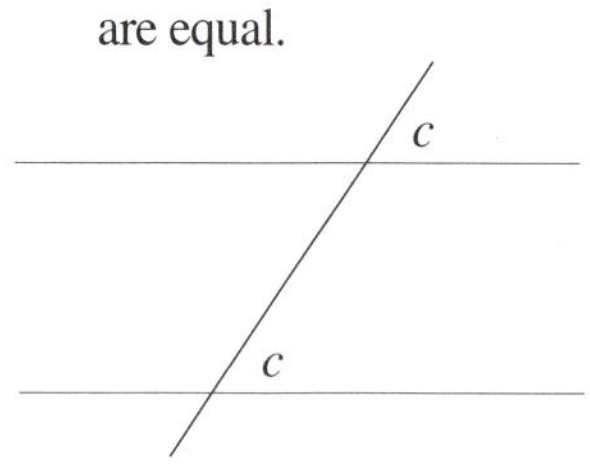

Interior angles on the same side of the transversal are supplementary.

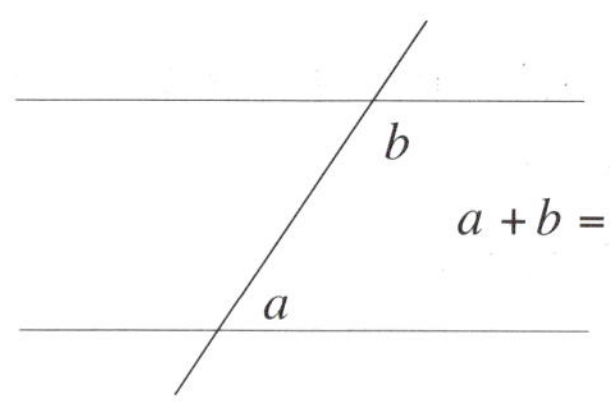

$a + b = 180°$

11. In a triangle, an exterior angle is equal to the sum of its remote interior angles and therefore greater than either of them.

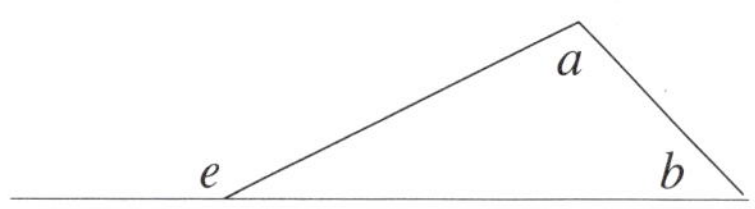

$e = a + b$ and $e > a$ and $e > b$

12. A central angle has by definition the same measure as its intercepted arc.

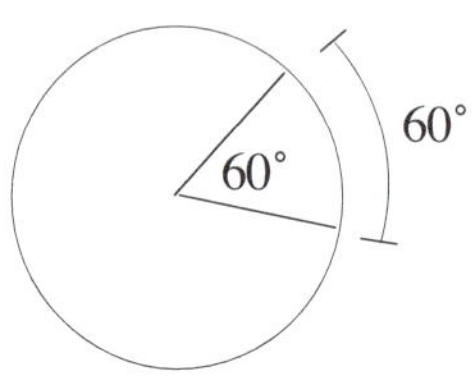

13. An inscribed angle has one-half the measure of its intercepted arc.

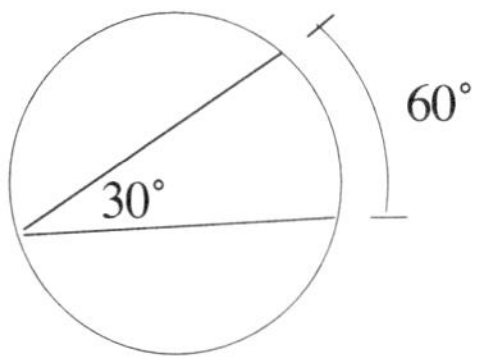

14. There are 180° in a straight angle.

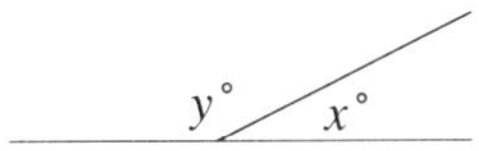

$x + y = 180°$

15. The angle sum of a triangle is 180°.

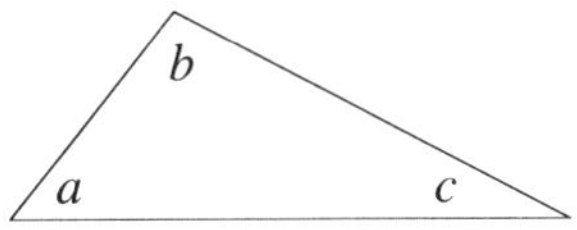

$a + b + c = 180°$

Example: In the triangle to the right, what is the degree measure of angle c ?

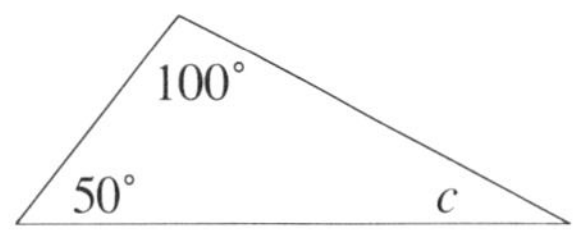

(A) 17
(B) 20
(C) 30
(D) 40
(E) 45

Since a triangle has 180°, we get $100 + 50 + c = 180$. Solving for c yields $c = 30$. Hence, the answer is (C).

16. To find the percentage increase, find the absolute increase and divide by the original amount.

Example: If a shirt selling for \$18 is marked up to \$20, then the absolute increase is $20 - 18 = 2$. Thus, the percentage increase is $\frac{increase}{original\ amount} = \frac{2}{18} = \frac{1}{9} \approx 11\%$.

17. Systems of simultaneous equations can most often be solved by merely adding or subtracting the equations.

Example: If $4x + y = 14$ and $3x + 2y = 13$, then $x - y =$

Solution: Merely subtract the second equation from the first:

$$\begin{array}{rr} & 4x + y = 14 \\ (-) & 3x + 2y = 13 \\ \hline & x - y = 1 \end{array}$$

18. When counting elements that are in overlapping sets, the total number will equal the number in one group plus the number in the other group minus the number common to both groups. Venn diagrams are very helpful with these problems.

Example: If in a certain school 20 students are taking math and 10 are taking history and 7 are taking both, how many students are taking either math or history?

Solution:

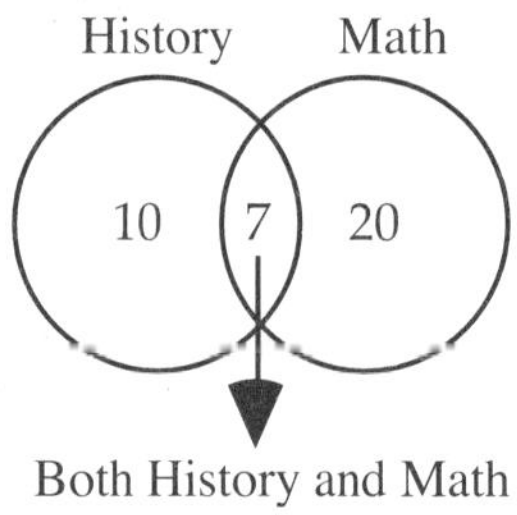

By the principle stated above, we add 10 and 20 and then subtract 7 from the result. Thus, there are $(10 + 20) - 7 = 23$ students.

19. The number of integers between two integers <u>inclusive</u> is one more than their difference.

For example: The number of integers between 49 and 101 inclusive is $(101 - 49) + 1 = 53$. To see this more clearly, choose smaller numbers, say, 9 and 11. The difference between 9 and 11 is 2. But there are three numbers between them inclusive—9, 10, and 11—one more than their difference.

20. Rounding Off: The convention used for rounding numbers is *"if the following digit is less than five, then the preceding digit is not changed. But if the following digit is greater than or equal to five, then the preceding digit is increased by one."*

Example: **65**,439 —> **65**,000 (following digit is 4)
5.**567**1 —> 5.**57**00 (dropping the unnecessary zeros gives 5.57)

21. Writing a number as a product of a power of 10 and a number $1 \le n < 10$ is called scientific notation. This notation has the following form: $n \times 10^c$, where $1 \le n < 10$ and c is an integer.

Example: $326,000,000 = 3.26 \times 10^8$
Notice that the exponent is the number of significant places that the decimal is moved[*], not the number zeros. Students often use 6 as the exponent in the above example because there are 6 zeros.

Example: $0.00007 = 7 \times 10^{-5}$
Notice that for a small number the exponent is negative and for a large number the exponent is positive.

---

[*] Although no decimal is shown in the number 326,000,000, you can place a decimal at the end of the number and add as many trailing zeros as you like without changing the value of the number: 326,000,000 = 326,000,000.00 . . . .

Number Theory

This broad category is a popular source for GRE questions. At first, students often struggle with these problems since they have forgotten many of the basic properties of arithmetic. So, before we begin solving these problems, let's review some of these basic properties.

- ***"The remainder is r when p is divided by k"* means $p = kq + r$; the integer q is called the quotient. For instance, *"The remainder is 1 when 7 is divided by 3"* means $7 = 3 \cdot 2 + 1$. Dividing both sides of $p = kq + r$ by k gives the following alternative form $p/k = q + r/k$.**

Example 1: The remainder is 57 when a number is divided by 10,000. What is the remainder when the same number is divided by 1,000?

(A) 5
(B) 7
(C) 43
(D) 57
(E) 570

Since the remainder is 57 when the number is divided by 10,000, the number can be expressed as $10{,}000n + 57$, where n is an integer. Rewriting 10,000 as 1,000(10) yields

$$1{,}000(10)n + 57 =$$

$$1{,}000(10n) + 57 =$$

Now, since n is an integer, $10n$ is an integer. Letting $10n = q$, we get

$$1{,}000q + 57 =$$

Hence, the remainder is still 57 (by the $p = kq + r$ form) when the number is divided by 1,000. The answer is (D).

Method II (Alternative form)
Since the remainder is 57 when the number is divided by 10,000, the number can be expressed as $10{,}000n + 57$. Dividing this number by 1,000 yields

$$\frac{10{,}000n + 57}{1000} =$$

$$\frac{10{,}000n}{1{,}000} + \frac{57}{1{,}000} =$$

$$10n + \frac{57}{1{,}000}$$

Hence, the remainder is 57 (by the alternative form $p/k = q + r/k$), and the answer is (D).

- **A number n is even if the remainder is zero when n is divided by 2: $n = 2z + 0$, or $n = 2z$.**

- **A number n is odd if the remainder is one when n is divided by 2: $n = 2z + 1$.**

➢ **The following properties for odd and even numbers are very useful—you should memorize them:**

$$\begin{aligned} &even \times even = even \\ &odd \times odd = odd \\ &even \times odd = even \end{aligned}$$

$$\begin{aligned} &even + even = even \\ &odd + odd = even \\ &even + odd = odd \end{aligned}$$

Example 2: If n is a positive integer and $(n + 1)(n + 3)$ is odd, then $(n + 2)(n + 4)$ must be a multiple of which one of the following?

(A) 3 (B) 5 (C) 6 (D) 8 (E) 16

$(n + 1)(n + 3)$ is odd only when both $(n + 1)$ and $(n + 3)$ are odd. This is possible only when n is even. Hence, $n = 2m$, where m is a positive integer. Then,

$(n + 2)(n + 4) =$
$(2m + 2)(2m + 4) =$
$2(m + 1)2(m + 2) =$
$4(m + 1)(m + 2) =$
$4 \times$ (product of two consecutive positive integers, one which must be even) =
$4 \times$ (an even number), and this equals a number that is at least a multiple of 8

Hence, the answer is (D).

➢ **Consecutive integers are written as $x, x + 1, x + 2, \ldots$**

➢ **Consecutive even or odd integers are written as $x, x + 2, x + 4, \ldots$**

➢ **The integer zero is neither positive nor negative, but it is even: $0 = 2 \times 0$.**

➢ **A *prime number* is an integer that is divisible only by itself and 1.**

The prime numbers are 2, 3, 5, 7, 11, 13, 17, 19, 23, 29, 31, 37, 41, . . .

➢ **A number is divisible by 3 if the sum of its digits is divisible by 3.**

For example, 135 is divisible by 3 because the sum of its digits $(1 + 3 + 5 = 9)$ is divisible by 3.

➢ **A *common multiple* is a multiple of two or more integers.**

For example, some common multiples of 2 and 5 are 0, 10, 20, 40, and 50.

➢ **The *least common multiple* (LCM) of two integers is the smallest positive integer that is a multiple of both.**

For example, the LCM of 4 and 10 is 20. The standard method of calculating the LCM is to prime factor the numbers and then form a product by selecting each factor the greatest number of times it occurs. For 4 and 10, we get

$4 = 2^2$
$10 = 2 \bullet 5$

In this case, select 2^2 instead of 2 because it has the greater number of factors of 2, and select 5 by default since there are no other factors of 5. Hence, the LCM is $2^2 \bullet 5 = 4 \bullet 5 = 20$.

For another example, let's find the LCM of 8, 36, and 54. Prime factoring yields

$8 = 2^3$
$36 = 2^2 \bullet 3^2$
$54 = 2 \bullet 3^3$

In this case, select 2^3 because it has more factors of 2 than 2^2 or 2 itself, and select 3^3 because is has more factors of 3 than 3^2 does. Hence, the LCM is $2^3 \bullet 3^3 = 8 \bullet 27 = 216$.

A shortcut for finding the LCM is to just keep adding the largest number to itself until the other numbers divide into it evenly. For 4 and 10, we would add 10 to itself: 10 + 10 = 20. Since 4 divides evenly in to 20, the LCM is 20. For 8, 36, and 54, we would add 54 to itself: 54 + 54 + 54 + 54 = 216. Since both 8 and 36 divide evenly into 216, the LCM is 216.

➤ **The absolute value of a number, $| \, |$, is always positive. In other words, the absolute value symbol eliminates negative signs.**

For example, $|-7| = 7$ and $|-\pi| = \pi$. Caution, the absolute value symbol acts only on what is inside the symbol, $| \, |$. For example, $-|-(7-\pi)| = -(7-\pi)$. Here, only the negative sign inside the absolute value symbol but outside the parentheses is eliminated.

Example 3: The number of prime numbers divisible by 2 plus the number of prime numbers divisible by 3 is

(A) 0 (B) 1 (C) 2 (D) 3 (E) 4

A prime number is divisible by no other numbers, but itself and 1. Hence, the only prime number divisible by 2 is 2 itself; and the only prime number divisible by 3 is 3 itself. Hence, The number of prime numbers divisible by 2 is one, and the number of prime numbers divisible by 3 is one. The sum of the two is 1 + 1 = 2, and the answer is (C).

Example 4: If $15x + 16 = 0$, then $15|x|$ equals which one of the following?

(A) 15 (B) $-16x$ (C) $15x$ (D) 16 (E) $16x$

Solving the given equation $15x + 16 = 0$ yields $x = -16/15$.

Substituting this into the expression $15|x|$ yields

$$15|x| = 15\left|\frac{-16}{15}\right| = 15\left(\frac{16}{15}\right) = 16$$

The answer is (D).

➤ **The product (quotient) of positive numbers is positive.**

➤ **The product (quotient) of a positive number and a negative number is negative.**

For example, $-5(3) = -15$ and $\frac{6}{-3} = -2$.

➤ **The product (quotient) of an even number of negative numbers is positive.**

For example, $(-5)(-3)(-2)(-1) = 30$ is positive because there is an even number, 4, of positives. $\frac{-9}{-2} = \frac{9}{2}$ is positive because there is an even number, 2, of positives.

➤ **The product (quotient) of an odd number of negative numbers is negative.**

For example, $(-2)(-\pi)(-\sqrt{3}) = -2\pi\sqrt{3}$ is negative because there is an odd number, 3, of negatives. $\frac{(-2)(-9)(-6)}{(-12)\left(-18/2\right)} = -1$ is negative because there is an odd number, 5, of negatives.

➤ **The sum of negative numbers is negative.**

For example, $-3 - 5 = -8$. Some students have trouble recognizing this structure as a sum because there is no plus symbol, +. But recall that subtraction is defined as negative addition. So $-3 - 5 = -3 + (-5)$.

➤ **A number raised to an even exponent is greater than or equal to zero.**

For example, $(-\pi)^4 = \pi^4 \geq 0$, and $x^2 \geq 0$, and $0^2 = 0 \cdot 0 = 0 \geq 0$.

Problem Set E:

➤ **Easy**

1. If x is divisible by both 3 and 4, then the number x must be a multiple of which one of the following?
   (A) 8
   (B) 12
   (C) 15
   (D) 18
   (E) 21

2. The last digit of the positive even number n equals the last digit of n^2. Which one of the following could be n ?
   (A) 12
   (B) 14
   (C) 15
   (D) 16
   (E) 17

3. Which one of the following is divisible by both 2 and 3?
   (A) 1005
   (B) 1296
   (C) 1351
   (D) 1406
   (E) 1414

4.

Column A	Column B
The number of prime numbers divisible by 2	The number of prime numbers divisible by 3.

5. Which one of the following equals the product of exactly two prime numbers?
   (A) $11 \cdot 6$
   (B) $13 \cdot 22$
   (C) $14 \cdot 23$
   (D) $17 \cdot 21$
   (E) $13 \cdot 23$

6. Which one of the following equals the product of the smallest prime number greater than 21 and the largest prime number less than 16?
   (A) $13 \cdot 16$
   (B) $13 \cdot 29$
   (C) $13 \cdot 23$
   (D) $15 \cdot 23$
   (E) $16 \cdot 21$

7.

Column A		Column B
	a and b are the digits of a two-digit number ab, and $a = b + 3$.	
The two-digit number ab		40

8.

Column A	Column B
The last digit in the number 252^{56}	The last digit in the number 152^{56}

➢ Medium

9. Column A | p and q are two positive integers and $p/q = 7.15$ | Column B

q | | 15

10. Column A | X is a 3-digit number and Y is a 4-digit number. All the digits of X are greater than 4, and all the digits of Y are less than 5. | Column B

The sum of the digits of X | | The sum of the digits of Y

11. Column A | | Column B

The least number divisible by 2, 3, 4, 5, and 6 | | The least number that is a multiple of 2, 3, 4, 5 and 6

12. The number 3 divides a with a result of b and a remainder of 2. The number 3 divides b with a result of 2 and a remainder of 1. What is the value of a ?

(A) 13
(B) 17
(C) 21
(D) 23
(E) 27

13 The remainder when the positive integer m is divided by 7 is x. The remainder when m is divided by 14 is $x + 7$. Which one of the following could m equal?

(A) 45
(B) 53
(C) 72
(D) 85
(E) 100

14. Which one of the following choices does not equal any of the other choices?

(A) $5.43 + 4.63 - 3.24 - 2.32$
(B) $5.53 + 4.73 - 3.34 - 2.42$
(C) $5.53 + 4.53 - 3.34 - 2.22$
(D) $5.43 + 4.73 - 3.24 - 2.42$
(E) $5.43 + 4.73 - 3.14 - 2.22$

15. Each of the two positive integers a and b ends with the digit 2. With which one of the following numbers does $a - b$ end?

(A) 0
(B) 1
(C) 2
(D) 3
(E) 4

16. Column A | x is a two-digit number. The digits of the number differ by 6, and the squares of the digits differ by 60. | Column B

x | | 60

17. If each of the three nonzero numbers a, b, and c is divisible by 3, then abc must be divisible by which one of the following the numbers?

(A) 8
(B) 27
(C) 81
(D) 121
(E) 159

18. How many positive five-digit numbers can be formed with the digits 0, 3, and 5?

(A) 14
(B) 15
(C) 108
(D) 162
(E) 243

19. If n is a positive integer, which one of the following numbers must have a remainder of 3 when divided by any of the numbers 4, 5, and 6?

(A) $12n + 3$
(B) $24n + 3$
(C) $80n + 3$
(D) $90n + 2$
(E) $120n + 3$

20.

Column A	Column B
Least common multiple of the two positive integers m and n	mn

21. The number 3072 is divisible by both 6 and 8. Which one of the following is the first integer larger than 3072 that is also divisible by both 6 and 8?

(A) 3078
(B) 3084
(C) 3086
(D) 3090
(E) 3096

22. If the least common multiple of m and n is 24, then what is the first integer larger than 3070 that is divisible by both m and n?

(A) 3072
(B) 3078
(C) 3084
(D) 3088
(E) 3094

23.

Column A	Column B
The first number larger than 300 that is a multiple of both 6 and 8	324

24.

Column A		Column B
$a/5 - b/6$	a, b, and c are consecutive integers in increasing order of size.	$b/5 - c/6$

25. How many 3-digit numbers do not have an even digit or a zero?

(A) 20
(B) 30
(C) 60
(D) 80
(E) 125

26.

Column A		Column B
	The digits of a two-digit number x differ by 4.	
The positive difference between the squares of the digits of x		15

➤ Hard

27. Which one of the following is the minimum value of the sum of two integers whose product is 36?

(A) 37
(B) 20
(C) 15
(D) 13
(E) 12

28. If a and b are positive integers, and $x = 2 \cdot 3 \cdot 7 \cdot a$, and $y = 2 \cdot 2 \cdot 8 \cdot b$, and the values of both x and y lie between 120 and 130 (not including the two), then $a - b =$

(A) –2
(B) –1
(C) 0
(D) 1
(E) 2

29.

Column A		Column B
	m and n are two positive integers. $5m + 7n = 46$.	
m		n

30.

Column A		Column B
	a and b are the digits of a two-digit number ab, and $b = a + 3$.	
The positive two-digit number ab		The positive two-digit number ba

31. a, b, c, d, and e are five consecutive numbers in increasing order of size. Deleting one of the five numbers from the set decreased the sum of the remaining numbers in the set by 20%. Which one of the following numbers was deleted?

(A) a
(B) b
(C) c
(D) d
(E) e

32. A set has exactly five consecutive positive integers starting with 1. What is the percentage decrease in the average of the numbers when the greatest one of the numbers is removed from the set?

(A) 5
(B) 8.5
(C) 12.5
(D) 15.2
(E) 16.66

33. What is the maximum value of m such that 7^m divides into 14! evenly?

(A) 1
(B) 2
(C) 3
(D) 4
(E) 5

34. Column A — The sum of the multiples of 6 between 0 and 100

The sum of the positive integers from 1 through n can be calculated by the formula

$$\frac{n(n+1)}{2}$$

Column B — The sum of the multiples of 8 between 0 and 100

35. Column A — Smallest palindrome number greater than 233

A *palindrome number* is a number that reads the same forward or backward. For example, 787 is a palindrome number.

Column B — Smallest palindrome greater than 239

➢ **Very Hard**

36. $2ab5$ is a four-digit number divisible by 25. If the number formed from the two digits ab is a multiple of 13, then $ab =$

(A) 10
(B) 25
(C) 52
(D) 65
(E) 75

37. The positive integers m and n leave remainders of 2 and 3, respectively, when divided by 6. $m > n$. What is the remainder when $m - n$ is divided by 6?

(A) 1
(B) 2
(C) 3
(D) 4
(E) 5

38. The remainder when $m + n$ is divided by 12 is 8, and the remainder when $m - n$ is divided by 12 is 6. If $m > n$, then what is the remainder when mn divided by 6?

(A) 1
(B) 2
(C) 3
(D) 4
(E) 5

39. What is the remainder when $7^2 \cdot 8^2$ is divided by 6?

(A) 1
(B) 2
(C) 3
(D) 4
(E) 5

40. a, b, c, d, and e are five consecutive integers in increasing order of size. Which one of the following expressions is not odd?

(A) $a + b + c$
(B) $ab + c$
(C) $ab + d$
(D) $ac + d$
(E) $ac + e$

41. How many positive integers less than 500 can be formed using the numbers 1, 2, 3, and 5 for the digits?

(A) 48
(B) 52
(C) 66
(D) 68
(E) 84

42. What is the remainder when 3^7 is divided by 8?

(A) 1
(B) 2
(C) 3
(D) 5
(E) 7

43.

Column A		Column B
	The positive integers m and n leave a remainder of 2 and 3, respectively, when divided by 6. $m > n$.	
The remainder when $m + n$ is divided by 6		The remainder when $m - n$ is divided by 6

Answers and Solutions to Problem Set E

➢ Easy

1. We are given that x is divisible by 3 and 4. Hence, x must be a common multiple of 3 and 4. The least common multiple of 3 and 4 is 12. So, x is a multiple of 12. The answer is (B).

2. Numbers ending with 0, 1, 5, or 6 will have their squares also ending with the same digit.

For example,

10 ends with 0, and 10^2 (= 100) also ends with 0.
11 ends with 1, and 11^2 (= 121) also ends with 1.
15 ends with 5, and 15^2 (= 225) also ends with 5.
16 ends with 6, and 16^2 (= 256) also ends with 6.

Among the four numbers 0, 1, 5, or 6, even numbers only end with 0 or 6. Choice (D) has one such number. The answer is (D).

3. A number divisible by 2 ends with one of the digits 0, 2, 4, 6, or 8.

If a number is divisible by 3, then the sum of its digits is also divisible by 3.

Hence, a number divisible by both 2 and 3 will follow both of the above rules.

Choices (A) and (C) do not end with an even digit. Hence, eliminate them.

The sum of digits of Choice (B) is 1 + 2 + 9 + 6 = 18, which is divisible by 3. Also, the last digit is 6. Hence, choice (B) is correct.

Next, the sum of the digits of choices (D) and (E) are 1 + 4 + 0 + 6 (= 11) and 1 + 4 + 1 + 4 (= 10), respectively, and neither is divisible by 3. Hence, reject the two choices.

Hence, the answer is (B).

4. A prime number is divisible by no other numbers, but itself and 1. Hence, the only prime number divisible by 2 is 2 itself; and the only prime number divisible by 3 is 3 itself. Since the number of primes in each column is 1, the answer is (C).

5. Choice (A): $11 \cdot 6$ can be factored as $11 \cdot 2 \cdot 3$. The product of more than two primes. Reject.
Choice (B): $13 \cdot 22$ can be factored as $13 \cdot 2 \cdot 11$. The product of more than two primes. Reject.
Choice (C): $14 \cdot 23$ can be factored as $7 \cdot 2 \cdot 23$. The product of more than two primes. Reject.
Choice (D): $17 \cdot 21$ can be factored as $17 \cdot 3 \cdot 7$. The product of more than two primes. Reject.
Choice (E): $13 \cdot 23$ cannot be further factored and is itself the product of two primes. Accept.

The answer is (E).

6. The smallest prime number greater than 21 is 23, and the largest prime number less than 16 is 13. The product of the two is $13 \cdot 23$, which is listed in choice (C). The answer is (C).

7. Suppose b equals 0. Then $a = b + 3 = 0 + 3 = 3$. Hence, Column A equals $ab = 30$. Here, Column A is less than Column B, which equals 40.

Now, suppose b equals 1. Then $a = b + 3 = 1 + 3 = 4$. Hence, Column A equals $ab = 41$. Here Column A is greater than Column B, which equals 40.

Hence, we have a double case, and the answer is (D).

8. The last digit of the number 252 (in Column A) is 2, and the last digit of the number 152 (in Column B) is also 2. Hence, both numbers raised to the same power (here 56) should end with the same digit. So, 152^{56} should end with the same digit as 252^{56}. The answer is (C). A small example: the last digit of 6^2 (= 36) is 6 and the last digit of 16^2 (= 256) is also 6.

➢ Medium

9. We have that $p/q = 7.15$. Solving for p yields $p = 7.15q = 7q + 0.15q$ = 7(a positive integer) + $0.15q$ = (a positive integer) + $0.15q$. Now, p is a positive integer only when $0.15q$ is an integer. Now, $0.15q$ equals $15/100 \cdot q = 3q/20$ and would result in an integer only when the denominator of the fraction (i.e., 20) is canceled out by q. This happens only when q is a multiple of 20. Hence, q = 20, or 40, or 60, ... Any of these values is greater than 15. Hence, Column A is greater than Column B, and the answer is (A).

10. Column A: Since all the digits of the 3-digit number X are greater than 4, each digit must be greater than or equal to 5. Hence, the sum of the three digits must be greater than or equal to $3 \times 5 = 15$. Also, since the maximum value of each digit is 9, the maximum possible value of the sum of its digits is $3 \times 9 = 27$. Hence, we have the inequality $15 \le$ Column A ≤ 27.

Column B: Since all the digits in the 4-digit number Y are less than 5, each digit must be less than or equal to 4. Hence, the sum of the four digits must be less than or equal to $4 \times 4 = 16$. Also, the minimum value of the sum of the digits of a 4-digit number is 1 (for example, for 1000, the sum of the digits is 1). Hence, we have the inequality $1 \le$ Column B ≤ 16.

Since the inequality for Column A, $15 \le$ Column A ≤ 27, and the inequality for Column B, $1 \le$ Column B ≤ 16, have a common range (numbers between 15 and 16, inclusive, satisfy both inequalities), an inequality between the two columns cannot be derived. Hence, we cannot know which column is greater. The answer is (D).

11. Any number that is divisible by the five numbers 2, 3, 4, 5, and 6 must also be a multiple of all five numbers. So, Column A (refers to the set of numbers that are divisible by 2, 3, 4, 5, and 6) and Column B (refers to the set of numbers that are a multiple of 2, 3, 4, 5 and 6) refer to the same set. Hence, the least values of the two columns are the same. So, Column A equals Column B, and the answer is (C).

12. Since 3 divides b with a result of 2 and a remainder of 1, $b = 2 \cdot 3 + 1 = 7$. Since number 3 divides a with a result of b (which we now know equals 7) and a remainder of 2, $a = b \cdot 3 + 2 = 7 \cdot 3 + 2 = 23$. The answer is (D).

13. Choice (A): $45/7 = 6 + 3/7$, so $x = 3$. Now, $45/14 = 3 + 3/14$. The remainder is 3, not $x + 7$ (= 10). Reject.

Choice (B): $53/7 = 7 + 4/7$, so $x = 4$. Now, $53/14 = 3 + 11/14$. The remainder is 11, and equals $x + 7$ (= 11). Accept the choice.

Choice (C): $72/7 = 10 + 2/7$, so $x = 2$. Now, $72/14 = 5 + 2/14$. The remainder is 2, not $x + 7$ (= 9). Reject.

Choice (D): $85/7 = 12 + 1/7$, so $x = 1$. Now, $85/14 = 6 + 1/14$. The remainder is 1, not $x + 7$ (= 8). Reject.

Choice (E): $100/7 = 14 + 2/7$, so $x = 2$. Now, $100/14 = 7 + 2/14$. The remainder is 2, not $x + 7$ (= 9). Reject.

The answer is (B).

14. Choice (A) = 5.43 + 4.63 – 3.24 – 2.32 = 4.5.

Choice (B) = 5.53 + 4.73 – 3.34 – 2.42 = 4.5 = Choice (A). Reject choices (A) and (B).

Choice (C) = 5.53 + 4.53 – 3.34 – 2.22 = 4.5 = Choice (A). Reject choice (C).

Choice (D) = 5.43 + 4.73 – 3.24 – 2.42 = 4.5 = Choice (A). Reject choice (D).

Choice (E) = 5.43 + 4.73 – 3.14 – 2.22 = 4.8. Correct.

The answer is (E).

15. Since each of the two integers a and b ends with the same digit, the difference of the two numbers ends with 0. For example 642 – 182 = 460, and 460 ends with 0. Hence, the answer is (A).

16. Suppose a and b are the two digits of the number x, and let a represent the greater of the two. From the given information, we have

$$a - b = 6 \qquad (1)$$
$$a^2 - b^2 = 60$$

Applying the Difference of Squares formula, $a^2 - b^2 = (a - b)(a + b)$, to the second equation yields

$$(a - b)(a + b) = 60$$
$$6(a + b) = 60 \qquad \text{since } a - b = 6$$
$$a + b = 60/6 = 10 \qquad (2)$$

Adding equations (1) and (2) yields $2a = 16$. Dividing by 2 yields $a = 8$. Substituting this in equation (2) yields $8 + b = 10$. Solving for b yields $b = 2$. Hence, the two digits are 2 and 8. They can be arranged in any order. Hence, 28 and 82 are two feasible solutions. Now, 28 is less than Column B, and 82 is greater than Column B. Hence, we have a double case, and the answer is (D).

17. Since each one of the three numbers a, b, and c is divisible by 3, the numbers can be represented as $3p$, $3q$, and $3r$, respectively, where p, q, and r are integers. The product of the three numbers is $3p \cdot 3q \cdot 3r = 27(pqr)$. Since p, q, and r are integers, pqr is an integer and therefore abc is divisible by 27. The answer is (B).

18. Let the digits of the five-digit positive number be represented by 5 compartments:

Each of the last four compartments can be filled in 3 ways (with any one of the numbers 0, 3 and 5).

The first compartment can be filled in only 2 ways (with only 3 and 5, not with 0, because placing 0 in the first compartment would yield a number with fewer than 5 digits).

3 5	0 3 5	0 3 5	0 3 5	0 3 5

Hence, the total number of ways the five compartments can be filled in is $2 \cdot 3 \cdot 3 \cdot 3 \cdot 3 = 162$. The answer is (D).

19. Let m be a number that has a remainder of 3 when divided by any of the numbers 4, 5, and 6. Then $m - 3$ must be exactly divisible by all three numbers. Hence, $m - 3$ must be a multiple of the Least Common Multiple of the numbers 4, 5, and 6. The LCM is $3 \cdot 4 \cdot 5 = 60$. Hence, we can suppose $m - 3 = 60p$, where p is a positive integer. Replacing p with n, we get $m - 3 = 60n$. So, $m = 60n + 3$. Choice (E) is in the same format $120n + 3 = 60(2n) + 3$. Hence, the answer is (E).

We can also subtract 3 from each answer choice, and the correct answer will be divisible by 60:

Choice (A): If $n = 1$, then $(12n + 3) - 3 = 12n = 12$, not divisible by 60. Reject.
Choice (B): If $n = 1$, then $(24n + 3) - 3 = 24n = 24$, not divisible by 60. Reject.
Choice (C): If $n = 1$, then $(80n + 3) - 3 = 80n = 80$, not divisible by 60. Reject.
Choice (D): If $n = 1$, then $(90n + 2) - 3 = 90n - 1 = 89$, not divisible by 60. Reject.
Choice (E): $(120n + 3) - 3 = 120n$, divisible by 60 for any integer n. Hence, correct.

20. Suppose the two positive integers m and n do not have a common factor, apart from 1. Then the LCM of m and n is mn. For example, when $m = 14$ (= 2 • 7) and $n = 15$ (= 3 • 5), the LCM is 14 • 15. In this case, the LCM equals mn, and the Column A equals Column B.

Now, suppose the two integers m and n have at least one common factor (other than 1). Then the LCM uses the common factors only once, unlike mn. Hence, the LCM is less than mn. For example, suppose $m = 10$ (= 2 • 5) and $n = 14$ (= 2 • 7). Here, m and n have 2 as a common factor. The LCM of m and n is 2 • 5 • 7 = 70, and mn equals 2 • 5 • $\underline{2}$ • 7 = 140. The LCM did not use the underlined 2 in the evaluation. Hence, here, Column A is less than Column B.

Since this is a double case, the answer is (D).

21. Any number divisible by both 6 and 8 must be a multiple of the least common multiple of the two numbers, which is 24. Hence, any such number can be represented as $24n$. If 3072 is one such number and is represented as $24n$, then the next such number should be $24(n + 1) = 24n + 24 = 3072 + 24 = 3096$. The answer is (E).

22. Any number divisible by both m and n must be a multiple of the least common multiple of the two numbers, which is given to be 24. The first multiple of 24 greater than 3070 is 3072. Hence, the answer is (A).

23. The least common multiple of 6 and 8 is 24. Hence, a multiple of both 6 and 8 must also be a multiple of 24.

We know that a multiple of 24 exists once in every 24 consecutive numbers. For example, between the numbers 34 (not itself a multiple of 24) and 34 + 24, there must be exactly one multiple of 24. Similarly, exactly one multiple of 24 must exist between 300 (not itself a multiple of 24) and 300 + 24 (= 324, Column B), and this number is the first multiple of 24 larger than 300, Column A. So, Column A lies between 300 and 324 (Column B) and therefore is less than Column B. The answer is (B).

24. The consecutive integers a, b, and c in the increasing order of size can be expressed as a, $a + 1$, $a + 2$, respectively.

Now, Column A equals $\frac{a}{5} - \frac{b}{6} = \frac{a}{5} - \frac{a+1}{6} = \frac{6a - 5a - 5}{30} = \frac{a-5}{30}$.

Column B equals

$$\frac{b}{5} - \frac{c}{6} =$$

$$\frac{a+1}{5} - \frac{a+2}{6} =$$

$$\frac{6a + 6 - 5a - 10}{30} =$$

$$\frac{a-4}{30} =$$

$$\frac{a-4+1-1}{30} = \quad \text{by adding and subtracting 1 from the numerator}$$

$$\frac{a-5+1}{30} =$$

$$\frac{a-5}{30} + \frac{1}{30} =$$

$$\text{Column A} + \frac{1}{30}$$

Since Column B is 1/30 units greater than Column A, the answer is (B).

25. There are 5 digits that are not even or zero: 1, 3, 5, 7, and 9. Now, let's count all the three-digit numbers that can be formed from these five digits. The first digit of the number can be filled in 5 ways with any one of the mentioned digits. Similarly, the second and third digits of the number can be filled in 5 ways. Hence, the total number of ways of forming the three-digit number is 125 ($= 5 \times 5 \times 5$). The answer is (E).

26. Since the digits of the number differ by 4, the number can be any one of the numbers 15, 26, 37, 48, and 59 or any one of their reverse numbers. Hence, the positive difference between the digits varies from $5^2 - 1^2 = 24$ (for the numbers 15 or 51) to $9^2 - 5^2 = 56$ (for the numbers 59 or 95). Hence, Column A is a number between 24 and 56, inclusive, and therefore is always greater than 15, the value of Column B. Hence, the answer is (A).

➢ **Hard**

27. List all possible factors x and y whose product is 36, and calculate the corresponding sum $x + y$:

x	y	xy	$x + y$
1	36	36	37
2	18	36	20
3	12	36	15
4	9	36	13
6	6	36	12

From the table, the minimum sum is 12. The answer is (E).

28. We are given that $x = 2 \cdot 3 \cdot 7 \cdot a = 42a$ and $y = 2 \cdot 2 \cdot 8 \cdot b = 32b$.

We are given that the values of both x and y lie between 120 and 130 (not including the two).

The only multiple of 42 in this range is $42 \times 3 = 126$. Hence, $x = 126$ and $a = 3$. The only multiple of 32 in this range is $32 \times 4 = 128$. Hence, $y = 128$ and $b = 4$. Hence, $a - b = 3 - 4 = -1$. The answer is (B).

29. Usually, a system having a single constraint such as $5m + 7n = 46$ will not have a unique solution.

But the given system has two constraints:

1) $5m + 7n = 46$
2) m and n are positive integers

Hence, we might have a unique solution. Let's see:

Let $p = 5m$ (p is a multiple of 5) and $q = 7n$ (q is a multiple of 7) such that and $p + q = 46$. Subtracting q from both sides yields $p = 46 - q$ [(a positive multiple of 5) equals 46 – (a positive multiple of 7)]. Let's look at how many such solutions exist:

If $q = 7$, $p = 46 - 7 = 39$, not a multiple of 5. Reject.

If $q = 14$, $p = 46 - 14 = 32$, not a multiple of 5. Reject.

If $q = 21$, $p = 46 - 21 = 25$, a multiple of 5. Acceptable. So, $m = 25/5 = 5$ and $q = 21/7 = 3$.

If $q = 28$, $p = 46 - 28 = 18$, not a multiple of 5. Reject.

If $q = 35$, $p = 46 - 35 = 11$, not a multiple of 5. Reject.

If $q = 42$, $p = 46 - 42 = 4$, not a multiple of 5. Reject.

If $q \geq 49$, $p \leq 46 - 49 = -3$, not positive. Reject.

The solution is Column A = m = 5, and Column B = n = 3. Hence, Column A is greater and the answer is (A).

30. Since $b = a + 3$, the digit b is greater than the digit a and $a - b = -3$.

In a two-digit number, the leftmost digit (the tens-digit) is the "more significant" digit. Since $b > a$, the number ba (which has b in tens-digit position) is greater than the number ab (which has a in tens-digit position). Hence, Column B is greater than Column A, and the answer is (B).

Method II
Column A: $ab = 10a + b$.

Column B: $ba = 10b + a$.

Column A – Column B = $10a + b - (10b + a) = 9a - 9b = 9(a - b) = 9(-3) = -27$.

Hence, Column A is 27 units less than Column B. The answer is (B).

31. Since a, b, c, d, and e are consecutive numbers in the increasing order, we have $b = a + 1$, $c = a + 2$, $d = a + 3$ and $e = a + 4$. The sum of the five numbers is $a + (a + 1) + (a + 2) + (a + 3) + (a + 4) = 5a + 10$.

Now, we are given that the sum decreased by 20% when one number was deleted. Hence, the new sum should be $(5a + 10)(1 - 20/100) = (5a + 10)(1 - 1/5) = (5a + 10)(4/5) = 4a + 8$. Now, since New Sum = Old Sum – Dropped Number, we have $(5a + 10) = (4a + 8)$ + Dropped Number. Hence, the number dropped is $(5a + 10) - (4a + 8) = a + 2$. Since $c = a + 2$, the answer is (C).

32. The average of the five consecutive positive integers 1, 2, 3, 4, and 5 is $(1 + 2 + 3 + 4 + 5)/5 = 15/5 = 3$. After dropping 5 (the greatest number), the new average becomes $(1 + 2 + 3 + 4)/4 = 10/4 = 2.5$. The percentage drop in the average is

$$\frac{\text{Old average} - \text{New average}}{\text{Old average}} \cdot 100 =$$

$$\frac{3 - 2.5}{3} \cdot 100 =$$

$$\frac{100}{6} =$$

$$16.66\%$$

The answer is (E).

33. The term 14! equals the product of the numbers 1, 2, 3, 4, 5, 6, 7, 8, 9, 10, 11, 12, 13, and 14. Only two of these numbers are divisible by 7. The numbers are 7 and 14. Hence, 14! can be expressed as the product of $k \cdot 7 \cdot 14$, where k is not divisible by 7. Now, since there are two 7s in 14!, the numbers 7 and 7^2 divide 14! evenly. 7^3 and further powers of 7 leave a remainder when divided into 14!. Hence, the maximum value of m is 2. The answer is (B).

34. The sum of the multiples of 6 between 0 and 100 equals $6 + 12 + 18 + \dots + 96 = 6(1) + 6(2) + 6(3) + \dots + 6(16) = 6(1 + 2 + 3 + \dots + 16) = 6\left(\frac{16(16+1)}{2}\right) = 3 \cdot 16 \cdot 17$.

The sum of multiples of 8 between 0 and 100 equals $8 + 16 + 24 + \dots + 96 = 8(1) + 8(2) + 8(3) + \dots + 8(12) = 8(1 + 2 + 3 + \dots + 12) = 8\left(\frac{12(12+1)}{2}\right) = 4 \cdot 12 \cdot 13$.

Since $3 \cdot 16 \cdot 17$ ($= 48 \cdot 17$) is clearly greater than $4 \cdot 12 \cdot 13$ ($= 48 \cdot 13$), Column A is greater than Column B and the answer is (A).

35. A palindrome number reads the same forward or backward. There is no palindrome number between 233 through 239 since none of the numbers read the same both forward and backward. Hence, the palindrome number immediately after 233 is the same as the palindrome number immediate after 239. Hence, Column A and Column B refer to the same number, and the answer is (C).

➢ Very Hard

36. We have that the number $2ab5$ is divisible by 25. Any number divisible by 25 ends with the last two digits 00, 25, 50, or 75. So, $b5$ should equal 25 or 75. Hence, $b = 2$ or 7. Since a is now free to take any digit from 0 through 9, ab can have multiple values.

We also have that ab is divisible by 13. The multiples of 13 are 13, 26, 39, 52, 65, 78, and 91. Among these, the only number ending with 2 or 7 is 52. Hence, $ab = 52$. The answer is (C).

37. We are given that the numbers m and n, when divided by 6, leave remainders of 2 and 3, respectively. Hence, we can represent the numbers m and n as $6p + 2$ and $6q + 3$, respectively, where p and q are suitable integers.

Now, $m - n = (6p + 2) - (6q + 3) = 6p - 6q - 1 = 6(p - q) - 1$. A remainder must be positive, so let's add 6 to this expression and compensate by subtracting 6:

$$6(p - q) - 1 =$$

$$6(p - q) - 6 + 6 - 1 =$$

$$6(p - q) - 6 + 5 =$$

$$6(p - q - 1) + 5$$

Thus, the remainder is 5, and the answer is (E).

38. Since the remainder when $m + n$ is divided by 12 is 8, $m + n = 12p + 8$; and since the remainder when $m - n$ is divided by 12 is 6, $m - n = 12q + 6$. Here, p and q are integers. Adding the two equations yields $2m = 12p + 12q + 14$. Solving for m yields $m = 6p + 6q + 7 = 6(p + q + 1) + 1 = 6r + 1$, where r is a positive integer equaling $p + q + 1$. Now, let's subtract the equations $m + n = 12p + 8$ and $m - n = 12q + 6$. This yields $2n = (12p + 8) - (12q + 6) = 12(p - q) + 2$. Solving for n yields $n = 6(p - q) + 1 = 6t + 1$, where t is an integer equaling $p - q$.

Hence, we have

$$\begin{aligned} mn &= (6r + 1)(6t + 1) \\ &= 36rt + 6r + 6t + 1 \\ &= 6(6rt + r + t) + 1 \qquad \text{by factoring out 6} \end{aligned}$$

Hence, the remainder is 1, and the answer is (A).

39. $7^2 \cdot 8^2 = (7 \cdot 8)^2 = 56^2$.

The number immediately before 56 that is divisible by 6 is 54. Now, writing 56^2 as $(54 + 2)^2$, we have

$$\begin{aligned} 56^2 &= (54 + 2)^2 \\ &= 54^2 + 2(2)(54) + 2^2 \qquad \text{by the formula } (a + b)^2 = a^2 + 2ab + b^2 \\ &= 54[54 + 2(2)] + 2^2 \\ &= 6 \times 9[54 + 2(2)] + 4 \qquad \text{here, the remainder is 4} \end{aligned}$$

Since the remainder is 4, the answer is (D).

40. Choice (A): $a + b + c$: Suppose a is an even number. Then b, the integer following a, must be odd, and c, the integer following b, must be even. Hence, $a + b + c$ = sum of two even numbers (a and c) and an odd number (b). Since the sum of any number of even numbers with an odd number is odd (For example, if $a = 4$, then $b = 5$, $c = 6$, and $a + b + c$ equals $4 + 5 + 6 = 15$ (odd)), $a + b + c$ is odd. Reject.

Choice (B): $ab + c$: At least one of every two consecutive positive integers a and b must be even. Hence, the product ab is an even number. Now, if c is odd (which happens when a is odd), $ab + c$ must be odd. For example, if $a = 3$, $b = 4$, and $c = 5$, then $ab + c$ must equal $12 + 5 = 17$, an odd number. Reject.

Choice (C): $ab + d$: We know that ab being the product of two consecutive numbers must be even. Hence, if d happens to be an odd number (it happens when b is odd), then the sum $ab + d$ is also odd. For example, if $a = 4$, then $b = 5$, $c = 6$, and $d = 7$, then $ab + d = 3 \cdot 5 + 7 = 15 + 7 = 23$, an odd number. Reject.

Choice (D): $ac + d$: Suppose a is odd. Then c must also be odd, being a number 2 more than a. Hence, ac is the product of two odd numbers and must therefore be odd. Now, d is the integer following c and must be even. Hence, $ac + d$ = odd + even = odd. For example, if $a = 3$, then $b = 3 + 1 = 4$, $c = 4 + 1 = 5$ (odd) and $d = 5 + 1 = 6$ (even) and $ac + d = 3 \cdot 5 + 6 = 21$, an odd number. Reject.

Choice (E): $ac + e$: Suppose a is an odd number. Then both c and e must also be odd. Now, ac is product of two odd numbers and therefore must be odd. Summing this with another odd number e yields an even number. For example, if $a = 1$, then c must equal 3, and e must equal 5 and $ac + e$ must equal $1 \cdot 3 + 5 = 8$, an even number. Now, suppose a is an even number. Then both c and e must also be even. Hence,

$ac + e =$

(product of two even numbers) + (an even number) =

(even number) + (even number) =

an even number

For example, if $a = 2$, then c must equal 4, and e must equal 6 and the expression $ac + e$ equals 14, an even number. Hence, in any case, $ac + e$ is even. Correct.

The answer is (E).

41. A number less than 500 will be 1) a single-digit number, or 2) a double-digit number, or a 3) triple-digit number with left-most digit less than 5.

Let the compartments shown below represent the single, double and three digit numbers.

1 or 2 or 3 or 5 ------- (Total 4 ways)	+	1 or 2 or 3 or 5 ------- (Total 4 ways)	1 or 2 or 3 or 5 ------- (Total 4 ways)	+	1 or 2 or 3 or 5 not allowed ------- (Total 3 ways)	1 or 2 or 3 or 5 ------- (Total 4 ways)	1 or 2 or 3 or 5 ------- (Total 4 ways)
Compartment for single-digit number		Compartments for double-digit number.			Compartments for triple-digit number.		

$= 4 + (4 \times 4) + (3 \times 4 \times 4) = 4 + 16 + 48 = 68$

The compartment for the single-digit number can be filled in 4 ways (with any one of the numbers 1, 2, 3, and 5).

Each of the two compartments for the double-digit number can be filled in 4 ways (with any one of the 4 numbers 1, 2, 3, and 5) each. Hence, the two-digit number can be made in $4 \times 4 = 16$ ways.

Regarding the three-digit number, the left most compartment can be filled in 3 ways (with any one of the numbers 1, 2, and 3). Each of the remaining two compartments can be filled in 4 ways (with any one of the numbers 1, 2, 3, and 5) each. Hence, total number of ways of forming the three-digit number equals $3 \times 4 \times 4 = 48$.

Hence, the total number of ways of forming the number is $4 + 16 + 48 = 68$. The answer is (D).

42. We know that $3^7 = 3 \cdot 3^3 \cdot 3^3 = 3 \cdot 27 \cdot 27 = 3(27^2)$. The number immediately before 27 that is divisible by 8 is 24. Hence, replace 27 with 24 + 3. Then we have

$$3^7 = 3(27^2) = 3(24 + 3)^2 = 3(24^2 + 2 \cdot 24 \cdot 3 + 3^2)$$
$$= 3 \cdot 24^2 + 3 \cdot 2 \cdot 24 \cdot 3 + 3 \cdot 9$$

Now,

$$\frac{3^7}{8} = \frac{3 \cdot 24^2 + 3 \cdot 2 \cdot 24 \cdot 3 + 3 \cdot 9}{8}$$
$$= \frac{3 \cdot 24^2}{8} + \frac{3 \cdot 2 \cdot 24 \cdot 3}{8} + \frac{3 \cdot 9}{8}$$
$$= \text{Integer} + \text{Integer} + \frac{27}{8}$$
$$= \text{Integer} + \text{Integer} + \frac{24 + 3}{8}$$
$$= \text{Integer} + \text{Integer} + 3 + \frac{3}{8}$$

Hence, the remainder is 3, and the answer is (C).

43. We are given that the numbers m and n, when divided by 6, leave remainders of 2 and 3, respectively. Hence, we can represent the numbers m and n as $6p + 2$ and $6q + 3$, respectively, where p and q are suitable integers.

Now, $m + n = (6p + 2) + (6q + 3) = 6p + 6q + 5 = 6(p + q) + 5$; the remainder is 5. Hence, Column A equals 5.

Now, $m - n = (6p + 2) - (6q + 3) = 6p - 6q - 1 = 6(p - q) - 1$. A remainder must be positive, so let's add 6 to this expression and compensate by subtracting 6:

$$6(p - q) - 1 =$$

$$6(p - q) - 6 + 6 - 1 =$$

$$6(p - q) - 6 + 5 =$$

$$6(p - q - 1) + 5$$

Thus, the remainder is 5, and Column B equals 5.

Since both columns equal 5, the answer is (C).

Quantitative Comparisons

Quantitative comparisons make up one-half of the math portion of the GRE. This is good news because they are the easiest problems to improve on.

Generally, quantitative comparison questions require much less calculating than do multiple-choice questions. But they are trickier.

Substitution is very effective with quantitative comparison problems. But you must plug in all five major types of numbers: positives, negatives, fractions, 0, and 1. Test 0, 1, 2, –2, and 1/2, in that order.

GENERAL PRINCIPLES FOR SOLVING QUANTITATIVE COMPARISONS

The following principles can greatly simplify quantitative comparison problems.

You Can Add or Subtract the Same Term (Number) from Both Sides of a Quantitative Comparison Problem.

You Can Multiply or Divide Both Sides of a Quantitative Comparison Problem by the Same Positive Term (Number). (**Caution:** This cannot be done if the term can ever be negative or zero.)

You can think of a quantitative comparison problem as an inequality/equation. Your job is to determine whether the correct symbol with which to compare the columns is <, =, >, or that it cannot be determined. Therefore, all the rules that apply to solving inequalities apply to quantitative comparisons. That is, you can always add or subtract the same term to both columns of the problem. If the term is always positive, then you can multiply or divide both columns by it. (The term cannot be negative because it would then invert the inequality. And, of course, it cannot be zero if you are dividing.)

Example:

Column A	Column B
$\frac{1}{5}+\frac{1}{3}+\frac{1}{8}$	$\frac{1}{8}+\frac{1}{5}+\frac{1}{4}$

Don't solve this problem by adding the fractions in each column; that would be too time consuming—the LCD is 120! Instead, merely subtract $\frac{1}{5}$ and $\frac{1}{8}$ from both columns:

Column A	Column B
$\frac{1}{3}$	$\frac{1}{4}$

Now $\frac{1}{3}$ is larger than $\frac{1}{4}$, so Column A is larger than Column B.

If there are only numbers (i.e., no variables) in a quantitative comparison problem, then "not-enough-information" cannot be the answer. Hence (D), not-enough-information, cannot be the answer to the example above.

Example:

Column A	$y > 0$	Column B
$y^3 + y^4$		$y^4 - 2y^2$

First cancel y^4 from both columns:

Column A	$y > 0$	Column B
y^3		$-2y^2$

Since $y > 0$, we can divide both columns by y^2:

Column A	$y > 0$	Column B
y		-2

Now, we are given that $y > 0$. Hence, Column A is greater. The answer is (A).

Example:

Column A	$x > 1$	Column B
$\frac{1}{x}$		$\frac{1}{x-1}$

Since $x > 1$, $x - 1 > 0$. Hence, we can multiply both columns by $x(x - 1)$ to clear fractions. This yields

Column A	$x > 1$	Column B
$x - 1$		x

Subtracting x from both columns yields

Column A	$x > 1$	Column B
-1		0

In this form, it is clear that Column B is larger. The answer is (B).

Example:

Column A	n is a positive integer and $0 < x < 1$	Column B
$\frac{n^2}{x}$		n^2

Since we are given that n is positive, we may multiply both columns by $\frac{1}{n^2}$:

Column A	Column B
$\frac{n^2}{x} \cdot \frac{1}{n^2}$	$n^2 \cdot \frac{1}{n^2}$

Reducing yields

Column A	Column B
$\frac{1}{x}$	1

We are also given that $0 < x < 1$. So, we may multiply both columns by x to get

Column A	Column B
1	x

But again, we know that $0 < x < 1$. Hence, Column A is larger.

Watch out!

You Must Be Certain That the Quantity You Are Multiplying or Dividing by Can Never Be Zero or Negative. (There are no restrictions on adding or subtracting.)

The following example illustrates the false results that can occur if you don't guarantee that the number you are multiplying or dividing by is positive.

Column A	$0 \le x < 1$	Column B
x^3		x^2

Solution (Invalid): Dividing both columns by x^2 yields

Column A	Column B
x	1

We are given that $x < 1$, so Column B is larger. But this is a false result because when $x = 0$, the two original columns are equal:

Column A	Column B
$0^3 = 0$	$0^2 = 0$

Hence, the answer is actually (D), not-enough-information to decide.

Watch out!

Don't Cancel Willy-Nilly.

Some people are tempted to cancel the x^2s from both columns of the following problem:

Column A	Column B
$x^2 + 4x - 6$	$6 + 4x - x^2$

You cannot cancel the x^2s from both columns because they do not have the same sign. In Column A, x^2 is positive. Whereas in Column B, it is negative.

Strategy

You Can Square Both Sides of a Quantitative Comparison Problem to Eliminate Square Roots.

Example:

Column A	Column B
$\sqrt{3} + \sqrt{5}$	$\sqrt{8}$

Squaring both columns yields

Column A	Column B
$\left(\sqrt{3} + \sqrt{5}\right)^2$	$\left(\sqrt{8}\right)^2$

or

Column A	Column B
$3+2\sqrt{3}\sqrt{5}+5$	8

Reducing gives

Column A	Column B
$8+2\sqrt{3}\sqrt{5}$	8

Now, clearly Column A is larger.

Example:

Column A	Column B
$\dfrac{\sqrt{2}}{3}$	$\dfrac{2}{5}$

Multiplying both columns by 15 to clear fractions yields

Column A	Column B
$5\sqrt{2}$	6

Squaring both columns yields

Column A	Column B
$25\cdot 2$	36

Performing the multiplication in Column A yields

Column A	Column B
50	36

Hence, Column A is larger, and the answer is (A).

<u>SUBSTITUTION</u> (Special Cases)

We already studied this method in the section *Substitution*. Here, we will practice more and learn a couple of special cases.

A. In a problem with two variables, say, x and y, you must check the case in which x = y. (This often gives a double case.)

Example:

Column A		Column B
	x and y are positive.	
Average of x and y		Average of x^3 and y^3

Let $x = y = 1$. Then Column A becomes $\dfrac{1+1}{2}=1$. And Column B becomes $\dfrac{1^3+1^3}{2}=1$. In this case, the columns are equal. But if $x = y = 2$, then Column A becomes $\dfrac{2+2}{2}=2$ and Column B becomes $\dfrac{2^3+2^3}{2}=8$. In this case, the columns are unequal. This is a double case and therefore the answer is (D).

Example:

Column A	x and y are integers greater than or equal to 1.	Column B
2^{x+y}		$2^x + 2^y$

If $x \neq y$, then Column A is larger than Column B. (Plug in a few numbers until you are convinced.) But if $x = y = 1$, then the columns are equal: $2^{x+y} = 2^{1+1} = 2^2 = 4$ and $2^x + 2^y = 2^1 + 2^1 = 4$. Hence, there is not enough information to decide.

B. When you are given x < 0, you must plug in negative whole numbers, negative fractions, and –1. (Choose the numbers –1, –2, and –1/2, in that order.)

Example:

Column A	$k < 0$	Column B
$k^2\left(k+\frac{1}{2}\right)^2$		0

If k is –1 or –2, then Column A is larger since it is a product of squares. But if $k = -1/2$, then the two columns are equal: $k^2\left(k+\frac{1}{2}\right)^2 = \left(-\frac{1}{2}\right)^2\left(-\frac{1}{2}+\frac{1}{2}\right)^2 = \frac{1}{4}\cdot 0 = 0$. Hence, there is not enough information to decide and the answer is (D).

C. Sometimes you have to plug in the first three numbers (but never more than three) from a class of numbers.

Example:

x is both an integer and greater than 1. Let $\boxed{x}$ stand for the smallest positive integer factor of x not equal to 1.

Column A	Column B
$\boxed{x}$	$\boxed{x^3}$

Choose $x = 2$, 3, and 4. If $x = 2$, then $\boxed{x} = 2$ and $\boxed{x^3} = \boxed{8} = 2$. So, for this choice of x, the two columns are equal. If $x = 3$, then $\boxed{x} = 3$ and $\boxed{x^3} = \boxed{27} = 3$, again the columns are equal. Finally, If $x = 4$, then $\boxed{x} = 2$ and $\boxed{x^3} = \boxed{64} = 2$, still again the columns are equal. Hence, the answer is (C). Note, there is no need to check $x = 5$. The writers of the GRE do not change the results after the third number.

Problem Set F:

➢ **Easy**

1\. Column A | n is a positive integer | Column B

n | | The sum of two integers whose product is n

2\. Column A | $0 < x < 1$ | Column B

x^2 | | x

3\. Column A | | Column B

The count of the numbers between 100 and 300 that are divisible by both 5 and 6 | | The count of the numbers between 100 and 300 that are divisible by either 5 or 6

➢ **Medium**

4\. Column A | In the rectangular coordinate system shown, points A and E lie on the x-axis, and points B and D lie on the y-axis. Point C is the midpoint of the line AB, and point F is the midpoint of the line DE. | Column B

The slope of the line AB | | The slope of the line DE

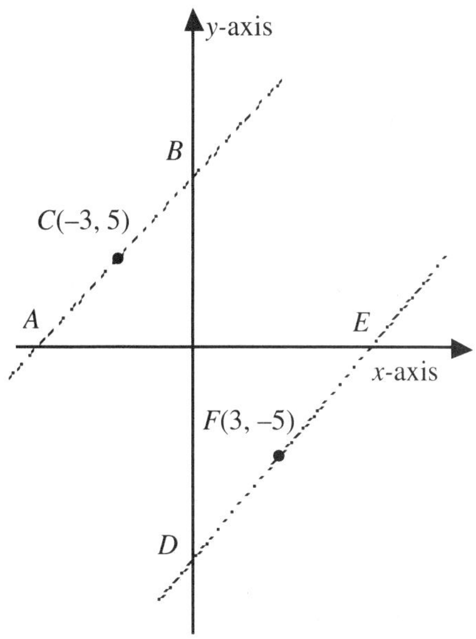

5\. Column A | $x = 3.635 \bullet 10^{16}$ | Column B

$\frac{x+1}{x-1}$ | | $\frac{x-1}{x+1}$

6\. Column A | $a + \frac{b}{10} > 0$
 $b + \frac{a}{10} < 0$ | Column B

a | | b

	Column A		Column B
7.	x	$\lvert x\rvert + \lvert y\rvert = 2$ and $x + y = 0$.	y
8.	$(4 - 5x) + (4 - 5x)^2$	$(4 - 5x)^2 = 1$	1
9.	$\dfrac{x+14}{y+14}$	The ratio of x to y is 3 : 4, and the ratio of $x + 7$ to $y + 7$ is 4 : 5.	$\dfrac{5}{6}$
10.	The product of x and y	x and y are two numbers	The product of 80% of x and 125% of y
11.	$\dfrac{\text{The area of } \Delta DEF}{\text{The area of } \Delta ABC}$	ABC and DEF are right triangles. Each side of ΔABC is twice the length of the corresponding side of ΔDEF.	$\dfrac{1}{2}$
12.	a_5	The nth term of the sequence $a_1, a_2, a_3, \ldots, a_n$ is defined as $a_n = -(a_{n-1})$. The first term a_1 equals -1.	1
13.	The probability that both Tom and John get the degree	Everyone who passes the test will be awarded a degree. The probability that Tom passes the test is 0.5, and the probability that John passes the test is 0.4. The two events are independent of each other.	The probability that at least one of them gets the degree
14.	The difference between two angles of a triangle		179.5°
15.	x	$2x + 1 > 3x + 2$ and $5x + 2 > 4x$	1

➤ Hard

	Column A		Column B
16.	k	$\dfrac{l}{m+n} = \dfrac{m}{n+l} = \dfrac{n}{l+m} = k$, where k is a real number.	1/3

17.

Column A		Column B
	The average of a set of six positive numbers is 30.	
The average of the numbers in the set after replacing the smallest number in the set with 0		The average of the remaining numbers in the set after removing the smallest number from the set

18.

Column A		Column B
	In the figure, *ABCD* and *BECD* are parallelograms.	
Area of parallelogram *ABCD*		Area of parallelogram *BECD*

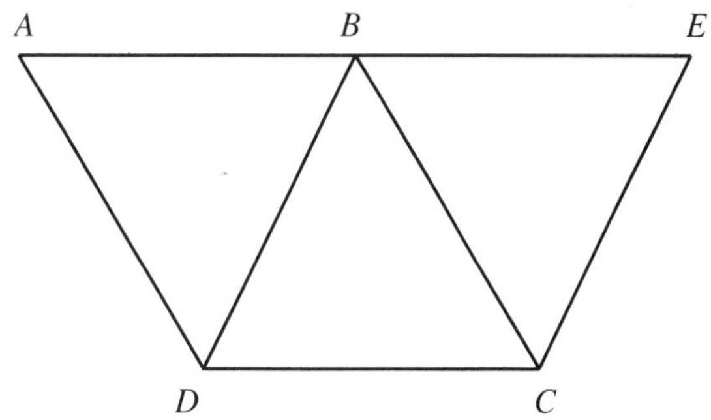

➢ Very Hard

19.

Column A		Column B
	A set has exactly five consecutive positive integers.	
The percentage decrease in the average of the numbers when one of the numbers is dropped from the set		20%

20.

Column A		Column B
	A tank is filled with x pounds of wheat. The tank has a hole at the bottom and each day 1% of the wheat is lost from the tank through the hole.	
Percentage of wheat lost in the first three days		3%

Answers and Solutions to Problem Set F

➢ Easy

1. Suppose $n = 6$. Now, the factors of 6 are 6 and 1, and 2 and 3.

In the first case, factors 6 and 1 sum to 7 (= 6 + 1), which is greater than 6 (= Column A). Here, Column B is greater.

In the second case, the factors 2 and 3 sum to 5 (= 2 + 3), which is less than 6 (= Column A). Here, Column A is greater. This is a double case, and therefore the answer is (D).

2. Since $0 < x < 1$, we know that x is positive. Now, multiplying both sides of the inequality $0 < x < 1$ by x yields

$$0 < x^2 < x$$

Hence, Column B is greater than Column A, and the answer is (B).

3. The least common multiple of 5 and 6 is the product of the two (since 5 and 6 have no common factors), which is $5 \cdot 6 = 30$. Hence, if a number is divisible by both 5 and 6, it must be a multiple of 30. For example, the numbers 30, 60, 90, … are divisible by both 5 and 6.

Now, if a number is divisible by either 5 or 6, the number must be a multiple of at least one of them. Since there are more multiples of 5 and 6 than there are of 30 between 100 and 300, the count in Column B is greater than the count in Column A.

The answer is (B).

➢ Medium

4. Let the coordinate representations of points A and E (which are on the x-axis) be $(a, 0)$, and $(e, 0)$, respectively. Also, let the coordinates of the points B and D (which are on the y-axis) be $(0, b)$ and $(0, d)$, respectively.

The formula for the midpoint of two points (x_1, y_1) and (x_2, y_2) in a coordinate system is $\left(\dfrac{x_1 + x_2}{2}, \dfrac{y_1 + y_2}{2}\right)$.

Hence, the midpoint of AB is $\left(\dfrac{a+0}{2}, \dfrac{0+b}{2}\right) = \left(\dfrac{a}{2}, \dfrac{b}{2}\right) = C(-3, 5)$. Equating the x- and y-coordinates on both sides yields $a/2 = -3$ and $b/2 = 5$. Solving for a and b yields $a = -6$ and $b = 10$.

Similarly, the midpoint of DE is $\left(\dfrac{e+0}{2}, \dfrac{0+d}{2}\right) = \left(\dfrac{e}{2}, \dfrac{d}{2}\right) = F(3, -5)$. Equating the x- and y-coordinates on both sides yields $e/2 = 3$ and $d/2 = -5$. Solving for e and d yields $e = 6$ and $d = -10$.

The slope of a line through two points (x_1, y_1) and (x_2, y_2) in a coordinate system is $\dfrac{y_2 - y_1}{x_2 - x_1}$.

Hence, Column A, which is the slope of AB, equals $\dfrac{b-0}{0-a} = -\dfrac{b}{a} = -\dfrac{10}{-6} = \dfrac{5}{3}$.

Column B, which is the slope of DE, equals $\dfrac{d-0}{0-e} = -\dfrac{d}{e} = -\dfrac{-10}{6} = \dfrac{5}{3}$.

Since both columns equal 5/3, the answer is (C).

5. Since x is a large positive number, both $x + 1$ and $x - 1$ are positive. Hence, we can clear fractions by multiplying both columns by $(x + 1)(x - 1)$, which yields

$$(x+1)^2 \qquad\qquad\qquad (x-1)^2$$

Performing the multiplication yields

$$x^2 + 2x + 1 \qquad\qquad\qquad x^2 - 2x + 1$$

Subtracting x^2 and 1 from both columns yields

$$2x \qquad\qquad\qquad -2x$$

Since x is a positive number, Column A is positive and Column B is negative.

Since all positive numbers are greater than all negative numbers, Column A is greater than Column B and the answer is (A).

6. Multiplying the given inequalities by 10 yields

$$10a + b > 0$$
$$10b + a < 0$$

Multiplying the second inequality by –1 and flipping the direction of the inequality yields

$$10a + b > 0$$
$$-a - 10b > 0$$

Adding these inequalities yields

$$9a - 9b > 0$$

Adding $9b$ to both sides of the inequality yields

$$9a > 9b$$

Finally, dividing both sides of the inequality by 9 yields

$$a > b$$

Hence, Column A is greater than Column B. The answer is (A).

7. Suppose $x = 1$ and $y = -1$. Then $|x| + |y| = |1| + |-1| = 1 + 1 = 2$, and $x + y = 1 + (-1) = 0$. Hence, $x = 1$ and $y = -1$ is a solution of the system of equations, and here x is greater than y.

Now, suppose x = –1 and y = 1. Then $|x| + |y| = |-1| + |1| = 1 + 1 = 2$, and $x + y = -1 + 1 = 0$. Hence, $x = -1$ and $y = 1$ is another solution of the system of equations, and here x is less than y.

Thus, we have a double case, and the answer is (D).

8. $(4 - 5x)^2$ equals 1 when either $4 - 5x = -1$ or $+1$. If $4 - 5x$ equals –1, then $(4 - 5x) + (4 - 5x)^2 = -1 + 1 = 0$; and if $4 - 5x$ equals 1, then $(4 - 5x) + (4 - 5x)^2 = 1 + 1 = 2$. In the first case, Column A is less than Column B; and in the second case, Column B is less than Column A. Hence, we have a double case, and the answer is (D).

9. Forming the two ratios yields $\frac{x}{y}=\frac{3}{4}$ and $\frac{x+7}{y+7}=\frac{4}{5}$. Let's solve this system of equations by the substitution method. Multiplying the first equation by y yields $x = 3y/4$. Substituting this into the second equation yields $\frac{\frac{3y}{4}+7}{y+7}=\frac{4}{5}$. Cross-multiplying yields

$$5(3y/4 + 7) = 4(y + 7)$$
$$15y/4 + 35 = 4y + 28$$
$$15y/4 - 4y = 28 - 35$$
$$-y/4 = -7$$
$$y = 28$$

and $x=\frac{3y}{4}=\frac{3\cdot 28}{4}=3\cdot 7=21$.

Plugging these values for x and y into the expression in Column A, $\frac{x+14}{y+14}$, yields

$$\frac{21+14}{28+14}=$$
$$\frac{35}{42}=$$
$$\frac{5}{6}=$$
Column B

Hence, the answer is (C).

10. Column A: The product of x and y is xy.

Column B: 80% of x is $\frac{80}{100}x=\frac{4}{5}x$.

And 125% of y is $\frac{125}{100}y=\frac{5}{4}y$.

The product of the two is $\left(\frac{4}{5}x\right)\left(\frac{5}{4}y\right)= xy =$ Column A.

Hence, the columns are equal, and the answer is (C).

11. We are given that each side of triangle *ABC* is twice the length of the corresponding side of triangle *DEF*. Hence, each leg of triangle *ABC* must be twice the length of the corresponding leg in triangle *DEF*. The formula for the area of a right triangle is 1/2 • (*product of the measures of the two legs*). Hence,

$$\text{Column A} =$$

$$\frac{\text{The area of } \Delta DEF}{\text{The area of } \Delta ABC} =$$

$$\frac{\frac{1}{2}(\text{leg 1 of } \Delta DEF)(\text{leg 2 of } \Delta DEF)}{\frac{1}{2}(\text{leg 1 of } \Delta ABC)(\text{leg 2 of } \Delta ABC)} =$$

$$\frac{\frac{1}{2}(\text{leg 1 of } \Delta DEF)(\text{leg 2 of } \Delta DEF)}{\frac{1}{2}(2 \bullet \text{leg 1 of } \Delta DEF)(2 \bullet \text{leg 2 of } \Delta DEF)} =$$

$$\frac{\frac{1}{2}}{\frac{1}{2}(2)(2)} =$$

$$\frac{1}{2 \cdot 2} =$$

$$\frac{1}{4} \text{ and this is less than } \frac{1}{2} \text{ (= Column B)}$$

Hence, the answer is (B).

12. The rule for the sequence is $a_n = -(a_{n-1})$. Putting $n = 2$ and 3 in the rule yields

$$a_2 = -(a_{2-1}) = -a_1 = -(-1) = 1 \qquad \text{(given that } a_1 = -1)$$
$$a_3 = -(a_{3-1}) = -a_2 = -1$$

Similarly, we get that *each* even numbered term (when n is even) equals 1 and *each* odd numbered term (when n is odd) equals –1. Since a_5 is an odd numbered term, it equals –1. Hence, Column A equals –1 and is less than 1 (= Column B). The answer is (B).

13. The case of both Tom and John getting a degree is just one of the cases in which at least one of them gets the degree (Column A is one of the cases of Column B). Hence, the probability of the former is less than the probability of the later (Column B is greater). Also, the probability of the remaining case (exactly one of the two passing) is not zero. So, Column A cannot equal Column B. The answer is (B).

14. Suppose the angles of the triangle measure 179.8°, 0.1°, and 0.1°. Then the difference between the first two angles of the triangle is 179.8 – 0.1 = 179.7 > Column B and the difference between the last two angles is 0.1 – 0.1 = 0 < Column B.

Hence, we have a double case, and the answer is (D).

15. We are given the two inequalities

$$2x + 1 > 3x + 2$$
$$5x + 2 > 4x$$

Subtracting $2x + 2$ from both sides of the top inequality and subtracting $4x + 2$ from both sides of the bottom inequality yields

$$-1 > x$$
$$x > -2$$

Combining these inequalities yields $-1 > x > -2$. Since any number between –1 and –2 is less than 1, Column B is greater than Column A. The answer is (B).

➢ Hard

16. The given equation is $\frac{l}{m+n}=\frac{m}{n+l}=\frac{n}{l+m}=k$. Forming the three equations yields $l=(m+n)k$, $m=(n+l)k$, $n=(l+m)k$. Summing these three equations yields

$$\begin{aligned}
l+m+n &= (m+n)k+(n+l)k+(l+m)k \\
&= k[(m+n)+(n+l)+(l+m)] \\
&= k(m+n+n+l+l+m) \\
&= k(2m+2n+2l) \\
&= 2k(m+n+l) \\
1 &= 2k \qquad \text{by canceling } m+n+l \text{ from each side} \\
1/2 &= k
\end{aligned}$$

Hence, Column A equals 1/2. Since 1/2 is greater than 1/3, Column A is greater than Column B, and the answer is (A).

17. Let a, b, c, d, e, and f be the numbers in the set, and let f be the smallest number in the set.

When the smallest number (f) in the set is replaced by 0, the numbers in the set are a, b, c, d, e, and 0. Column A equals the average of these six numbers, which equals $\frac{a+b+c+d+e+0}{6}=\frac{a+b+c+d+e}{6}$.

Instead, if the smallest number in the set is removed, the remaining numbers in the set would be a, b, c, d, and e. Now there are only 5 numbers in the set. Hence, Column B, which equals the average of the remaining numbers (five numbers) in the set, equals $\frac{a+b+c+d+e}{5}$.

Since all the numbers in the set are positive (given), the sum of the five numbers $a+b+c+d+e$ is also positive. Note that dividing a positive number by 5 yields a greater result than dividing it by 6. Hence, $\frac{a+b+c+d+e}{5}$ is greater than $\frac{a+b+c+d+e}{6}$. Thus, Column B is greater than Column A, and the answer is (B).

18. We know that a diagonal of a parallelogram cuts the parallelogram into two triangles of equal area. Since BD is a diagonal of parallelogram $ABCD$, the area of the parallelogram equals twice the area of either of the two equal triangles ΔABD or ΔDBC. Hence, Column A = 2(area of ΔDBC).

Similarly, since BC is a diagonal of parallelogram $BECD$, the area of the parallelogram $BECD$ is 2(area of ΔDBC).

Thus, both columns equal 2(area of ΔDBC), and the answer is (C).

➢ Very Hard

19. The average of the five consecutive positive integers, say, a, $a + 1$, $a + 2$, $a + 3$, and $a + 4$ is

$$\frac{a+(a+1)+(a+2)+(a+3)+(a+4)}{5} =$$
$$\frac{5a+10}{5} =$$
$$a+2$$

The average decrease is a maximum when the greatest number in the set is dropped. Hence, after dropping $a + 4$, the average of the remaining numbers a, $a + 1$, $a + 2$, and $a + 3$ is

$$\frac{a+(a+1)+(a+2)+(a+3)}{4} =$$
$$\frac{4a+6}{4} =$$
$$a+\frac{3}{2}$$

The percentage decrease in the average is

$$\frac{\text{Old Average} - \text{New Average}}{\text{Old Average}} \times 100 =$$
$$\frac{(a+2)-\left(a+\frac{3}{2}\right)}{a+2} \times 100 =$$
$$\frac{\frac{1}{2}}{a+2} \times 100$$

The percentage is a maximum when a takes the minimum possible value. Since a is a positive integer, the minimum value of a is 1. Hence, the maximum possible percentage equals

$$\frac{\frac{1}{2}}{1+2} \times 100 =$$
$$\frac{\frac{1}{2}}{3} \times 100 =$$
$$\frac{1}{2 \cdot 3} \times 100 =$$
$$\frac{100}{6} =$$
$$16.66\%$$

Hence, the maximum possible value of Column A is 16.66%, which is less than Column B. Hence, the answer is (B).

20. We are given that each day 1% of the remaining wheat in the tank is lost. The initial content in the tank is x pounds. By the end of the first day, the content remaining is

$$(\text{Initial content})(1 - \text{Loss percent}/100) = x(1 - 1/100) = 0.99x$$

Similarly, the content remaining after the end of the second day is $0.99x(1 - 1/100) = (0.99)(0.99)x$; and by the end of the third day, the content remaining is $(0.99)(0.99)(0.99)x$. Hence, Column A, which equals the net loss percentage in the three consecutive days, is

$$\frac{\text{Initial content} - \text{Final content}}{\text{Final content}} \bullet 100 =$$

$$\frac{x - 0.99 \bullet 0.99 \bullet 0.99x}{x} \bullet 100 =$$

$$100 - 99(0.99)(0.99) \approx 2.9, \text{ which is less than } 3$$

Hence, Column A < Column B and the answer is (B).

Hard Quantitative Comparisons

Most of the time, we have an intuitive feel for whether a problem is hard or easy. But on tricky problems (problems that appear easy but are actually hard) our intuition can fail us.

On the test, your first question will be of medium difficulty. If you answer it correctly, the next question will be a little harder. If you again answer it correctly, the next question will be harder still, and so on. If your math skills are strong and you are not making any mistakes, you should reach the medium-hard or hard problems by about the fifth problem. Although this is not very precise, it can be quite helpful. Once you have passed the fifth question, you should be alert to subtleties in any seemingly simple problems.

There are special techniques and strategies that apply to the hard problems only. Do not apply the methods of this section to the easy or medium quantitative comparison problems.

On Hard Quantitative Comparison Problems, The Obvious Answer (The Eye-Catcher) Will Almost Always Be Wrong. (If one expression looks at first glance to be larger than another, then it will not be.)

This is so because when people cannot solve a problem, they most often pick the answer-choice that "looks right." But if that were the answer, most people would answer it correctly and therefore it would not be a "hard" problem.

Example 1:

Column A		Column B
x^{10}	$x \geq 1$	x^{100}

One would expect x^{100} to be larger than x^{10}. But this is a hard problem and therefore what we expect will not be the answer. Now, clearly x^{100} cannot always be less than x^{10}. And just as clearly x^{100} cannot always be equal to x^{10}. Hence, the answer is (D)—not-enough-information. (A double case can also be obtained by substituting $x = 1$ and then $x = 2$.)

Example 2:

Column A	Column B
The number of distinct prime factors of x	The number of distinct prime factors of $4x$

We expect the number of prime factors of $4x$ to be larger than the number of prime factors of x. But that is the eye-catcher. Now, the number of prime factors of $4x$ cannot be less than the number of prime factors of x since $4x$ contains all the factors of x. So, the answer must be that either they are equal or there is not enough information. In fact, there is not enough information, as can be verified by plugging in the numbers $x = 2$ and then $x = 3$.

Example 3:

Column A	Column B
The area of a square with perimeter 12	The area of a parallelogram with perimeter 16

We expect the area of the parallelogram to be larger. After all, the parallelogram could be a square with perimeter 16, which of course has a larger area than a square with perimeter 12. But that would be too easy. Hence, there must be a parallelogram whose area is equal to or less than the area of the square. (See whether you can draw it. Hint: Look at the extreme cases.) Thus, we have a double case, and the answer is (D)—not-enough-information.

Note 1: When plugging in on quantitative comparison problems, be sure to check 0, 1, 2, –2, and 1/2, in that order.

Note 2: If there are only numbers in a quantitative comparison problem, i.e., no variables, then (D), not-enough-information, cannot be the answer.

Note 3: When drawing geometric figures, don't forget extreme cases.

<u>**Problem Set G:**</u> Eliminate the eye-catcher and then solve the following problems.

1.

Column A		Column B
	$x > 0$	
$1/2x$		$2x$

2.

Column A		Column B
	$x > 0$	
$x^3 + 1$		$x^4 + 1$

3.

Column A	Column B
The largest power of 3 that is a factor of $5 \cdot 3^2 + 3^2 \cdot 2$	The largest power of 3 that is a factor of $3 \cdot 2 + 7 \cdot 3$

4.

Column A		Column B
	x is an even integer.	
The number of distinct prime factors of $4x$		The number of distinct prime factors of x

5.

Column A	Column B
The average of three numbers if the greatest is 20	The average of three numbers if the greatest is 2

6.

Column A		Column B
	a and b are integers greater than zero.	
a/b		a^2

7.

Column A		Column B
	q is an integer greater than 1. Let $\boxed{q}$ stand for the smallest positive integer factor of q that is greater than 1.	
$\boxed{q}$		$\boxed{q^3}$

8.

Column A		Column B
	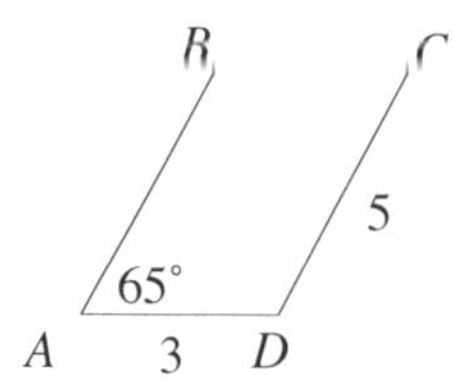	
15		The area of parallelogram *ABCD*

Answers and Solutions to Problem Set G

1. Intuitively, one expects $2x$ to be larger than the fraction $1/2x$. But that would be too easy to be the answer to a hard problem. Now, clearly, $1/2x$ cannot always be greater than $2x$, nor can it always be equal to $2x$. Hence, the answer is (D).

Let's also solve this problem by substitution. If $x = 1$, then $1/2x = 1/2$ and $2x = 2$. In this case, Column B is greater. But if $x = 1/2$, then $\frac{1}{2x} = \frac{1}{2\cdot\left(\frac{1}{2}\right)} = \frac{1}{1} = 1$ and $2x = 2\cdot\frac{1}{2} = 1$. In this case, the columns are equal. This is a double case and the answer is (D).

2. Intuitively, one expects $x^4 + 1$ to be larger than $x^3 + 1$. But this is a hard problem, so we can reject (B) as the answer. Now, if $x = 0$ or 1, then expressions are equal. However, for any other value of x, the expressions are unequal. Hence, the answer is (D).

3. At first glance, Column A appears larger than Column B since it has more 3's. But this is a hard problem, so that could not be the answer. Now, if we multiply out each expression, Column A becomes $63 = 3^2 \cdot 7$ and Column B becomes $27 = 3^3$. The power of 3^3 is larger than the power of 3^2. Hence, Column B is larger. The answer is (B).

4. We expect $4x$ to have more prime factors than x since $4x$ contains every factor of x. But as this is a hard problem, we eliminate (A). $4x$ contains every factor of x, so x cannot have more prime factors than $4x$. This eliminates (B). Now, $4x = 2^2x$. But we are given that x is even, so it already contains the prime factor 2. Hence, the 4 does not add any more distinct prime factors. So, the columns are equal. The answer is (C).

5. At first glance, Column A appears larger than Column B. However, the problem does not exclude negative numbers. Suppose the three numbers in Column A are –20, 0, and 20 and that the three numbers in Column B are 0, 1, and 2. Then the average for Column A would be $\frac{-20+0+20}{3} = \frac{0}{3} = 0$, and the average for Column B would be $\frac{0+1+2}{3} = \frac{3}{3} = 1$. In this case, Column B is larger. Clearly, there are also numbers for which Column A would be larger. Hence, the answer is (D).

6. Intuitively, we expect a^2 to be larger than the fraction a/b. So, that will not be the answer. Now, if $a = b = 1$, then both columns equal 1. However, if $a = b = 2$, then Column B is larger. Hence, the answer is (D).

7. The eye-catcher is Column A since we are looking for the smallest factor and q is smaller than q^3. Let's use substitution to solve this problem. Since $q > 1$, we need to look at only 2, 3, and 4 (see Substitution Special Cases). If $q = 2$, then $\boxed{q} = \boxed{2} = 2$ and $\boxed{q^3} = \boxed{2^3} = \boxed{8} = 2$. In this case, the two columns are equal. If $q = 3$, then $\boxed{q} = 3$ and $\boxed{q^3} = 3$. In this case, the two columns are again equal. If $q = 4$, then $\boxed{q} = 2$ and $\boxed{q^3} = 2$. Once again, the two columns are equal. Hence, the answer is (C).

8. If the parallelogram were a rectangle, then its area would be 15 and the columns would be equal. But as the rectangle is tilted to the right, its area decreases:

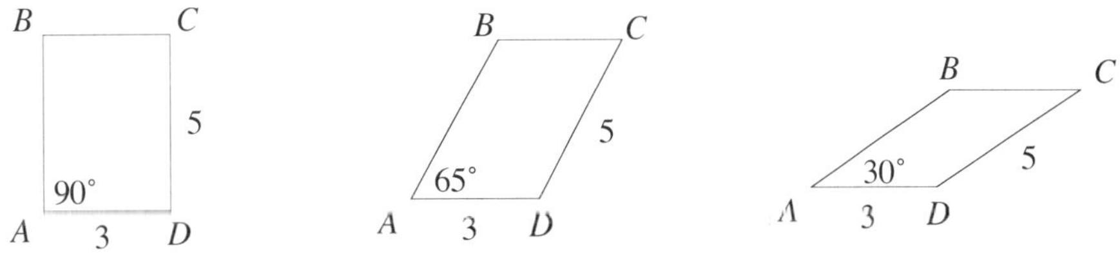

Area = 15 Area ≈ 13.5 Area = 7.5

The answer is (A).

Eliminate Answer-Choices That Are Too Easily Derived or Too Ordinary.

Example 1:

Column A	$x \cdot y = 3$	Column B
$x + y$		4

The numbers 3 and 1 are solutions to the equation $x \cdot y = 3$ because $3 \cdot 1 = 3$. So for this choice of x and y, Column A equals Column B, since $3 + 1 = 4$. But that is too easy: Everyone will notice 1 and 3 as solutions of the equation $x \cdot y = 3$. Hence, there must be another pair of numbers whose product is 3 and whose sum is not 4. In fact, there are an infinite number of pairs. For example, $9 \cdot \frac{1}{3} = 3$, but $9 + \frac{1}{3} \neq 4$. This is a double case and therefore the answer is (D).

Example 2:

Column A		Column B
The greatest number of regions into which two straight lines will divide the shaded region.		4

Most people will draw one or the other of the two drawings below:

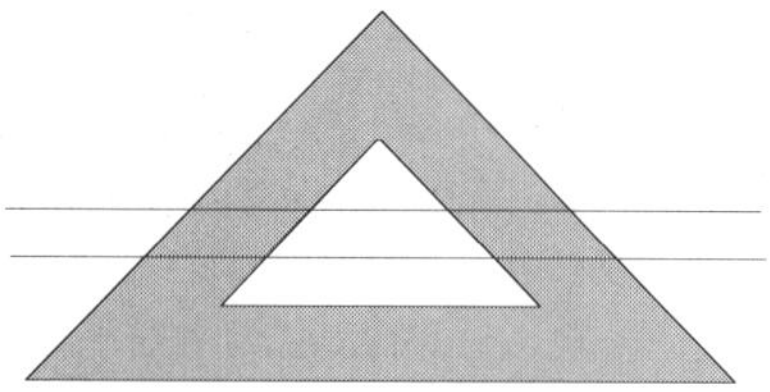

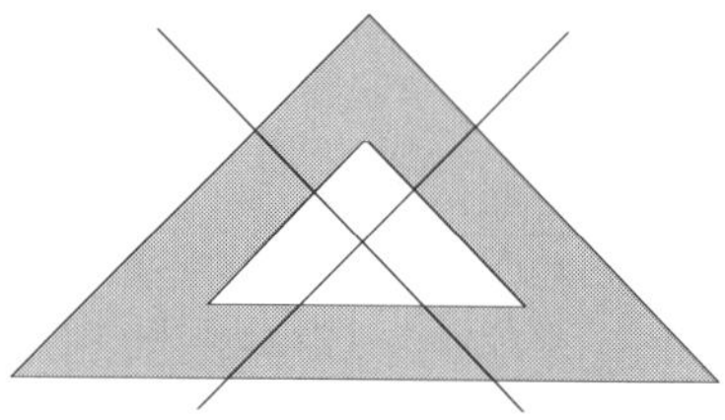

In each case, four separate shaded regions are formed. But these drawings are too ordinary, too easy. There must be a way to draw the lines to form more than four regions. Try to draw it before looking at the answer below.

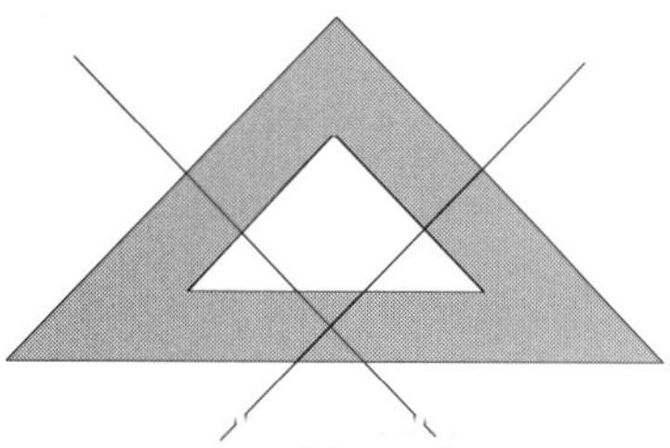

The lines must intersect in the shaded region.

Problem Set H: Eliminate the eye-catcher and then solve the following problems.

1.

Column A	Column B
Volume of a cylinder with a height of 10	Volume of a cone with a height of 10

2.

Column A	Column B
The greatest possible number of points common to a triangle and a circle	3

3. On the final exam in History 101, the average score for the girls was 72 and for the boys, 70.

Column A	Column B
The average score for the class	71

4.

Column A	Column B
Perimeter of a rectangle with an area of 10	Perimeter of a triangle with an area of 10

5. Line segments *AB* and *CD* are both parallel and congruent. The mid-point of AB is M.

Column A	Column B
The length of segment *CM*	The length of segment *DM*

6. Let $\boxed{x}$ denote the greatest integer less than or equal to *x*.

Column A	Column B
$\boxed{3.1} + \boxed{-3.1}$	0

7.

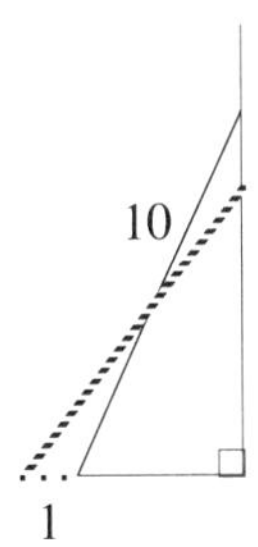

A 10-foot ladder is leaning against a vertical wall. The top of the ladder touches the wall at a point 8 feet above the ground. The base of the ladder slips 1 foot away from the wall.

Column A	Column B
1	The distance the top of the ladder slides down the wall

Answers and Solutions to Problem Set H

1. Since we are not given the radius of the cylinder, we can make the cylinder very narrow or very broad by taking the radius to very small or very large. The same can be done with the cone. Hence, we have a double case, and the answer is (D).

2. There are six possible points of intersection as shown in the diagram below:

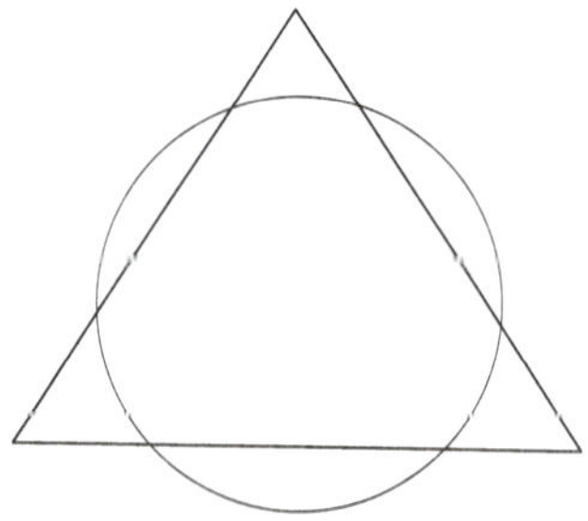

The answer is (A).

3. The eye-catcher is that the two columns are equal. That won't be the answer to this hard problem. Now, if there are more girls in the class, then the average will be closer to 72 than to 70. On the other hand, if there are more boys in the class, then the average will be closer to 70. This is a double case, and therefore the answer is (D).

4. The eye-catcher is Column A since one expects the perimeter of a rectangle to be longer than that of a triangle of similar size. However, by making the base of the triangle progressively longer, we can make the perimeter of the triangle as long as we want. The following diagram displays a rectangle and a triangle with the same area, yet the triangle's perimeter is longer than the rectangle's:

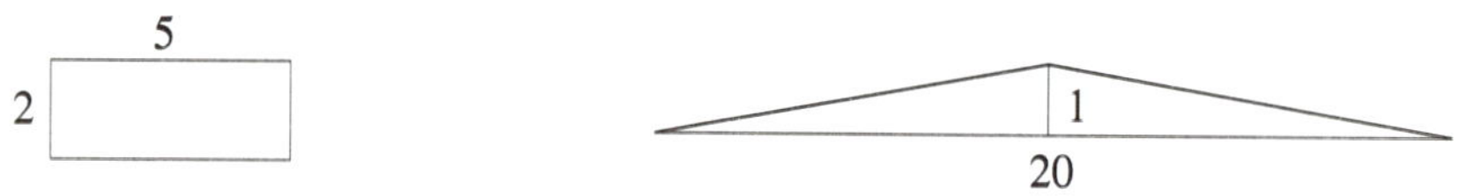

The answer is (D).

5. Most people will draw the figure as follows:

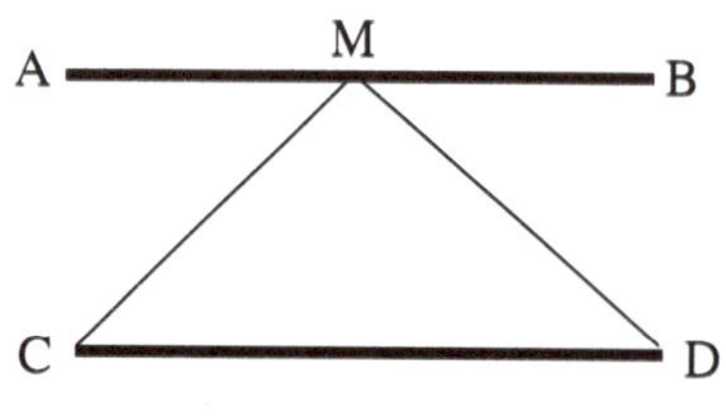

In this drawing, CM equals DM. But that is too ordinary. There must be a way to draw the lines so that the lengths are not equal. One such drawing is as follows:

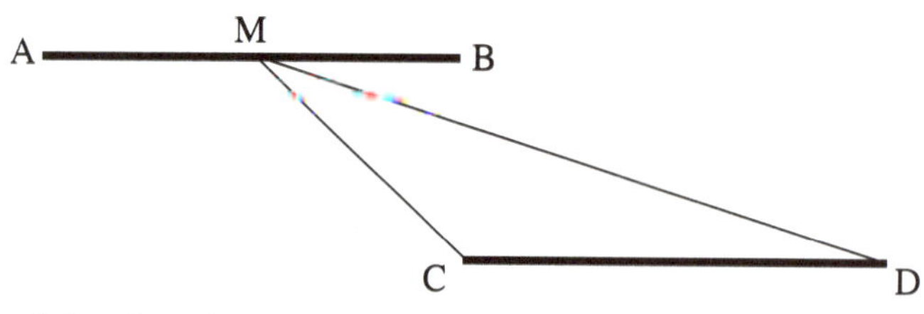

This is a double case, and therefore the answer is (D). (Note: When drawing a geometric figure, be careful not to assume more than what is given. In this problem, we are told <u>only</u> that the two lines are parallel and congruent; we cannot assume that they are aligned.)

6. The eye-catcher is that the columns are equal: 3.1 – 3.1 = 0. But that won't be the answer to this hard problem. Now, $\boxed{x}$ denotes the greatest integer less than or equal to x. That is, $\boxed{x}$ is the first integer smaller than x. Further, if x is an integer, then $\boxed{x}$ is equal to x itself. Therefore, $\boxed{3.1} = 3$, and $\boxed{-3.1} = -4$ (not –3). Hence, $\boxed{3.1} + \boxed{-3.1} = 3 + (-4) = -1$. Therefore, Column B is larger. The answer is (B).

7. We can immediately eliminate (C) because that would be too easy. Let y be the distance the top of the ladder slides down the wall, let h be the height of the new resting point of the top of the ladder, and x be the original distance of the bottom of the ladder from the wall:

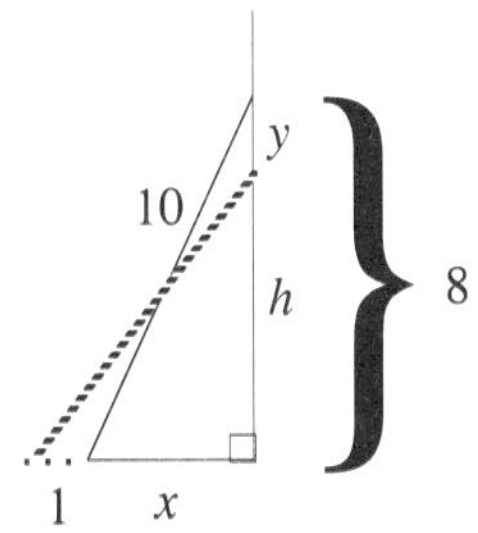

Applying the Pythagorean Theorem to the original triangle yields	$x^2 + 8^2 = 10^2$
Solving this equation for x yields	$x = 6$
Hence, the base of the final triangle is	$1 + 6 = 7$
Applying the Pythagorean Theorem to the final triangle yields	$h^2 + 7^2 = 10^2$
Solving this equation for h yields	$h = \sqrt{51}$

Adding this information to the drawing yields

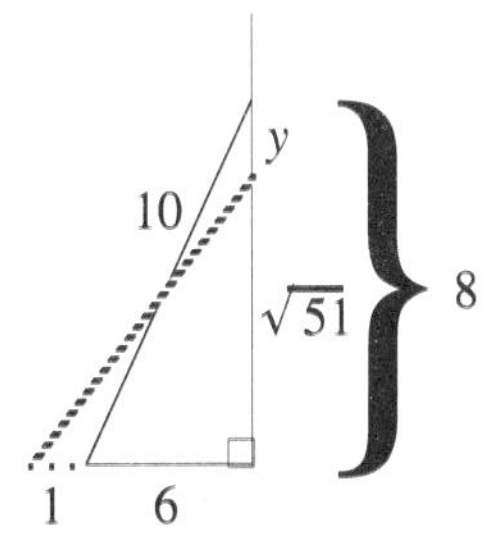

From the drawing, $y = 8 - \sqrt{51} < 8 - 7 = 1$, since $\sqrt{51} \approx 7.1$. Hence, Column A is larger, and the answer is (A).

Geometry

About one-fourth of the math problems on the GRE involve geometry. (There are no proofs.) Unfortunately, the figures on the GRE are usually not drawn to scale. Hence, in most cases, you cannot solve a problem or check your work by "eyeballing" the drawing.

Following is a discussion of the basic properties of geometry. You probably know many of these properties. Memorize any that you do not know.

Lines & Angles

When two straight lines meet at a point, they form an angle. The point is called the vertex of the angle, and the lines are called the sides of the angle.

The angle to the right can be identified in three ways:

1. $\angle x$
2. $\angle B$
3. $\angle ABC$ or $\angle CBA$

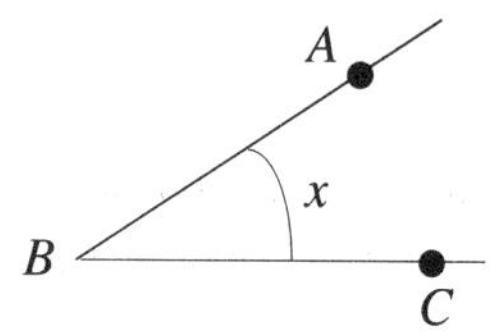

When two straight lines meet at a point, they form four angles. The angles opposite each other are called vertical angles, and they are congruent (equal). In the figure to the right, $a = b$, and $c = d$.

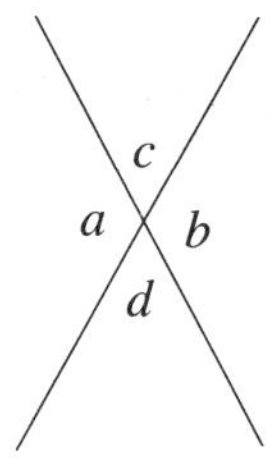

$a = b$ and $c = d$

Angles are measured in degrees, °. By definition, a circle has 360°. So an angle can be measured by its fractional part of a circle. For example, an angle that is $\frac{1}{360}$ of the arc of a circle is 1°. And an angle that is $\frac{1}{4}$ of the arc of a circle is $\frac{1}{4} \times 360 = 90°$.

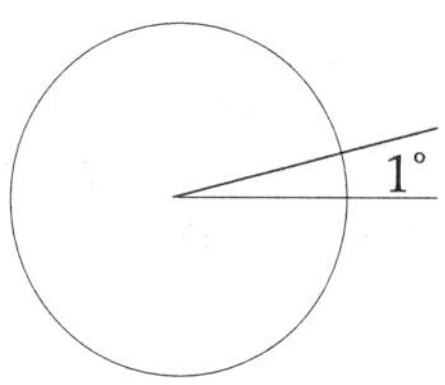

1/360 of an arc of a circle

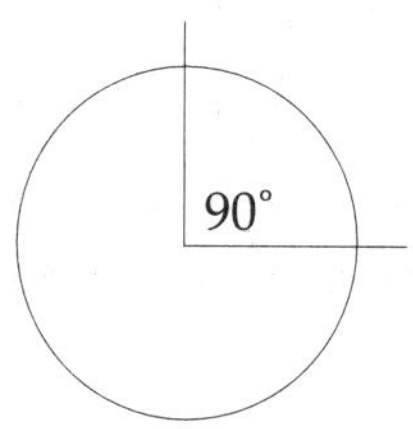

1/4 of an arc of a circle

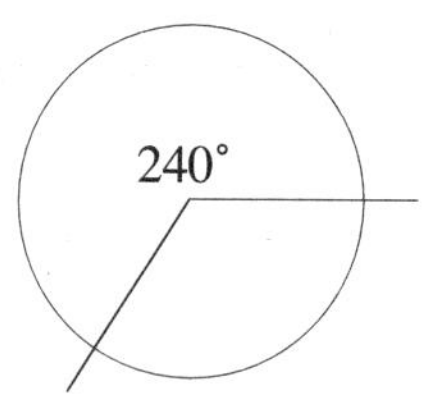

2/3 of an arc of a circle

There are four major types of angle measures:

An **acute angle** has measure less than 90°:

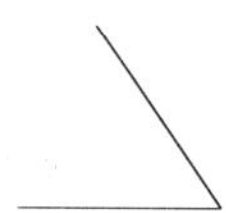

A **right angle** has measure 90°:

An **obtuse angle** has measure greater than 90°:

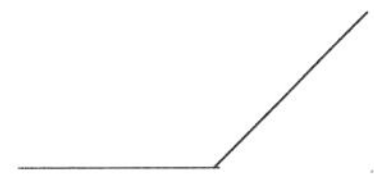

A **straight angle** has measure 180°:

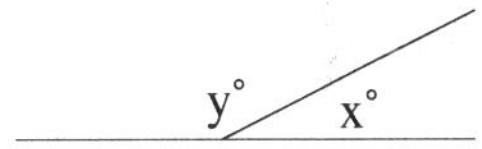

x + y = 180°

Example: In the figure, AOB is a straight line. What is the average of the four numbers a, b, c, d ?

(A) 45
(B) 360/7
(C) 60
(D) 90
(E) 120

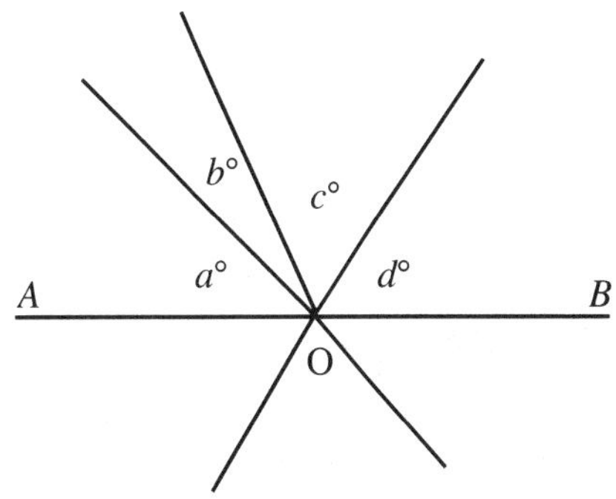

In the figure, AOB is a straight line, and a straight angle measures 180°. Hence, the sum of the angles a, b, c, and d is 180, and the average of the four is their sum divided by 4: 180/4 = 45. The answer is (A).

Example: In the figure, lines l, m, and n intersect at O. Which one of the following must be true about the value of a ?

(A) $a < 5/8$
(B) $a = 5/7$
(C) $a = 5/6$
(D) $a = 1$
(E) $a > 1$

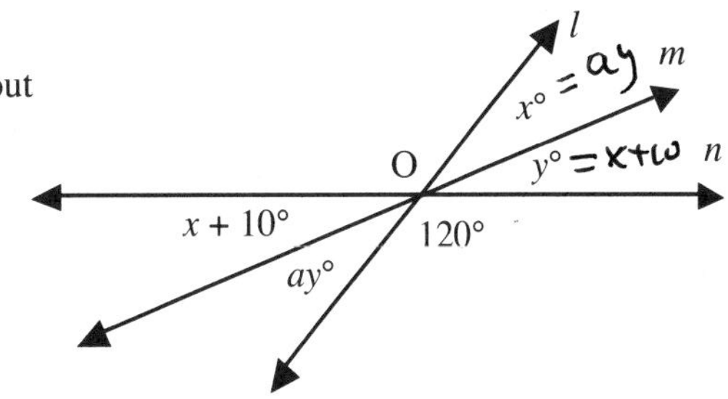

Equating vertical angles yields $x = ay$ and $y = x + 10$. Plugging the second equation into the first yields $x = a(x + 10)$. Solving for x yields $x = \frac{10a}{1-a}$. Also, $y = x + 10 = \frac{10a}{1-a} + 10 = \frac{10a + 10 - 10a}{1-a} = \frac{10}{1-a}$. Now, we know that the angle made by any point on a line is 180°. Hence, the angle made by point O on line n is 180°. Hence, $120 + ay + x + 10 = 180$. Simplifying yields $ay + x = 50$. Substituting the known results $x = \frac{10a}{1-a}$ and $y = \frac{10}{1-a}$ into this equation yields $\frac{10a}{1-a} + \frac{10a}{1-a} = 50$. Hence, $2\left(\frac{10a}{1-a}\right) = 50$. Multiplying both sides by $(1 - a)$ yields $20a = 50(1 - a)$. Distributing the 50 yields $20a = 50 - 50a$. Adding $50a$ to both sides yields $70a = 50$. Finally, dividing both sides by 70 yields $a = 5/7$. The answer is (B).

Two angles are supplementary if their angle sum is 180°:

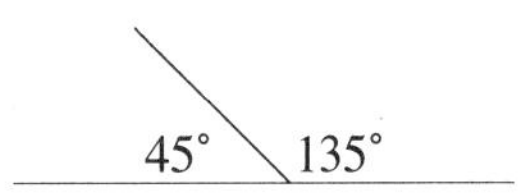

45 + 135 = 180

Two angles are complementary if their angle sum is 90°:

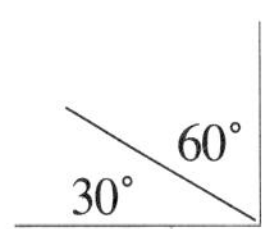

30 + 60 = 90

Perpendicular lines meet at right angles:

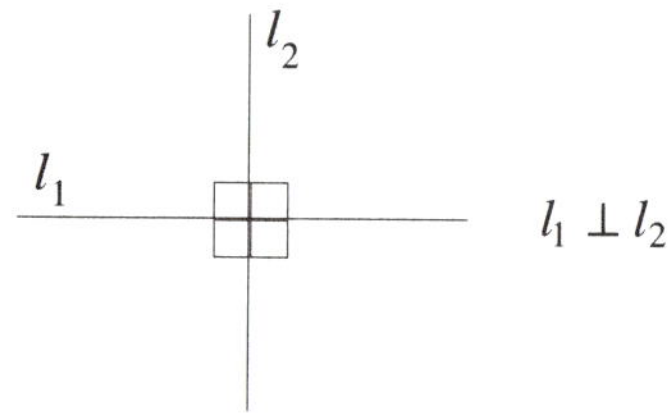

Two lines in the same plane are parallel if they never intersect. Parallel lines have the same slope.

When parallel lines are cut by a transversal, three important angle relationships exist:

Alternate interior angles are equal.

Corresponding angles are equal.

Interior angles on the same side of the transversal are supplementary.

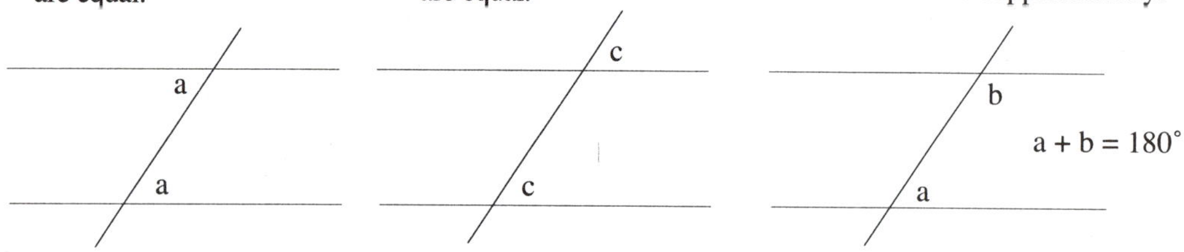

The shortest distance from a point to a line is along a new line that passes through the point and is perpendicular to the original line.

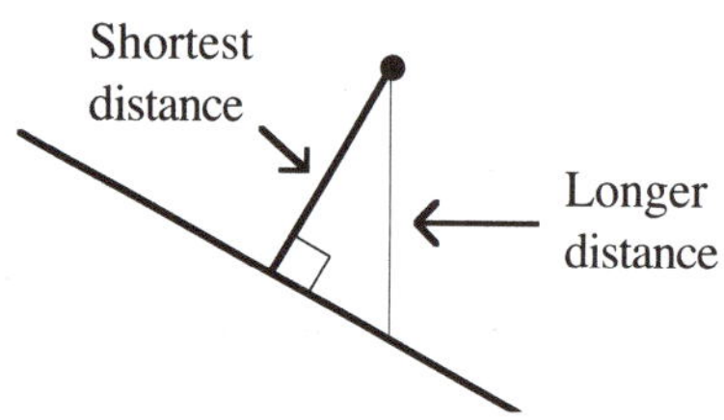

Triangles

A triangle containing a right angle is called a *right triangle*. The right angle is denoted by a small square:

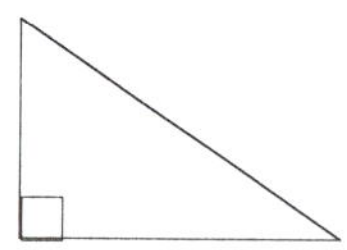

A triangle with two equal sides is called *isosceles*. The angles opposite the equal sides are called the base angles, and they are congruent (equal). A triangle with all three sides equal is called *equilateral*, and each angle is 60°. A triangle with no equal sides (and therefore no equal angles) is called *scalene*:

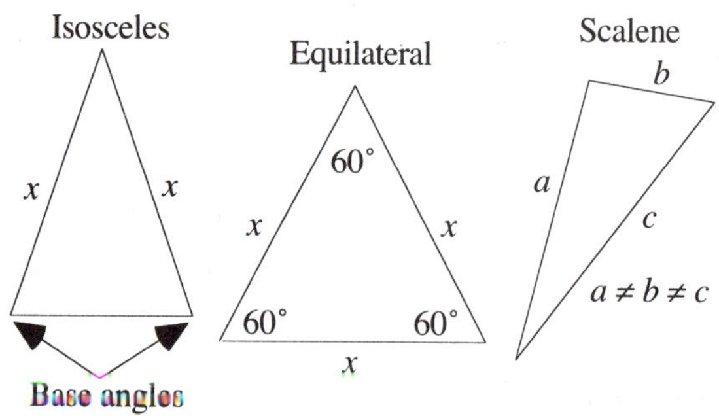

The altitude to the base of an isosceles or equilateral triangle bisects the base and bisects the vertex angle:

Isosceles:

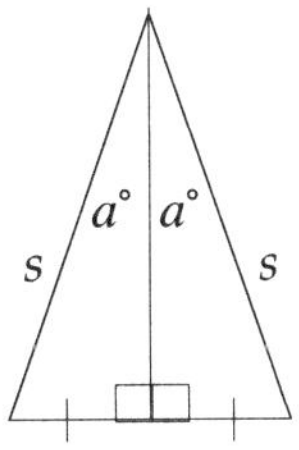

Equilateral:

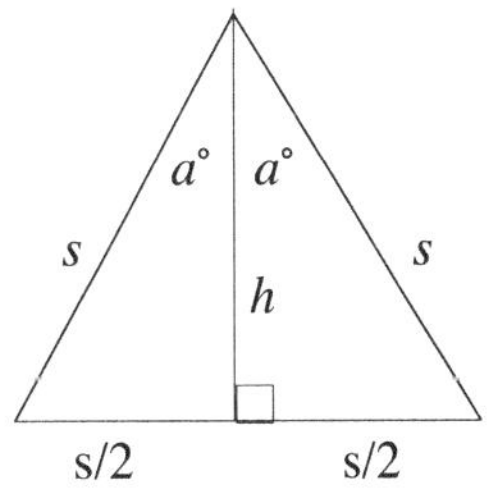

$$h = \frac{s\sqrt{3}}{2}$$

The angle sum of a triangle is 180°:

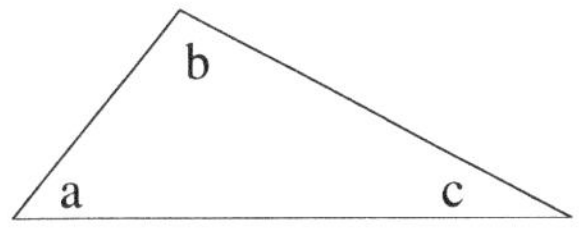

$$a + b + c = 180°$$

Example: In the figure, lines AB and DE are parallel. What is the value of x?

(A) 22.5
(B) 45
(C) 60
(D) 67.5
(E) 70

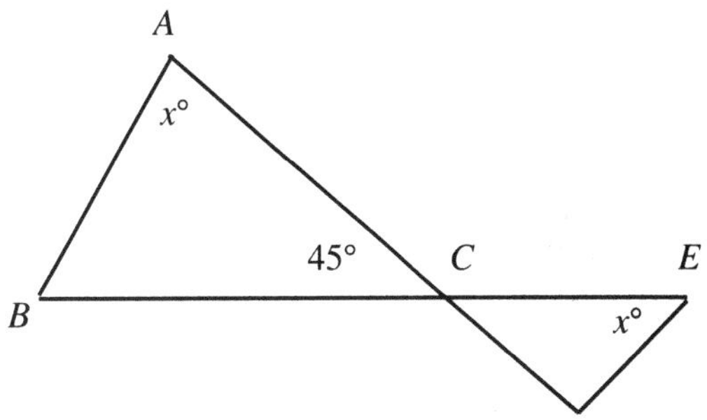

Since lines AB and DE are parallel, we can equate the alternate interior angles at A and D to get $\angle A = \angle D = x$ (from the figure). Also, equating vertical angles ACB and DCE yields $\angle ACB = \angle DCE = 45$ (from the figure). Now, we know that the angle sum of a triangle is 180°. Hence, $\angle DCE + \angle CED + \angle EDC = 180$. Plugging the known values into this equation yields $45 + x + x = 180$. Solving this equation for x yields $x = 67.5$. The answer is (D).

The area of a triangle is $\frac{1}{2}bh$, where b is the base and h is the height. Sometimes the base must be extended in order to draw the altitude, as in the third drawing directly below:

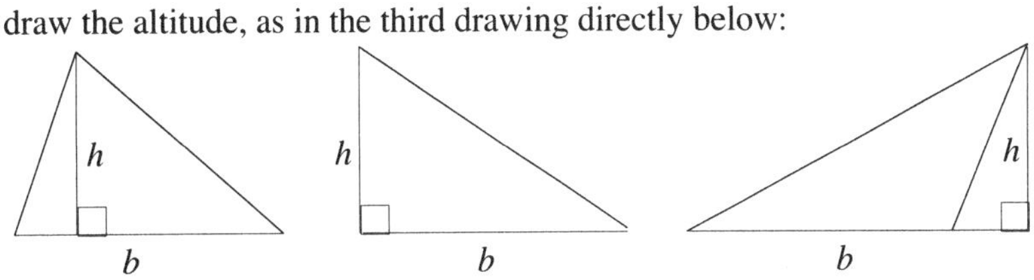

$$A = \frac{1}{2}bh$$

In a triangle, the longer side is opposite the larger angle, and vice versa:

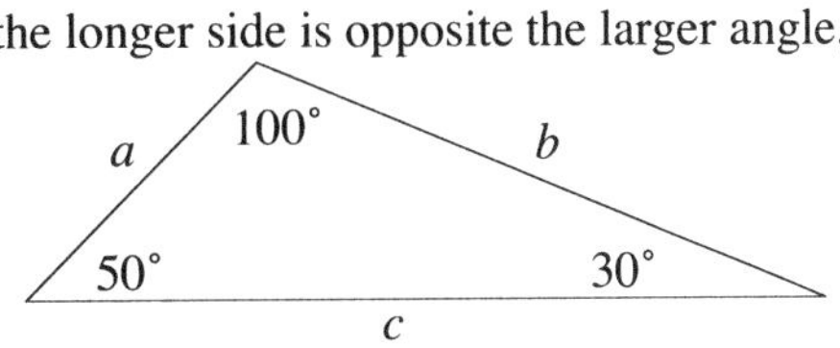

50° is larger than 30°, so side b is longer than side a.

Pythagorean Theorem (right triangles only): The square of the hypotenuse is equal to the sum of the squares of the legs.

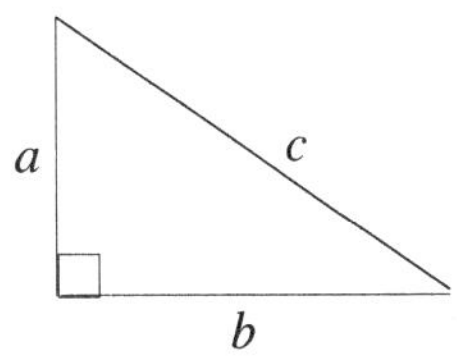

$$c^2 = a^2 + b^2$$

Pythagorean triples: The numbers 3, 4, and 5 can always represent the sides of a right triangle and they appear very often: $5^2 = 3^2 + 4^2$. Another, but less common, Pythagorean Triple is 5, 12, 13: $13^2 = 5^2 + 12^2$.

Two triangles are similar (same shape and usually different sizes) if their corresponding angles are equal. If two triangles are similar, their corresponding sides are proportional:

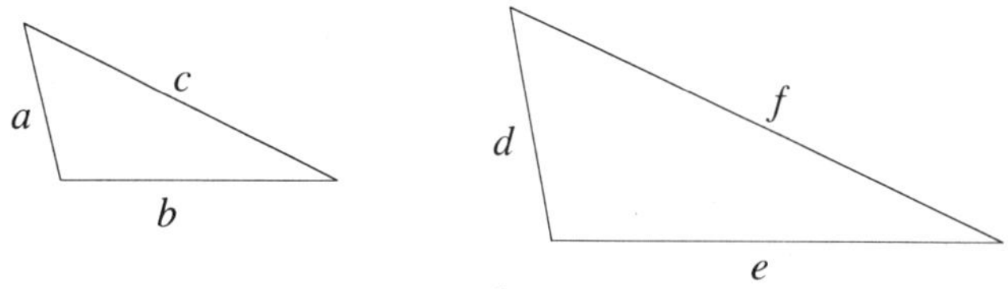

$$\frac{a}{d} = \frac{b}{e} = \frac{c}{f}$$

If two angles of a triangle are congruent to two angles of another triangle, the triangles are similar.

In the figure to the right, the large and small triangles are similar because both contain a right angle and they share $\angle A$..

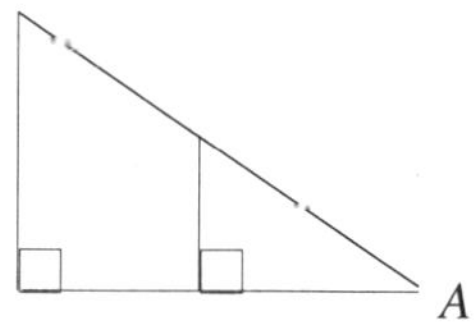

Two triangles are congruent (identical) if they have the same size and shape.

In a triangle, an exterior angle is equal to the sum of its remote interior angles and is therefore greater than either of them:

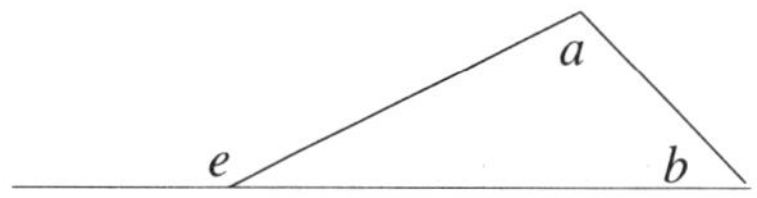

$e = a + b$ and $e > a$ and $e > b$

In a triangle, the sum of the lengths of any two sides is greater than the length of the remaining side:

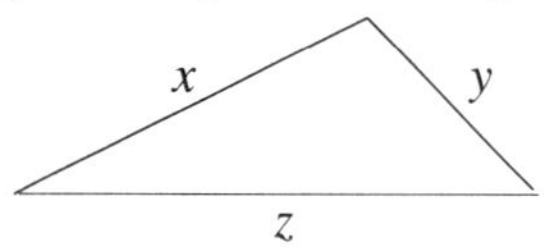

$x + y > z$

$y + z > x$

$x + z > y$

Example: Two sides of a triangle measure 4 and 12. Which one of the following could equal the length of the third side?

(A) 5
(B) 7
(C) 9
(D) 17
(E) 20

Each side of a triangle is shorter than the sum of the lengths of the other two sides, and, at the same time, longer than the difference of the two. Hence, the length of the third side of the triangle in the question is greater than the difference of the other two sides (12 – 4 = 8) and smaller than their sum (12 + 4 = 16). Since only choice (C) lies between the values, the answer is (C).

In a 30°–60°–90° triangle, the sides have the following relationships:

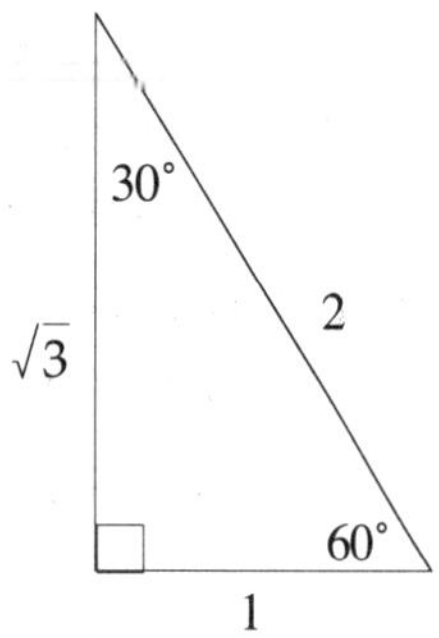

In general —>

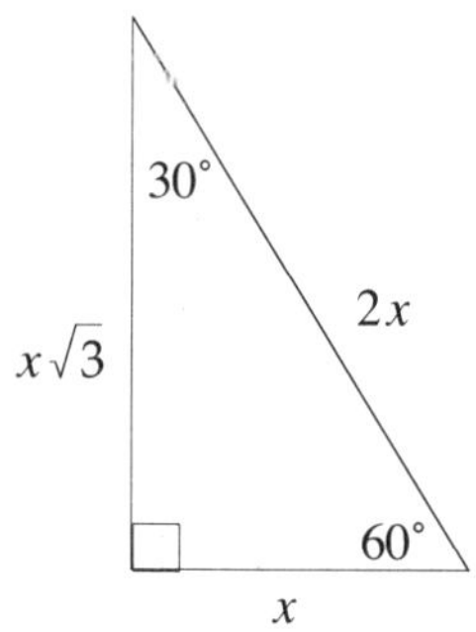

In a 45°–45°–90° triangle, the sides have the following relationships:

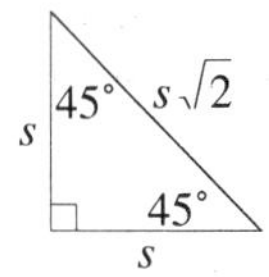

Quadrilaterals

A *quadrilateral* is a four-sided closed figure, where each side is a straight line.

The angle sum of a quadrilateral is 360°. You can view a quadrilateral as being composed of two 180-degree triangles:

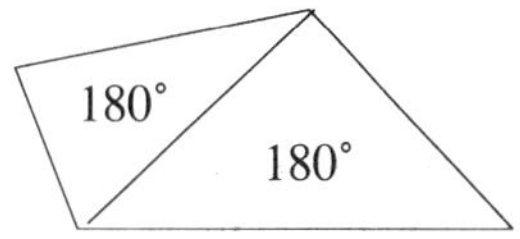

A *parallelogram* is a quadrilateral in which the opposite sides are both parallel and congruent. Its area is *base* × *height*:

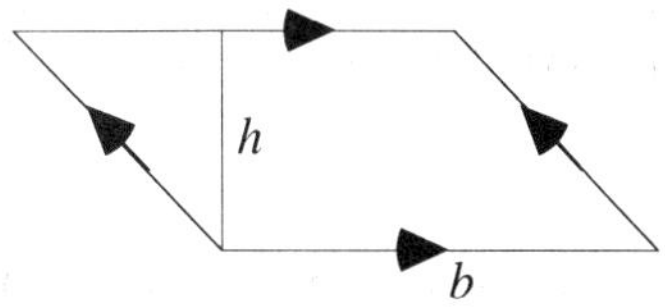

$A = bh$

The diagonals of a parallelogram bisect each other:

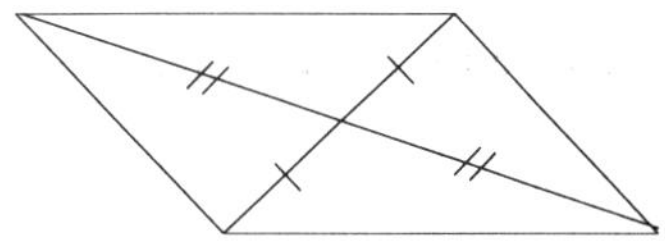

A parallelogram with four right angles is a *rectangle*. If w is the width and l is the length of a rectangle, then its area is and its perimeter is $P = 2w + 2l$.

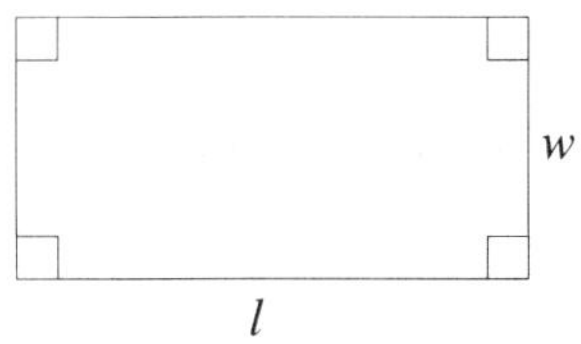

$A = l \cdot w$

$P = 2w + 2l$

Example: In the figure, the area of quadrilateral *ABCD* is 75. What is the area of parallelogram *EFGH* ?

(A) 96
(B) 153
(C) 157
(D) 165
(E) 171

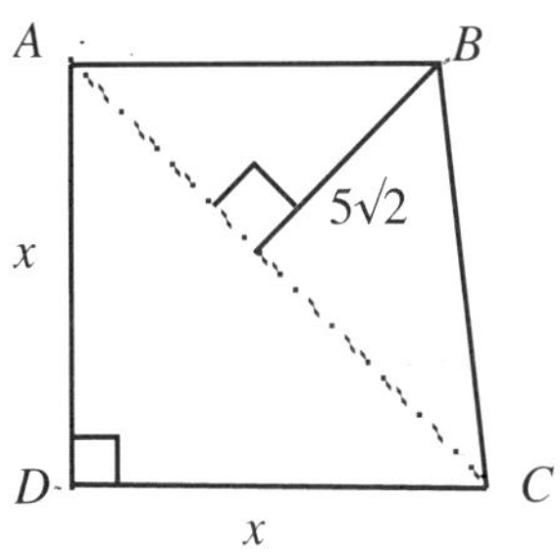

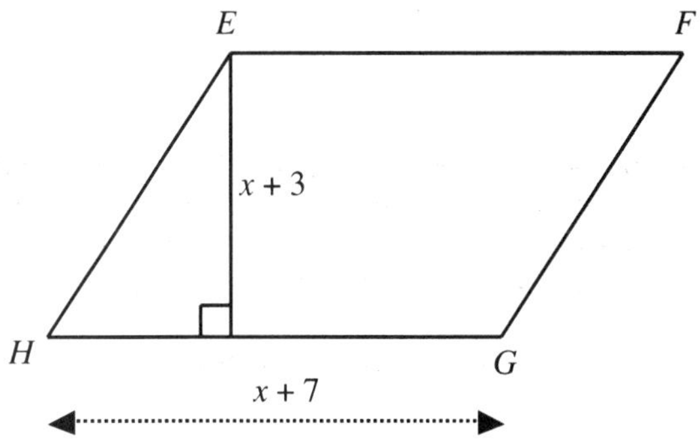

In the figure, ΔACD is right angled. Hence, by The Pythagorean Theorem, $AC^2 = AD^2 + DC^2 = x^2 + x^2 = 2x^2$. By square rooting the sides, we have $AC = x\sqrt{2}$.

The formula for the area of a triangle is 1/2 × *base* × *height*. Hence, the area of the right-triangle *ACD* is $1/2 \cdot x \cdot x$, and the area of triangle *ABC* is $\frac{1}{2} \cdot AC \cdot$ (altitude from B on AC) $= \frac{1}{2} \cdot x\sqrt{2} \cdot 5\sqrt{2} = 5x$. Now, the area of quadrilateral *ABCD*, which is given to be 75, is the sum of areas of the two triangles: $x^2/2 + 5x$. Hence, $x^2/2 + 5x = 75$. Multiplying both sides by 2 yields $x^2 + 10x = 150$.

Now, the formula for the area of a parallelogram is *base* × *height*. Hence, the area of *EFGH* is $(x + 3)(x + 7) = x^2 + 10x + 21 = (x^2 + 10x) + 21 = 150 + 21$ (since $x^2 + 10x = 150$) $= 171$. The answer is (E).

If the opposite sides of a rectangle are equal, it is a square and its area is $A = s^2$ and its perimeter is $P = 4s$, where s is the length of a side:

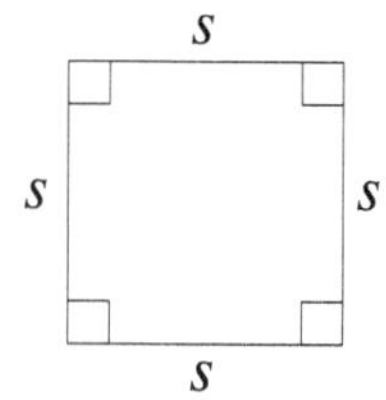

$$A = s^2$$
$$P = 4s$$

The diagonals of a square bisect each other and are perpendicular to each other:

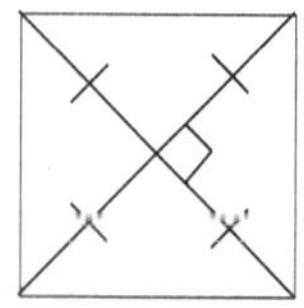

A quadrilateral with only one pair of parallel sides is a *trapezoid*. The parallel sides are called *bases*, and the non-parallel sides are called *legs*:

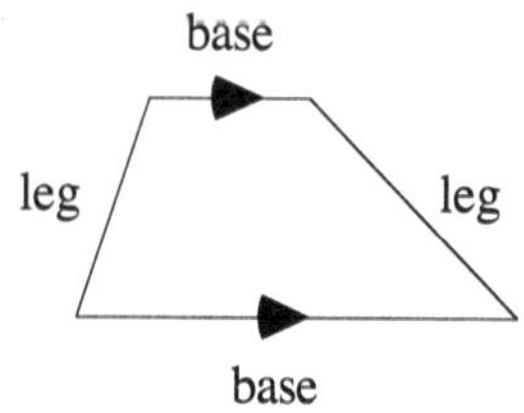

The area of a trapezoid is the average of the two bases times the height:

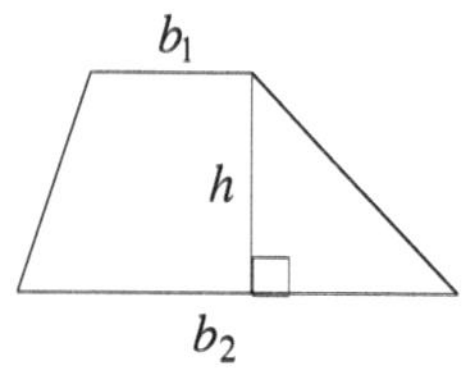

$$A = \left(\frac{b_1 + b_2}{2}\right)h$$

Volume

The volume of a rectangular solid (a box) is the product of the length, width, and height. The surface area is the sum of the area of the six faces:

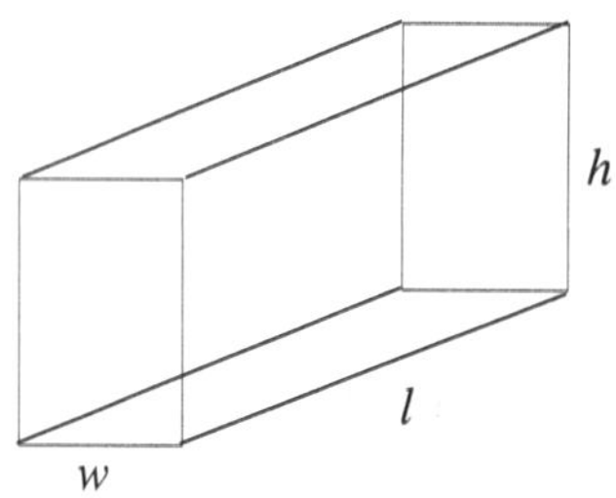

$$V = l \cdot w \cdot h$$
$$S = 2wl + 2hl + 2wh$$

If the length, width, and height of a rectangular solid (a box) are the same, it is a cube. Its volume is the cube of one of its sides, and its surface area is the sum of the areas of the six faces:

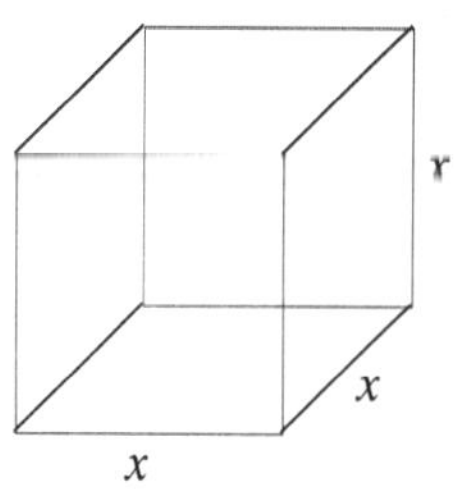

$$V = x^3$$
$$S = 6x^2$$

Example: The length, width, and depth of a rectangular box are 6 feet, 5 feet, and 7 feet, respectively. A hose supplies water at a rate of 6 cubic feet per minute. How much time in minutes would it take to fill a conical box whose volume is three times the volume of the rectangle box?

(A) 105
(B) 125
(C) 205
(D) 235
(E) 322

The volume of a rectangular tank is *length* × *width* × *depth* = 6 feet × 5 feet × 7 feet. Hence, the volume of the conical box, which is 3 times the volume of rectangular box, is 3(6 × 5 × 7). Now, the time taken to fill a tank equals the (volume of the tank) ÷ (the rate of filling) = 3(6 × 5 × 7) feet/6 cubic feet per minute = 105 minutes. The answer is (A).

The volume of a cylinder is $V = \pi r^2 h$, and the lateral surface (excluding the top and bottom) is $S = 2\pi rh$, where r is the radius and h is the height:

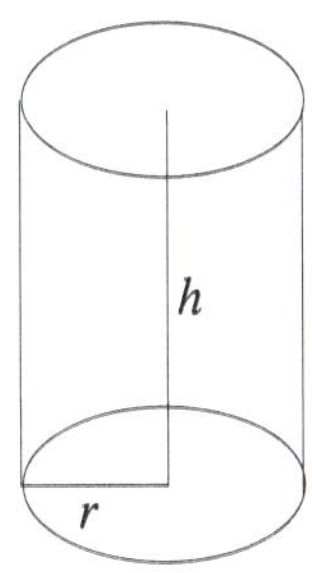

$$V = \pi r^2 h$$

$$S = 2\pi rh + 2\pi r^2$$

Circles

A circle is a set of points in a plane equidistant from a fixed point (the center of the circle). The perimeter of a circle is called the *circumference*.

A line segment from a circle to its center is a *radius*.

A line segment with both end points on a circle is a *chord.*

A chord passing though the center of a circle is a *diameter.*

A diameter can be viewed as two radii, and hence a diameter's length is twice that of a radius.

A line passing through two points on a circle is a *secant.*

A piece of the circumference is an *arc*.

The area bounded by the circumference and an angle with vertex at the center of the circle is a *sector.*

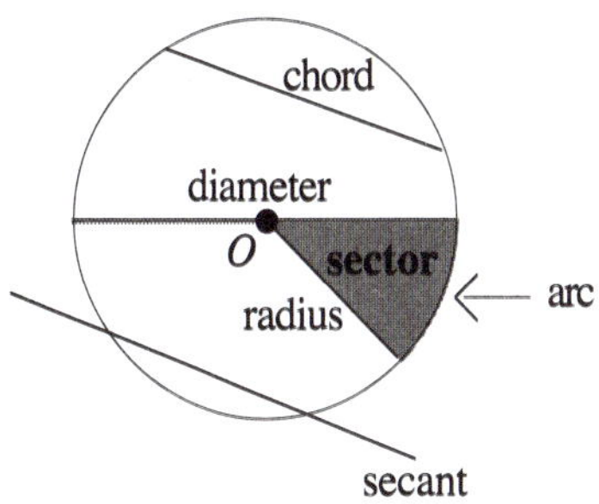

A tangent line to a circle intersects the circle at only one point. The radius of the circle is perpendicular to the tangent line at the point of tangency:

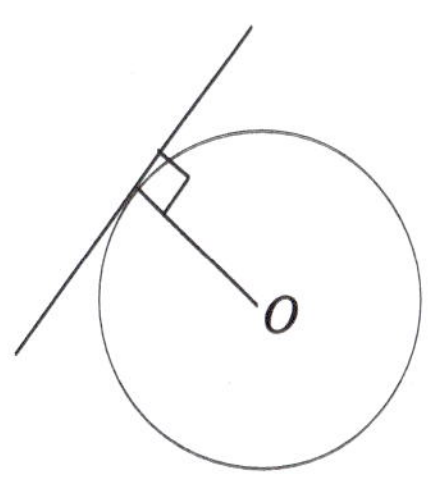

Two tangents to a circle from a common exterior point of the circle are congruent:

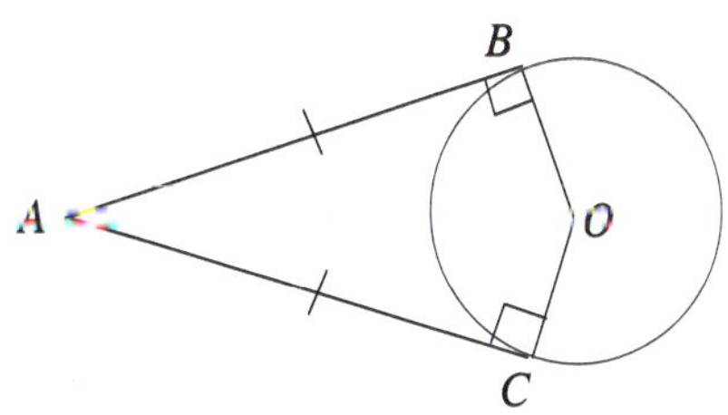

$AB \cong AC$

An angle inscribed in a semicircle is a right angle:

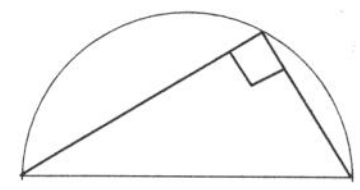

A central angle has by definition the same measure as its intercepted arc:

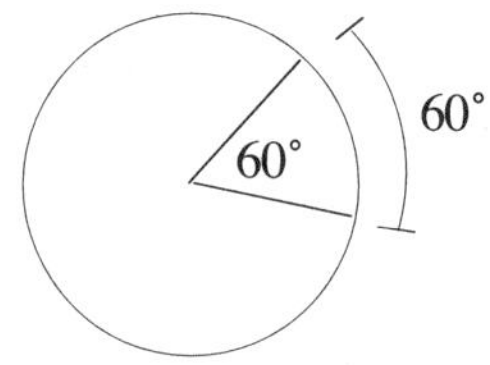

An inscribed angle has one-half the measure of its intercepted arc:

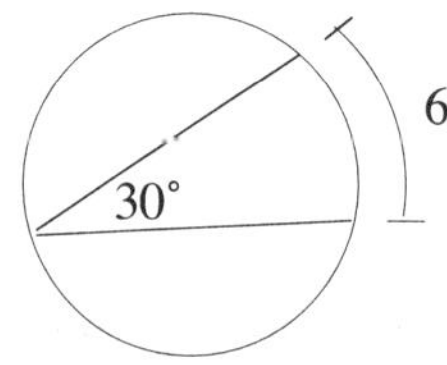

The area of a circle is πr^2, and its circumference (perimeter) is $2\pi r$, where r is the radius:

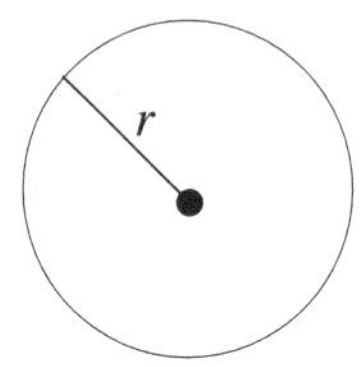

$A = \pi r^2$

$C = 2\pi r$

On the GRE, $\pi \approx 3$ is a sufficient approximation for π. You don't need $\pi \approx 3.14$.

Example: In the circle shown in the figure, the length of the arc ACB is 3 times the length of the arc AB. What is the length of the line segment AB ?

(A) 3
(B) 4
(C) 5
(D) $2\sqrt{3}$
(E) $3\sqrt{2}$

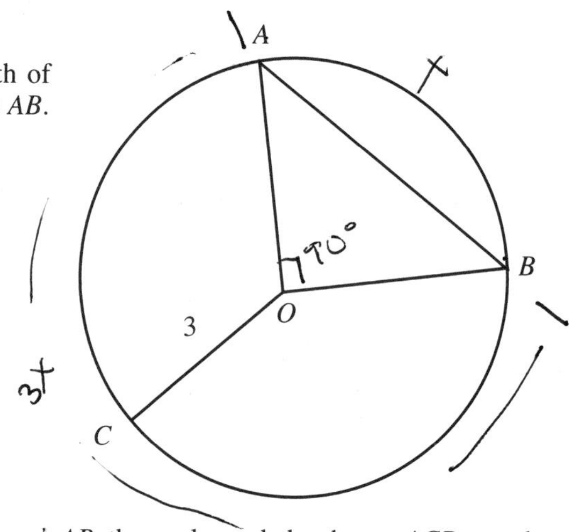

Since the length of the arc ACB is 3 times the length of the arc AB, the angle made by the arc ACB must be three times the angle made by the arc AB. Now, the two arcs together make 360° at the center of the circle. Hence, the smaller angle, the angle made by the arc AB, must equal one-quarter of the full angle, which is 360°. One quarter of 360° is 90°. Hence, $\angle AOB = 90°$. Hence, triangle AOB is a right triangle, and applying The Pythagorean Theorem to the triangle yields

$$AB^2 = OA^2 + OB^2$$
$$= 3^2 + 3^2 = 9 + 9 = 18 \qquad OA = OB = \text{radius of circle} = OC = 3 \text{ (from the figure)}$$
$$AB = \sqrt{18} = 3\sqrt{2}$$

The answer is (E).

Shaded Regions

To find the area of the shaded region of a figure, subtract the area of the unshaded region from the area of the entire figure.

Example: In the figure, $ABCD$ is a rectangle. What is the area of the shaded region in the figure?

(A) 18
(B) 20
(C) 24
(D) 28
(E) 32

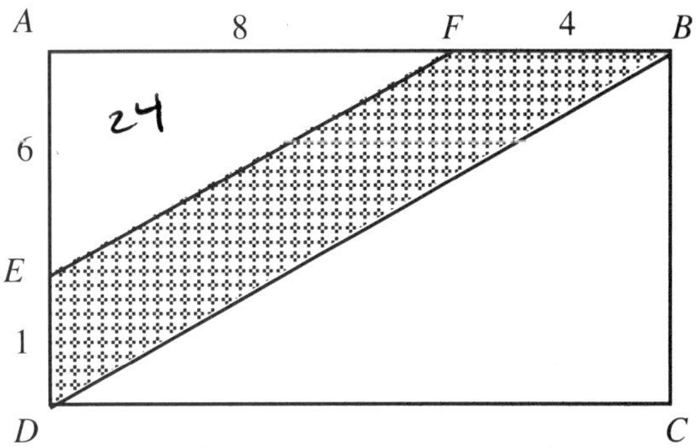

The figure is not drawn to scale.

From the figure, the area of the shaded region equals

$$(\text{Area of } \Delta ABD) - (\text{Area of } \Delta AFE)$$

Now, the area of ΔABD, by the formula $1/2 \times \textit{base} \times \textit{height}$, equals

$$1/2 \times AB \times AD = (1/2)(AF + FB)(AE + ED) = (1/2)(8 + 4)(6 + 1) = (1/2)(12)(7) = 6 \times 7 = 42$$

and the area of ΔAFE equals

$$1/2 \times AF \times AE = 1/2 \times 8 \times 6 = 4 \times 6 = 24$$

Hence, the area of the shaded region equals

$$(\text{Area of } \Delta ABD) - (\text{Area of } \Delta AFE) =$$
$$42 - 24 =$$
$$18$$

The answer is (A).

Example: What is the area of shaded region in the figure?

(A) $10\pi + 27\sqrt{3}$
(B) $10\pi + \frac{27}{4}\sqrt{3}$
(C) $30\pi + 27\sqrt{3}$
(D) $30\pi + 9\sqrt{3}$
(E) $36\pi + 27\sqrt{3}$

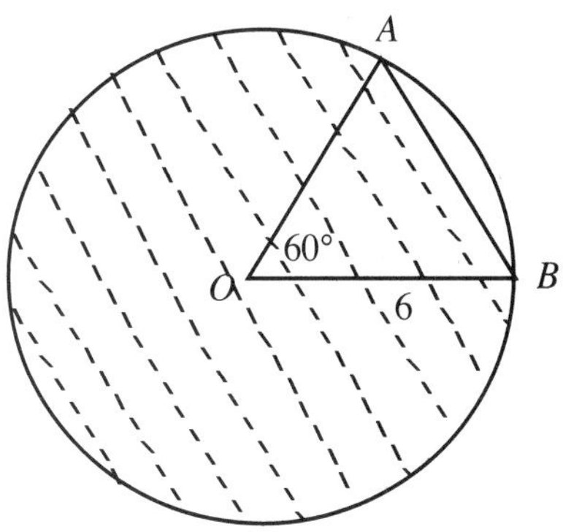

From the figure, we have

The area of the Shaded region = (Area of Circle) – (Area of Sector AOB) + (Area of ΔAOB).

By the formula of the area of a circle, the area of the circle in the figure is $\pi \times \textit{radius}^2 = \pi(6)^2 = 36\pi$.

The formula for the area of a sector is

$$(\text{Angle made by sector}/360°) \times (\text{Area of the circle}) =$$

$$60/360 \times 36\pi =$$

$$1/6 \times 36\pi =$$

$$6\pi$$

Also, since OA and OB are radii, angles opposite them are equal. Hence, AOB is an isosceles triangle with one angle ($\angle OAB =$) 60°. An isosceles triangle with one angle measuring 60° is always an equilateral triangle.

Now, the formula for the area of an equilateral triangle is $\frac{\sqrt{3}}{4} \bullet side^2$. Hence, the area of ΔAOB is

$$\frac{\sqrt{3}}{4} \bullet 6^2 = \frac{\sqrt{3}}{4} \bullet 36 = 9\sqrt{3}$$

Hence, the area of the shaded region is $36\pi - 6\pi + 9\sqrt{3} = 30\pi + 9\sqrt{3}$. The answer is (D).

"Birds-Eye" View

Most geometry problems on the GRE require straightforward calculations. However, some problems measure your insight into the basic rules of geometry. For this type of problem, you should step back and take a "birds-eye" view of the problem. The following example will illustrate.

Example: In the figure to the right, O is both the center of the circle with radius 2 and a vertex of the square $OPRS$. What is the length of diagonal PS ?

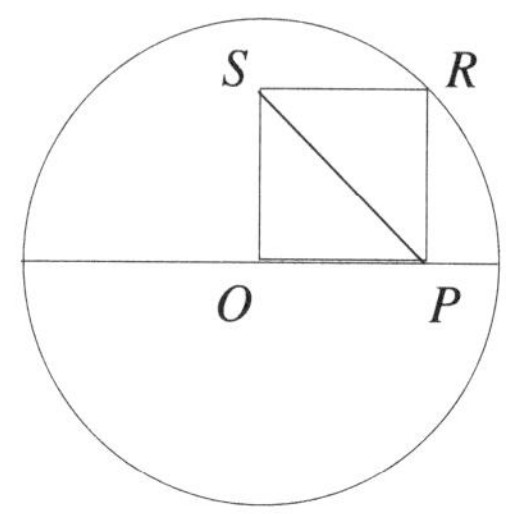

(A) 1/2
(B) $\frac{\sqrt{2}}{2}$
(C) 4
(D) 2
(E) $2\sqrt{5}$

The diagonals of a square are equal. Hence, line segment OR (not shown) is equal to SP. Now, OR is a radius of the circle and therefore $OR = 2$. Hence, $SP = 2$ as well, and the answer is (D).

Problem Set I:

➢ **Easy**

1. If p is the circumference of the circle Q and the area of the circle is 25π, what is the value of p ?

   (A) 25
   (B) 10π
   (C) 35
   (D) 15π
   (E) 25π

2. A rectangular field is 3.2 yards long. A fence marking the boundary is 11.2 yards in length. What is the area of the field in square yards?

   (A) 4.68
   (B) 7.68
   (C) 9.28
   (D) 11.28
   (E) 12.38

3. *ABCD* is a square and one of its sides *AB* is also a chord of the circle as shown in the figure. What is the area of the square?

   (A) 3
   (B) 9
   (C) 12
   (D) $12\sqrt{2}$
   (E) 18

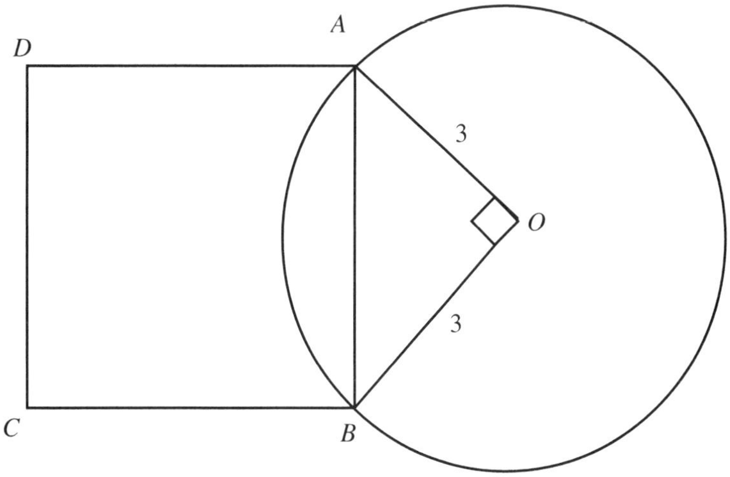

4. In the figure, lines l and m are parallel. If $y - z = 60$, then what is the value of x?

(A) 60
(B) 75
(C) 90
(D) 120
(E) 135

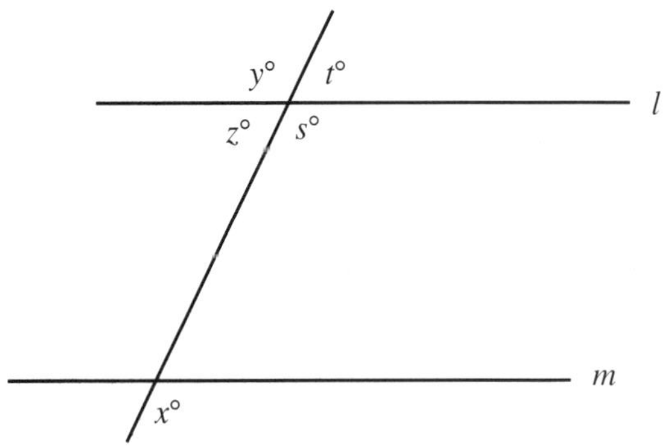

5. What is the value of y in the figure?

(A) 20
(B) 30
(C) 35
(D) 45
(E) 50

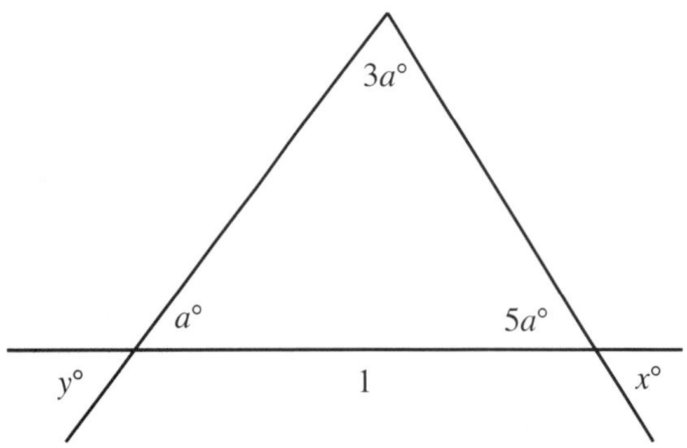

6. In the triangle, what is the value of x?

(A) 25
(B) 55
(C) 60
(D) 77
(E) 85

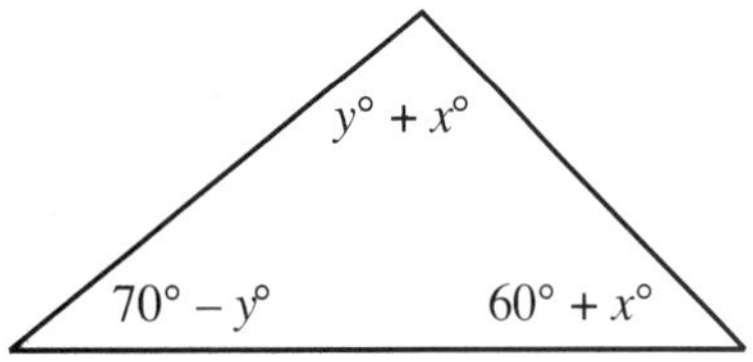

7. In the figure, what is the value of a ?

(A) 16
(B) 18
(C) 36
(D) 54
(E) 72

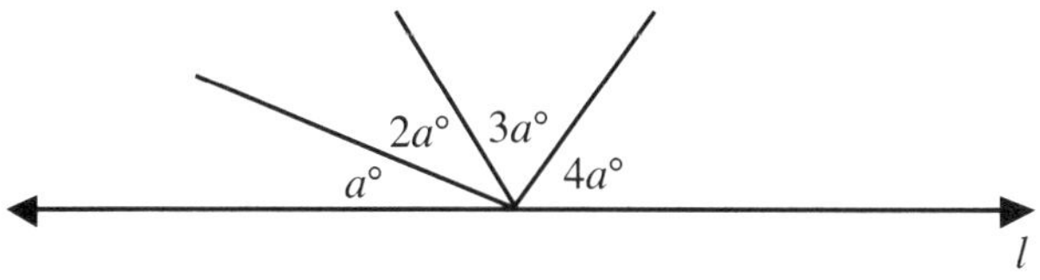

8. In the figure, what is the average of the five angles shown inside the circle?

(A) 36
(B) 45
(C) 60
(D) 72
(E) 90

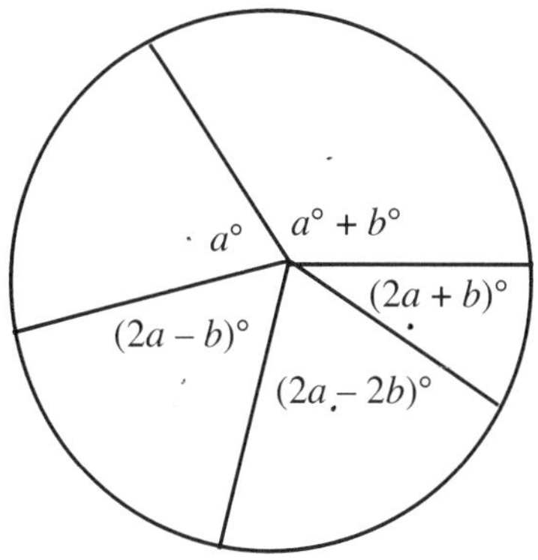

9. In the figure, O is the center of the circle. What is average of the numbers a, b, c, and d ?

(A) 45
(B) 60
(C) 90
(D) 180
(E) 360

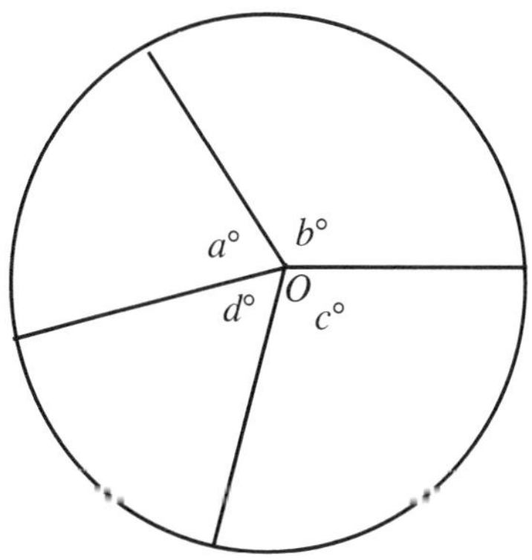

10. Column A | Column B

The perimeter of ΔABC | The circumference of the circle

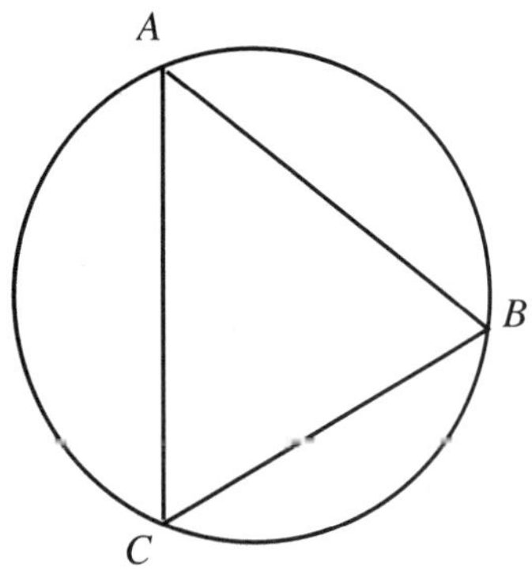

11. Column A | Column B

The perimeter of quadrilateral *ABCD* | The circumference of the circle

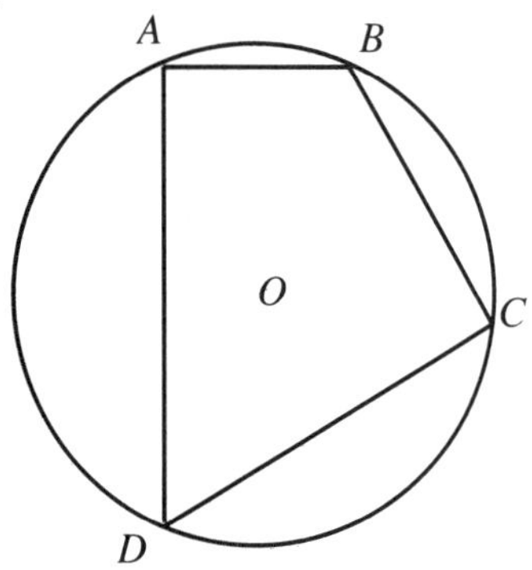

12. Column A | A circle is inscribed in a triangle. | Column B

The perimeter of the triangle | The circumference of the circle

➢ **Medium**

13. Column A | In ΔABC, $\angle B = 72°$ | Column B

$\angle A$ | 72°

14. In the figure, AD and BC are lines intersecting at O. What is the value of a ?

(A) 15
(B) 30
(C) 45
(D) 60
(E) 135

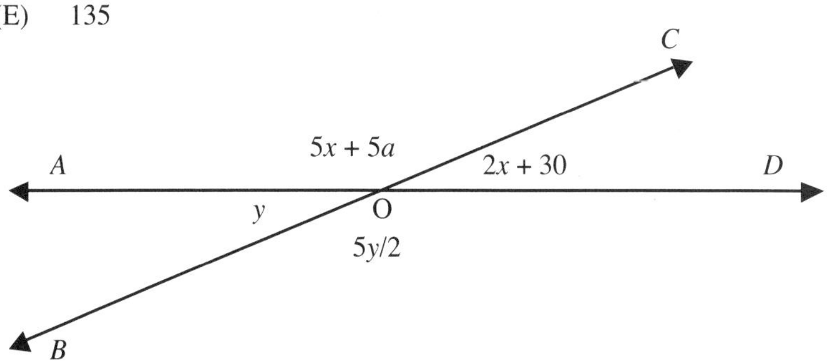

15. From the figure, which one of the following must be true?

(A) $y = z$
(B) $y < z$
(C) $y \le z$
(D) $y > z$
(E) $y \ge z$

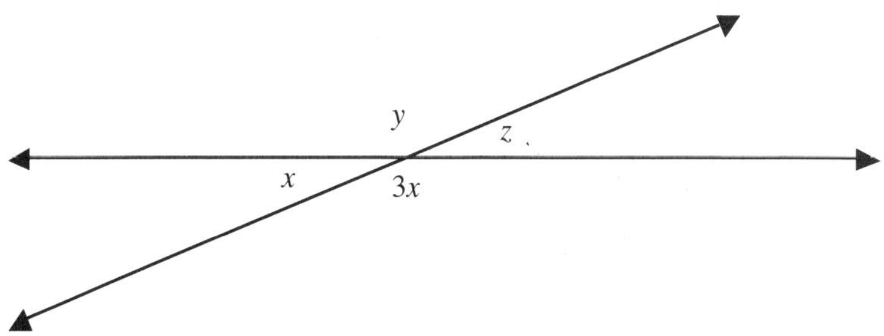

16. From the figure, which one of the following could be the value of b ?

(A) 20
(B) 30
(C) 60
(D) 75
(E) 90

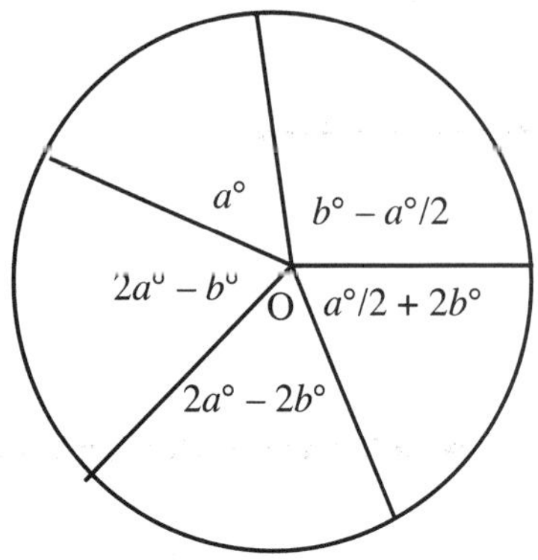

17. AD is the longest side of the right triangle ABD shown in the figure. What is the length of longest side of ΔABC ?

(A) 2
(B) 3
(C) $\sqrt{41}$
(D) 9
(E) 41

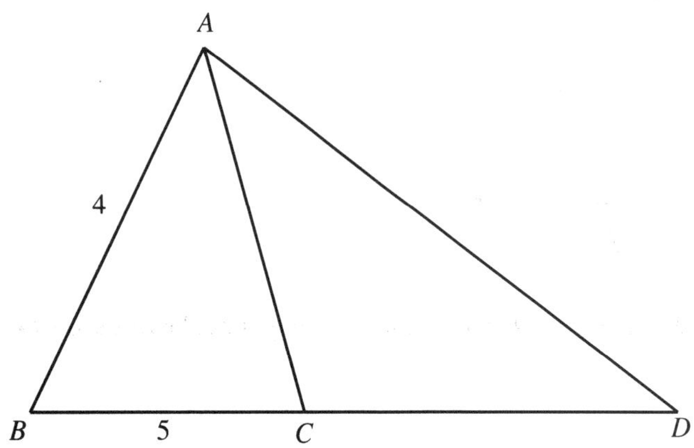

The figure is not drawn to scale

18. In the figure, what is the value of a ?

(A) 30
(B) 45
(C) 60
(D) 72
(E) 90

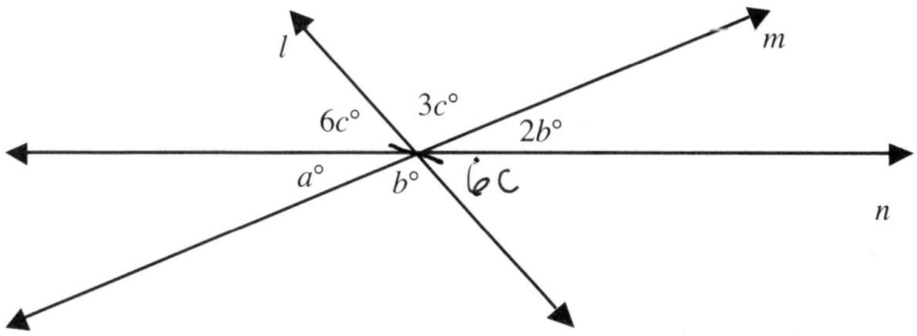

19. In the figure, lines l and m are parallel. Which one of the following, if true, makes lines p and q parallel?

(A) $a = b$
(B) $a = c$
(C) $c = d$
(D) $d = b$
(E) $b = c$

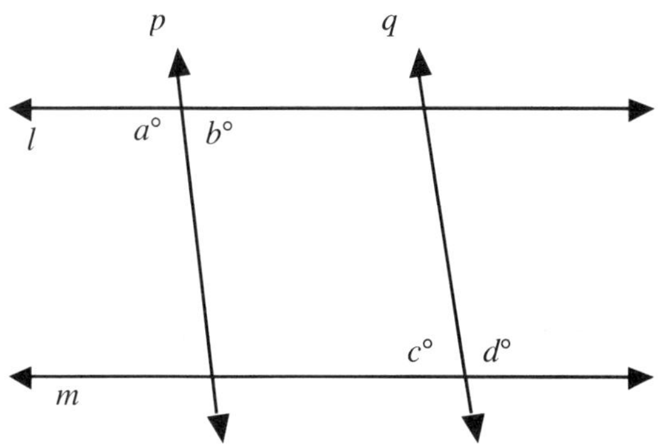

20. A, B, C, and D are points on a line such that point B bisects line AC and point A bisects line CD. What is the ratio of AB to CD ?

(A) 1/4
(B) 1/3
(C) 1/2
(D) 2/3
(E) 3/4

21. If A, B, C, D, and E are points in a plane such that line CD bisects $\angle ACB$ and line CB bisects right angle $\angle ACE$, then $\angle DCE =$

(A) 22.5°
(B) 45°
(C) 57.5°
(D) 67.5°
(E) 72.5°

22. From the figure, which of the following must be true?

(I) $x + y = 90$
(II) x is 35 units greater than y
(III) x is 35 units less than y

(A) I only
(B) II only
(C) III only
(D) I and II only
(E) I and III only

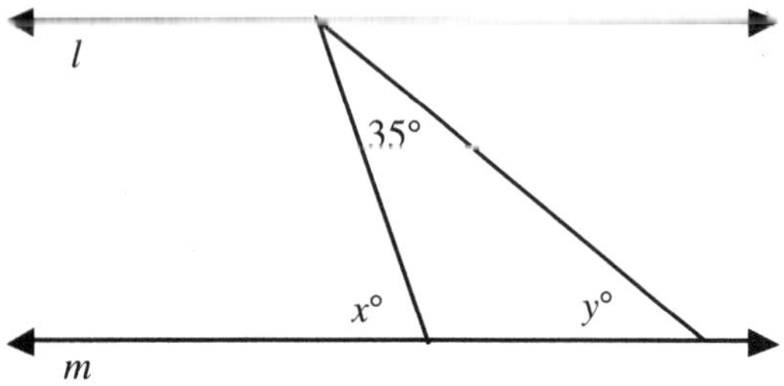

23. In the figure, triangles ABC and ABD are right triangles. What is the value of x?

(A) 20
(B) 30
(C) 50
(D) 70
(E) 90

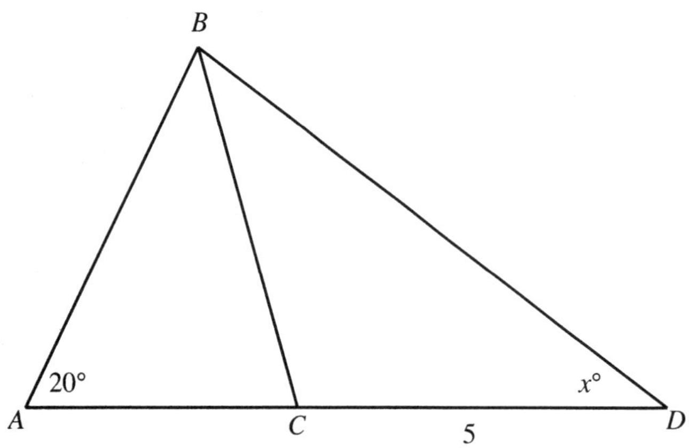

24. In the figure, ABC is a right triangle. What is the value of y ?

(A) 20
(B) 30
(C) 50
(D) 70
(E) 90

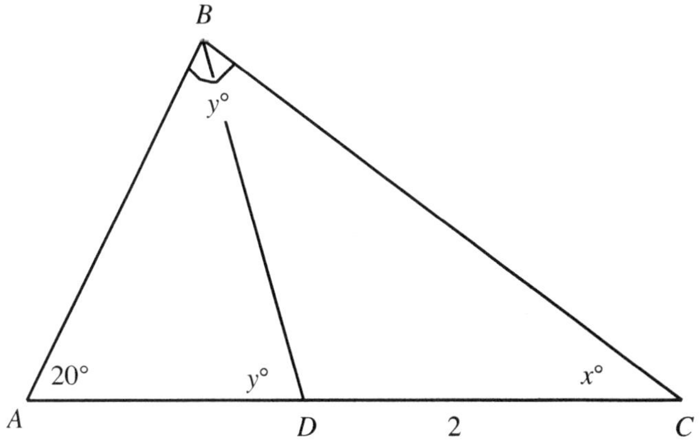

25. The following are the measures of the sides of five different triangles. Which one of them represents a right triangle?

(A) $\sqrt{3}$, $\sqrt{4}$, $\sqrt{5}$
(B) 1, 5, 4
(C) 7, 3, 4
(D) $\sqrt{3}$, $\sqrt{7}$, $\sqrt{4}$
(E) 4, 8, 10

26. ΔABC is a right-angled isosceles triangle, and $\angle B$ is the right angle in the triangle. If AC measures $7\sqrt{2}$, then which one of the following would equal the lengths of AB and BC, respectively?

(A) 7, 7
(B) 9, 9
(C) 10, 10
(D) 11, 12
(E) 7, 12

27. In the figure, if $AB = 8$, $BC = 6$, $AC = 10$ and $CD = 9$, then $AD =$

(A) 12
(B) 13
(C) 15
(D) 17
(E) 24

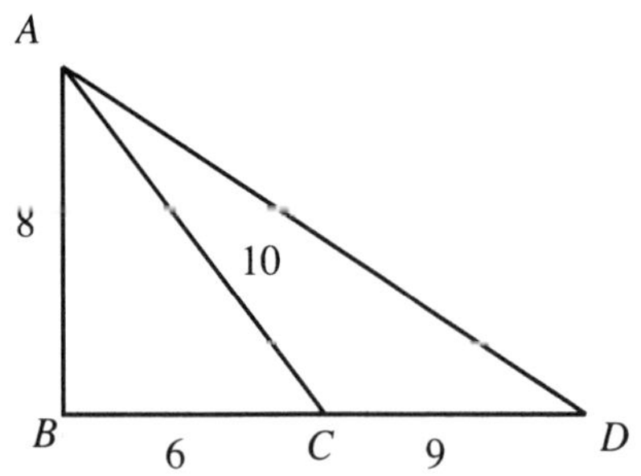

28. The average length of the sides of ΔABC is 12. What is the perimeter of ΔABC ?

(A) 4
(B) 6
(C) 12
(D) 24
(E) 36

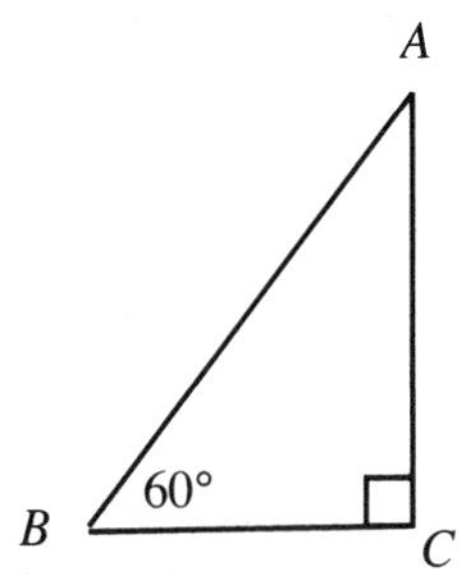

29. Which one of the following is true regarding the triangle shown in figure?

(A) $x > y > z$
(B) $x < y < z$
(C) $x = y = z$
(D) $2x = 3y/2 = z$
(E) $x/2 = 2y/3 = z$

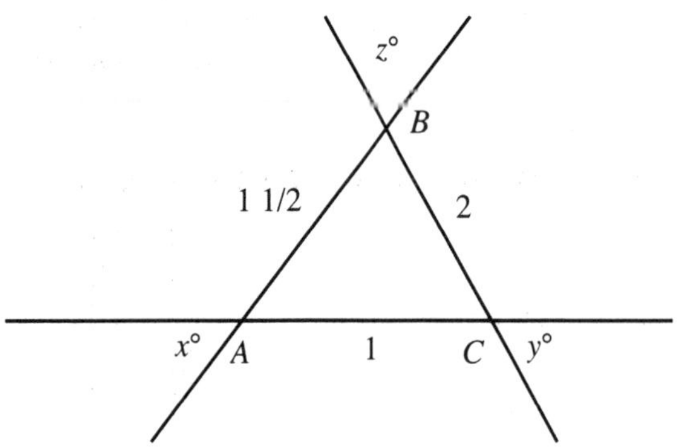

30. In the figure, *ABCD* is a rectangle, and the area of ΔACE is 10. What is the area of the rectangle?

(A) 18
(B) 22.5
(C) 36
(D) 44
(E) 45

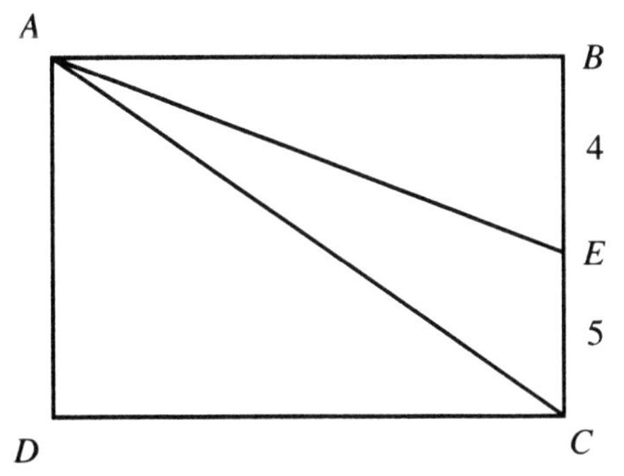

31. In the figure, what is the area of ΔABC ?

(A) 2
(B) $\sqrt{2}$
(C) 1
(D) $\frac{1}{\sqrt{2}}$
(E) 1/2

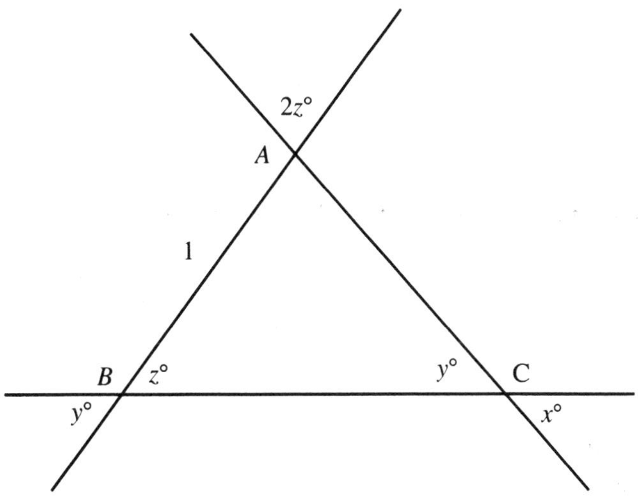

32. In the figure, O is the center of the circle. Which one of the following must be true about the perimeter of the triangle shown?

(A) Always less than 10
(B) Always greater than 40
(C) Always greater than 30
(D) Always less than 30
(E) Less than 40 and greater than 20

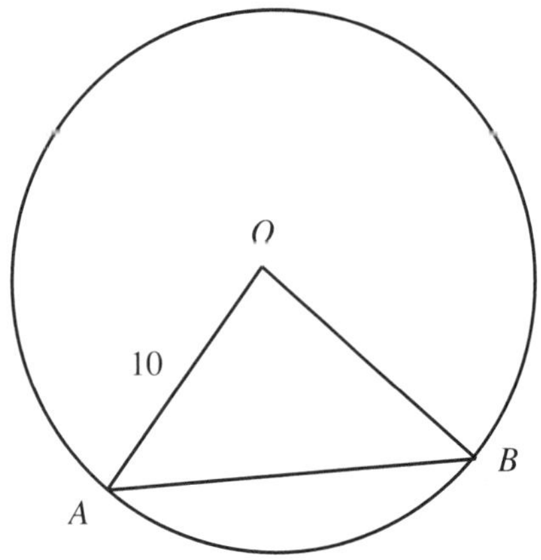

33. Which of the following must be true?

(I) The area of triangle P.
(II) The area of triangle Q.
(III) The area of triangle R.

(A) I = II = III
(B) I < II < III
(C) I > II < III
(D) III < I < II
(E) III > I > II

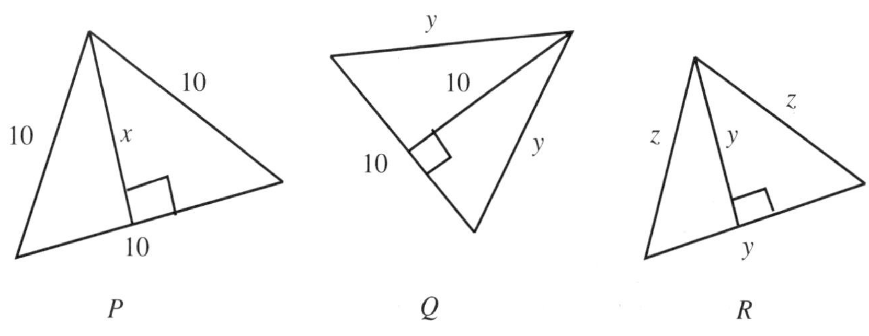

34. In the figure, *ABCD* is a parallelogram. Which one of the following is true?

(A) $x < y$
(B) $x > q$
(C) $x > p$
(D) $y > p$
(E) $y > q$

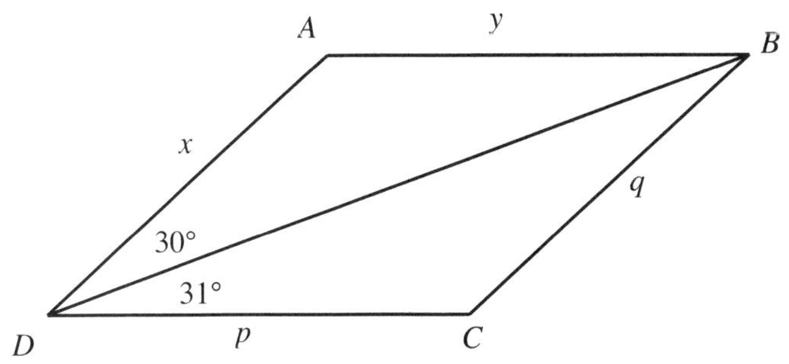

35. In the figure, the areas of parallelograms *EBFD* and *AECF* are 3 and 2, respectively. What is the area of rectangle *ABCD* ?

(A) 3
(B) 4
(C) 5
(D) $4\sqrt{3}$
(E) 7

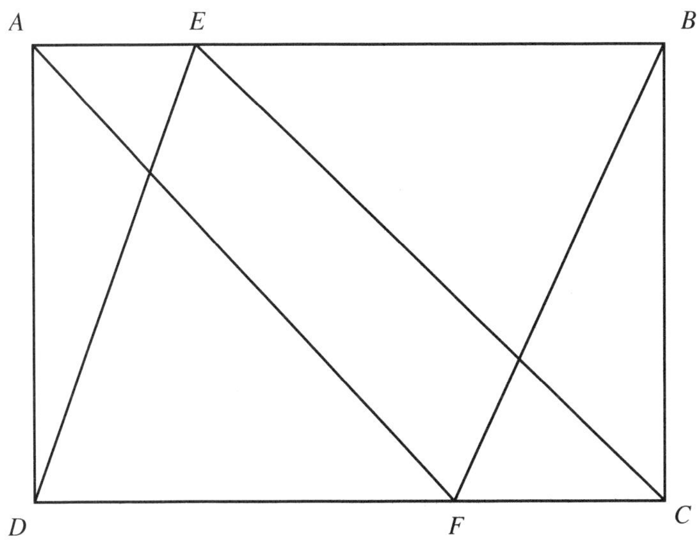

36. In the figure, the area of rectangle *ABCD* is 100. What is the area of the square *EFGH* ?

(A) 256
(B) 275
(C) 309
(D) 399
(E) 401

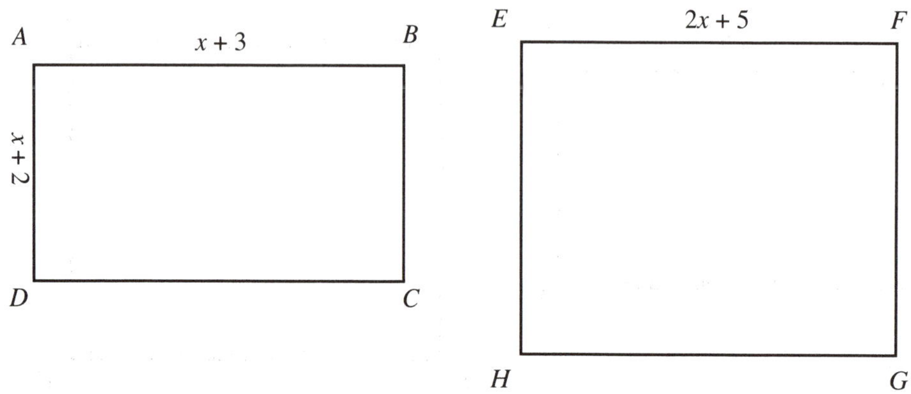

37. In the figure, the area of rectangle *EFGH* is 3 units greater than the area of rectangle *ABCD*. What is the value of ab if $a + b = 8$?

(A) 9
(B) 12
(C) 15
(D) 18
(E) 21

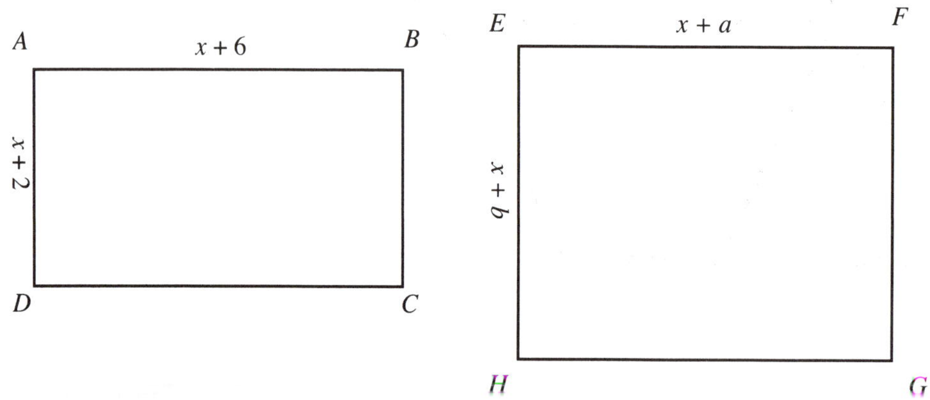

38. In the figure, the area of rectangle *ABCD* is 45. What is the area of the square *EFGH* ?

(A) 20
(B) 40
(C) 50
(D) 65
(E) 70

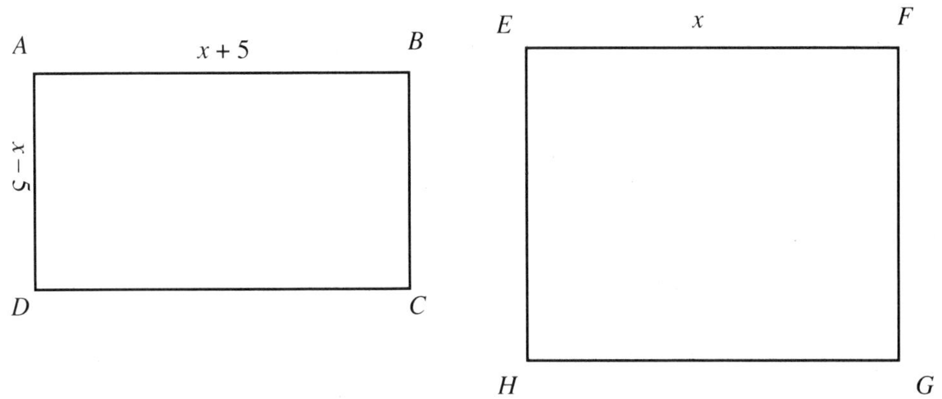

39. In the figure, *ABCD* is a parallelogram, what is the value of *b* ?

(A) 46
(B) 48
(C) 72
(D) 84
(E) 96

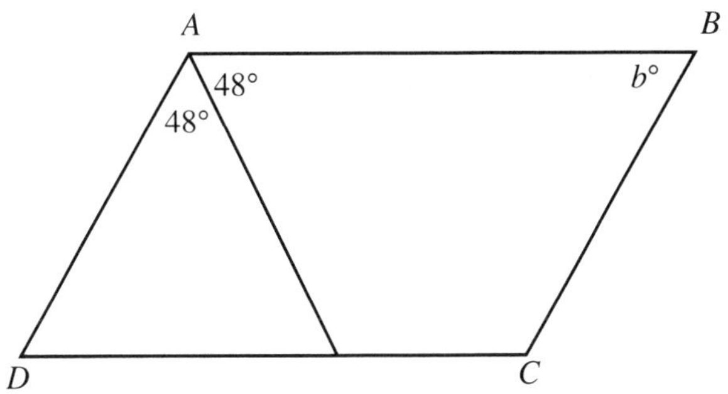

40. In the figure, AB and CD are the diameters of the circle. What is the value of x?

(A) 16°
(B) 18°
(C) 26°
(D) 32°
(E) 58°

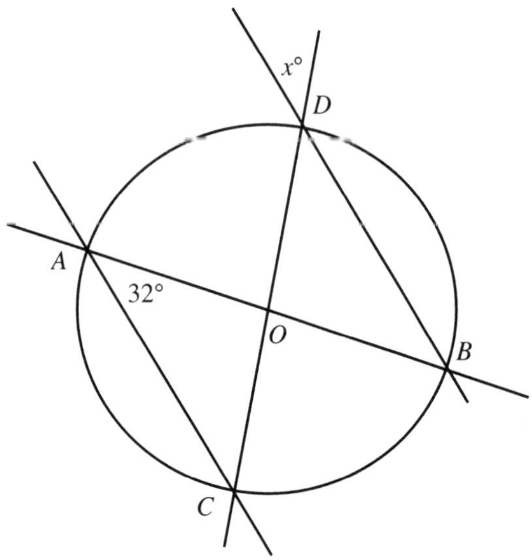

41. In the figure, *a*, *b*, *c*, *d*, *e*, *f*, *g*, *h*, *i*, and *j* are chords of the circle. Which two chords are parallel to each other?

(A) *a* and *f*
(B) *b* and *g*
(C) *c* and *h*
(D) *d* and *i*
(E) *e* and *j*

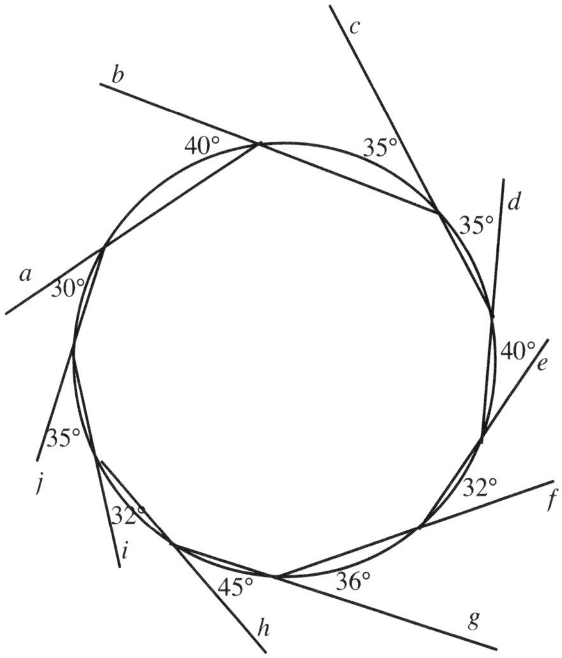

The figure is not drawn to scale.

42. *A* and *B* are centers of two circles that touch each other externally, as shown in the figure. What is the area of the circle whose diameter is *AB* ?

(A) 4π
(B) $25\pi/4$
(C) 9π
(D) 16π
(E) 25π

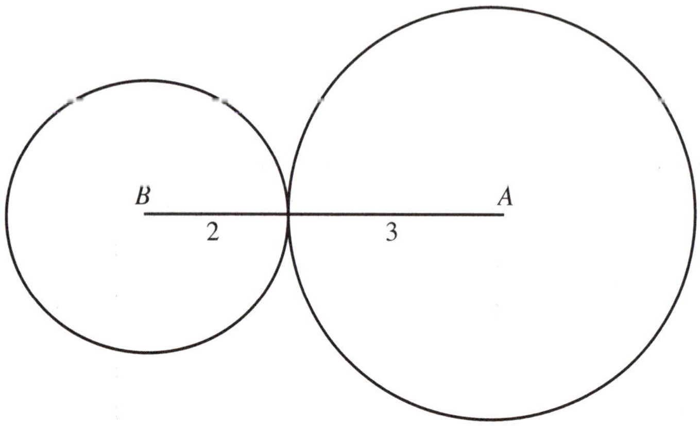

43. In the figure, which one of the following angles is the greatest?

(A) $\angle A$
(B) $\angle B$
(C) $\angle C$
(D) $\angle D$
(E) $\angle CDB$

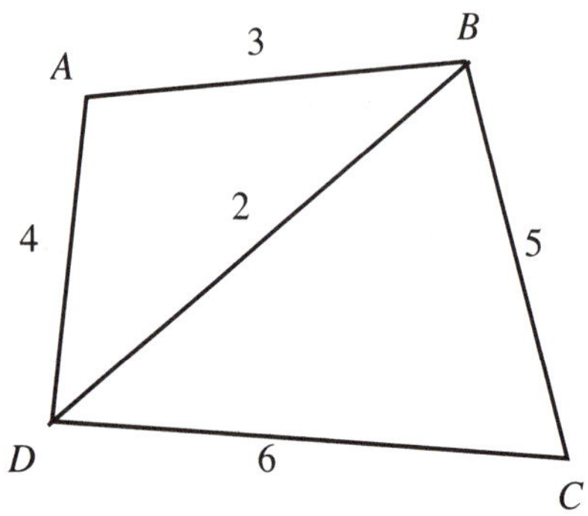

The figure is not drawn to scale.

44. In the figure, A, B, C, and D are points on a line in that order. If $AC = 5$, $BD = 10$, and $AD = 13$, then what is the length of BC ?

(A) 2
(B) 8
(C) 15
(D) 18
(E) 28

45. The side length of a square inscribed in a circle is 2. What is the area of the circle?

(A) π
(B) $\sqrt{2}\pi$
(C) 2π
(D) $2\sqrt{2}\pi$
(E) π^2

46. In the figure shown, if $\angle A = 60°$, $\angle B = \angle C$, and $BC = 20$, then $AB =$

(A) 20
(B) $10\sqrt{2}$
(C) $10\sqrt{3}$
(D) $20\sqrt{2}$
(E) $20\sqrt{3}$

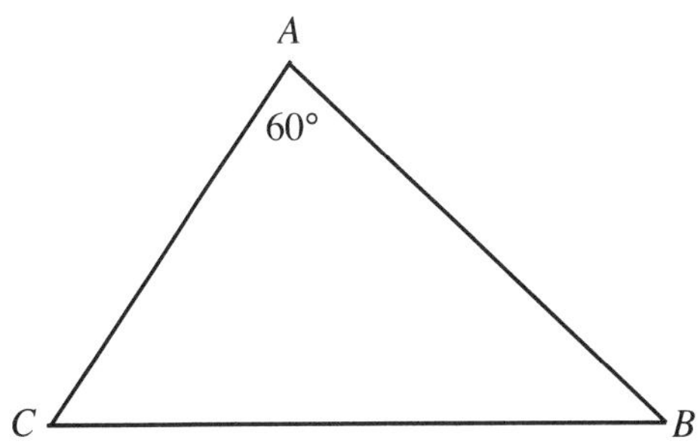

47. Which of the following could be the four angles of a parallelogram?

(I) 50°, 130°, 50°, 130°
(II) 125°, 50°, 125°, 60°
(III) 60°, 110°, 60°, 110°

(A) I only
(B) II only
(C) I and II only
(D) I and III only
(E) I, II and III

48. In the figure, $ABCD$ is a rectangle and E is a point on the side AB. If $AB = 10$ and $AD = 5$, what is the area of the shaded region in the figure?

(A) 25
(B) 30
(C) 35
(D) 40
(E) 45

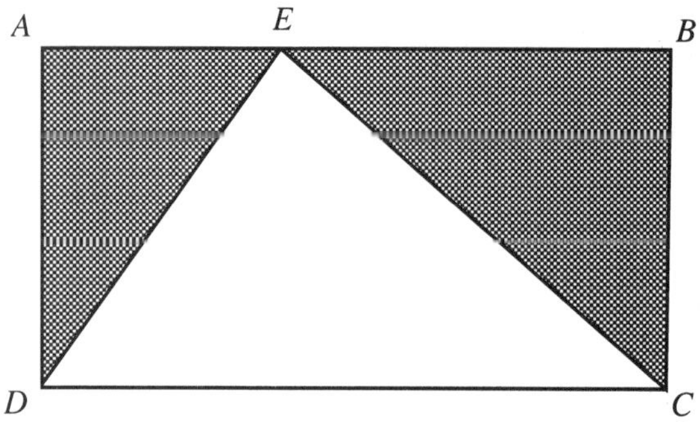

49. What is the area of the equilateral triangle if the base $BC = 6$?

(A) $9\sqrt{3}$
(B) $18\sqrt{3}$
(C) $26\sqrt{3}$
(D) $30\sqrt{3}$
(E) $36\sqrt{3}$

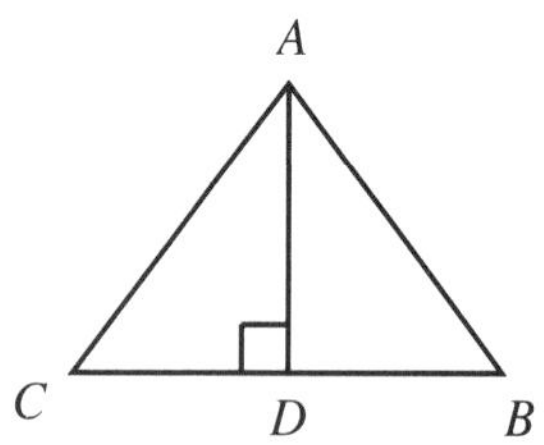

50. In the figure, $ABCD$ is a square, and OB is a radius of the circle. If BC is a tangent to the circle and $PC = 2$, then what is the area of the square?

(A) 16
(B) 20
(C) 25
(D) 36
(E) 42

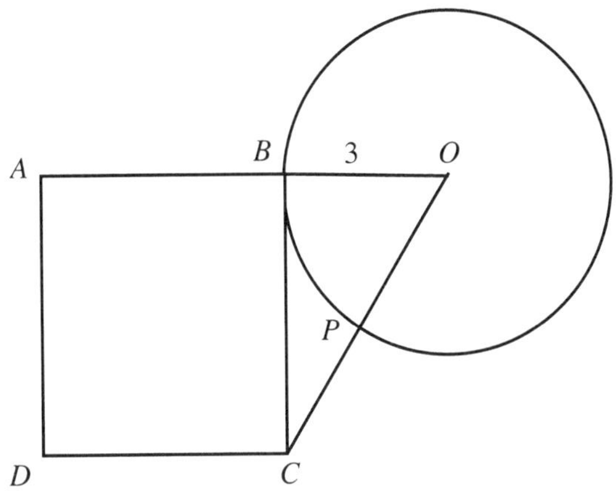

51. In the figure shown, AB is a diameter of the circle and O is the center of the circle. If $A = (3, 4)$, then what is the circumference of the circle?

(A) 3
(B) 4
(C) 4π
(D) 5π
(E) 10π

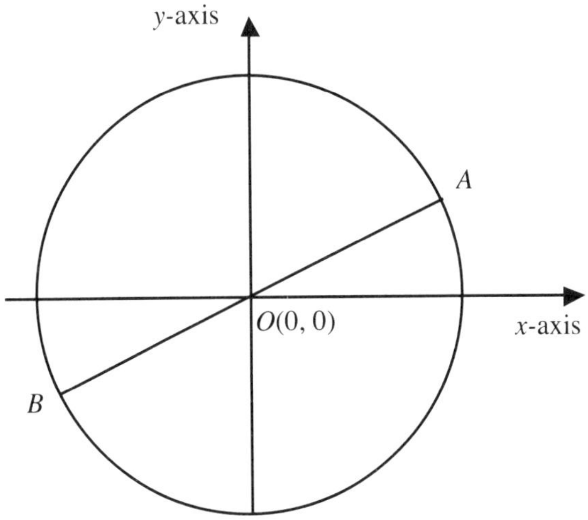

52. A, B, and C are three unequal faces of a rectangular tank. The tank contains a certain amount of water. When the tank is based on the face A, the height of the water is half the height of the tank. The dimensions of the side B are 3 ft × 4 ft and the dimensions of side C are 4 ft × 5 ft. What is the measure of the height of the water in the tank in feet?

(A) 2
(B) 2.5
(C) 3
(D) 4
(E) 5

53. In the figure, if line CE bisects $\angle ACB$, then $x =$

(A) 45
(B) 50
(C) 55
(D) 65
(E) 70

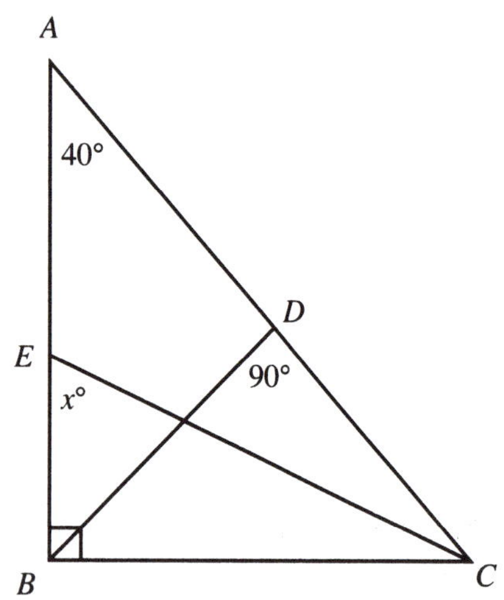

54. In the figure, $\angle P =$

(A) 15°
(B) 30°
(C) 35°
(D) 40°
(E) 50°

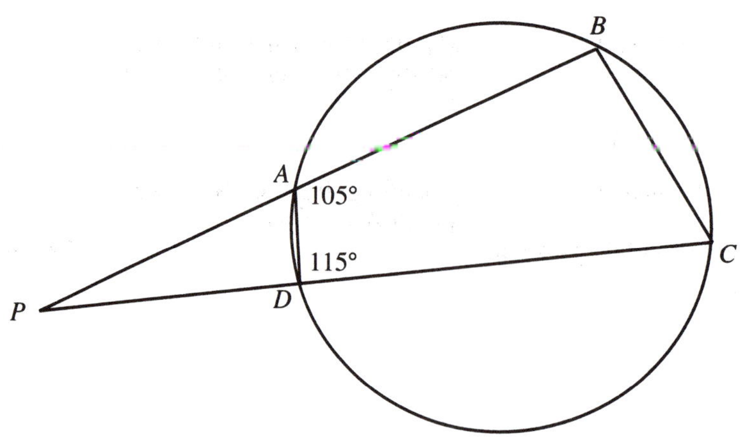

55. The length of a rectangular banner is 3 feet 2 inches, and the width is 2 feet 4 inches. Which one of the following equals the area of the banner?

(A) 5 sq. feet
(B) 5 1/2 sq. feet
(C) 6 1/3 sq. feet
(D) 7 sq. feet
(E) 7 7/18 sq. feet

56. In the figure, if $\angle A = \angle C$, then $x =$

(A) 30
(B) 60
(C) 90
(D) 150
(E) 180

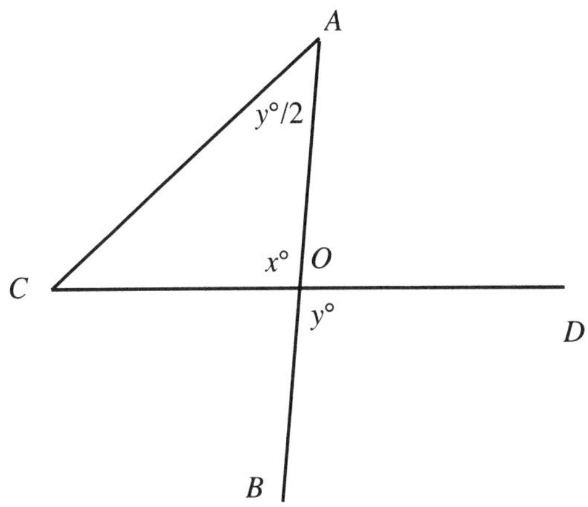

<u>The dimensions in the figure may be different from what they appear to be.</u>

57.

Column A		Column B
	The width and length of a rectangle are 6 inches and 12 inches, respectively.	
The area of the rectangle when its length is decreased by 4 inches and its width is not changed		The area of the rectangle when its width is decreased by 4 inches and its length is not changed

58.

Column A		Column B
	$\lvert a\rvert$ is the distance point a is from the origin on the number line. $x \neq 0$.	
$\lvert x\rvert + \lvert -2\rvert$		$\lvert x - 2\rvert$

59.

Column A		Column B
	A regular polygon of 24 sides is inscribed in a circle.	
The perimeter of the polygon		The circumference of the circle

> **Hard**

60. In the figure, lines l and k are parallel. Which one of the following must be true?

(A) $a < b$
(B) $a \le b$
(C) $a = b$
(D) $a \ge b$
(E) $a > b$

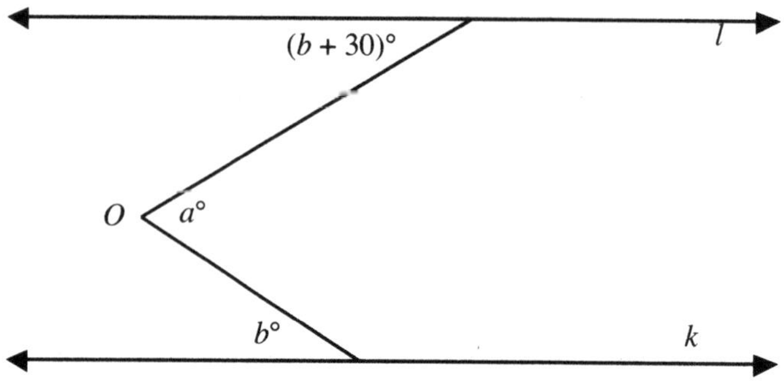

61. In the figure, lines l and k are parallel. If a is an acute angle, then which one of the following must be true?

(A) $b > 10$
(B) $b > 15$
(C) $b < 20$
(D) $b < 30$
(E) $b > 45$

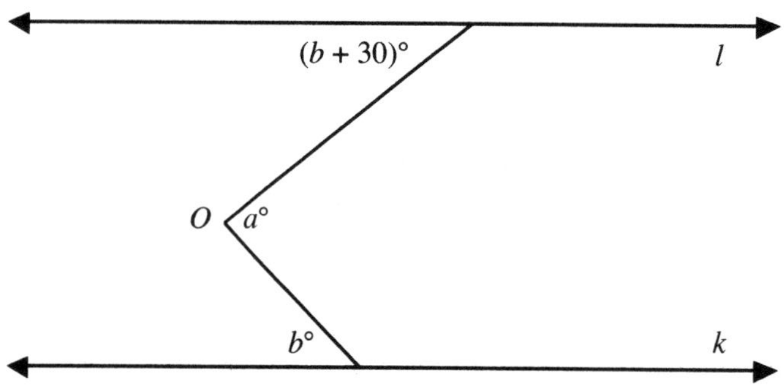

62. The diagonal length of a square is 14.1 sq. units. What is the area of the square, rounded to the nearest integer?

(A) 96
(B) 97
(C) 98
(D) 99
(E) 100

63. In the figure, ABC and ADC are right triangles. Which of the following could be the lengths of AD and DC, respectively?

(I) $\sqrt{3}$ and $\sqrt{4}$
(II) 4 and 6
(III) 1 and $\sqrt{24}$
(IV) 1 and $\sqrt{26}$

(A) I and II only
(B) II and III only
(C) III and IV only
(D) IV and I only
(E) I, II and III only

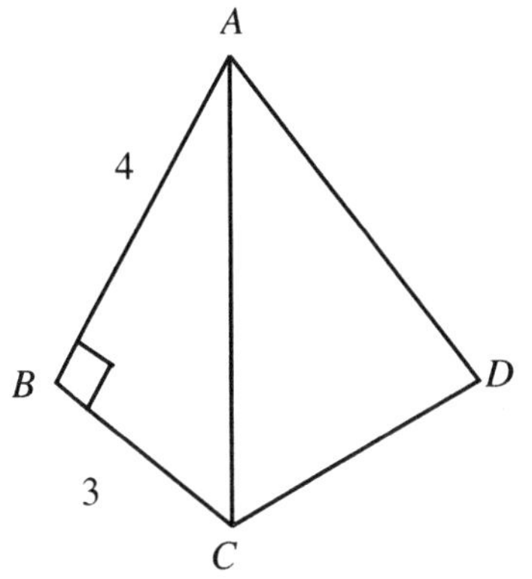

64. In the figure, what is the value of x ?

(A) 15
(B) 30
(C) 45
(D) 60
(E) 75

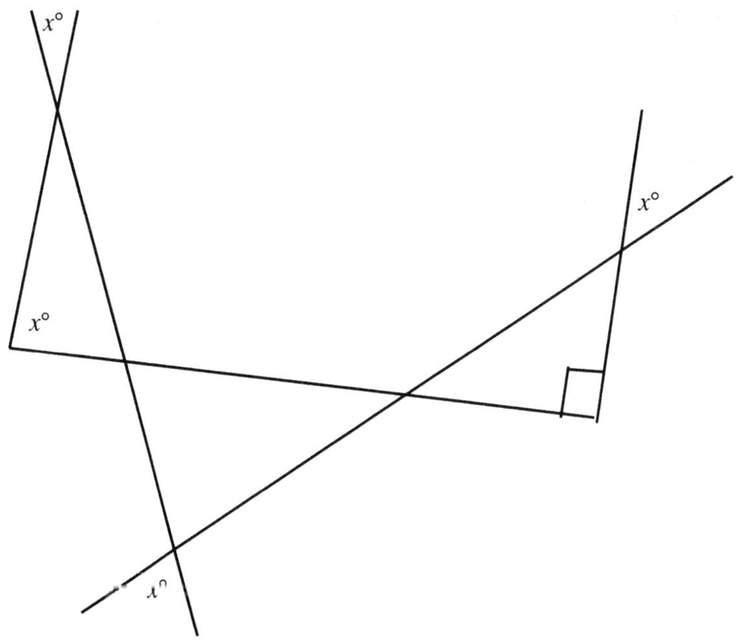

65. In the figure, what is the value of x?

(A) 10°
(B) 30°
(C) 45°
(D) 60°
(E) 75°

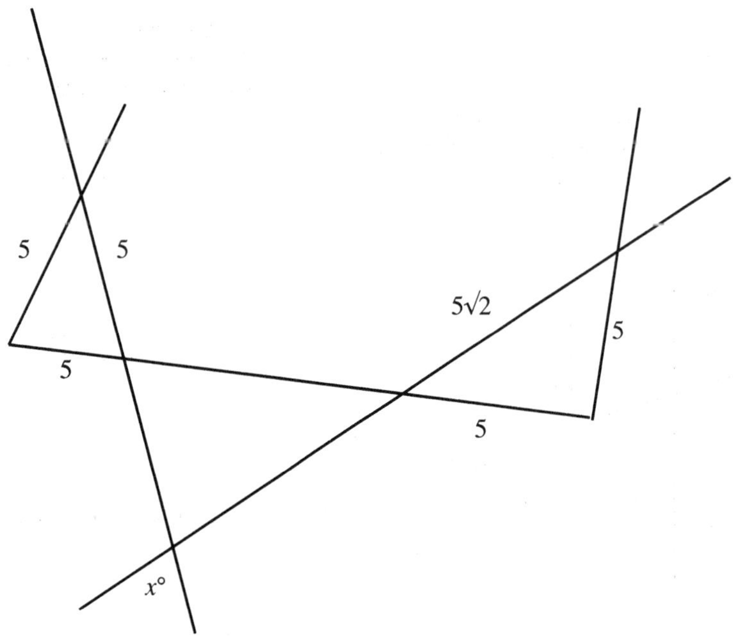

66. If $ABCD$ is a square and the area of ΔAFG is 10, then what is the area of ΔAEC?

(A) 5
(B) $\dfrac{10}{\sqrt{2}}$
(C) $\dfrac{10}{\sqrt{3}}$
(D) 10
(E) 20

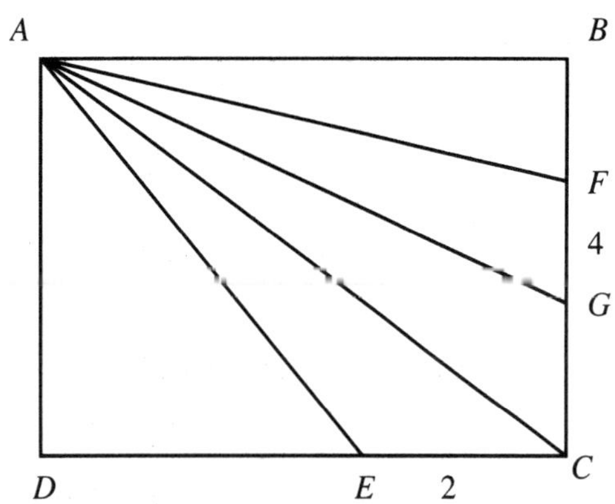

67. In the figure, $ABCD$ is a rectangle, and F and E are points on AB and BC, respectively. The area of ΔDFB is 9 and the area of ΔBED is 24. What is the perimeter of the rectangle?

(A) 18
(B) 23
(C) 30
(D) 42
(E) 48

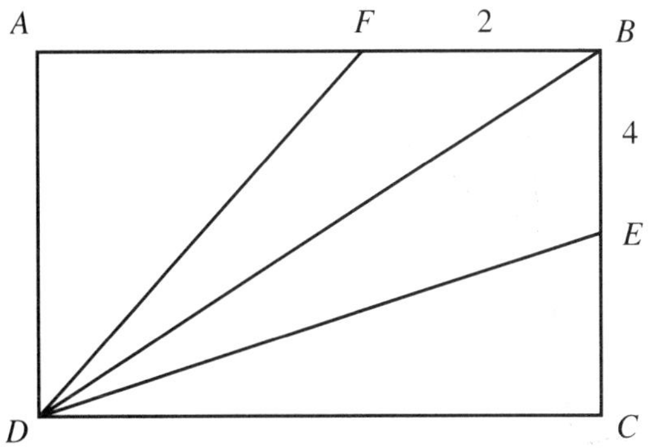

68. In the figure, $ABCD$ is a rectangle. The area of quadrilateral $EBFD$ is one-half the area of the rectangle $ABCD$. Which one of the following is the value of AD ?

(A) 5
(B) 6
(C) 7
(D) 12
(E) 15

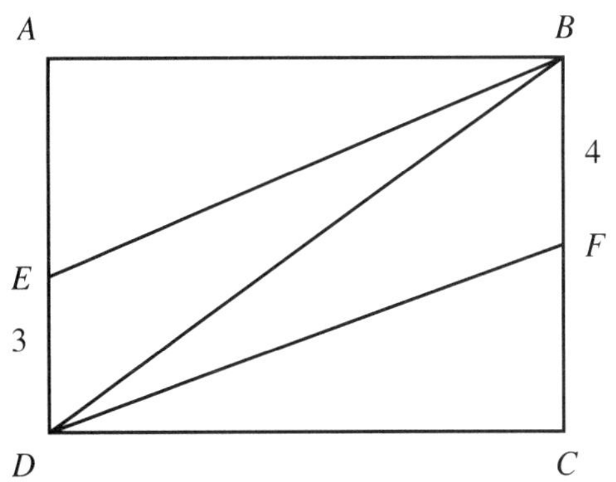

69. In the figure, $ABCD$ is a rectangle, and the area of quadrilateral $AFCE$ is equal to the area of ΔABC. What is the value of x ?

(A) 5
(B) 6
(C) 7
(D) 12
(E) 15

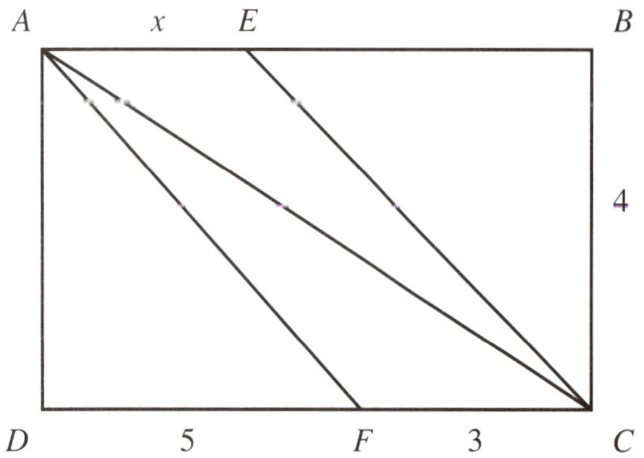

70. In the figure, $ABCD$ is a rectangle. Points E and F cut the sides BC and CD of the rectangle respectively such that $EC = 3$, $FC = 4$, and $AD = 12$, and the areas of the crossed and the shaded regions in the figure are equal. Which one of the following equals the perimeter of rectangle $ABCD$?

(A) 42
(B) 50
(C) 56
(D) 64
(E) 72

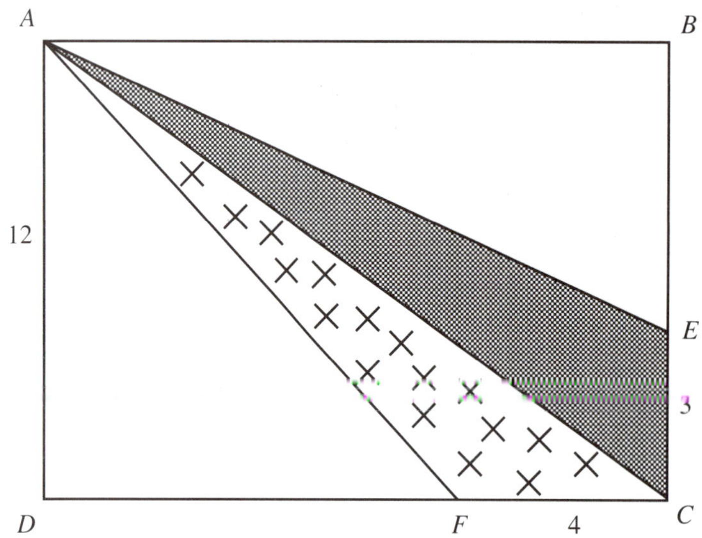

71. What is the perimeter of ΔABC shown in the figure?

(A) $2 + 4\sqrt{2}$
(B) $4 + 2\sqrt{2}$
(C) 8
(D) $4 + 4\sqrt{2}$
(E) $4 + 4\sqrt{3}$

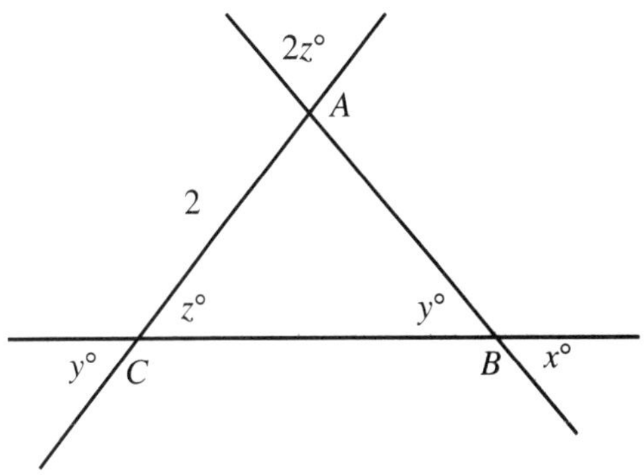

72. In the figure, what is the value of x?

(A) 90
(B) 95
(C) 100
(D) 105
(E) 115

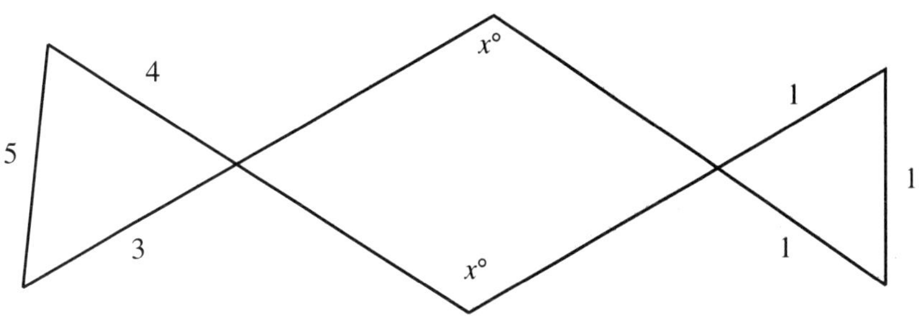

73. In the figure, $ABCD$ is a rectangle. Points P, Q, R, S, and T cut side AB of the rectangle such that $AP = 3$, $PQ = QR = RS = ST = 1$. E is a point on AD such that $AE = 3$. Which one of the following line segments is parallel to the diagonal BD of the rectangle?

(A) EP
(B) EQ
(C) ER
(D) ES
(E) ET

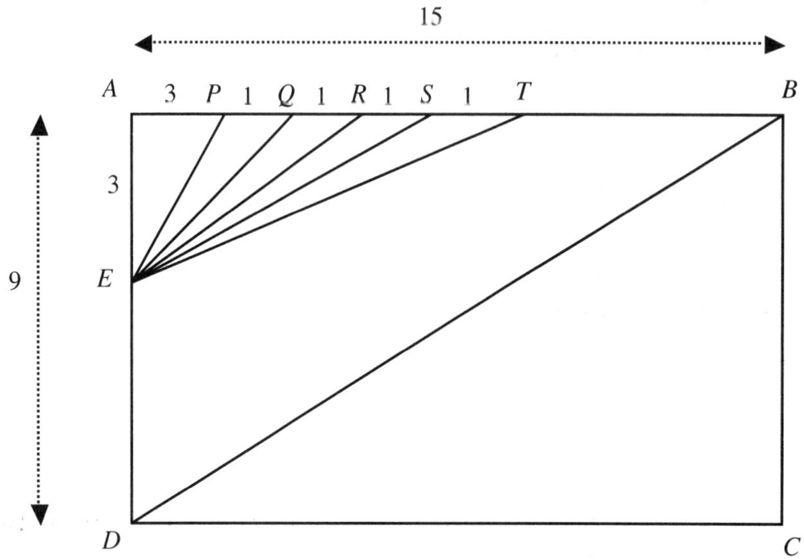

The figure is not drawn to scale.

74. In the, figure A, B and C are points on the circle. What is the value of x ?

(A) 45
(B) 55
(C) 60
(D) 65
(E) 70

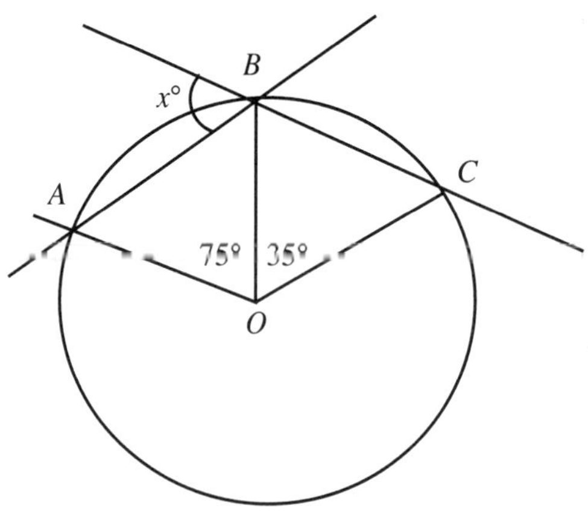

75. In the figure, *ABCD* and *PQRS* are two rectangles inscribed in the circle as shown and $AB = 4$, $AD = 3$, and $QR = 4$. What is the value of l ?

(A) 3/2
(B) 8/3
(C) 3
(D) 4
(E) 5

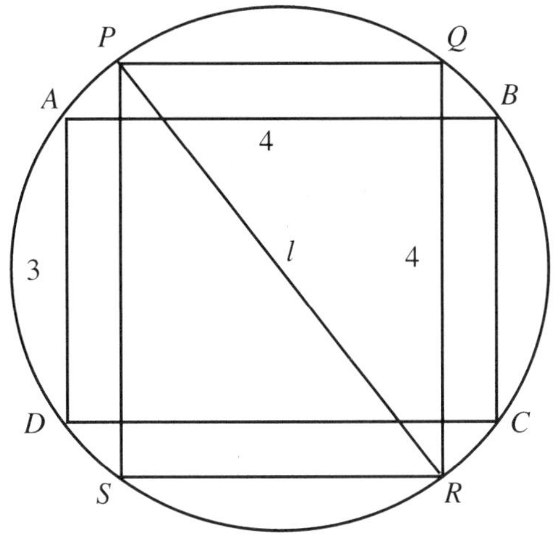

76. In the figure, *ABCD* and *PQRS* are rectangles inscribed in the circle shown in the figure. If $AB = 5$, $AD = 3$, and $QR = 4$, then what is the value of l ?

(A) 3
(B) 4
(C) 5
(D) $\sqrt{15}$
(E) $3\sqrt{2}$

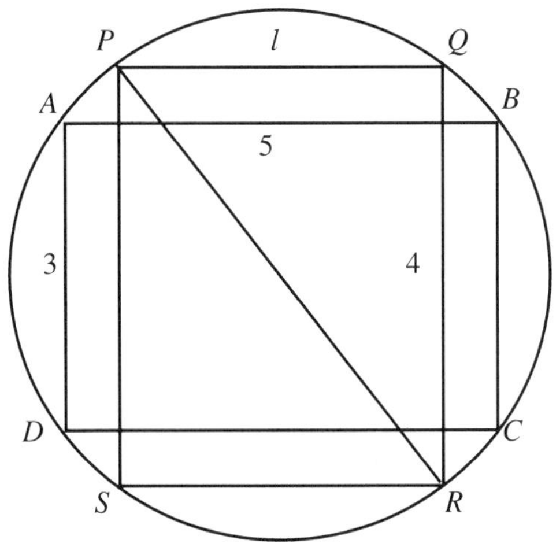

77. In the figure, *ABCD* is a rectangle inscribed in the circle shown. What is the length of the arc *DC* ?

(A) $\pi/4$
(B) $2\pi/3$
(C) $\pi/2$
(D) $3\pi/4$
(E) $4\pi/3$

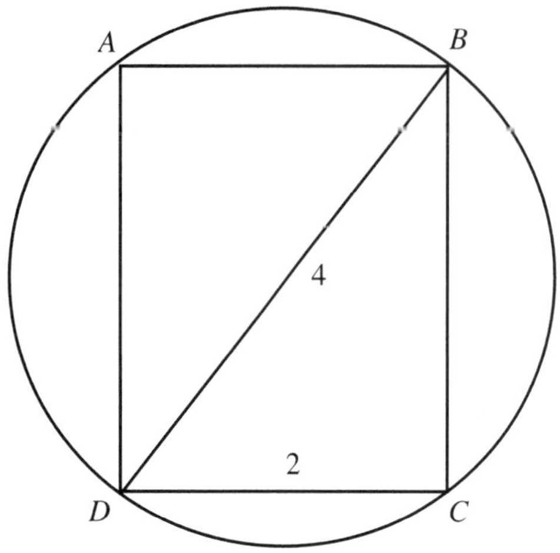

78. Which one of the following relations is true regarding the angles of the quadrilateral shown in figure?

(A) $\angle A = \angle C$
(B) $\angle B > \angle D$
(C) $\angle A < \angle C$
(D) $\angle B = \angle D$
(E) $\angle A = \angle B$

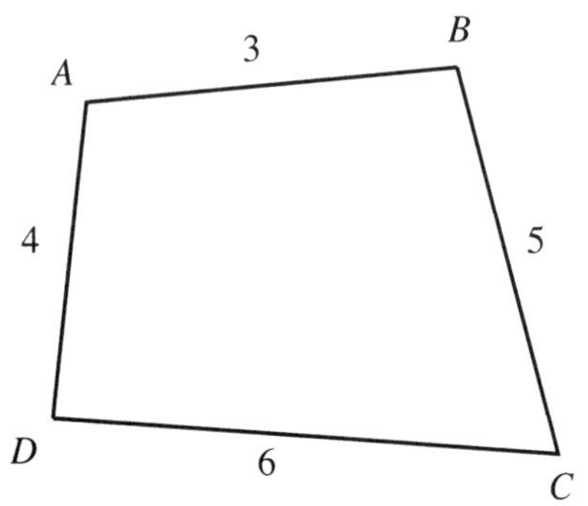

The figure is not drawn to scale.

79. In a triangle with sides of lengths 3, 4, and 5, the smallest angle is 36.87°. In the figure, *O* is the center of the circle of radius 5. *A* and *B* are two points on the circle, and the distance between the points is 6. What is the value of *x* ?

(A) 36.87
(B) 45
(C) 53.13
(D) 116.86
(E) 126.86

80. Which of following indicates that ΔABC is a right triangle?

(I) The angles of ΔABC are in the ratio 1 : 2 : 3.
(II) One of the angles of ΔABC equals the sum of the other two angles.
(III) ΔABC is similar to the right triangle ΔDEF.

(A) I only
(B) II only
(C) III only
(D) I and II only
(E) I, II, and III

81. In the figure, O is the center of the circle of radius 3, and $ABCD$ is a square. If $PC = 3$ and the side BC of the square is a tangent to the circle, then what is the area of the square $ABCD$?

(A) 25
(B) 27
(C) 36
(D) 42
(E) 56

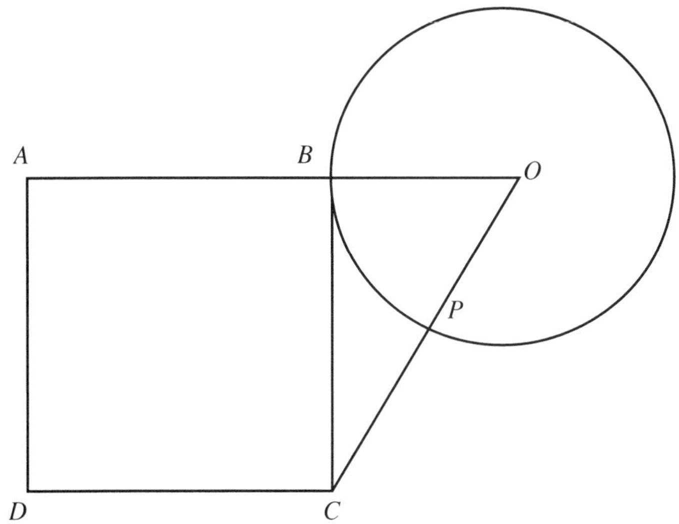

82. In the figure shown, $ABCDEF$ is a regular hexagon and AOF is an equilateral triangle. The perimeter of ΔAOF is $2a$ feet. What is the perimeter of the hexagon in feet?

(A) $2a$
(B) $3a$
(C) $4a$
(D) $6a$
(E) $12a$

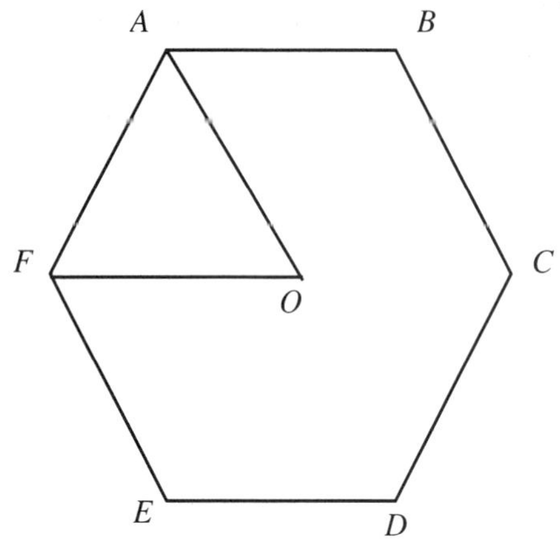

83. The area of the base of a tank is 100 sq. ft. It takes 20 seconds to fill the tank with water poured at rate of 25 cubic feet per second. What is the height in feet of the rectangular tank?

(A) 0.25
(B) 0.5
(C) 1
(D) 5
(E) 25

84. Point A is 10 miles West of Point B. Point B is 30 miles North of Point C. Point C is 20 miles East of Point D. What is the distance between points A and D?

(A) $10\sqrt{10}$ miles
(B) $10\sqrt{20}$ miles
(C) $20\sqrt{10}$ miles
(D) $30\sqrt{10}$ miles
(E) $30\sqrt{20}$ miles

85. Water is poured into an empty cylindrical tank at a constant rate. In 10 minutes, the height of the water increased by 7 feet. The radius of the tank is 10 feet. What is the rate at which the water is poured?

(A) $11/8\ \pi$ cubic feet per minute.
(B) $11/3\ \pi$ cubic feet per minute.
(C) $7/60\ \pi$ cubic feet per minute.
(D) 11π cubic feet per minute.
(E) 70π cubic feet per minute.

86. The length of a rectangle is increased by 25%. By what percentage should the width be decreased so that the area of the rectangle remains unchanged?

(A) 20
(B) 25
(C) 30
(D) 33.33
(E) 50

87. In the rectangular coordinate system shown, *ABCD* is a parallelogram. If the coordinates of the points *A*, *B*, *C*, and *D* are (0, 2), (a, b), $(a, 2)$, and (0, 0), respectively, then $b =$

(A) 0
(B) 2
(C) 3
(D) 4
(E) 5

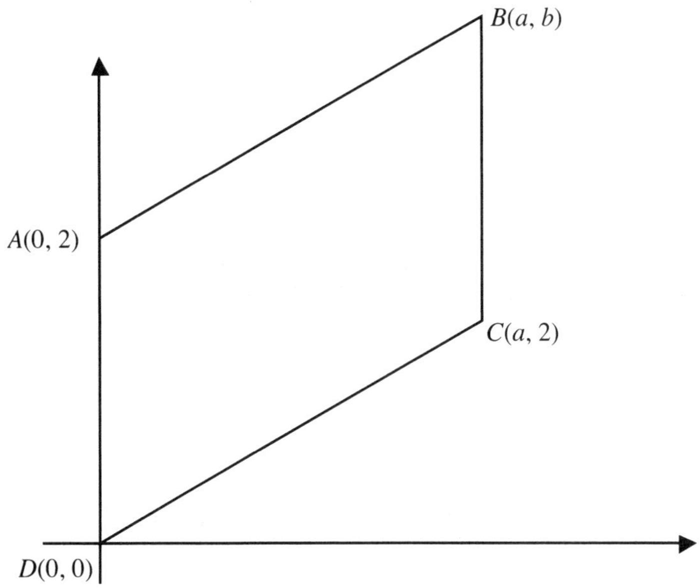

88. In the figure, if $AB = 10$, what is the length of the side CD ?

(A) 5
(B) $5\sqrt{3}$
(C) $\frac{10}{\sqrt{3}}$
(D) 10
(E) $10\sqrt{3}$

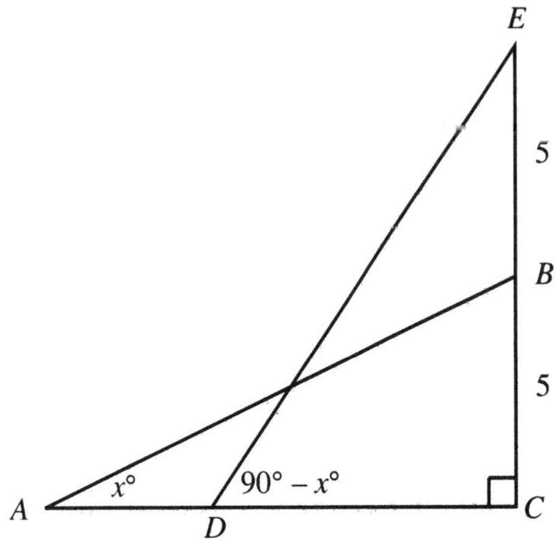

89. In the figure, what is the area of ΔABC if $EC/CD = 3$?

(A) 12
(B) 24
(C) 81
(D) 121.5
(E) 143

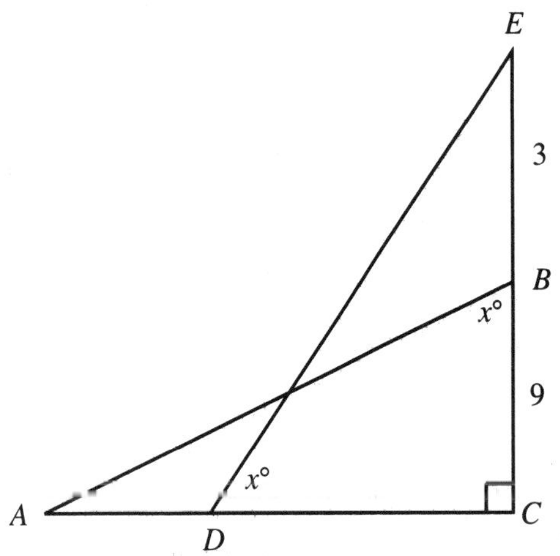

90. *AB* and *CD* are chords of the circle, and *E* and *F* are the midpoints of the chords, respectively. The line *EF* passes through the center *O* of the circle. If *EF* = 17, then what is radius of the circle?

(A) 10
(B) 12
(C) 13
(D) 15
(E) 25

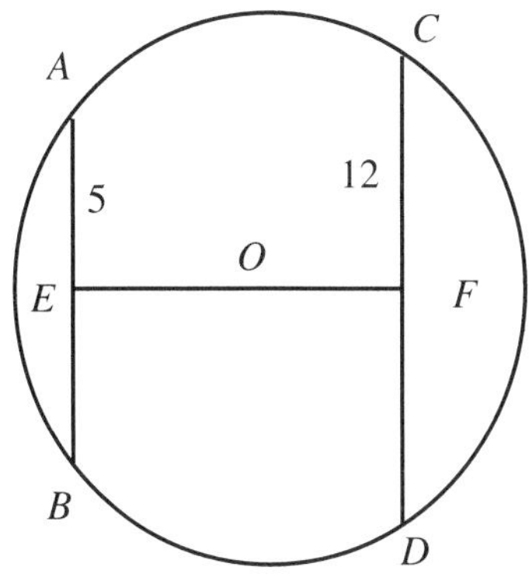

91. Column A | The perimeter of rectangle *ABCD* is 5/2 times as long as the side *AB*. | Column B

Length of side *AB* | | Length of side *BC*

92. A closed rectangular tank contains a certain amount of water. When the tank is placed on its 3 ft by 4 ft side, the height of the water in the tank is 5 ft. When the tank is placed on another side of dimensions 4 ft by 5 ft, what is the height, in feet, of the surface of the water above the ground?

(A) 2
(B) 3
(C) 4
(D) 5
(E) 6

93. Column A | Column B

The clockwise angle made by the hour hand and the minute hand at 12:15pm | The counterclockwise angle made by the hour hand and the minute hand at 12:45pm

94. *AC*, a diagonal of the rectangle *ABCD*, measures 5 units. The area of the rectangle is 12 sq. units. What is the perimeter of the rectangle?

(A) 7
(B) 14
(C) 17
(D) 20
(E) 28

95. In the figure, ΔABC is inscribed in the circle. The triangle does not contain the center of the circle O. Which one of the following could be the value of x in degrees?

(A) 35
(B) 70
(C) 85
(D) 90
(E) 105

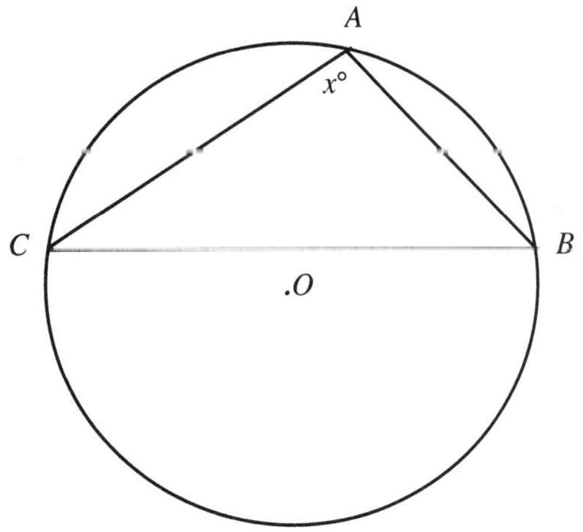

96. In the figure, $ABCD$ is a square, and BC is tangent to the circle with radius 3. If $PC = 2$, then what is the area of square $ABCD$?

(A) 9
(B) 13
(C) 16
(D) 18
(E) 25

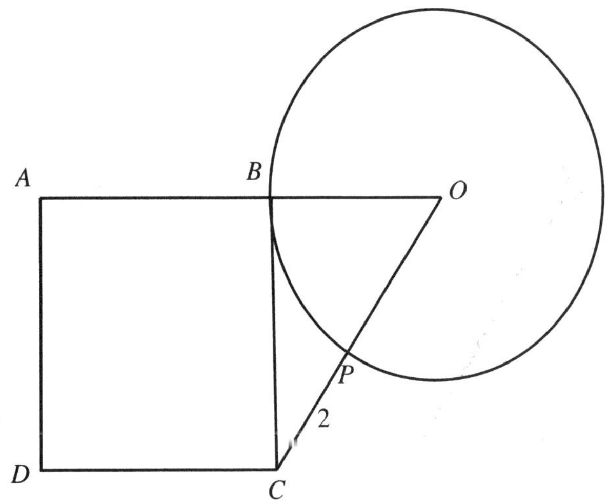

97. In the figure, *ABCD* is a square and *BCP* is an equilateral triangle. What is the measure of x ?

(A) 7.5
(B) 15
(C) 30
(D) 45
(E) 60

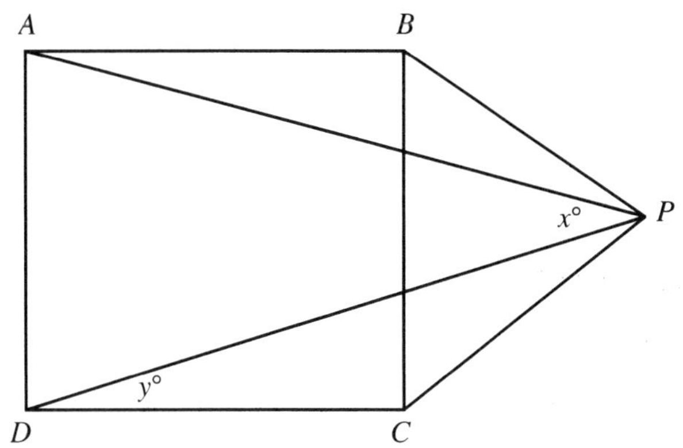

98. Column A | A circular park is enlarged uniformly such that it now occupies 21% more land. | Column B

Column A: The percentage increase in the radius of the park due to the enlargement

Column B: The percentage increase in the area of the park due to the enlargement

99. In the figure, *ABCD* and *ABEC* are parallelograms. The area of the quadrilateral *ABED* is 6. What is the area of the parallelogram *ABCD* ?

(A) 2
(B) 4
(C) 4.5
(D) 5
(E) 6

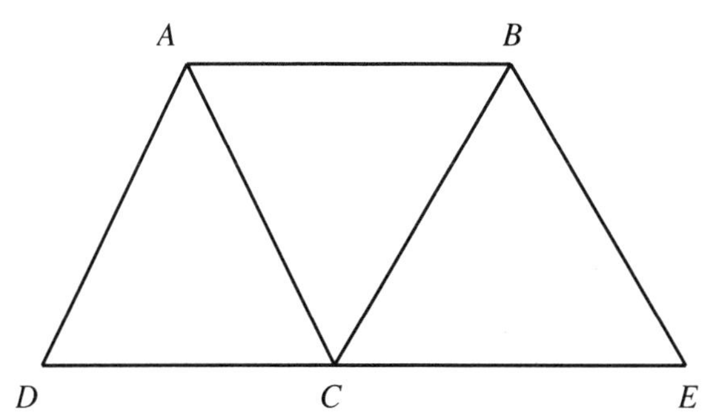

100. In the figure, lines l_1, l_2, and l_3 are parallel to one another. Line-segments *AC* and *DF* cut the three lines. If *AB* = 3, *BC* = 4, and *DE* = 5, then which one of the following equals *DF* ?

(A) 3/30
(B) 15/7
(C) 20/3
(D) 6
(E) 35/3

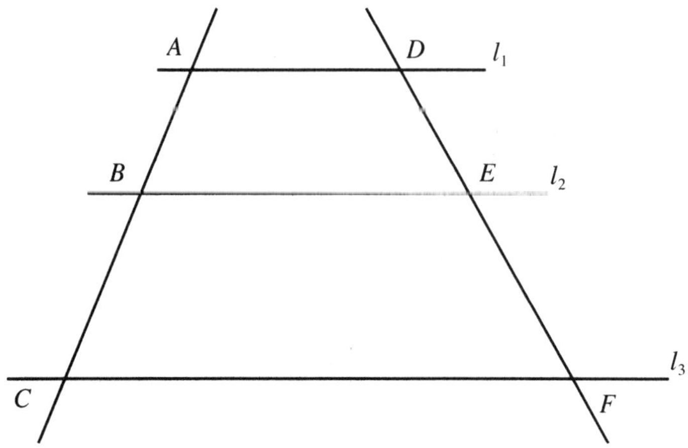

101. In the figure, *AB* is parallel to *CD*. What is the value of *x* ?

(A) 36
(B) 45
(C) 60
(D) 75
(E) Cannot be determined

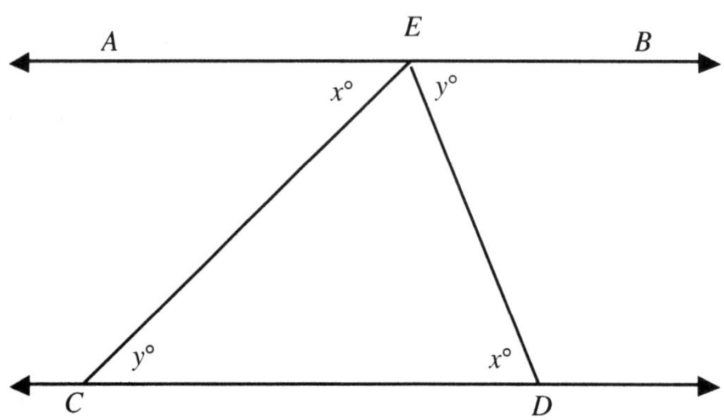

102. Column A | In the figure, P and Q are centers of the two circles of radii 3 and 4, respectively. A and B are the points at which a common tangent touches each circle. | Column B

AB | | PQ

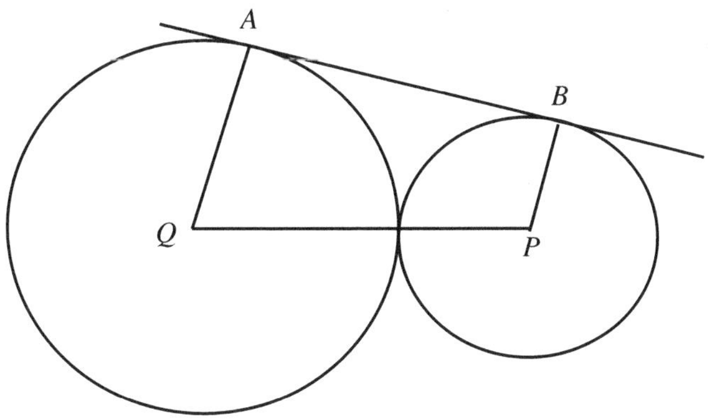

103. Column A | In the figure, ΔABC is inscribed in the circle. The triangle does not contain the center of the circle O. | Column B

x | | 90

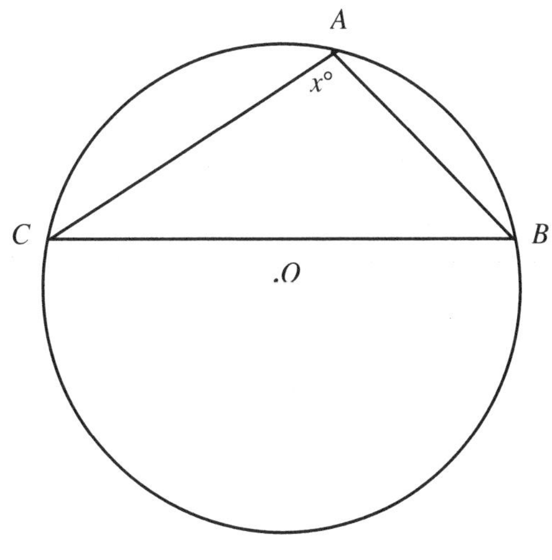

104. Column A | | Column B

Measure of the largest angle (in degrees) of a triangle with sides of length 5, 6, and 7 | | 60

➢ Very Hard

105. In the figure, if $y = 60$, then what is the value of z ?

(A) 20
(B) 30
(C) 55
(D) 75
(E) 90

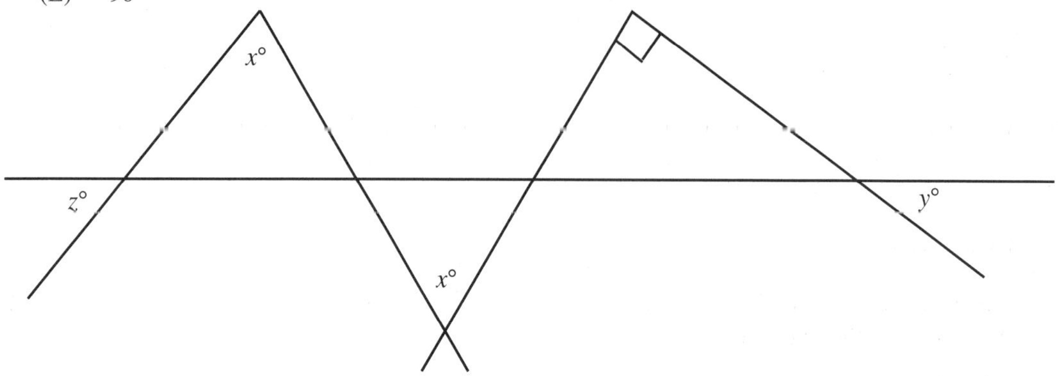

106. In the figure, *ABCD* is a rectangle and *AF* is parallel to *BE.* If $x = 5$, and $y = 10$, then what is the area of ΔAFD ?

(A) 2.5
(B) 5
(C) 12.5
(D) 50
(E) $50 + 5y$

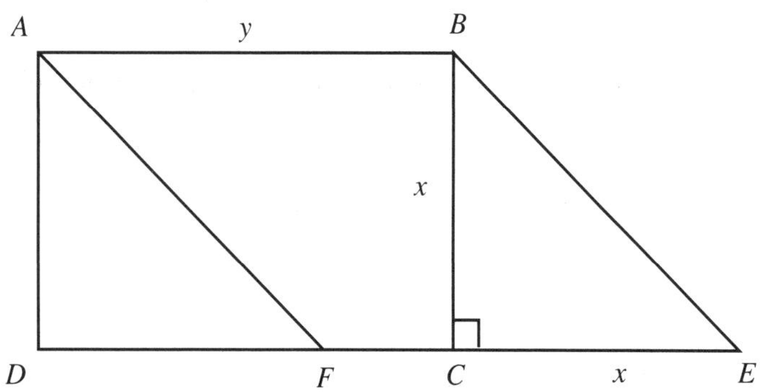

Answers and Solutions to Problem Set I

➢ Easy

1. If p is the circumference of the circle Q and the area of the circle is 25π, what is the value of p ?

   (A) 25
   (B) 10π
   (C) 35
   (D) 15π
   (E) 25π

The area of the circle $Q = \pi \cdot radius^2 = 25\pi$. Solving the equation for the radius yields $radius = \sqrt{\frac{25\pi}{\pi}} = 5$.

Now, the circumference of the circle $Q = 2\pi \cdot radius = 2\pi \cdot 5 = 10\pi$.

The answer is (B).

2. A rectangular field is 3.2 yards long. A fence marking the boundary is 11.2 yards in length. What is the area of the field in square yards?

   (A) 4.68
   (B) 7.68
   (C) 9.28
   (D) 11.28
   (E) 12.38

Let l be the length of the rectangle. Then $l = 3.2$ yards. We are given that the length of the fence required (perimeter) for the field is 11.2 yards. The formula for the perimeter of a rectangle is $2(length + width)$. Hence, the perimeter of the field is $2(l + w) = 11.2$, or $2(3.2 + w) = 11.2$. Solving for w yields $w = 2.4$. The formula for the area of a rectangle is $length \times width$. Hence, the area of the rectangle is $lw = 3.2 \times 2.4 = 7.68$. The answer is (B).

3. $ABCD$ is a square and one of its sides AB is also a chord of the circle as shown in the figure. What is the area of the square?

   (A) 3
   (B) 9
   (C) 12
   (D) $12\sqrt{2}$
   (E) 18

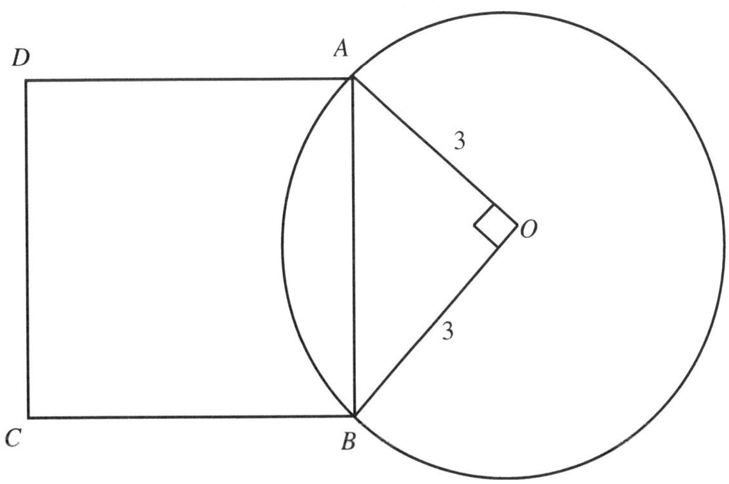

Side AB is the hypotenuse of the ΔAOB. Hence, by The Pythagorean Theorem, we have $AB^2 = AO^2 + BO^2 = 3^2 + 3^2 = 18$. Hence, the area of the square $ABCD$ equals $side^2 = AB^2 = 18$. The answer is (E).

4. In the figure, lines l and m are parallel. If $y - z = 60$, then what is the value of x?

   (A) 60
   (B) 75
   (C) 90
   (D) 120
   (E) 135

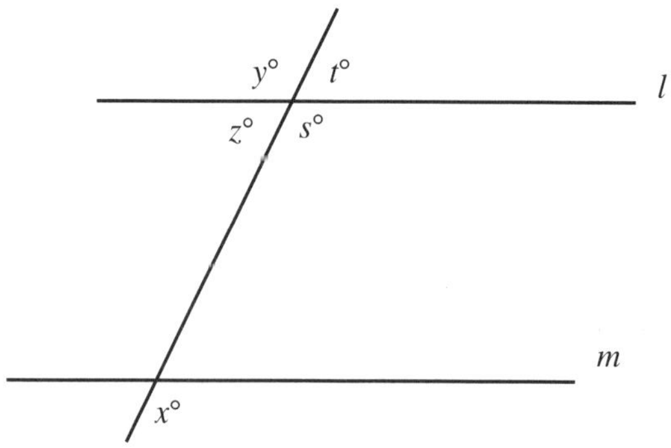

Since the angle made by a line is 180°, $z + y = 180$. Also, we are given that $y - z = 60$. Adding the equations yields

$$z + y + y - z = 180 + 60$$
$$2y = 240$$
$$y = 120$$

Since the lines l and m are parallel, the alternate exterior angles x and y are equal. Hence, x equals 120. The answer is (D).

5. What is the value of y in the figure?

   (A) 20
   (B) 30
   (C) 35
   (D) 45
   (E) 50

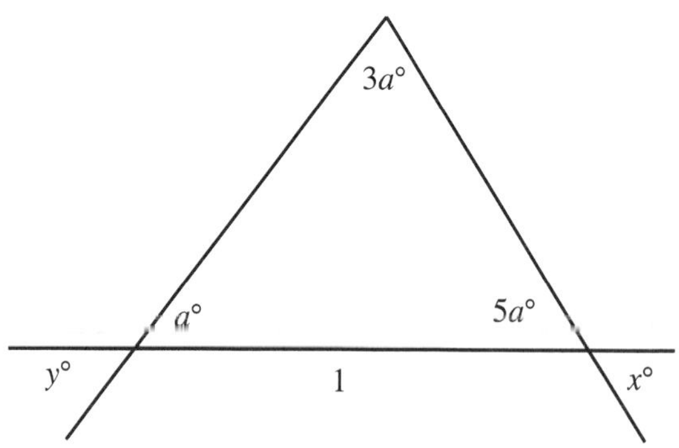

Summing the angles of the triangle in the figure to 180° yields $a + 3a + 5a = 180$. Solving this equation for a yields $a = 180/9 = 20$. Angles y and a in the figure are vertical and therefore are equal. So, $y = a = 20$. The answer is (A).

6. In the triangle, what is the value of x ?

(A) 25
(B) 55
(C) 60
(D) 77
(E) 85

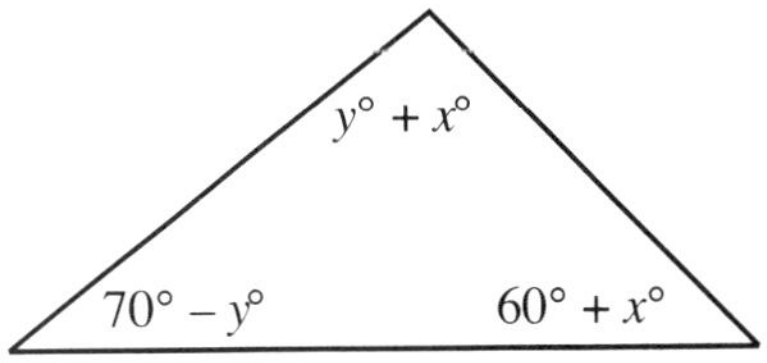

The angle sum of a triangle is 180°. Hence, $(y + x) + (60 + x) + (70 - y) = 180$. Simplifying the equation yields $2x + 130 = 180$. Solving for x yields $x = 25$. The answer is (A).

7. In the figure, what is the value of a ?

(A) 16
(B) 18
(C) 36
(D) 54
(E) 72

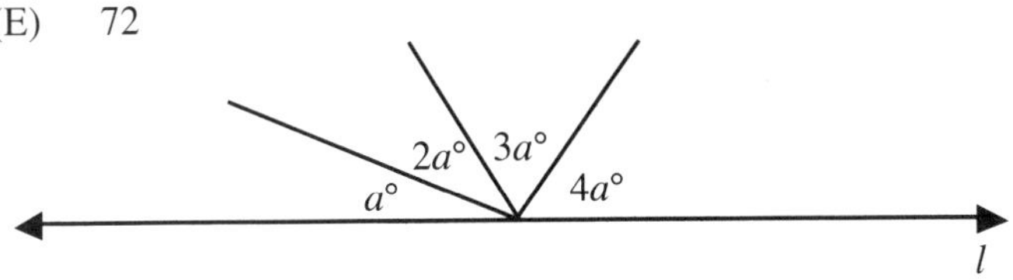

The angle made by a line is 180°. Hence, from the figure, we have

$$a + 2a + 3a + 4a = 180$$
$$10a = 180$$
$$a = 18$$

The answer is (B).

8. In the figure, what is the average of the five angles shown inside the circle?

(A) 36
(B) 45
(C) 60
(D) 72
(E) 90

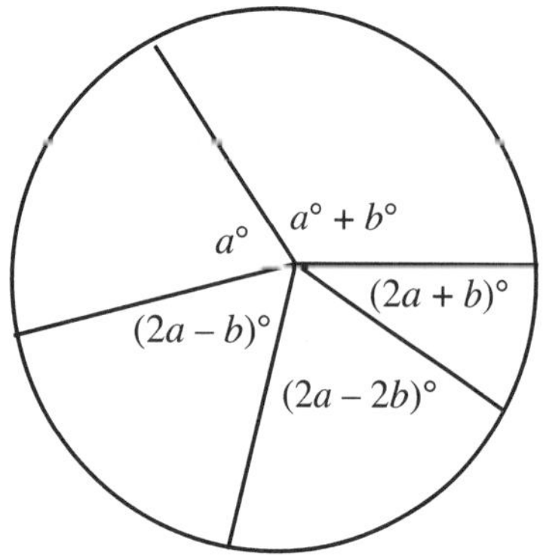

The average of the five angles is

$$\frac{\text{Sum of the five angles}}{5} = \frac{360}{5} = 72$$

The answer is (D).

9. In the figure, O is the center of the circle. What is average of the numbers a, b, c, and d ?

(A) 45
(B) 60
(C) 90
(D) 180
(E) 360

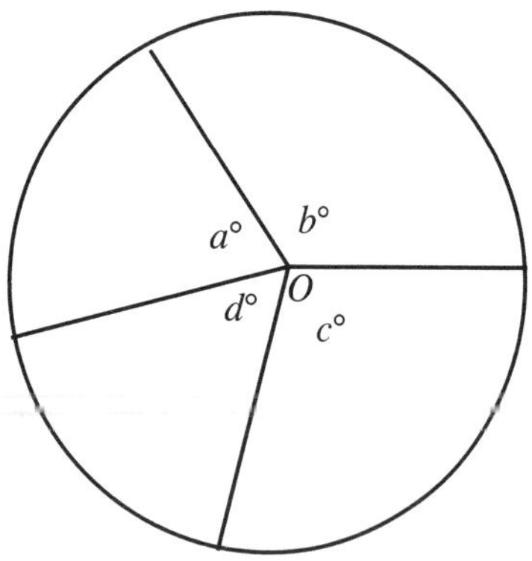

Since the angle around a point has 360°, the sum of the four angles a, b, c, and d is 360 and their average is 360/4 = 90. The answer is (C).

10. Column A | Column B

The perimeter of ΔABC | The circumference of the circle

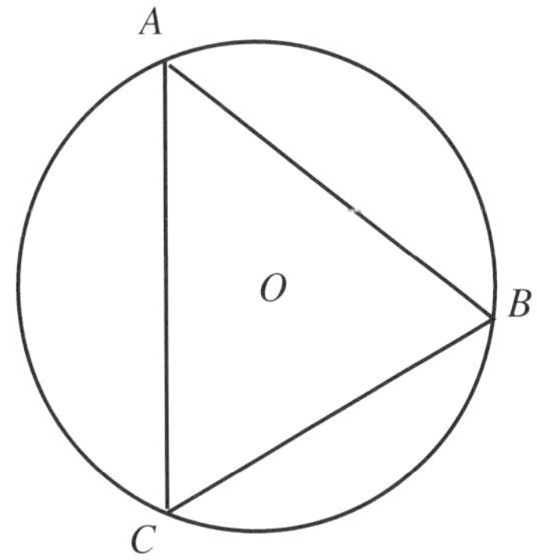

Since the shortest distance between two points is a straight line, the length of a chord is always less than the length of the arc it makes (subtends) on the circle. Hence, from the figure, we have

$$AB < \text{arc } AB$$
$$BC < \text{arc } BC$$
$$CA < \text{arc } CA$$

Adding the three inequalities yields $AB + BC + CA < \text{arc } AB + \text{arc } BC + \text{arc } CA$. The left side of the inequality is the perimeter of triangle ABC (which Column A equals), and the right side is the circumference of the circle (which Column B equals). Hence, Column A is less than Column B, and the answer is (B).

11. Column A | Column B

The perimeter of quadrilateral *ABCD* | The circumference of the circle

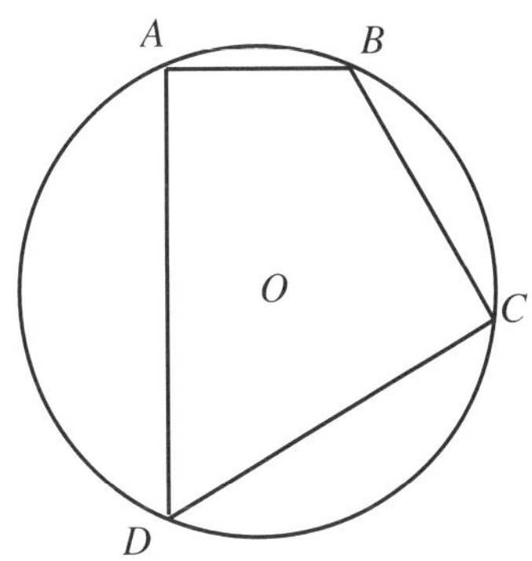

Since the shortest distance between two points is a straight line, the length of a chord is always shorter than the length of the arc that it makes on the circle. Hence, from the figure, we have

$$AB < \text{arc } AB$$
$$BC < \text{arc } BC$$
$$CD < \text{arc } CD$$
$$DA < \text{arc } DA$$

Adding the four inequalities yields $AB + BC + CD + DA < \text{arc } AB + \text{arc } BC + \text{arc } CD + \text{arc } DA$. The left side of the inequality is the perimeter of quadrilateral *ABCD* (which Column A equals), and the right side is the circumference of the circle (which Column B equals). Hence, Column A is less than Column B, and the answer is (B).

12. Column A — A circle is inscribed in a triangle. — Column B

Column A		Column B
The perimeter of the triangle		The circumference of the circle

Let ABC be the triangle, and let P, Q, and R be the points at which the inscribed circle touches the sides of the triangle.

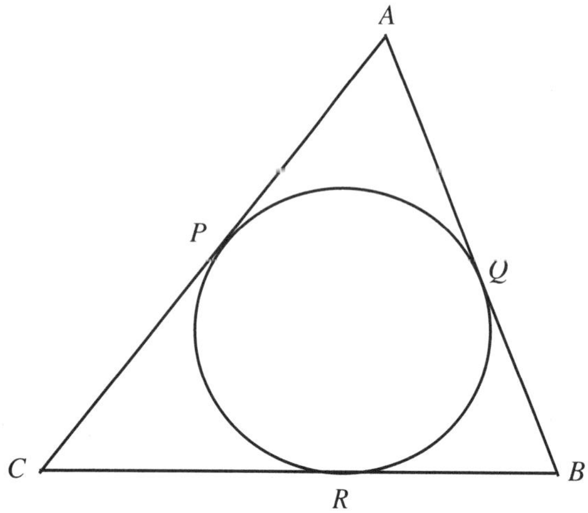

From the figure, it is clear that to go from one point on the circle, say, point P to another point, say, point Q, the shortest available path is the arc PQ. Hence, arc $PQ < PA + AQ$. Similarly, arc $QR < QB + BR$, and arc $RP < RC + CP$. Summing the three inequalities yields arc PQ + arc QR + arc $RP < (PA + AQ) + (QB + BR) + (RC + CP)$. The right side of the inequality is the perimeter of the triangle ABC (which Column A equals), and the left side is the circumference of the circle (which Column B equals). Hence, Column A is greater than Column B, and the answer is (A).

➢ **Medium**

13. Column A — In ΔABC, $\angle B = 72°$ — Column B

Column A		Column B
$\angle A$		72°

Summing the angles of ΔABC to 180° yields $\angle A + \angle B + \angle C = 180$. Substituting the value of $\angle B$ (= 72°) into this equation yields $\angle A + 72 + \angle C = 180$. Solving the equation for $\angle A$ yields $\angle A = 180 - 72 - \angle C = 108 - \angle C$ = Column A.

Now, Suppose $\angle C = 1°$. Then $\angle A = 108° - 1° = 107°$. In this case, $\angle A$ is greater than 72°. Now, Suppose $\angle C = 107°$. Then $\angle A = 108° - 107° = 1°$. In this case, $\angle A$ is less than 72°.

Hence, we have a double case, and the answer is (D).

14. In the figure, *AD* and *BC* are lines intersecting at *O*. What is the value of *a* ?

    (A) 15
    (B) 30
    (C) 45
    (D) 60
    (E) 135

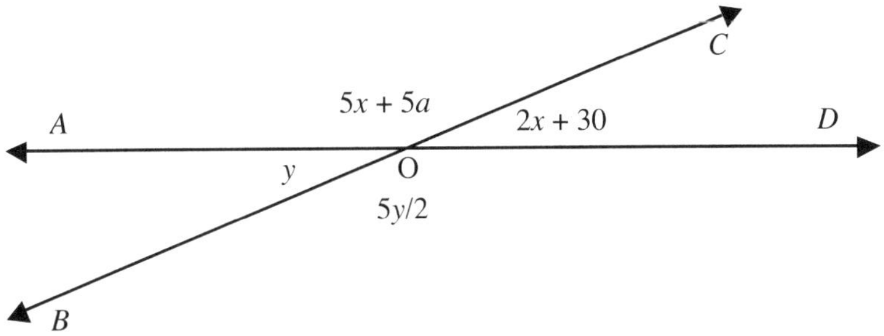

Equating vertical angles $\angle AOB$ and $\angle COD$ in the figure yields $y = 2x + 30$. Also, equating vertical angles $\angle AOC$ and $\angle BOD$ yields $5y/2 = 5x + 5a$. Multiplying this equation by 2/5 yields $y = 2x + 2a$. Subtracting this equation from the equation $y = 2x + 30$ yields $2a = 30$. Hence, $a = 30/2 = 15$, and the answer is (A).

15. From the figure, which one of the following must be true?

    (A) $y = z$
    (B) $y < z$
    (C) $y \leq z$
    (D) $y > z$
    (E) $y \geq z$

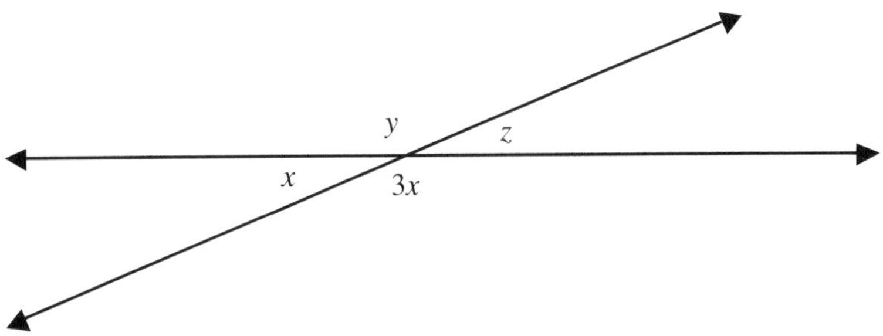

Equating the two pairs of vertical angles in the figure yields $y = 3x$ and $x = z$. Replacing x in first equation with z yields $y = 3z$. This equation says that y is 3 times as large as z. Hence, $y > z$. The answer is (D).

16. From the figure, which one of the following could be the value of b ?

(A) 20
(B) 30
(C) 60
(D) 75
(E) 90

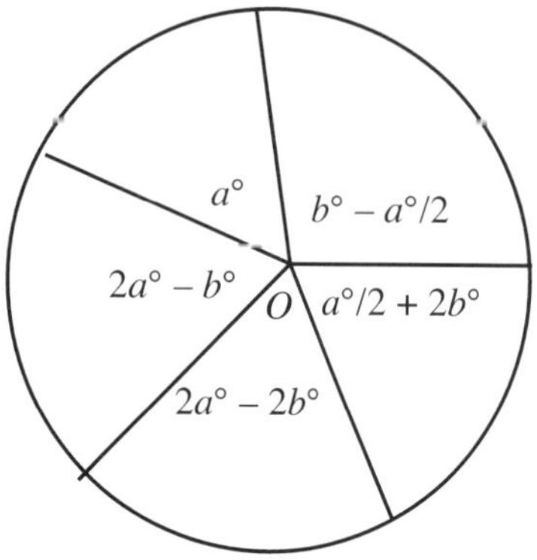

In the figure, angles $b° - a°/2$ and $2a° - 2b°$ must be positive. Hence, we have the inequalities, $b - a/2 > 0$ and $2a - 2b > 0$.

Adding $a/2$ to both sides of first inequality and $2b$ to both sides of second inequality yields the following two inequalities:

$$b > a/2$$
$$2a > 2b$$

Dividing the second inequality by 2 yields $a > b$.

Now, summing angles around point O to 360° yields $a + (b - a/2) + (a/2 + 2b) + (2a - 2b) + (2a - b) = 360$. Simplifying this yields $5a = 360$, and solving yields $a = 360/5 = 72$.

Substituting this value in the inequalities $b > a/2$ and $a > b$ yields

$$b > a/2 = 72/2 = 36, \text{ and } 72 > b$$

Combining the inequalities $b > 36$ and $72 > b$ yields $36 < b < 72$. The only choice in this range is (C), so the answer is (C).

17. AD is the longest side of the right triangle ABD shown in the figure. What is the length of longest side of ΔABC ?

(A) 2
(B) 3
(C) $\sqrt{41}$
(D) 9
(E) 41

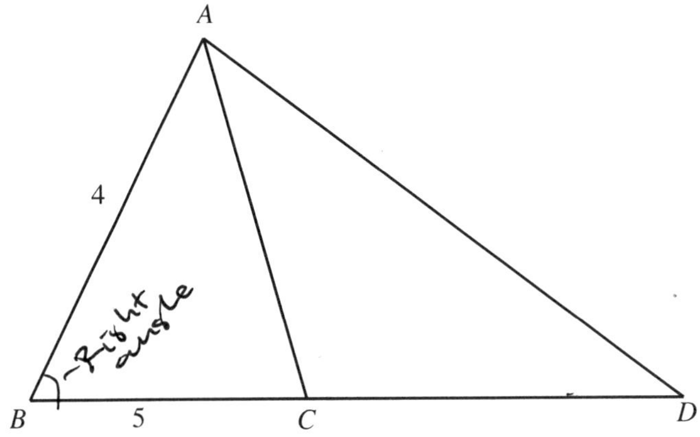

The figure is not drawn to scale

In a right triangle, the angle opposite the longest side is the right angle. Since AD is the longest side of the right triangle ABD, $\angle B$ must be a right angle and ΔABC must be a right triangle. Applying The Pythagorean Theorem to the right triangle ABC yields

$$AC^2 = AB^2 + BC^2$$
$$AC = \sqrt{AB^2 + BC^2}$$
$$AC = \sqrt{4^2 + 5^2}$$
$$AC = \sqrt{16 + 25}$$
$$AC = \sqrt{41}$$

The answer is (C).

18. In the figure, what is the value of a ?

(A) 30
(B) 45
(C) 60
(D) 72
(E) 90

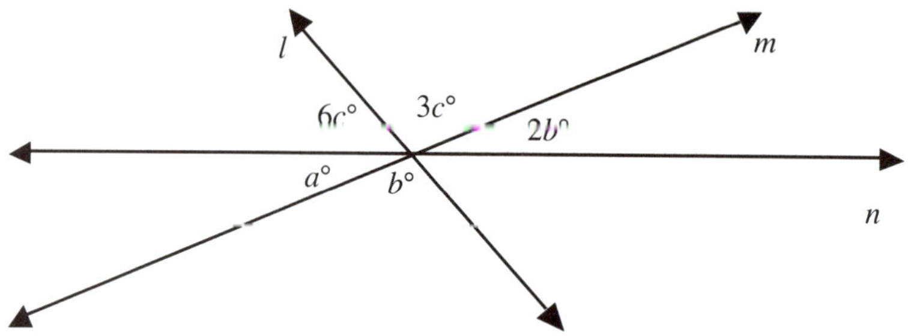

Equating vertical angles in the figure yields $a = 2b$ and $b = 3c$. From the first equation, we have $b = a/2$. Plugging this into the second equation yields $a/2 = 3c$, from which we can derive $c = a/6$. Since the angle made by a line is 180°, we have for line l that $b + a + 6c = 180$. Replacing b with $a/2$ and c with $a/6$ in this equation yields $a/2 + a + 6(a/6) = a/2 + a + a = 180$. Summing the left-hand side yields $5a/2 = 180$, and multiplying both sides by 2/5 yields $a = 180(2/5) = 72$. The answer is (D).

19. In the figure, lines l and m are parallel. Which one of the following, if true, makes lines p and q parallel?

(A) $a = b$
(B) $a = c$
(C) $c = d$
(D) $d = b$
(E) $b = c$

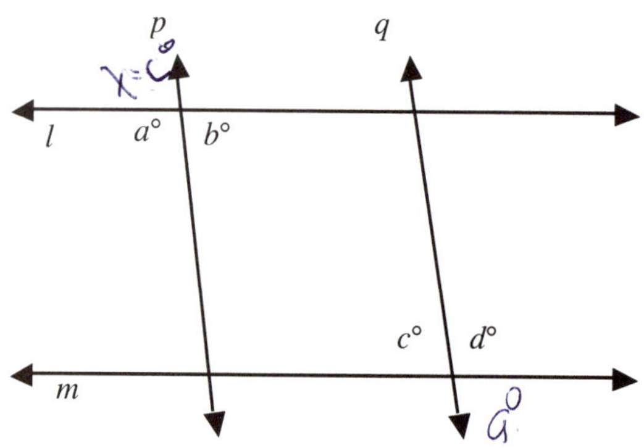

Superimposing parallel line m on line l yields a figure like this:

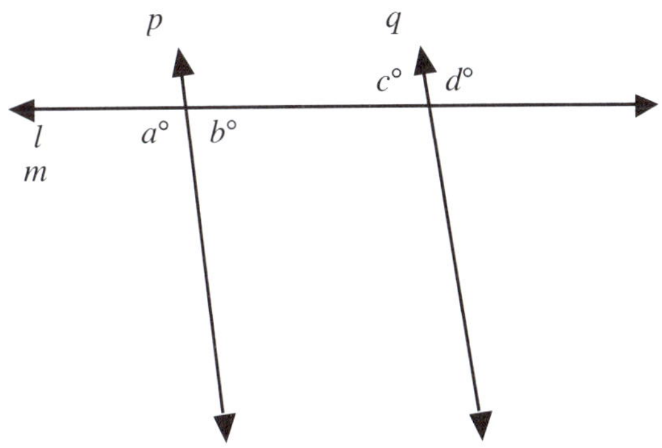

Now, when two lines (here p and q) cut by a transversal (here l) are parallel, we have

(I) Corresponding angles are equal: No corresponding angles are listed in the figure.
(II) Alternate interior angles are equal: $b = c$. In choice (E).
(III) Alternate exterior angles are equal: $a = d$. Not listed in any answer-choice.
(IV) Interior angles are supplementary.
(V) Exterior angles are supplementary.

The answer is (E).

20. A, B, C, and D are points on a line such that point B bisects line AC and point A bisects line CD. What is the ratio of AB to CD ?

(A) 1/4
(B) 1/3
(C) 1/2
(D) 2/3
(E) 3/4

Drawing the figure given in the question, we get

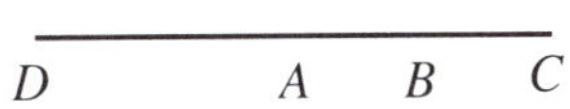

$AB = BC$
$CA = DA$

Suppose AB equal 1 unit. Since point B bisects line segment AC, AB equals half AC. Hence, AC equals twice $AB = 2(AB) = 2(1 \text{ unit}) = 2$ units. Again, since point A bisects line segment DC, DC equals twice $AC = 2(AC) = 2(2 \text{ units}) = 4$ units. Hence, $AB/DC = 1 \text{ unit}/4 \text{ units} = 1/4$. The answer is (A).

21. If A, B, C, D, and E are points in a plane such that line CD bisects $\angle ACB$ and line CB bisects right angle $\angle ACE$, then $\angle DCE =$

(A) 22.5°
(B) 45°
(C) 57.5°
(D) 67.5°
(E) 72.5°

Drawing the figure given in the question yields

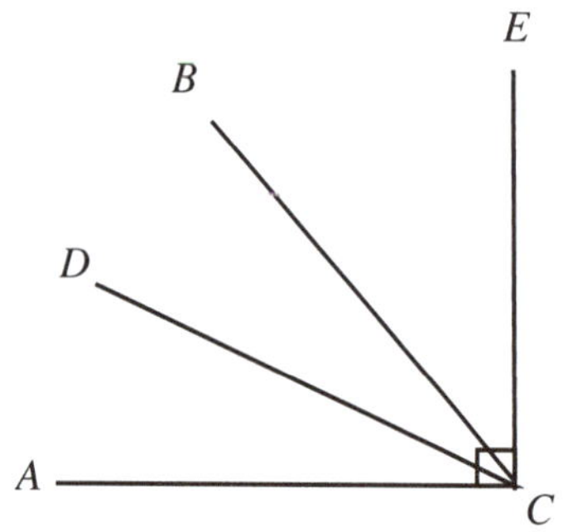

Figure not drawn to scale.

CD bisects $\angle ACB$
CB bisects $\angle ACE$

We are given that CB bisects the right-angle $\angle ACE$. Hence, $\angle ACB = \angle BCE = \angle ACE/2 = 90°/2 = 45°$. Also, since CD bisects $\angle ACB$, $\angle ACD = \angle DCB = \angle ACB/2 = 45°/2 = 22.5°$. Now, $\angle DCE = \angle DCB + \angle BCE = 22.5° + 45° = 67.5°$. The answer is (D).

22. From the figure, which of the following must be true?

(I) $x + y = 90$
(II) x is 35 units greater than y
(III) x is 35 units less than y

(A) I only
(B) II only
(C) III only
(D) I and II only
(E) I and III only

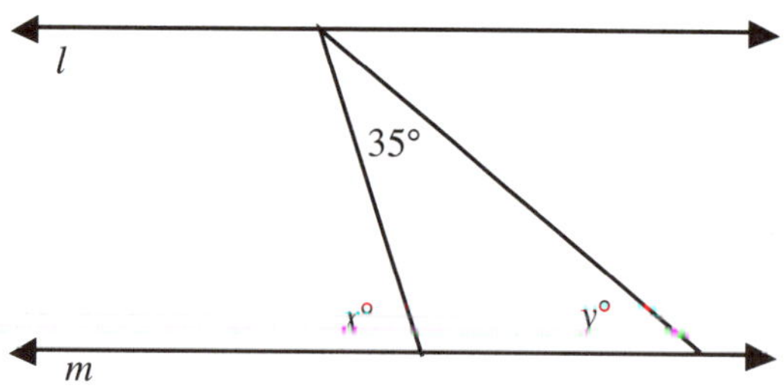

Angle x is an exterior angle of the triangle and therefore equals the sum of the remote interior angles, 35 and y. That is, $x = y + 35$. This equation says that x is 35 units greater than y. So, (II) is true and (III) is false. Now, if x is an obtuse angle ($x > 90$), then $x + y$ is greater than 90. Hence, $x + y$ need not equal 90. So, (I) is not necessarily true. The answer is (B).

23. In the figure, triangles *ABC* and *ABD* are right triangles. What is the value of x ?

(A) 20
(B) 30
(C) 50
(D) 70
(E) 90

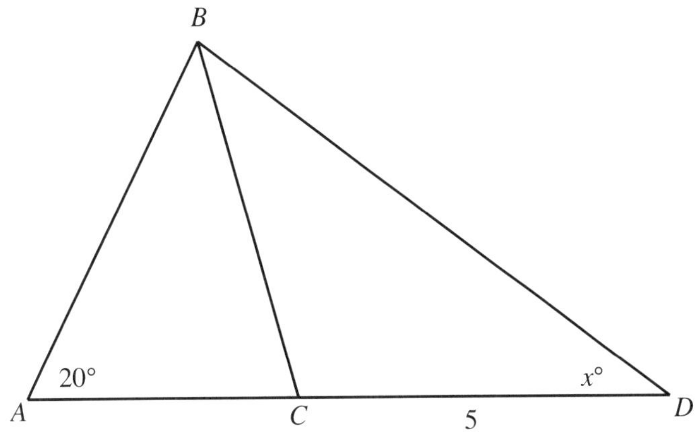

In triangle *ABC*, $\angle A$ is 20°. Hence, the right angle in ΔABC is either $\angle ABC$ or $\angle BCA$. If $\angle ABC$ is the right angle, then $\angle ABD$, of which $\angle ABC$ is a part, would be greater than the right angle (90°) and *ABD* would be an obtuse angle, not a right triangle. So, $\angle ABC$ is not 90° and therefore $\angle BCA$ must be a right angle. Since *AD* is a line, $\angle ACB + \angle BCD = 180°$. Solving for $\angle BCD$ yields $\angle BCD = 180 - 90 = 90$. Hence, ΔBCD is also a right triangle, with right angle at *C*. Since there can be only one right angle in a triangle, $\angle D$ is not a right angle. But, we are given that ΔABD is right angled, and from the figure $\angle A$ equals 20°, which is not a right angle. Hence, the remaining angle $\angle ABD$ is right angled. Now, since the sum of the angles in a triangle is 180°, in ΔABC, we have $20 + 90 + x = 180$. Solving for x yields $x = 70$. The answer is (D).

24. In the figure, *ABC* is a right triangle. What is the value of y ?

(A) 20
(B) 30
(C) 50
(D) 70
(E) 90

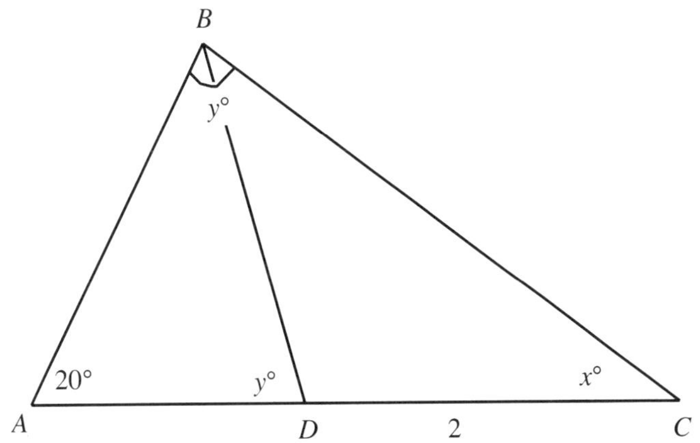

In the given right triangle, ΔABC, $\angle A$ is 20°. Hence, $\angle A$ is not the right angle in the triangle. Hence, either of the other two angles, $\angle C$ or $\angle B$, must be right angled.

Now, $\angle BDA$ (= $\angle ABC$ = $y°$, from the figure) is an exterior angle to ΔBCD and therefore equals the sum of the remote interior angles $\angle C$ and $\angle DBC$. Clearly, the sum is larger than $\angle C$ and therefore if $\angle C$ is a right angle, $\angle BDA$ (= $\angle ABC$) must be larger than a right angle, so $\angle BDA$, hence, $\angle ABC$ must be obtuse. But a triangle cannot accommodate a right angle and an obtuse angle simultaneously because the angle sum of the triangle would be greater than 180°. So, $\angle C$ is not a right angle and therefore the other angle, $\angle B$, is a right angle. Hence, $y° = \angle B = 90°$. The answer is (E).

25. The following are the measures of the sides of five different triangles. Which one of them represents a right triangle?

(A) $\sqrt{3}$, $\sqrt{4}$, $\sqrt{5}$
(B) 1, 5, 4
(C) 7, 3, 4
(D) $\sqrt{3}$, $\sqrt{7}$, $\sqrt{4}$
(E) 4, 8, 10

A right triangle must satisfy The Pythagorean Theorem: the square of the longest side of the triangle is equal to the sum of the squares of the other two sides. Hence, we look for the answer-choice that satisfies this theorem:

Choice (A): $\left(\sqrt{5}\right)^2 \neq \left(\sqrt{3}\right)^2 + \left(\sqrt{4}\right)^2$. Reject.
Choice (B): $5^2 \neq 1^2 + 4^2$. Reject.
Choice (C): $7^2 \neq 3^2 + 4^2$. Reject.
Choice (D): $\left(\sqrt{7}\right)^2 = \left(\sqrt{3}\right)^2 + \left(\sqrt{4}\right)^2$. Correct.
Choice (E): $10^2 \neq 8^2 + 4^2$. Reject.

The answer is (D).

26. ΔABC is a right-angled isosceles triangle, and $\angle B$ is the right angle in the triangle. If AC measures $7\sqrt{2}$, then which one of the following would equal the lengths of AB and BC, respectively?

(A) 7, 7
(B) 9, 9
(C) 10, 10
(D) 11, 12
(E) 7, 12

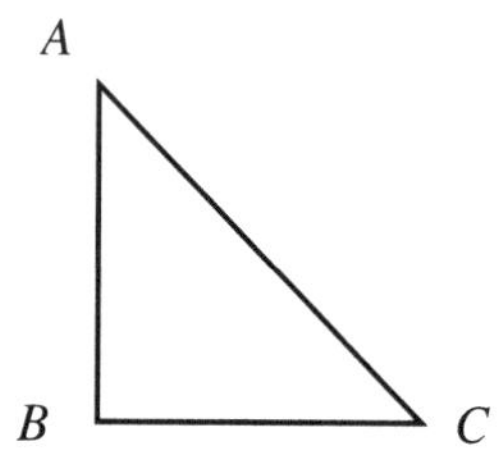

In a right-angled isosceles triangle, the sides of the right angle are equal. Now, in the given right-angled isosceles triangle ΔABC, $\angle B$ is given to be the right angle. Hence, the sides of the angle, AB and BC, are equal. Applying The Pythagorean Theorem to the triangle yields

$AB^2 + BC^2 = AC^2$
$BC^2 + BC^2 = \left(7\sqrt{2}\right)^2$ since $AB = BC$
$2(BC)^2 = 7^2 \times 2$
$BC^2 = 7^2$
$BC = 7$ by square rooting both sides

Hence, $AB = BC = 7$. The answer is (A).

27. In the figure, if $AB = 8$, $BC = 6$, $AC = 10$ and $CD = 9$, then $AD =$

(A) 12
(B) 13
(C) 15
(D) 17
(E) 24

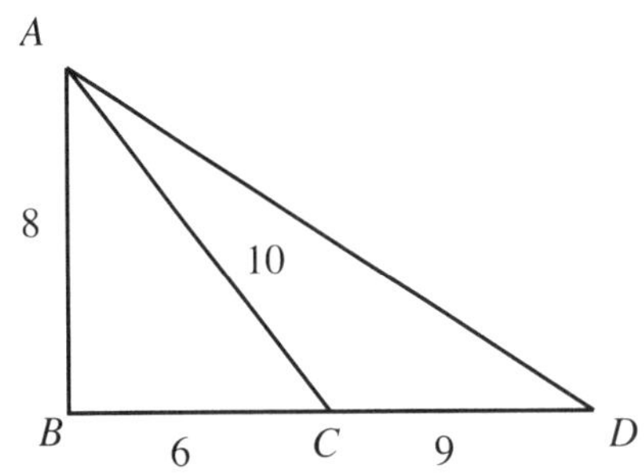

The lengths of the three sides of ΔABC are $AB = 8$, $BC = 6$, and $AC = 10$. The three sides satisfy The Pythagorean Theorem: $AC^2 = BC^2 + AB^2$ ($10^2 = 6^2 + 8^2$). Hence, triangle ABC is right angled and $\angle B$, the angle opposite the longest side AC (hypotenuse), is a right angle. Now, from the figure, this angle is part of ΔADB, so ΔADB is also right angled. Applying The Pythagorean Theorem to the triangle yields

$$
\begin{aligned}
AD^2 &= AB^2 + BD^2 \\
&= AB^2 + (BC + CD)^2 && \text{from the figure, } BD = BC + CD \\
&= 8^2 + (6 + 9)^2 \\
&= 8^2 + 15^2 \\
&= 289 \\
&= 17^2 \\
AD &= 17 && \text{by square rooting}
\end{aligned}
$$

The answer is (D).

28. The average length of the sides of ΔABC is 12. What is the perimeter of ΔABC ?

(A) 4
(B) 6
(C) 12
(D) 24
(E) 36

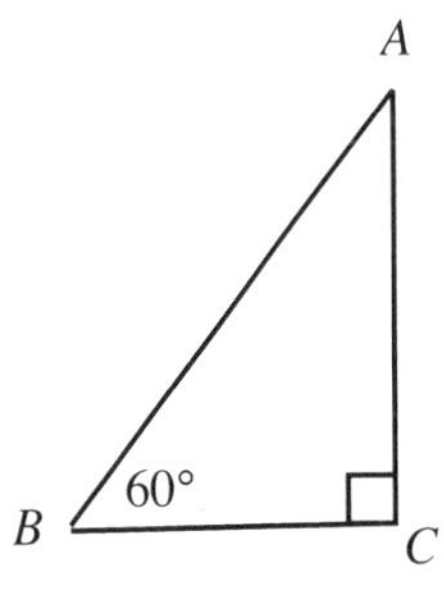

The perimeter of a triangle equals the sum of the lengths of the sides of the triangle.

The average length of the sides of the triangle equals (the sum of the lengths of the three sides)/3.

Hence, the perimeter of a triangle equals three times the average length of the sides of the triangle.

Now, we are given that the average length of the triangle is 12. Hence, the perimeter of the triangle equals $3 \times 12 = 36$. The answer is (E).

29. Which one of the following is true regarding the triangle shown in the figure?

(A) $x > y > z$
(B) $x < y < z$
(C) $x = y = z$
(D) $2x = 3y/2 = z$
(E) $x/2 = 2y/3 = z$

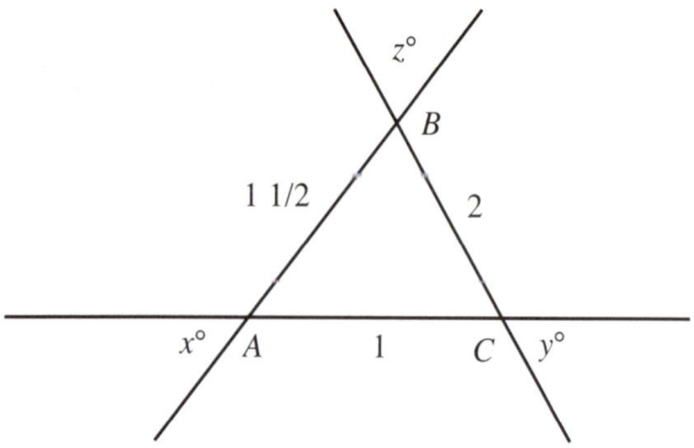

From the figure, we have the following inequality between the sides of the triangle: BC (= 2) > AB (= 1 1/2) > AC (= 1). In a triangle, the longer the side, the bigger the angle opposite it. Hence, we have the following inequality between the angles of the triangle: $\angle A > \angle C > \angle B$. Replacing the angles in the inequality with their respective vertical angles in the figure yields $x > y > z$. The answer is (A).

30. In the figure, *ABCD* is a rectangle, and the area of ΔACE is 10. What is the area of the rectangle?

(A) 18
(B) 22.5
(C) 36
(D) 44
(E) 45

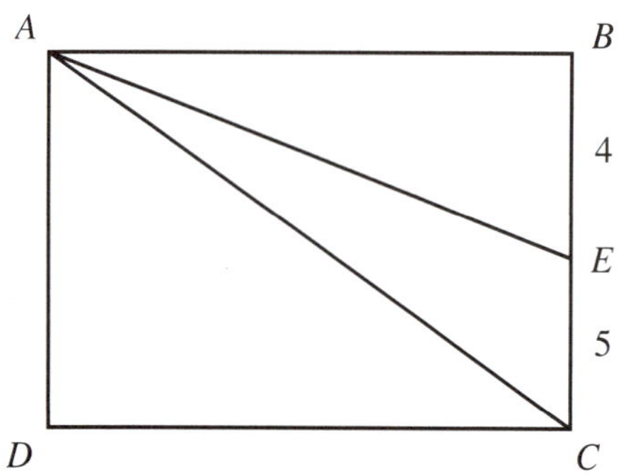

The formula for the area of a triangle is 1/2 × *base* × *height*. Hence, the area of ΔACE (which is given to equal 10) is $1/2 \times CE \times AB$. Hence, we have

$1/2 \times CE \times AB = 10$
$1/2 \times 5 \times AB = 10$ (from the figure, $CE = 5$)
$AB = 4$

Now, the formula for the area of a rectangle is *length* × *width*. Hence, the area of the rectangle $ABCD = BC \times AB$

$= (BE + EC) \times (AB)$ from the figure, $BC = BE + EC$
$= (4 + 5) \times 4$ from the figure, $BE = 4$ and $EC = 5$
$= 9 \times 4$
$= 36$

The answer is (C).

31. In the figure, what is the area of ΔABC ?

(A) 2
(B) $\sqrt{2}$
(C) 1
(D) $\frac{1}{\sqrt{2}}$
(E) 1/2

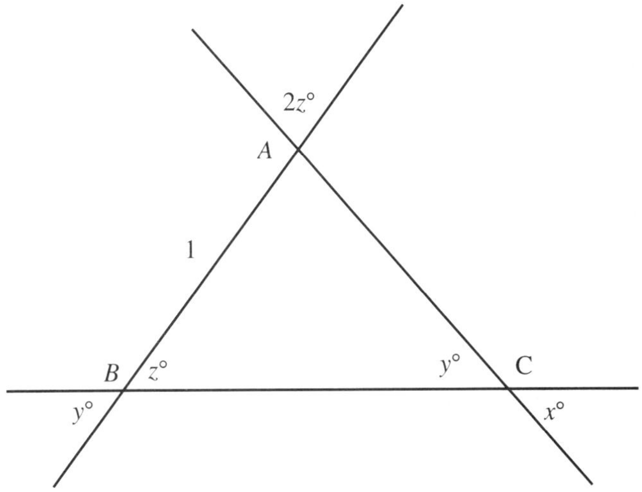

Equating vertical angles at point B in the figure yields $y = z$ and $\angle A = 2z$. So, the triangle is isosceles ($\angle B = \angle C = y$). Now, the angle sum of a triangle is 180°, so $2z + z + z = 180$. Solving for z yields $z = 180/4 = 45$. Hence, we have $2z = 2 \times 45 = 90$. So, ΔABC is a right-angled isosceles triangle with right angle at vertex A and equal angles at $\angle B$ and $\angle C$ (both equaling 45°). Since sides opposite equal angles in a triangle are equal, sides AB and AC are equal and each is 1 unit (given that AB is 1 unit). Now, the area of the right triangle ABC is $1/2 \times AB \times AC = 1/2 \times 1 \times 1 = 1/2$. The answer is (E).

32. In the figure, O is the center of the circle. Which one of the following must be true about the perimeter of the triangle shown?

(A) Always less than 10
(B) Always greater than 40
(C) Always greater than 30
(D) Always less than 30
(E) Less than 40 and greater than 20

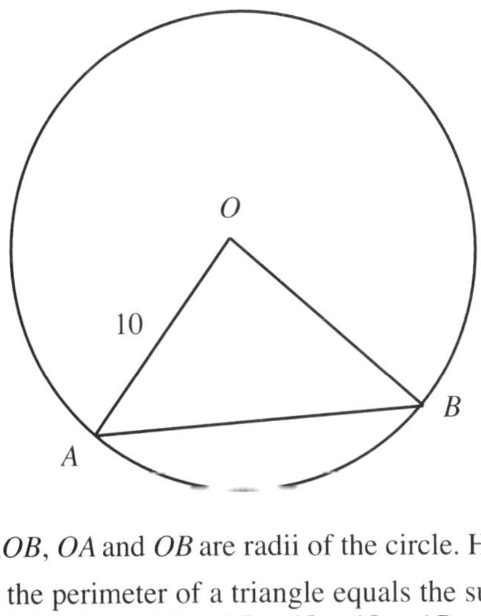

In ΔAOB, OA and OB are radii of the circle. Hence, both equal 10 (since $OA = 10$ in the figure).

Now, the perimeter of a triangle equals the sum of the lengths of the sides of the triangle. Hence, perimeter of $\Delta AOB = OA + OB + AB = 10 + 10 + AB = 20 + AB$.

In a triangle, the length of any side is less than the sum of the lengths of the other two sides and is greater than their difference. So, AB is less than $AO + OB$ (= 10 + 10 = 20) and is greater than 10 – 10 = 0. So, we have $0 < AB < 20$. Adding 20 to each part of this inequality yields $20 < 20 + AB < 20 + 20$, or 20 < Perimeter < 40. Hence, the answer is (E).

33. Which of the following must be true?

    (I) The area of triangle P.
    (II) The area of triangle Q.
    (III) The area of triangle R.

    (A) I – II – III
    (B) I < II < III
    (C) I > II > III
    (D) III < I < II
    (E) III > I > II

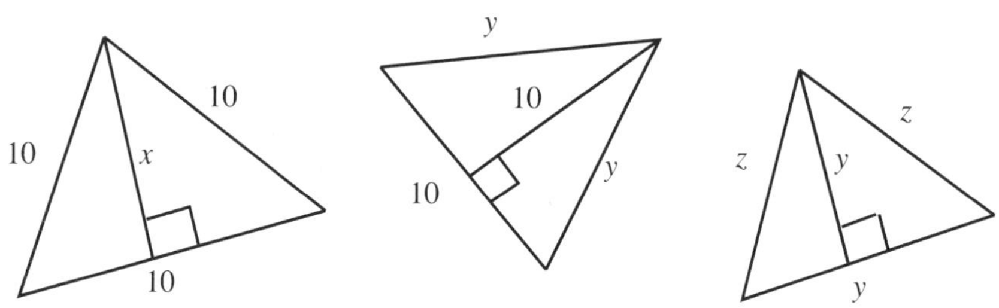

P *Q* *R*

In the figure, triangle P is equilateral, with each side measuring 10 units. So, as in any equilateral triangle, the altitude (x here) is shorter than any of the other sides of the triangle. Hence, x is less than 10. Now,

I = Area of Triangle P
= 1/2 × *base* × *height*
= 1/2 × 10 × x
= $5x$ and this is less than 50, since x is less than 10

Triangle Q has both base and altitude measuring 10 units, and the area of the triangle is 1/2 × *base* × *height* = 1/2 × 10 × 10 = 50. So, II = 50. We have one more detail to pick up: y, being the hypotenuse in the right triangle in figure Q, is greater than any other side of the triangle. Hence, y is greater than 10, the measure of one leg of the right triangle.

Triangle R has both the base and the altitude measuring y units. Hence,

II = Area of the Triangle R
= 1/2 × *base* × *height*
= 1/2 × y × y
= $(1/2)y^2$, and this is greater than $1/2 \times 10^2$ (since $y > 10$), which equals 50

Summarizing, the three results I < 50, II = 50, and III > 50 into a single inequality yields I < II < III. The answer is (B).

34. In the figure, *ABCD* is a parallelogram. Which one of the following is true?

(A) $x < y$
(B) $x > q$
(C) $x > p$
(D) $y > p$
(E) $y > q$

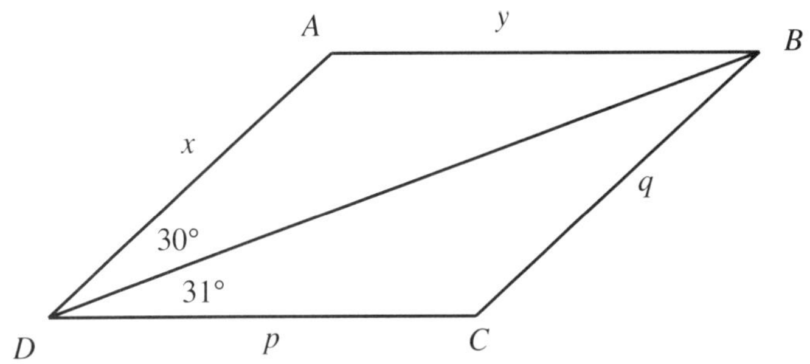

Since *ABCD* is a parallelogram, opposite sides are equal. So, $x = q$ and $y = p$. Now, line *BD* is a transversal cutting opposite sides *AB* and *DC* in the parallelogram. So, the alternate interior angles $\angle ABD$ and $\angle BDC$ both equal 31°. Hence, in ΔABD, $\angle B$ (which equals 31°) is greater than $\angle D$ (which equals 30°, from the figure). Since the sides opposite greater angles in a triangle are greater, we have $x > y$. But, $y = p$ (we know). Hence, $x > p$, and the answer is (C).

35. In the figure, the areas of parallelograms *EBFD* and *AECF* are 3 and 2, respectively. What is the area of rectangle *ABCD* ?

(A) 3
(B) 4
(C) 5
(D) $4\sqrt{3}$
(E) 7

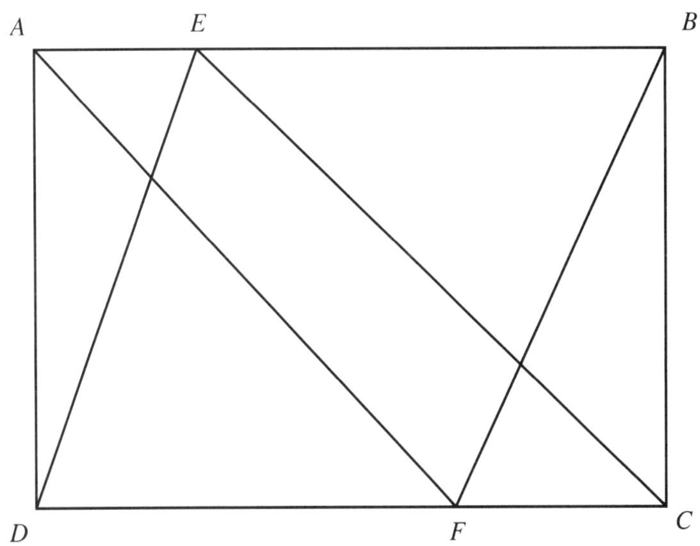

The area of the rectangle *ABCD* is *length* × *width* = *DC* × *AD*.

The area of the parallelogram *AECF* is *FC* × *AD* (*base* × *height*). Also, the area of parallelogram *EBFD* is *DF* × *AD*. Now, summing the areas of the two parallelograms *AECF* and *EBFD* yields *FC* × *AD* + *DF* × *AD* = (*FC* + *DF*)(*AD*) = *DC* × *AD* = the area of the rectangle *ABCD*. Hence, the area of rectangle *ABCD* equals the sum of areas of the two parallelograms, which is 2 + 3 = 5. The answer is (C).

36. In the figure, the area of rectangle $ABCD$ is 100. What is the area of the square $EFGH$?

(A) 256
(B) 275
(C) 309
(D) 399
(E) 401

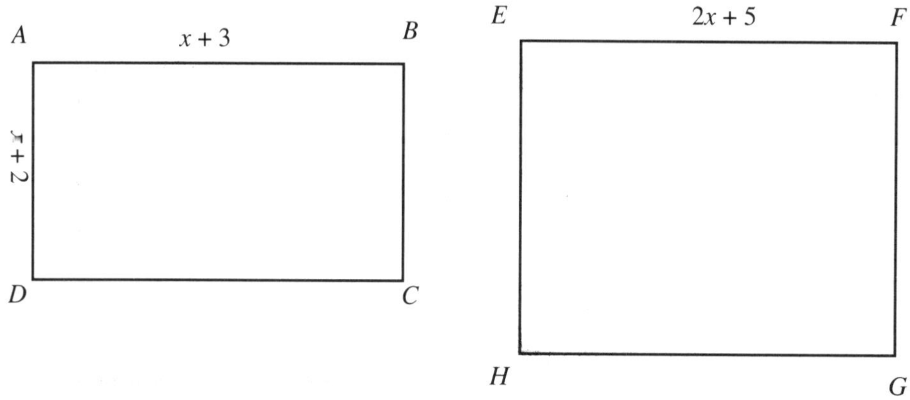

The area of the rectangle $ABCD$, *length* × *width*, is $(x + 3)(x + 2) = x^2 + 5x + 6 = 100$ (given the area of the rectangle $ABCD$ equals 100). Now, subtracting 6 from both sides yields $x^2 + 5x = 94$.

The area of square $EFGH$ is $(2x + 5)^2 = 4x^2 + 20x + 25 = 4(x^2 + 5x) + 25 = 4(94) + 25 = 401$. The answer is (E).

37. In the figure, the area of rectangle $EFGH$ is 3 units greater than the area of rectangle $ABCD$. What is the value of ab if $a + b = 8$?

(A) 9
(B) 12
(C) 15
(D) 18
(E) 21

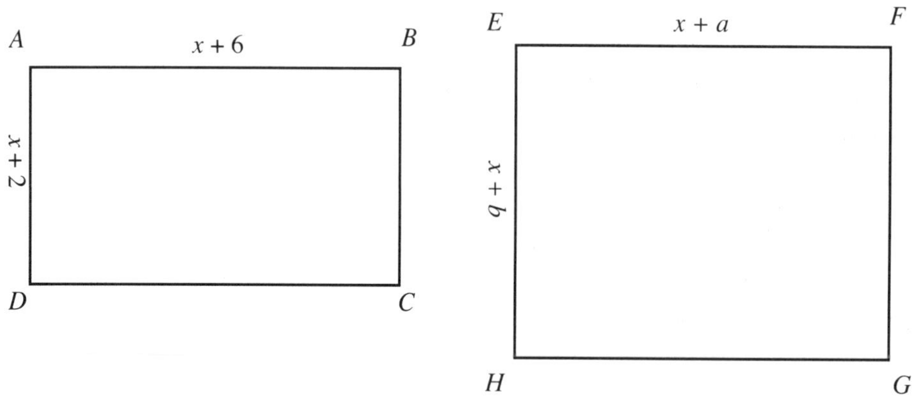

The formula for the area of a rectangle is *length* × *width*. Hence, the area of rectangle $ABCD$ is $(x + 6)(x + 2) = x^2 + 8x + 12$, and the area of the rectangle $EFGH$ is $(x + a)(x + b) = x^2 + (a + b)x + ab = x^2 + 8x + ab$ (given that $a + b = 8$). Now, we are given that the area of the rectangle $EFGH$ is 3 units greater than the area of the rectangle $ABCD$. Hence, we have

$$x^2 + 8x + ab = (x^2 + 8x + 12) + 3$$
$$ab = 12 + 3 \qquad \text{(by canceling } x^2 \text{ and } 8x \text{ from both sides)}$$
$$ab = 15$$

The answer is (C).

38. In the figure, the area of rectangle *ABCD* is 45. What is the area of the square *EFGH* ?

(A) 20
(B) 40
(C) 50
(D) 65
(E) 70

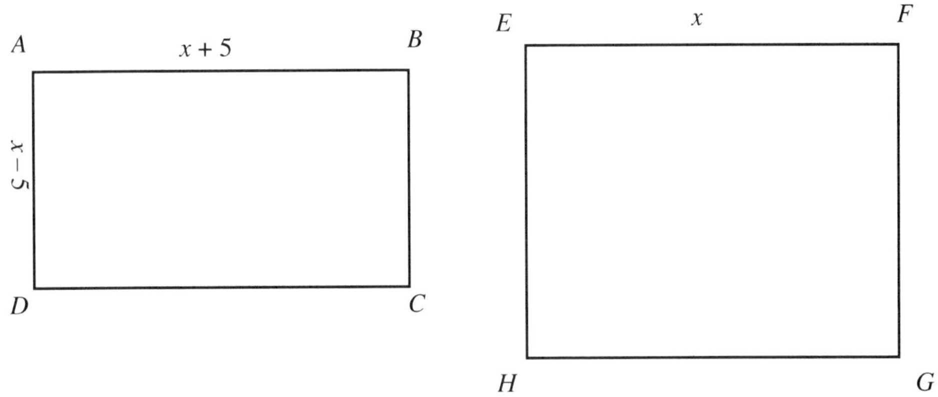

The formula for the area of the rectangle is *length* × *width*. Hence, the area of rectangle *ABCD* is $AB \times AD = (x + 5)(x - 5) = x^2 - 5^2$. We are given that the area is 45, so $x^2 - 5^2 = 45$. Solving the equation for x^2 yields $x^2 = 45 + 25 = 70$.

Now, the formula for the area of a square is $side^2$. Hence, the area of square *EFGH* is $EF^2 = x^2$. Now, as shown earlier, x^2 equals 70. The answer is (E).

39. In the figure, *ABCD* is a parallelogram, what is the value of *b* ?

(A) 46
(B) 48
(C) 72
(D) 84
(E) 96

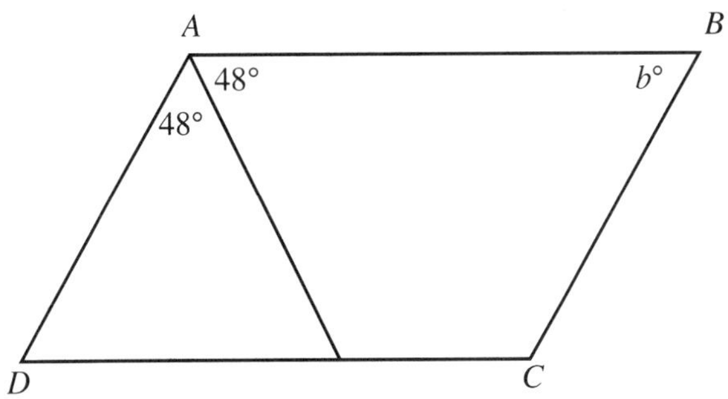

From the figure, $\angle A = 48 + 48 = 96$. Since the sum of any two adjacent angles of a parallelogram equals 180°, we have $\angle A + b = 180$ or $96 + b = 180$. Solving for *b* yields $b = 180 - 96 = 84$. The answer is (D).

40. In the figure, *AB* and *CD* are the diameters of the circle. What is the value of x ?

(A) 16°
(B) 18°
(C) 26°
(D) 32°
(E) 58°

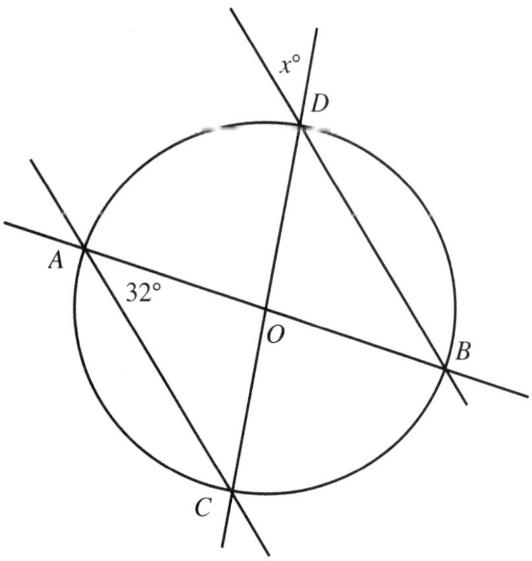

OA and *OC* are radii of the circle and therefore equal. Hence, the angles opposite the two sides in ΔAOC are equal: $\angle C = \angle A = 32°$ (from the figure). Now, summing the angles of the triangle to 180° yields $\angle A + \angle C + \angle AOC = 180$ or $32 + 32 + \angle AOC = 180$. Solving the equation for $\angle AOC$, we have $\angle AOC = 180 - (32 + 32) = 180 - 64 = 114$.

Since $\angle BOD$ and $\angle AOC$ vertical angles, they are equal. Hence, we have $\angle BOD = \angle AOC = 114$.

Now, *OD* and *OB* are radii of the circle and therefore equal. Hence, the angles opposite the two sides in ΔBOD are equal: $\angle B = \angle D$. Now, summing the angles of the triangle to 180° yields $\angle B + \angle D + \angle BOD = 180$ or $\angle D + \angle D + \angle BOD = 180$ or $2\angle D + 114 = 180$. Solving the equation yields $\angle D = 32$. Since $\angle D$ and angle $x°$ are vertical angles, x also equals 32.

The answer is (D).

41. In the figure, *a*, *b*, *c*, *d*, *e*, *f*, *g*, *h*, *i*, and *j* are chords of the circle. Which two chords are parallel to each other?

(A) *a* and *f*
(B) *b* and *g*
(C) *c* and *h*
(D) *d* and *i*
(E) *e* and *j*

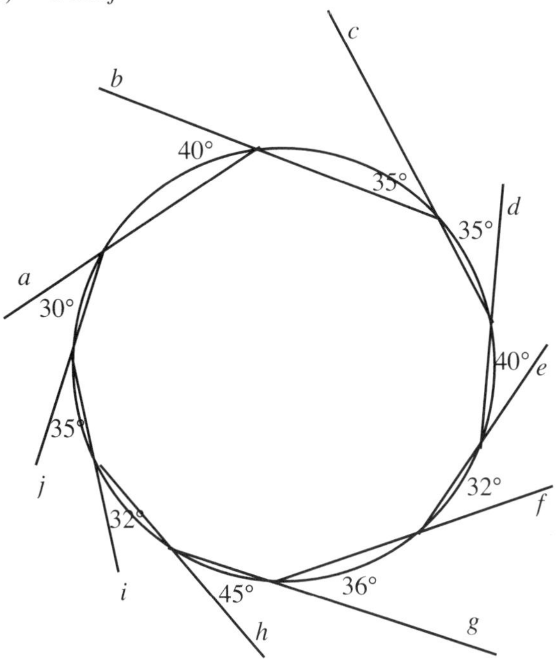

The figure is not drawn to scale.

Two lines are parallel to each other when the angle swept between them is 180°. Use this property to determine the answer:

Choice (A): Angle swept between lines *a* and *f* equals Angle between *a* and *b* + Angle between *b* and *c* + Angle between *c* and *d* + Angle between *d* and *e* + Angle between *e* and *f* = 40 + 35 + 35 + 40 + 32 = 182 ≠ 180. Hence, *a* and *f* are not parallel. Reject.

Choice (B): Angle swept between lines *b* and *g* equals Angle between *b* and *c* + Angle between *c* and *d* + Angle between *d* and *e* + Angle between *e* and *f* + Angle between *f* and *g* = 35 + 35 + 40 + 32 + 36 = 178 ≠ 180. Hence, *b* and *g* are not parallel. Reject.

Choice (C): Angle swept between lines *c* and *h* equals Angle between *c* and *d* + Angle between *d* and *e* + Angle between *e* and *f* + Angle between *f* and *g* + Angle between *g* and *h* = 35 + 40 + 32 + 36 + 45 = 188 ≠ 180. Hence, *c* and *h* are not parallel. Reject.

Choice (D): Angle swept between lines *d* and *i* equals Angle between *d* and *e* + Angle between *e* and *f* + Angle between *f* and *g* + Angle between *g* and *h* + Angle between *h* and *i* = 40 + 32 + 36 + 45 + 32 = 185 ≠ 180. Hence, *d* and *i* are not parallel. Reject.

Choice (E): Angle swept between lines *e* and *j* equals Angle between *e* and *f* + Angle between *f* and *g* + Angle between *g* and *h* + Angle between *h* and *i* + Angle between *i* and *j* = 32 + 36 + 45 + 32 + 35 = 180. Hence, *e* and *j* are parallel. Correct.

The answer is (E).

42. A and B are centers of two circles that touch each other externally, as shown in the figure. What is the area of the circle whose diameter is AB ?

(A) 4π
(B) $25\pi/4$
(C) 9π
(D) 16π
(E) 25π

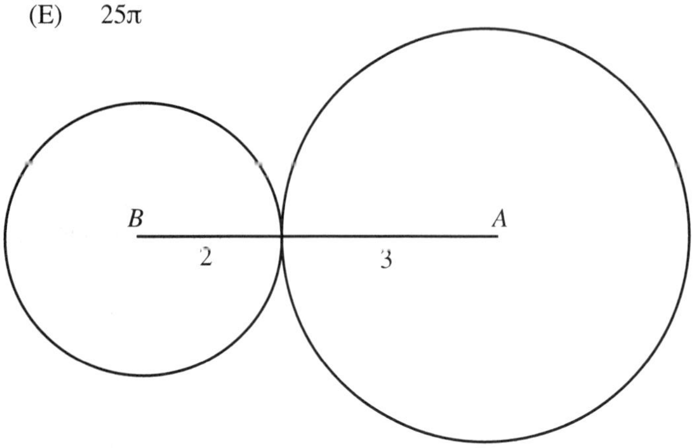

Since the two circles touch each other, the distance between their centers, AB, equals the sum of the radii of the two circles, which is $2 + 3 = 5$. Hence, the area of a circle with diameter AB (or radius $= AB/2$) is

$$\pi \times radius^2 =$$

$$\pi(AB/2)^2 =$$

$$\pi(5/2)^2 =$$

$$25\pi/4$$

The answer is (B).

43. In the figure, which one of the following angles is the greatest?

(A) $\angle A$
(B) $\angle B$
(C) $\angle C$
(D) $\angle D$
(E) $\angle CDB$

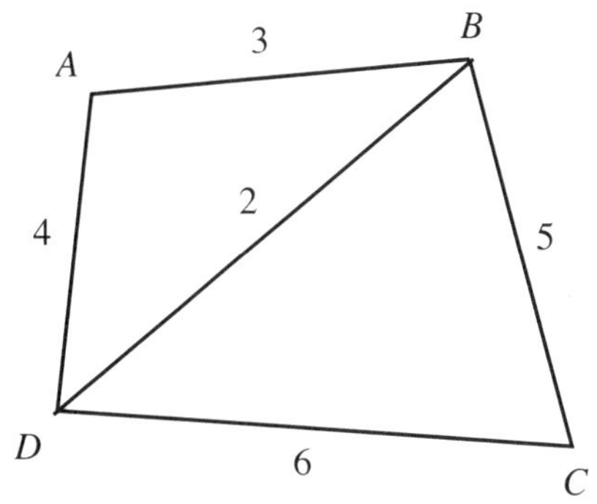

The figure is not drawn to scale.

In ΔABD, $AD = 4$, $AB = 3$, and $BD = 2$ (from the figure). Forming the inequality relation for the side lengths yields $AD > AB > BD$. Since in a triangle, the angle opposite the longer side is greater, we have a similar inequality for the angles opposite the corresponding sides: $\angle ABD > \angle BDA > \angle A$.

Similarly, in ΔBCD, $DC = 6$, $BC = 5$, and $BD = 2$. Forming the inequality for the side lengths yields $DC > BC > BD$. Also the angles opposite the corresponding sides follow the relation $\angle DBC > \angle CDB > \angle C$.

Now, summing the two known inequalities $\angle ABD > \angle BDA > \angle A$ and $\angle DBC > \angle CDB > \angle C$ yields $\angle ABD + \angle DBC > \angle BDA + \angle CDB > \angle A + \angle C$; $\angle B > \angle D > \angle A + \angle C$. From this inequality, clearly $\angle B$ is the greatest angle. Hence, the answer is (B).

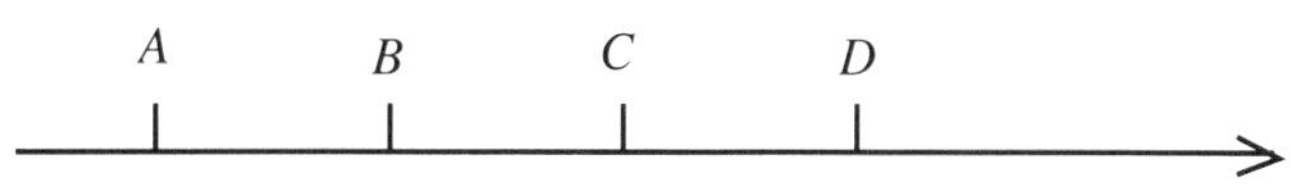

44. In the figure, A, B, C, and D are points on a line in that order. If $AC = 5$, $BD = 10$, and $AD = 13$, then what is the length of BC ?

(A) 2
(B) 8
(C) 15
(D) 18
(E) 28

From the number line, we have the equations

$$AC = AB + BC$$
$$BD = BC + CD$$
$$AD = AB + BC + CD$$

Adding the first two equations yields

$$AC + BD = (AB + BC) + (BC + CD) = AB + 2BC + CD$$

Subtracting the third equation from this equation yields

$$AC + BD - AD = AB + 2BC + CD - (AB + BC + CD)$$
$$= BC$$

Hence, $BC = AC + BD - AD = 5 + 10 - 13 = 2$ (given $AC = 5$, $BD = 10$, and $AD = 13$).

The answer is (A).

45. The side length of a square inscribed in a circle is 2. What is the area of the circle?

(A) π
(B) $\sqrt{2}\pi$
(C) 2π
(D) $2\sqrt{2}\pi$
(E) π^2

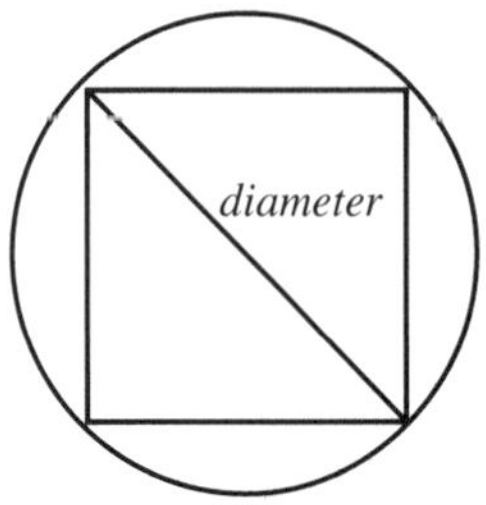

The diagonal of a square inscribed in a circle is a diameter of the circle. The formula for the diagonal of a square is $\sqrt{2} \times side$. Hence, the diameter of the circle inscribing the square of side length 2 is $\sqrt{2} \times 2 = 2\sqrt{2}$. Since *radius* = *diameter*/2, the radius of the circle is $\frac{2\sqrt{2}}{2} = \sqrt{2}$. Hence, the area of the circle is $\pi \cdot radius^2 = \pi\left(\sqrt{2}\right)^2 = 2\pi$. The answer is (C).

46. In the figure shown, if $\angle A = 60°$, $\angle B = \angle C$, and $BC = 20$, then $AB =$

(A) 20
(B) $10\sqrt{2}$
(C) $10\sqrt{3}$
(D) $20\sqrt{2}$
(E) $20\sqrt{3}$

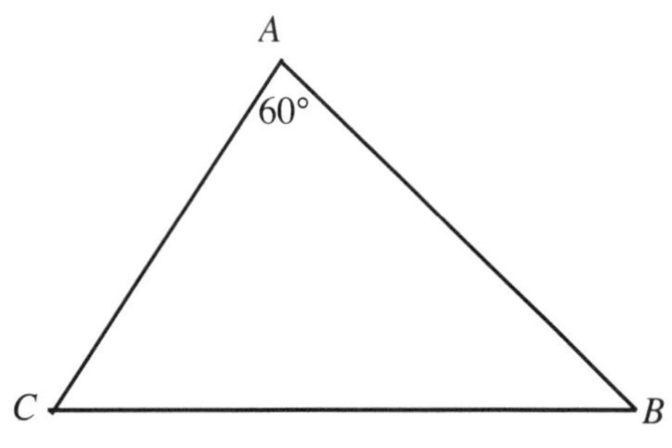

Summing the angles of ΔABC to 180° yields

$\angle A + \angle B + \angle C = 180$
$60 + \angle B + \angle B = 180$ since $\angle A = 60°$ and $\angle B = \angle C$
$2\angle B = 120$ by subtracting 60 from both sides
$\angle B = 60$

Hence, $\angle A = \angle B = \angle C = 60°$. Since the three angles of ΔABC are equal, the three sides of the triangle must also be equal. Hence, $AB = BC = 20$. The answer is (A).

47. Which of the following could be the four angles of a parallelogram?

(I) 50°, 130°, 50°, 130°
(II) 125°, 50°, 125°, 60°
(III) 60°, 110°, 60°, 110°

(A) I only
(B) II only
(C) I and II only
(D) I and III only
(E) I, II and III

A quadrilateral is a parallelogram if it satisfies two conditions:

1) The opposite angles are equal.
2) The angles sum to 360°.

Now, in (I), opposite angles are equal (one pair of opposite angles equals 50°, and the other pair of opposite angles equals 130°). Also, all the angles sum to 360° (= 50° + 130° + 50° + 130° = 360°). Hence, (I) is true.

In (II), not all opposite angles are equal (50° ≠ 60°). Hence, (II) is not a parallelogram.

In (III), the angle sum is not equal to 360° (60° + 110° + 60° + 110° = 340° ≠ 360°). Hence, (III) does not represent a quadrilateral.

Hence, only (I) is true, and the answer is (A).

48. In the figure, *ABCD* is a rectangle and *E* is a point on the side *AB*. If *AB* = 10 and *AD* = 5, what is the area of the shaded region in the figure?

(A) 25
(B) 30
(C) 35
(D) 40
(E) 45

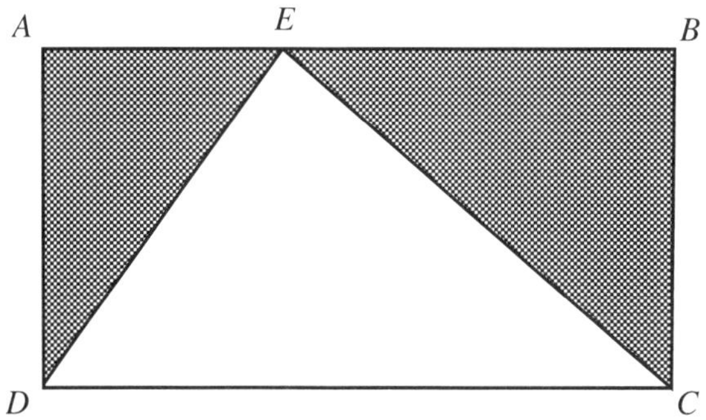

The shaded region contains two triangles ΔAED and ΔEBC. Hence, the area of the shaded region equals the sum of the areas of these two triangles.

Now, the formula for the area of a triangle is 1/2 × *base* × *height*. Hence, the area of ΔAED = $1/2 \times AE \times AD$, and the area of $\Delta EBC = 1/2 \times EB \times BC = 1/2 \times EB \times AD$ (since $BC = AD$).

Hence, the area of the shaded region equals $1/2 \times AE \times AD + 1/2 \times EB \times AD = 1/2 \times AD(AE + EB) = 1/2 \times AD \times AB$ [since $AB = AE + EB$] $= 1/2 \times 5 \times 10 = 25$.

The answer is (A).

49. What is the area of the equilateral triangle if the base $BC = 6$?

(A) $9\sqrt{3}$
(B) $18\sqrt{3}$
(C) $26\sqrt{3}$
(D) $30\sqrt{3}$
(E) $36\sqrt{3}$

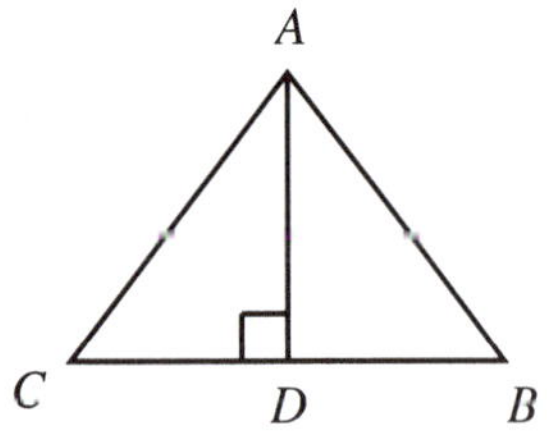

Since the side BC of the equilateral triangle measures 6 (given), the other side AC also measures 6. Since he altitude AD bisects the base (this is true in all equilateral or isosceles triangles), $CD = BD = (1/2)BC = 1/2 \times 6 = 3$. Applying The Pythagorean Theorem to ΔADC yields $AD^2 = AC^2 - CD^2 = 6^2 - 3^2 = 36 - 9 = 27$, or $AD = 3\sqrt{3}$. Hence, the area of the triangle is $1/2 \times base \times height = 1/2 \times BC \times AD = 1/2 \times 6 \times 3\sqrt{3} = 9\sqrt{3}$. The answer is (A).

50. In the figure, $ABCD$ is a square, and OB is a radius of the circle. If BC is a tangent to the circle and $PC = 2$, then what is the area of the square?

(A) 16
(B) 20
(C) 25
(D) 36
(E) 42

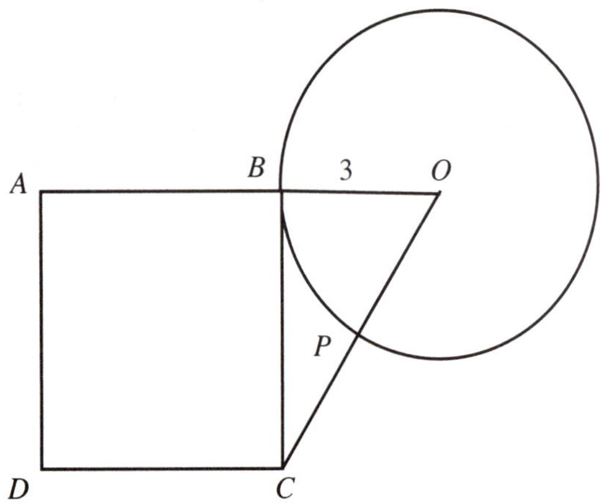

Side BC in the square is also a tangent to the circle shown. Since, at the point of tangency, a tangent is perpendicular to the radius of a circle, $\angle CBO$ is a right angle. Hence, ΔCBO is right angled; and by The Pythagorean Theorem, we have $OC^2 = OB^2 + BC^2$.

Now, we have $PC = 2$. Hence, $OC = OP + PC = Radius + PC = 3 + 2 = 5$. Putting the results in the known equation $OB^2 + BC^2 = OC^2$ yields $3^2 + BC^2 = 5^2$, or $BC^2 = 5^2 - 3^2 = 25 - 9 = 16$.

Now, the area of the square is $side^2 = BC^2 = 16$. The answer is (A).

51. In the figure shown, AB is a diameter of the circle and O is the center of the circle. If $A = (3, 4)$, then what is the circumference of the circle?

(A) 3
(B) 4
(C) 4π
(D) 5π
(E) 10π

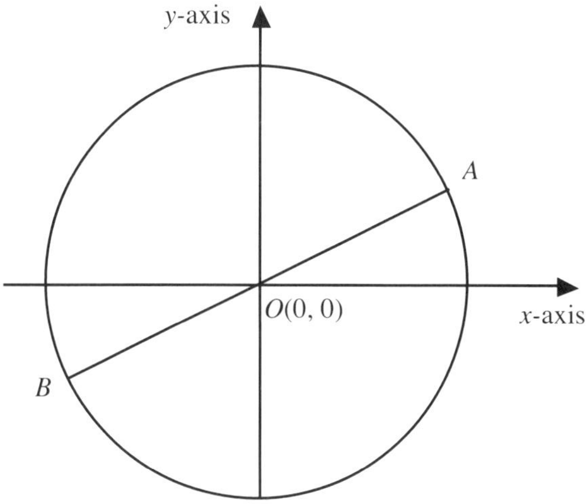

Since AB is a diameter of the circle, the point $A(3, 4)$ is on the circle. The center of the circle is $O(0, 0)$. Since the radius of a circle equals the distance between a point on the circle and the center, AO is a *radius*. By the formula for the distance between two points, $AO = \sqrt{(3-0)^2 + (4-0)^2} = \sqrt{3^2 + 4^2} = \sqrt{25} = 5$.

The formula for the circumference of a circle is $2\pi \times radius$. Hence, the circumference is $2\pi \times 5 = 10\pi$. The answer is (E).

52. A, B, and C are three unequal faces of a rectangular tank. The tank contains a certain amount of water. When the tank is based on the face A, the height of the water is half the height of the tank. The dimensions of the side B are 3 ft × 4 ft and the dimensions of side C are 4 ft × 5 ft. What is the measure of the height of the water in the tank in feet?

(A) 2
(B) 2.5
(C) 3
(D) 4
(E) 5

Draw a rectangular tank as given (based on face A).

Mapping the corresponding given values to the sides of the faces yields

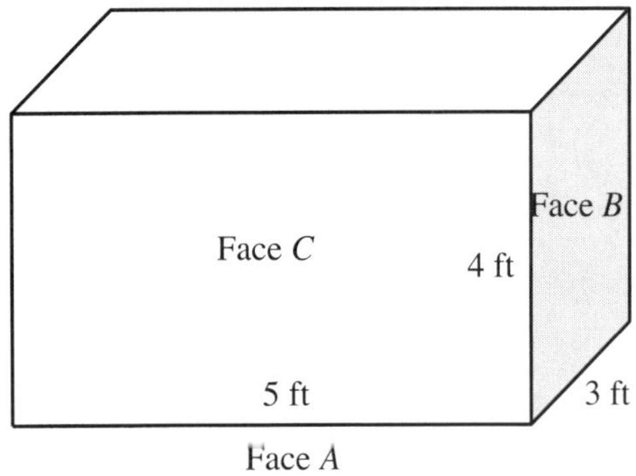

The height of the tank is 4 ft, and the height of the water is half the height of the tank, which is 4/2 = 2 ft. The answer is (A).

53. In the figure, if line *CE* bisects $\angle ACB$, then $x =$

(A) 45
(B) 50
(C) 55
(D) 65
(E) 70

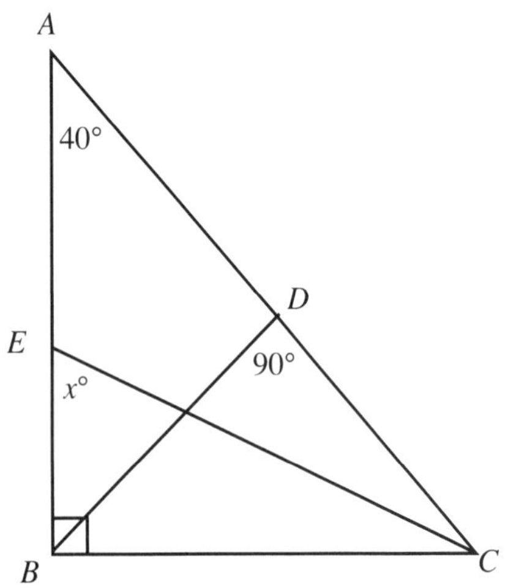

Summing the angles of ΔABC to 180° yields

$\angle A + \angle B + \angle C = 180$
$40 + 90 + \angle C = 180$ by substituting known values
$\angle C = 180 - (40 + 90) = 50$

We are given that *CE* bisects $\angle ACB$. Hence, $\angle ECB = \angle ACE$ = one half of the full angle $\angle ACB$, which equals $1/2 \cdot 50 = 25$. Now, summing the angles of ΔECB to 180° yields

$\angle BEC + \angle ECB + \angle CBE = 180$
$x + 25 + 90 = 180$
$x = 180 - 25 - 90 = 65$

The answer is (D).

54. In the figure, $\angle P =$

(A) 15°
(B) 30°
(C) 35°
(D) 40°
(E) 50°

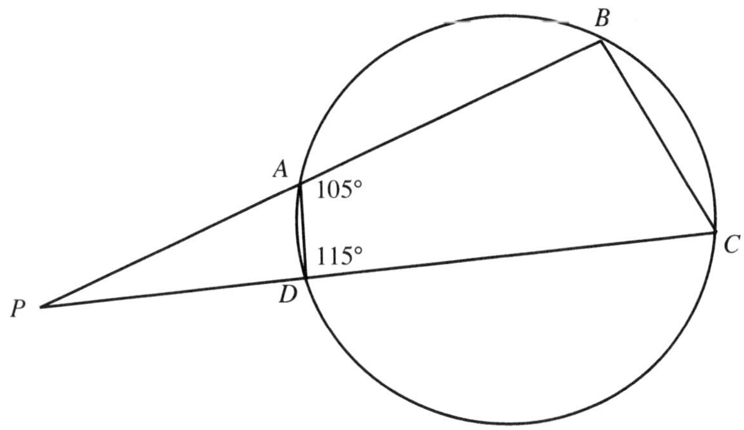

Since the angle in a line is 180°, we have $\angle PAD + \angle DAB = 180$, or $\angle PAD + 105 = 180$ (from the figure, $\angle DAB = 105°$). Solving for $\angle PAD$ yields $\angle PAD = 75$.

Since the angle in a line is 180°, we have $\angle ADP + \angle ADC = 180$, or $\angle ADP + 115 = 180$ (from the figure, $\angle ADC = 115°$). Solving for $\angle ADP$ yields $\angle ADP = 65$.

Now, summing the angles of the triangle *PAD* to 180° yields

$$\angle P + \angle PAD + \angle ADP = 180$$

$$\angle P + 75 + 65 = 180$$

$$\angle P = 40$$

The answer is (D).

55. The length of a rectangular banner is 3 feet 2 inches, and the width is 2 feet 4 inches. Which one of the following equals the area of the banner?

(A) 5 sq. feet
(B) 5 1/2 sq. feet
(C) 6 1/3 sq. feet
(D) 7 sq. feet
(E) 7 7/18 sq. feet

First, let's convert all the measurements to feet. There are 12 inches in a foot, so 2 inches equals 2/12 = 1/6 feet, and 4 inches equals 4/12 = 1/3 feet.

Hence, 3 feet 2 inches equals 3 1/6 feet, and 2 feet 4 inches equals 2 1/3 feet.

Now, the area of a rectangle is length × width. Hence,

$$\text{Area} = 3\ 1/6 \times 2\ 1/3 = 19/6 \times 7/3 = 133/18 = 7\ 7/18$$

The answer is (E).

56. In the figure, if $\angle A = \angle C$, then $x =$

(A) 30
(B) 60
(C) 90
(D) 150
(E) 180

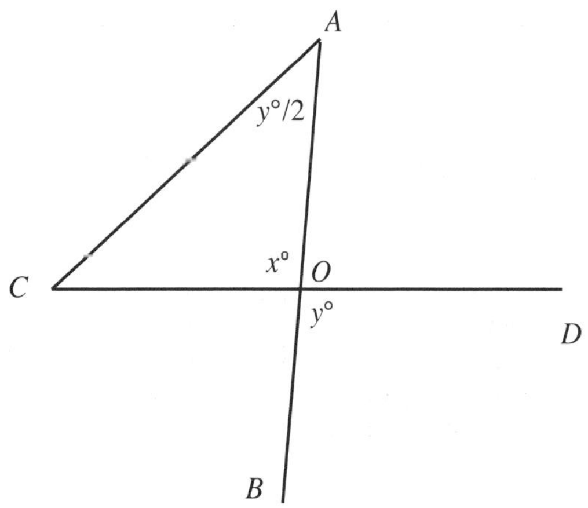

The dimensions in the figure may be different from what they appear to be.

From the figure, $\angle A = y/2$. Since we are given that $\angle A = \angle C$, $\angle C$ also equals $y/2$. Also, in figure, $\angle O = x = y$ (vertical angles are equal). Summing the angles of ΔAOC to 180 yields

$$\angle A + \angle O + \angle C = 180$$
$$y/2 + x + y/2 = 180$$
$$x/2 + x + x/2 = 180 \qquad \text{(since } x = y\text{)}$$
$$2x = 180$$
$$x = 180/2 = 90$$

The answer is (C).

57.

Column A	The width and length of a rectangle are 6 inches and 12 inches, respectively.	Column B
The area of the rectangle when its length is decreased by 4 inches and its width is not changed		The area of the rectangle when its width is decreased by 4 inches and its length is not changed

The formula for the area of a rectangle is *length* × *width*. The original length and width of the rectangle are 12 inches and 6 inches, respectively.

Column A: After the length of the rectangle is decreased by 4 inches, the new length becomes $12 - 4 = 8$ inches and the new area becomes $8 \cdot 6 = 48$.

Column B: After the width of the rectangle is decreased by 4 inches, the new width becomes $6 - 4 = 2$ inches and the new area becomes $12 \cdot 2 = 24$.

Since Column A is greater than Column B, the answer is (A).

58. Column A | $|a|$ is the distance point a is from the origin on the number line. $x \neq 0$. | Column B

$|x| + |-2|$ | $|x - 2|$

Suppose $x = -2$. Then Column A equals $|x| + |-2| = |-2| + |-2| = 2 + 2 = 4$, and Column B equals $|x - 2| = |-2 - 2| = |-4| = 4$. Here, Column A equals Column B.

Now, suppose $x = 2$. Then Column A equals $|x| + |-2| = |2| + |-2| = 2 + 2 = 4$, and Column B equals $|x - 2| = |2 - 2| = |0| = 0$. Here, Column A is greater than Column B.

This is a double case, and therefore the answer is (D).

59. Column A | A regular polygon of 24 sides is inscribed in a circle. | Column B

The perimeter of the polygon | The circumference of the circle

Each side of the regular polygon represents a chord of the circle, and each chord subtends a unique arc on the circle. The length of a chord is always less than the length of the arc it subtends. Hence, the sum of the lengths of the 24 sides (which are the 24 chords of the circle) is less than the sum of the lengths of the arcs that each one of the cords subtends. Hence, the perimeter of the polygon is less than the circumference of the circle, and therefore Column A is less than Column B. The answer is (B).

➢ Hard

60. In the figure, lines l and k are parallel. Which one of the following must be true?

(A) $a < b$
(B) $a \le b$
(C) $a = b$
(D) $a \ge b$
(E) $a > b$

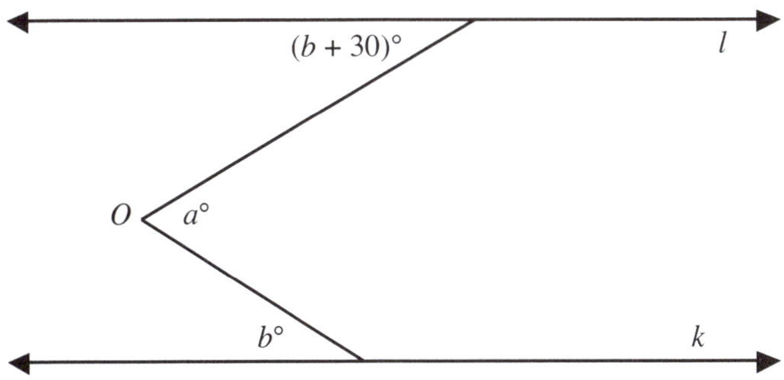

Draw line m passing through O and parallel to both line l and line k.

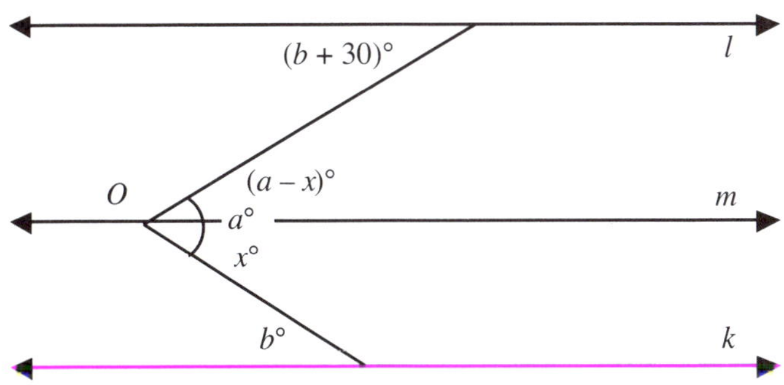

Now, observe that angle x is only part of angle a, and $x = b$ since they are alternate interior angles. Since x is only part of angle a, $a > x$ and $a > b$. The answer is (E).

61. In the figure, lines l and k are parallel. If a is an acute angle, then which one of the following must be true?

(A) $b > 10$
(B) $b > 15$
(C) $b < 20$
(D) $b < 30$
(E) $b > 45$

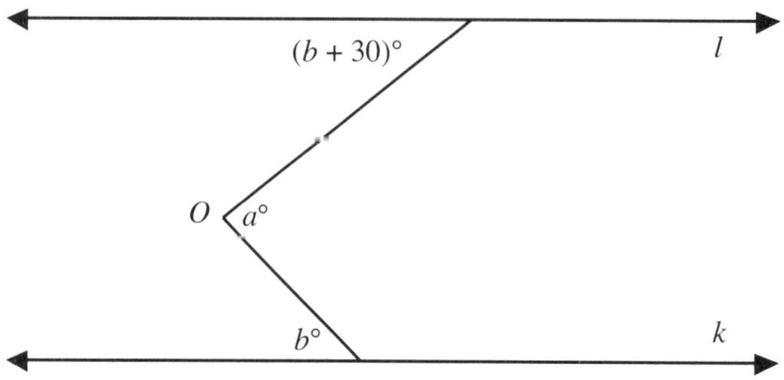

Draw a line parallel to both of the lines l and k and passing through O.

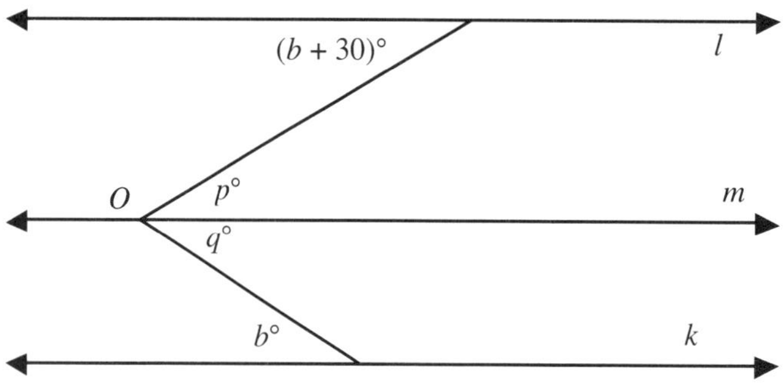

We are given that a is an acute angle. Hence, $a < 90$. Since angles p and $b + 30$ are alternate interior angles, they are equal. Hence, $p = b + 30$. Similarly, angles q and b are alternate interior angles. Hence, $q = b$. Since angle a is the sum of its sub-angles p and q, $a = p + q = (b + 30) + b = 2b + 30$. Solving this equation for b yields $b = (a - 30)/2 = a/2 - 15$. Now, dividing both sides of the inequality $a < 90$ by 2 yields $a/2 < 45$. Also, subtracting 15 from both sides of the inequality yields $a/2 - 15 < 30$. Since $a/2 - 15 = b$, we have $b < 30$. The answer is (D).

62. The diagonal length of a square is 14.1 sq. units. What is the area of the square, rounded to the nearest integer? ($\sqrt{2}$ is approximately 1.41.)

(A) 96
(B) 97
(C) 98
(D) 99
(E) 100

If a is the length of a side of the square, then a diagonal divides the square into two congruent (equal) right triangles. Applying The Pythagorean Theorem to either triangle yields $diagonal^2 = side^2 + side^2 = a^2 + a^2 = 2a^2$. Taking the square root of both sides of this equation yields $diagonal = a\sqrt{2}$. We are given that the diagonal length is 14.1. Hence, $a\sqrt{2} = 14.1$ or $a = \dfrac{14.1}{\sqrt{2}}$. Now, the area, a^2, equals $\left(\dfrac{14.1}{\sqrt{2}}\right)^2 = \dfrac{14.1^2}{\left(\sqrt{2}\right)^2} =$

$\dfrac{14.1^2}{2} = \dfrac{198.81}{2} = 99.4$. The number 99.4 is nearest to 99. Hence, the answer is (D).

Note: If you had approximated $\frac{14.1}{\sqrt{2}}$ with 10, you would have mistakenly gotten 100 and would have answered (E). Approximation is the culprit. Prefer doing it in the last.

63. In the figure, *ABC* and *ADC* are right triangles. Which of the following could be the lengths of *AD* and *DC*, respectively?

(I) $\sqrt{3}$ and $\sqrt{4}$
(II) 4 and 6
(III) 1 and $\sqrt{24}$
(IV) 1 and $\sqrt{26}$

(A) I and II only
(B) II and III only
(C) III and IV only
(D) IV and I only
(E) I, II and III only

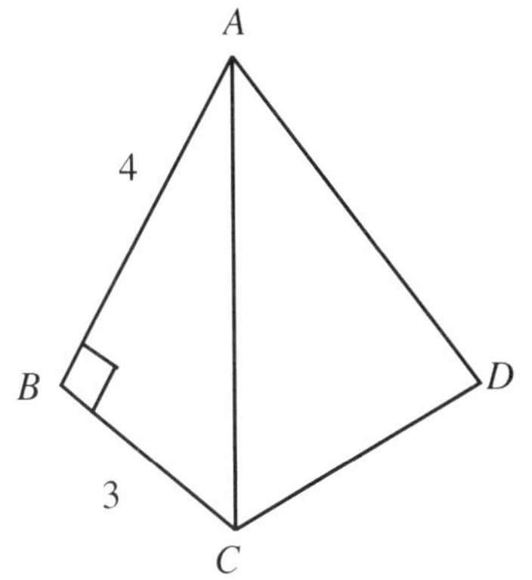

Figure not drawn to scale

From the figure, we have that $\angle B$ is a right angle in ΔABC. Applying The Pythagorean Theorem to the triangle yields $AC^2 = AB^2 + BC^2 = 4^2 + 3^2 = 25$. Hence, $AC = \sqrt{25} = 5$.

Now, we are given that ΔADC is a right-angled triangle. But, we are not given which one of the three angles of the triangle is right-angled. We have two possibilities: either the common side of the two triangles, *AC*, is the hypotenuse of the triangle, or it is not.

In the case *AC* is the hypotenuse of the triangle, we have by The Pythagorean Theorem,

$$AC^2 = AD^2 + DC^2$$
$$5^2 = AD^2 + DC^2$$

This equation is satisfied by III since $5^2 = 1^2 + \left(\sqrt{24}\right)^2$. Hence, III is possible.

In the case *AC* is not the hypotenuse of the triangle and, say, *DC* is the hypotenuse, then by applying The Pythagorean Theorem to the triangle, we have

$$AD^2 + AC^2 = DC^2$$
$$AD^2 + 5^2 = DC^2$$

This equation is satisfied by IV: $5^2 + 1^2 = \left(\sqrt{26}\right)^2$.

Hence, we conclude that III and IV are possible. The two are available in choice (C). Hence, the answer is (C).

64. In the figure, what is the value of x ?

(A) 15
(B) 30
(C) 45
(D) 60
(E) 75

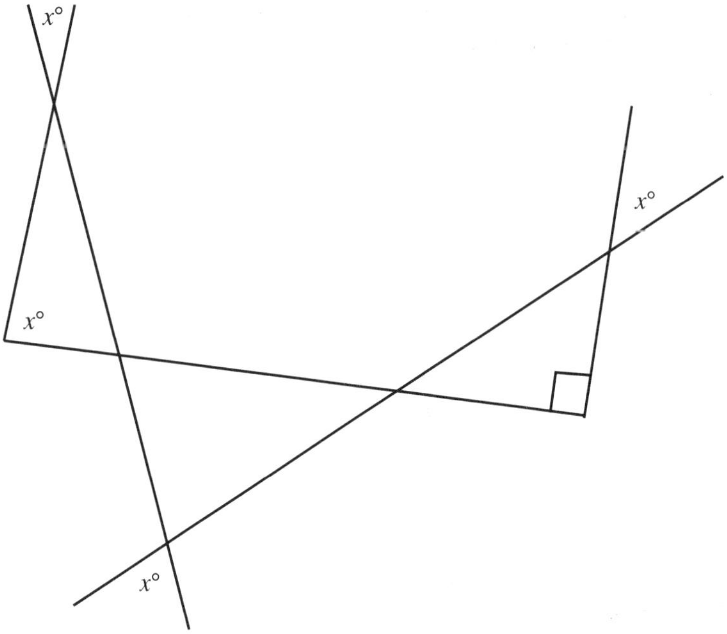

Let's name the vertices as shown in the figure

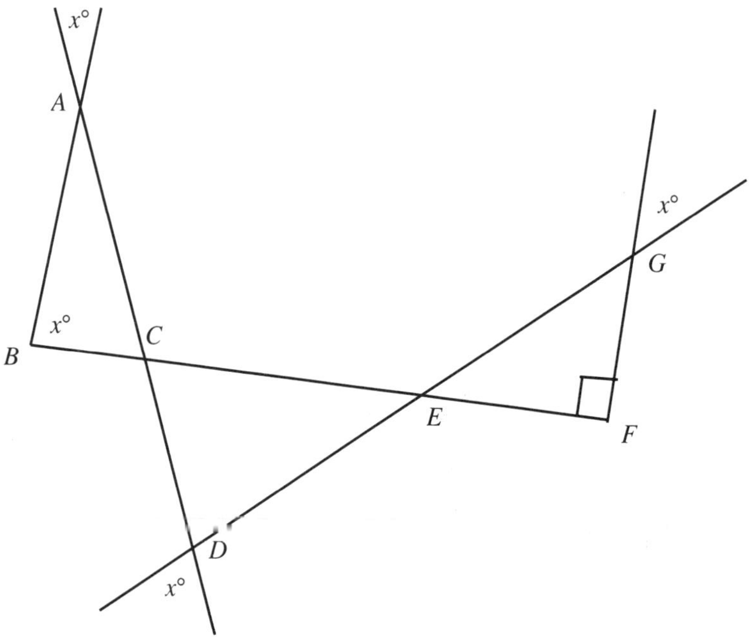

Let's start evaluating the unknown angles of the triangles in the figure from left most located triangle through the right most located triangle. Then the value of x can be derived by summing the angles of right most located triangle to 180°. This is done as follows:

In the first triangle from the left: ΔABC, we have from the figure that $\angle B = x°$ and $\angle A = x°$ (vertical angles are equal). Summing the angles of the triangle to 180° yields $x + x + \angle C = 180$. Solving the equation for $\angle C$ yields $\angle C = 180 - 2x$.

In the second triangle from the left: ΔCED, we have from the figure that $\angle D = x°$ (vertical angles are equal), and we have $\angle C$ [in ΔCED] = $\angle C$ in ΔACB (They are vertical angles) = $180 - 2x$ (Known result). Now, summing the angles of the triangle to 180° yields $(180 - 2x) + x + \angle E = 180$. Solving the equation for $\angle E$ yields $\angle E = x$.

In the third triangle from the left: ΔGFE, we have from the figure that $\angle G = x°$ (vertical angles are equal) and $\angle E$ [in ΔGFE] = $\angle E$ in ΔCED (Vertical angles are equal) = $x°$ (Known result). Now, we also have $\angle F = 90°$ (From the figure). Now, summing the three angles of the triangle to 180° yields $x + x + 90 = 180$. Solving for x yields $2x = 90$ or $x = 45$. Hence, the answer is (C).

65. In the figure, what is the value of x ?

(A) 10°
(B) 30°
(C) 45°
(D) 60°
(E) 75°

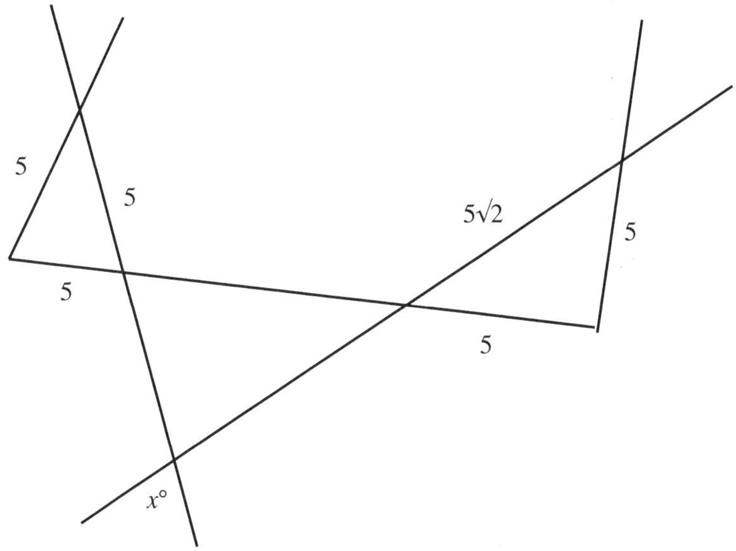

Let's name the vertices of the figure as shown

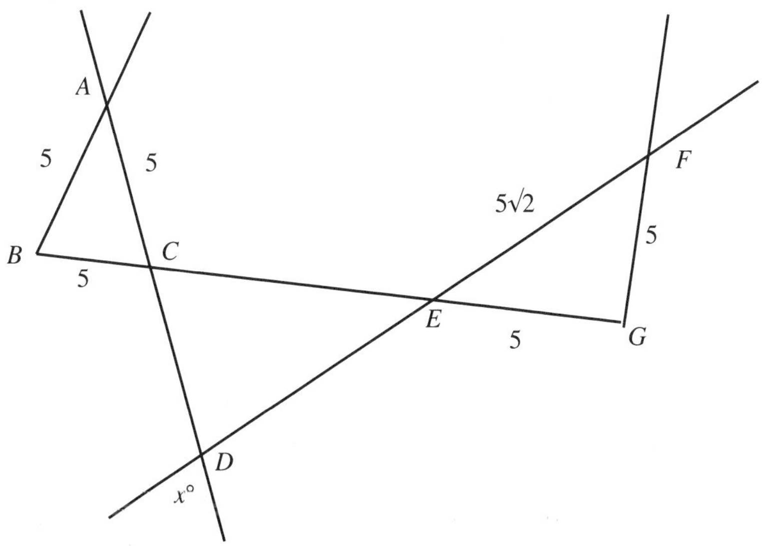

In ΔABC, all the sides are equal (each equals 5). Hence, the triangle is equilateral, and $\angle A = \angle B = \angle C = 60°$.

Also, ΔEFG is a right-angled isosceles triangle (since $EG = FG = 5$), and The Pythagorean Theorem is satisfied ($EG^2 + FG^2 = 5^2 + 5^2 = 50 = EF^2 = \left(5\sqrt{2}\right)^2$. Hence, $\angle E = \angle F = 45°$ (Angles opposite equal sides of an isosceles right triangle measure 45° each).

Now, in ΔCED, we have:

$\angle D = x°$	vertical angles, from the figure
$\angle C = \angle C$ in ΔABC	vertical angles, from the figure
$= 60°$	we know
$\angle E$ in $\Delta CED = \angle E$ in ΔEFG	vertical angles
$= 45°$	we know

Now, summing these three angles of ΔCED to 180° yields $60 + 45 + x = 180$. Solving this equation for x yields $x = 75$. The answer is (E).

66. If $ABCD$ is a square and the area of ΔAFG is 10, then what is the area of ΔAEC ?

 (A) 5
 (B) $\dfrac{10}{\sqrt{2}}$
 (C) $\dfrac{10}{\sqrt{3}}$
 (D) 10
 (E) 20

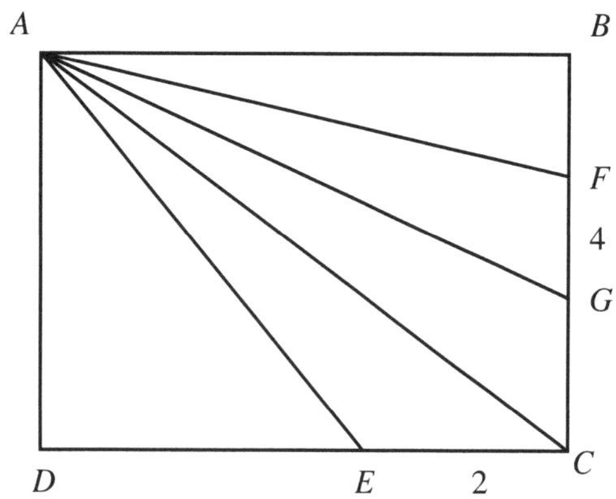

The formula for the area of a triangle is 1/2 × *base* × *height*. By the formula, the area of ΔAFG (which is given to be 10) is $1/2 \times FG \times AB$. Hence, we have

$1/2 \times 4 \times AB = 10$	given that the area of $\Delta AFG = 10$
$AB = 5$	

Also, by the same formula, the area of ΔAEC is

$1/2 \times EC \times DA$	
$= 1/2 \times 2 \times DA$	from the figure, $EC = 2$ units
$= 1/2 \times 2 \times AB$	$ABCD$ is a square. Hence, side DA = side AB
$= 1/2 \times 2 \times 5$	
$= 5$	

The answer is (A).

67. In the figure, $ABCD$ is a rectangle, and F and E are points on AB and BC, respectively. The area of ΔDFB is 9 and the area of ΔBED is 24. What is the perimeter of the rectangle?

(A) 18
(B) 23
(C) 30
(D) 42
(E) 48

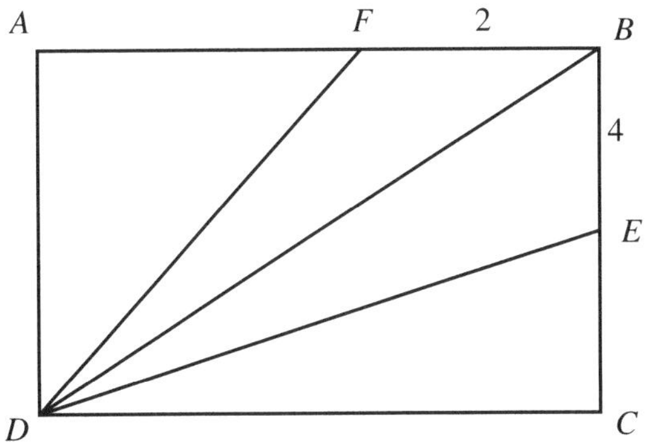

The formula for the area of a triangle is $1/2 \times base \times height$. Hence, the area of ΔDFB is $1/2 \times FB \times AD$. We are given that the area of ΔDFB is 9. Hence, we have

$$1/2 \times FB \times AD = 9$$

$$1/2 \times 2 \times AD = 9$$

$$AD = 9$$

Similarly, the area of ΔBED is $1/2 \times BE \times DC$, and we are given that the area of the triangle is 24. Hence, we have

$$1/2 \times BE \times DC = 24$$

$$1/2 \times 4 \times DC = 24 \qquad \text{from the figure, } BE = 4$$

$$DC = 12$$

Now, the formula for the perimeter of a rectangle is

2 × (the sum of the lengths of any two adjacent sides of the rectangle)

Hence, the perimeter of the rectangle $ABCD = 2(AD + DC) = 2(9 + 12) = 2 \times 21 = 42$. The answer is (D).

68. In the figure, $ABCD$ is a rectangle. The area of quadrilateral $EBFD$ is one-half the area of the rectangle $ABCD$. Which one of the following is the value of AD ?

(A) 5
(B) 6
(C) 7
(D) 12
(E) 15

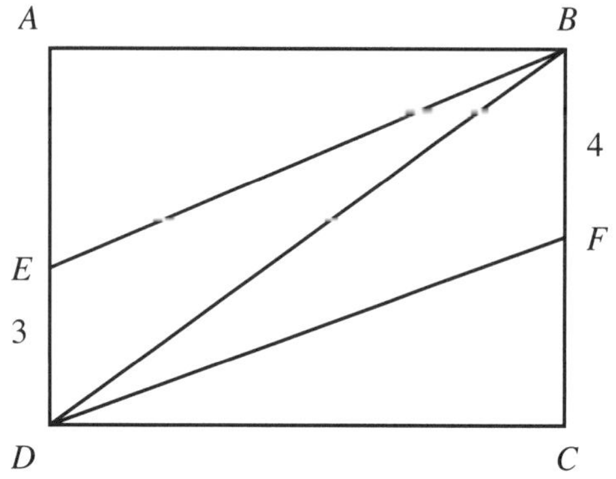

From the figure, it is clear that the area of quadrilateral $EBFD$ equals the sum of the areas of the triangles ΔEBD and ΔDBF. Hence, the area of the quadrilateral $EBFD$

= area of ΔEBD + area of ΔDBF	
$= 1/2 \times ED \times AB + 1/2 \times BF \times DC$	area of a triangle equals $1/2 \times base \times height$
$= 1/2 \times 3 \times AB + 1/2 \times 4 \times DC$	from the figure, $ED = 3$ and $BF = 4$
$= (3/2)AB + 2DC$	
$= (3/2)AB + 2AB$	opposite sides AB and DC must be equal
$= (7/2)AB$	

Now, the formula for the area of a rectangle is $length \times width$. Hence, the area of the rectangle $ABCD$ equals $AD \times AB$. Since we are given that the area of quadrilateral $EBFD$ is half the area of the rectangle $ABCD$, we have

$$\frac{1}{2}(AD \times AB) = \frac{7}{2}AB$$

$$AD \times AB = 7AB$$

$$AD = 7$$ (by canceling AB from both sides)

The answer is (C).

69. In the figure, $ABCD$ is a rectangle, and the area of quadrilateral $AFCE$ is equal to the area of ΔABC. What is the value of x ?

(A) 5
(B) 6
(C) 7
(D) 12
(E) 15

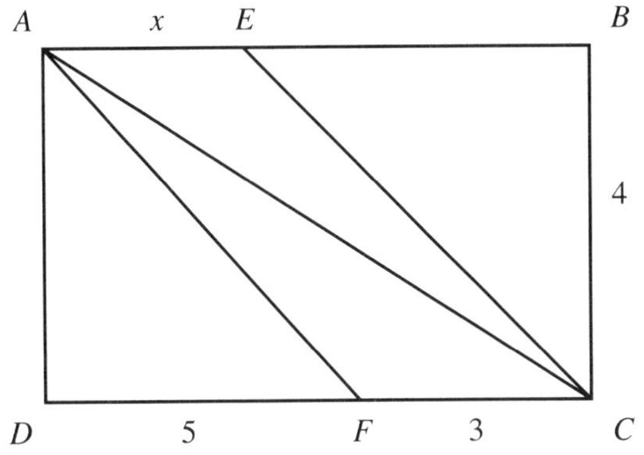

The formula for the area of a triangle is $1/2 \times base \times height$. Hence, the areas of triangles AEC, ACF, and ABC are

The area of $\Delta AEC = (1/2)(x)(BC) = (1/2)(x)(4) = 2x$

The area of $\Delta ACF = (1/2)(3)(AD)$

$= (1/2)(3)(BC)$ — $AD = BC$, since opposite sides a rectangle are equal
$= (3/2)BC = (3/2)(4) = 6$

The area of $\Delta ABC = (1/2)(AB)(AD)$

$= (1/2)(CD)(BC)$ — $AB = CD$ and $AD = BC$ since opposite sides in a rectangle are equal

$= (1/2)(DF + FC)(BC)$
$= (1/2)(5 + 3)(4)$
$= (1/2)(8)(4)$
$= 16$

Now, the area of the quadrilateral $AECF$ equals (the area of ΔAEC) + (the area of ΔACF) = $2x + 6$, and the area of $\Delta ABC = 16$. Since we are given that the area of the triangle ABC equals the area of the quadrilateral $AFCE$, $2x + 6 = 16$. Solving for x in this equation yields $x = 5$. The answer is (A).

70. In the figure, $ABCD$ is a rectangle. Points E and F cut the sides BC and CD of the rectangle respectively such that $EC = 3$, $FC = 4$, and $AD = 12$, and the areas of the crossed and the shaded regions in the figure are equal. Which one of the following equals the perimeter of rectangle $ABCD$?

(A) 42
(B) 50
(C) 56
(D) 64
(E) 72

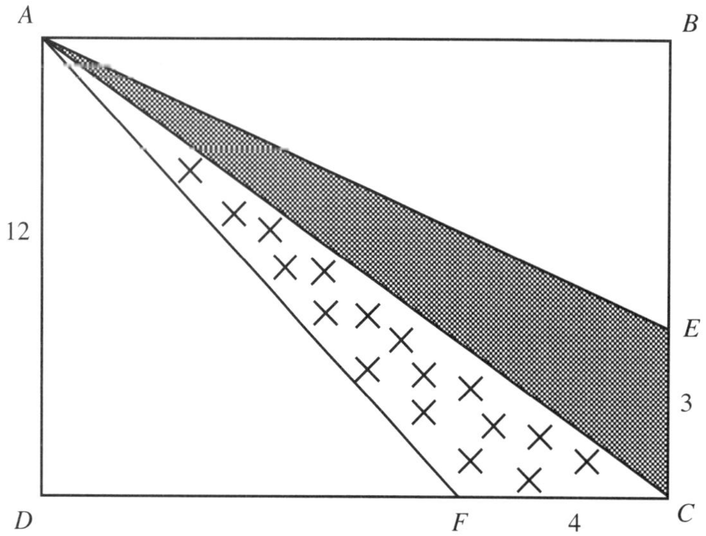

The formula for the area of a triangle is $1/2 \times base \times height$. Hence, the area of ΔAFC (the crossed region) equals $1/2 \times FC \times AD$, and the area of ΔAEC (the shaded region) equals $1/2 \times EC \times AB$. We are given that the two areas are equal. Hence, we have the equation

$1/2 \times EC \times AB = 1/2 \times FC \times AD$

$1/2 \times 3 \times AB = 1/2 \times 4 \times 12$ from the figure, $EC = 3$, $FC = 4$, and $AD = 12$

$(3/2)AB = 24$

$AB = (2/3)(24) = 16$

Now, the perimeter of a rectangle is two times the sum of the lengths of any two adjacent sides of the rectangle. Hence, the perimeter of $ABCD$ is $2(AB + AD) = 2(16 + 12) = 56$. The answer is (C).

71. What is the perimeter of ΔABC shown in the figure?

(A) $2 + 4\sqrt{2}$
(B) $4 + 2\sqrt{2}$
(C) 8
(D) $4 + 4\sqrt{2}$
(E) $4 + 4\sqrt{3}$

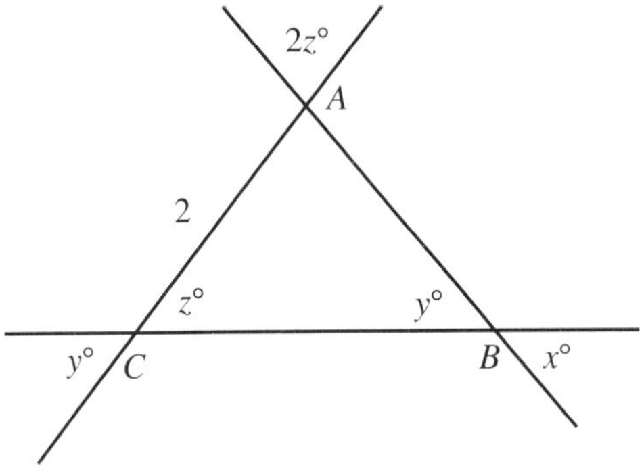

Equating the vertical angles at points A and C in the figure yields $\angle A = 2z$ and $y = z$. Summing the angles of the triangle to 180° yields

$\angle A + \angle B + \angle C = 180$	
$2z + y + z = 180$	we know that $\angle A = 2z$, $\angle B = y$, and $\angle C = z$
$2z + z + z = 180$	we know that $y = z$
$4z = 180$	
$z = 180/4 = 45$	

So, $\angle A = 2z = 2(45) = 90$, $\angle B = y = z = 45$ and $\angle C = z = 45$. Hence, ΔABC is a right triangle. Also, since angles $\angle C$ and $\angle B$ are equal (equal to 45), the sides opposite these two angles, AB and AC, must be equal. Since AC equals 2 (from the figure), AB also equals 2. Now, applying The Pythagorean Theorem to the triangle yields

$BC^2 = AB^2 + AC^2$	
$= 2^2 + 2^2$	given that $AB = AC = 2$
$= 4 + 4$	
$= 8$	
$BC = 2\sqrt{2}$	square rooting both sides

Now, the perimeter of $\Delta ABC = AB + BC + CA = 2 + 2\sqrt{2} + 2 = 4 + 2\sqrt{2}$. The answer is (B).

72. In the figure, what is the value of x ?

(A) 90
(B) 95
(C) 100
(D) 105
(E) 115

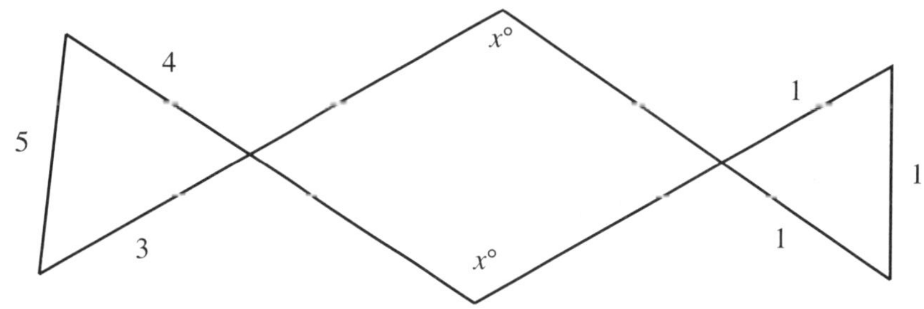

Let's name the vertices of the figure as shown below

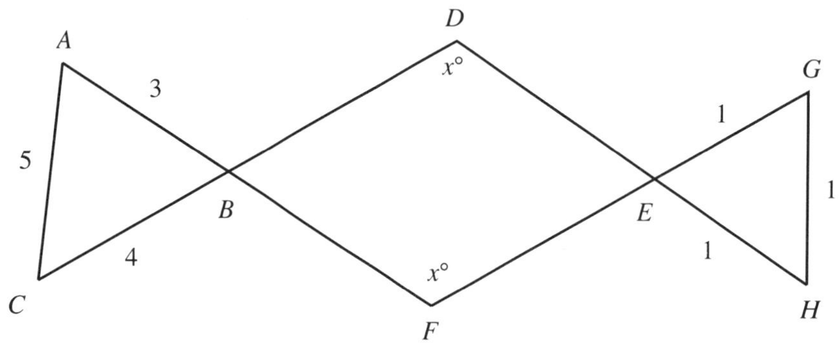

Now, in ΔABC, $AB = 3$, $BC = 4$, and $AC = 5$. Hence, AC^2 equals $5^2 = 25$, and $AB^2 + BC^2$ equals $3^2 + 4^2 = 9 + 16 = 25$. Hence, $AC^2 = 25 = AB^2 + BC^2$. From this, it is clear that the triangle satisfies The Pythagorean Theorem. Hence, by the theorem, ABC must be a right triangle, AC the longest side the hypotenuse, and the angle opposite the side, $\angle B$ must be the right angle. Hence, $\angle B = 90°$.

Now, in ΔEGH, from the figure, we have each side (EG, GH, and EH) of the triangle measures 1 unit. Hence, all the sides of the triangle are equal, and the triangle is equilateral. Since in any equilateral triangle, each angle measures 60°, we have $\angle E = 60°$.

Now, from the figure, we have

$\angle B$ in the quadrilateral $= \angle B$ in ΔABC	vertical angles are equal
$= 90°$	known result

Also, we have

$\angle E$ in the quadrilateral $= \angle E$ in ΔEGH	vertical angles are equal
$= 60°$	

Now, summing the angles of the quadrilateral to 360° yields $\angle B + \angle E + \angle D + \angle F = 360°$. Hence, we have $90 + 60 + x + x = 360$. Solving for x yields $2x = 210$ or $x = 210/2 = 105$. The answer is (D).

73. In the figure, $ABCD$ is a rectangle. Points P, Q, R, S, and T cut side AB of the rectangle such that $AP = 3$, $PQ = QR = RS = ST = 1$. E is a point on AD such that $AE = 3$. Which one of the following line segments is parallel to the diagonal BD of the rectangle?

(A) EP
(B) EQ
(C) ER
(D) ES
(E) ET

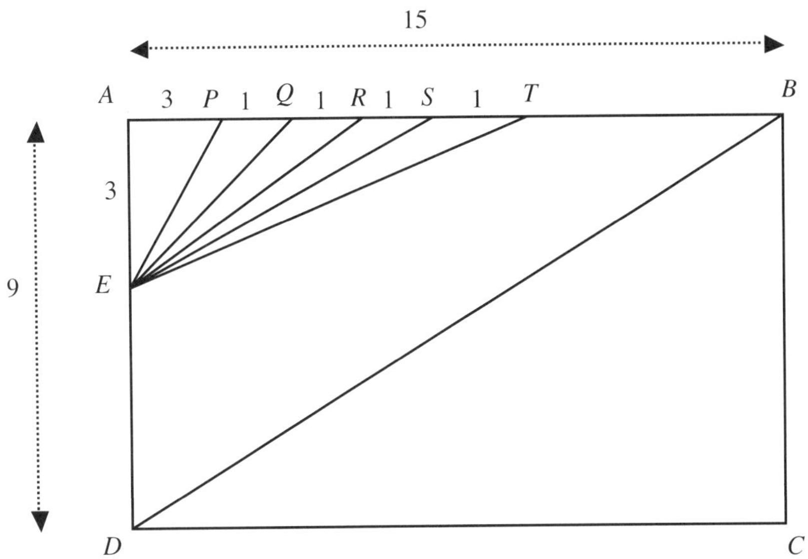

The figure is not drawn to scale.

Let X represent the point on AB that makes EX parallel to the diagonal BD. Then ΔAXE and ΔABD must be similar because

$\angle EAX = \angle DAB$	common angles
$\angle AEX = \angle ADB$	corresponding angles are equal

Hence, the sets of corresponding angles are equal and the triangles are similar.

Hence, the corresponding sides of the triangles must be in the same ratio. This yields the following equations:

$AE/AD = AX/AB$
$3/9 = AX/15$
$AX = 15 \times 3/9 = 15 \times 1/3 = 5$

Hence, the point X is 5 units away from the point A on the side AB. Now,

$AP = 3 \neq AX$, which equals 5.
$AQ = AP + PQ = 3 + 1 = 4 \neq AX$, which equals 5.
$AR = AP + PQ + QR = 3 + 1 + 1 = 5 = AX$, which equals 5.

Hence, point R coincides with point X and therefore just like EX, ER must be parallel to BD. Hence, the answer is (C).

74. In the, figure *A*, *B* and *C* are points on the circle. What is the value of x ?

(A) 45
(B) 55
(C) 60
(D) 65
(E) 70

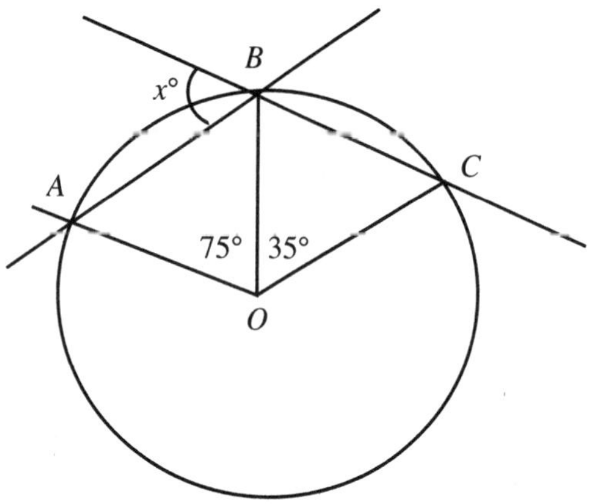

OA and *OB* are radii of the circle. Hence, angles opposite them in ΔAOB are equal: $\angle OAB = \angle ABO$. Summing the angles of ΔAOB to 180° yields $\angle OAB + \angle ABO + \angle AOB = 180$ or $2\angle ABO + 75° = 180$ [since $\angle OAB = \angle OBA$); $\angle OBA = (180 - 75)/2 = 105/2$].

Similarly, *OB* equals *OC* (radii of a circle are equal) and angles opposite them in ΔBOC are equal: $\angle OBC = \angle BCO$. Summing angles of the triangle to 180° yields $\angle OBC + \angle BCO + 35 = 180$ or $2\angle OBC + 35 = 180$ [since $\angle OBC = \angle BCO$; $\angle OBC = (180 - 35)/2 = 145/2$].

Now, since an angle made by a line is 180°, we have

$$x + \angle ABO + \angle OBC = 180$$

$$x + 105/2 + 145/2 = 180$$

$$x + 250/2 = 180$$

$$x + 125 = 180$$

$$x = 180 - 125 = 55$$

The answer is (B).

75. In the figure, *ABCD* and *PQRS* are two rectangles inscribed in the circle as shown and $AB = 4$, $AD = 3$, and $QR = 4$. What is the value of l ?

(A) 3/2
(B) 8/3
(C) 3
(D) 4
(E) 5

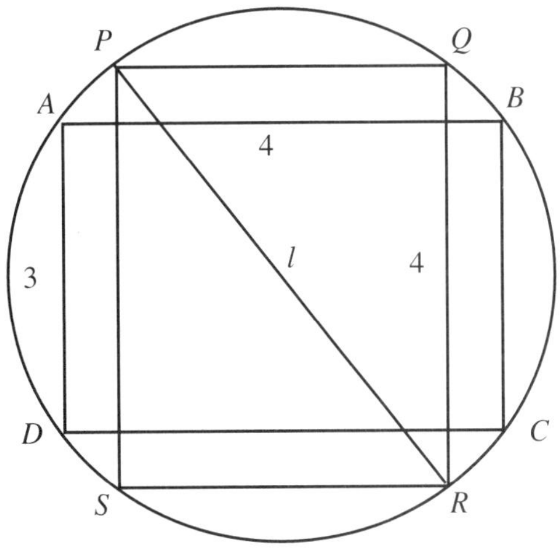

PQRS is a rectangle inscribed in the circle. Hence, diagonal *PR* must pass through the center of the circle. So, *PR* is a diameter of the circle.

Similarly, *BD* is a diagonal of rectangle *ABCD*, which is also inscribed in the same circle. Hence, the two diagonals must be diameters and equal. So, we have $PR = BD$.

Now, in the figure, let's join the opposite vertices *B* and *D* of the rectangle *ABCD*:

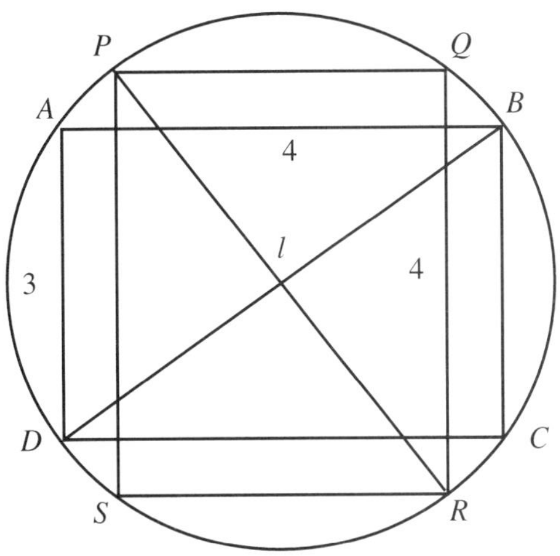

Applying The Pythagorean Theorem to the right triangle *ABD* yields $BD^2 = AB^2 + AD^2 = 4^2 + 3^2 = 16 + 9 = 25$. By square rooting, we get $BD = \sqrt{25} = 5$. Hence, *PR* also equals 5. Since, from the figure, l equals *PR*, l equals 5. The answer is (E).

76. In the figure, $ABCD$ and $PQRS$ are rectangles inscribed in the circle shown in the figure. If $AB = 5$, $AD = 3$, and $QR = 4$, then what is the value of l ?

(A) 3
(B) 4
(C) 5
(D) $\sqrt{15}$
(E) $3\sqrt{2}$

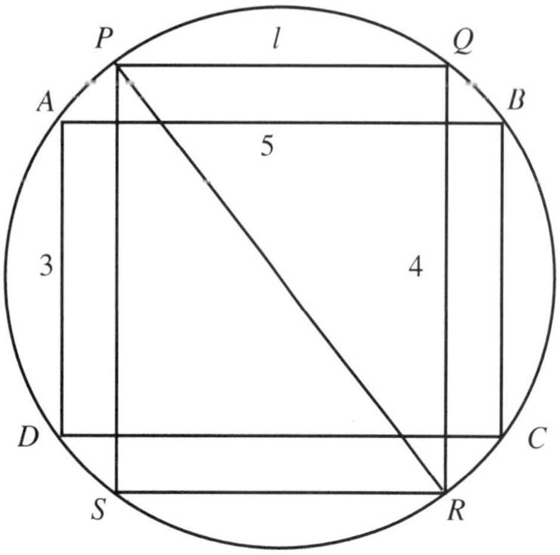

$PQRS$ is a rectangle inscribed in the circle. Hence, its diagonal PR must pass through the center of the circle. So, PR is a diameter of the circle.

Similarly, AC is a diagonal of the rectangle $ABCD$, which is also inscribed in the same circle. Hence, AC must also be a diameter of the circle. Since the diameters of a circle are equal, $PR = AC$.

Applying The Pythagorean Theorem to ΔABC yields $AC^2 = AB^2 + BC^2 = 5^2 + 3^2 = 25 + 9 = 34$. Hence, PR^2 also equals 34. Now, applying The Pythagorean Theorem to ΔPQR yields

$$PQ^2 + QR^2 = PR^2$$
$$l^2 + 4^2 = 34$$
$$l^2 + 16 = 34$$
$$l^2 = 18$$
$$l = \sqrt{18}$$
$$l = 3\sqrt{2}$$

The answer is (E).

77. In the figure, *ABCD* is a rectangle inscribed in the circle shown. What is the length of the arc *DC* ?

(A) $\pi/4$
(B) $2\pi/3$
(C) $\pi/2$
(D) $3\pi/4$
(E) $4\pi/3$

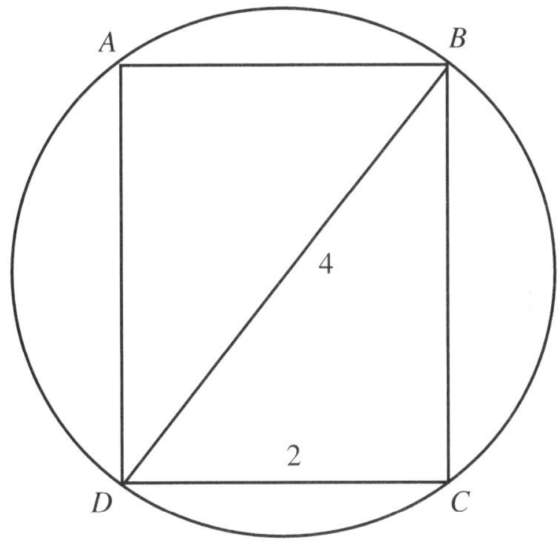

In the figure, *BD* is a diagonal of the rectangle inscribed in the circle. Hence, *BD* is a diameter of the circle. So, the midpoint of the diagonal must be the center of the circle, and the radius must equal half the length of the diameter: $BD/2 = 4/2 = 2$. Now, joining the center of the circle, say, *O* to the point *C* yields the following figure:

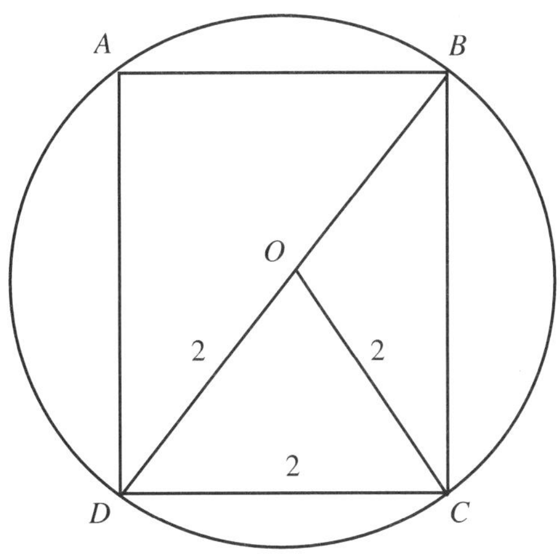

Now, in ΔDOC since *OD* and *OC* are radii of the circle, both equal 2. So, $OD = OC = 2$. Since *DC* also equals 2 (in figure given), $OD = OC = DC = 2$ (all three sides are equal). Hence, the triangle is equilateral and each angle must equal 60°, including $\angle DOC$. Now, the circumference of the given circle equals $2\pi \times radius = 2\pi(2) = 4\pi$. The fraction of the complete angle that the arc *DC* makes in the circle is $60°/360° = 1/6$. The arc length would also be the same fraction of the circumference of the circle. Hence, the arc length equals $1/6 \times 4\pi = 2\pi/3$. The answer is (B).

78. Which one of the following relations is true regarding the angles of the quadrilateral shown in the figure?

(A) $\angle A = \angle C$
(B) $\angle B > \angle D$
(C) $\angle A < \angle C$
(D) $\angle B = \angle D$
(E) $\angle A = \angle B$

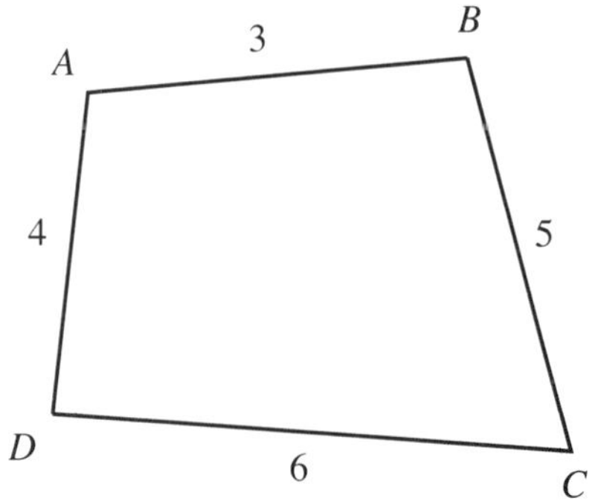

The figure is not drawn to scale.

Joining the opposite vertices B and D on the quadrilateral yields the following figure:

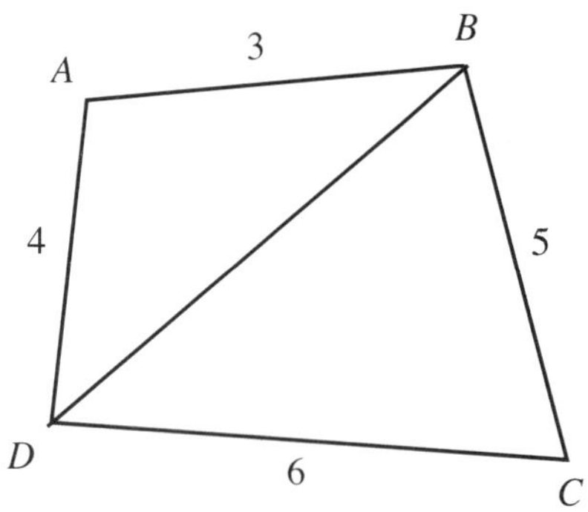

Since the angle opposite the longer side in a triangle is greater, we have

AD (= 4) > AB(= 3) (from the figure). Hence, $\angle ABD > \angle BDA$ and

CD (= 6) > BC(= 5) (from the figure). Hence, $\angle DBC > \angle CDB$.

Adding the two known inequalities $\angle ABD > \angle BDA$ and $\angle DBC > \angle CDB$ yields

$\angle ABD + \angle DBC > \angle BDA + \angle CDB$

$\angle B > \angle D$ Since from the figure, $\angle ABD + \angle DBC$ equals $\angle ABC$ (= $\angle B$) and $\angle BDA + \angle CDB$ equals $\angle CDA$ (= $\angle D$)

Hence, the answer is (B).

79. In a triangle with sides of lengths 3, 4, and 5, the smallest angle is 36.87°. In the figure, *O* is the center of the circle of radius 5. *A* and *B* are two points on the circle, and the distance between the points is 6. What is the value of *x* ?

(A) 36.87
(B) 45
(C) 53.13
(D) 116.86
(E) 126.86

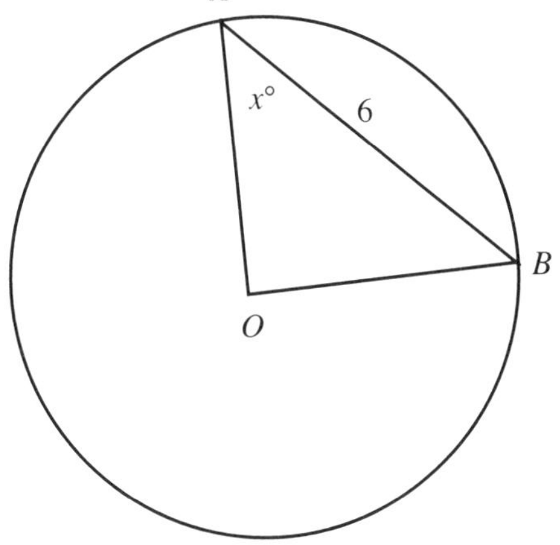

We are given that the smallest angle of any triangle of side lengths 3, 4, and 5 is 36.87°. The smallest angle is the angle opposite the smallest side, which measures 3.

Now, in ΔAOB, $OA = OB$ = radius of the circle = 5 (given). Hence, the angles opposite the two sides in the triangle are equal and therefore the triangle is isosceles. Just as in any isosceles triangle, the altitude on the third side *AB* must divide the side equally. Say, the altitude cuts *AB* at *J*. Then we have $AJ = JB$, and both equal $AB/2 = 3$.

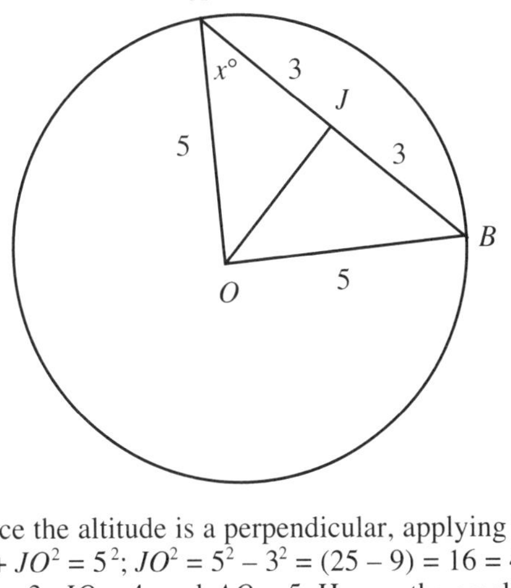

Since the altitude is a perpendicular, applying The Pythagorean Theorem to ΔAOJ yields $AJ^2 + JO^2 = OA^2$; $3^2 + JO^2 = 5^2$; $JO^2 = 5^2 - 3^2 = (25 - 9) = 16 = 4^2$. Square rooting both sides yields $JO = 4$. Hence, in ΔAOJ, $AJ = 3$, $JO = 4$, and $AO = 5$. Hence, the smallest angle, the angle opposite the smallest side, $\angle AOJ$ equals 36.87°. Summing the angles of ΔAOJ to 180° yields $x + \angle AJO + \angle JOA = 180$; $x + 90 + 36.87 = 180$. Solving this equation yields $x = 180 - (90 + 36.87) = 180 - 126.87 = 53.13$. The answer is (C).

80. Which of following indicates that ΔABC is a right triangle?

    (I) The angles of ΔABC are in the ratio 1 : 2 : 3.
    (II) One of the angles of ΔABC equals the sum of the other two angles.
    (III) ΔABC is similar to the right triangle ΔDEF.

    (A) I only
    (B) II only
    (C) III only
    (D) I and II only
    (E) I, II, and III

From I, we have that the ratio of the three angles of the triangle is 1 : 2 : 3. Let $k°$, $2k°$, and $3k°$ be the three angles. Summing the three angles to 180° yields $k + 2k + 3k = 180$; $6k = 180$; $k = 180/6 = 30$. Now that we have the value of k, we can calculate the three angles and determine whether the triangle is a right triangle or not. Hence, I determines whether ΔABC is a right triangle.

From II, we have that one angle of the triangle equals the sum of the other two angles. Let $\angle A = \angle B + \angle C$. Summing the angles of the triangle to 180° yields $\angle A + \angle B + \angle C = 180$; $\angle A + \angle A = 180$ (since $\angle B + \angle C = \angle A$); $2\angle A = 180$; $\angle A = 180/2 = 90$. Hence, the triangle is a right triangle. Hence, II determines whether ΔABC is a right triangle.

From III, we have that ΔABC is similar to the right triangle ΔDEF. Hence, ΔABC is a triangle of same type as ΔDEF. So, III determines whether ΔABC is a right triangle.

Hence, the answer is (E).

81. In the figure, O is the center of the circle of radius 3, and $ABCD$ is a square. If $PC = 3$ and the side BC of the square is a tangent to the circle, then what is the area of the square $ABCD$?

(A) 25
(B) 27
(C) 36
(D) 42
(E) 56

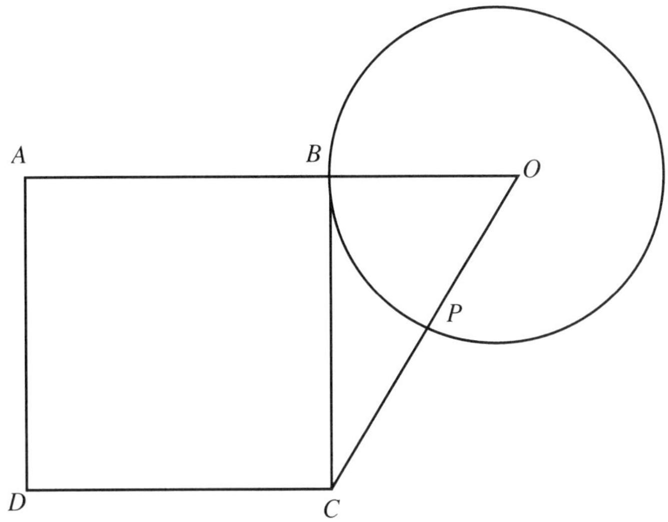

We are given that the radius of the circle is 3 units and BC, which is tangent to the circle, is a side of the square $ABCD$.

Since we have that $PC = 3$ (given), OC equals

$$[OP \text{ (radius of circle)}] + [PC \text{ (= 3 units, given)}] =$$

$$\text{radius} + 3 =$$

$$3 + 3 =$$

$$6$$

Now, since BC is tangent to the circle (given), $\angle OBC = 90°$, a right triangle. Hence, applying The Pythagorean Theorem to the triangle yields

$$OC^2 = BC^2 + OB^2$$

$$6^2 = BC^2 + 3^2$$

$$BC^2 = 6^2 - 3^2 = 36 - 9 = 27$$

Since the area of a square is $(\textit{side length})^2$, the area of square $ABCD$ is $BC^2 = 27$. The answer is (B).

82. In the figure shown, *ABCDEF* is a regular hexagon and *AOF* is an equilateral triangle. The perimeter of ΔAOF is $2a$ feet. What is the perimeter of the hexagon in feet?

(A) $2a$
(B) $3a$
(C) $4a$
(D) $6a$
(E) $12a$

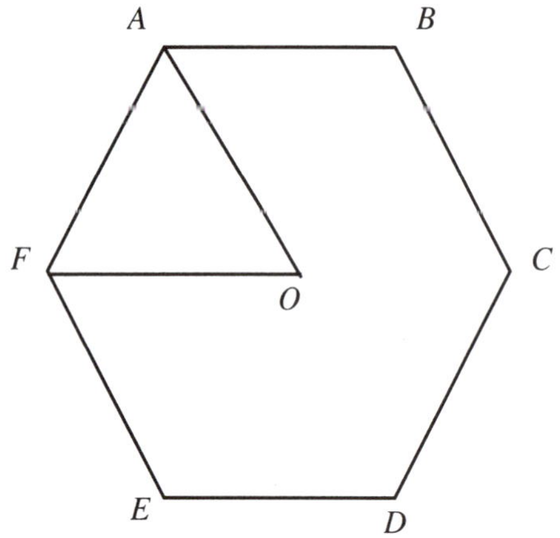

We are given that *AOF* is an equilateral triangle. In an equilateral triangle, all three sides are equal and therefore the perimeter of the triangle equals (*number of sides*) × (*side length*) = $3AF$ (where *AF* is one side of the equilateral triangle). Now, we are given that the perimeter of ΔAOF is $2a$. Hence, $3AF = 2a$, or $AF = 2a/3$.

We are given that *ABCDEF* is a regular hexagon. In a regular hexagon, all six sides are equal and therefore the perimeter of the hexagon equals (*number of sides*) × (*side length*) = $6AF$ (where *AF* is also one side of the hexagon). Substituting $AF = 2a/3$ into this formula yields

$$6AF = 6(2a/3) = 4a$$

The answer is (C).

83. The area of the base of a tank is 100 sq. ft. It takes 20 seconds to fill the tank with water poured at rate of 25 cubic feet per second. What is the height in feet of the rectangular tank?

(A) 0.25
(B) 0.5
(C) 1
(D) 5
(E) 25

The formula for the volume of a rectangular tank is (*area of base*) × *height*. Hence, we have

$$\textit{volume} = (\textit{area of base}) \times \textit{height}$$
or
$$\textit{height} = \textit{volume}/(\textit{area of base})$$

We are given that it takes 20 seconds to fill the tank at the rate of 25 cubic feet per second. Hence, the volume of the tank = rate × time = 25 cubic feet × 20 seconds = 500 cubic feet. Using this in the equation for height yields *height* = *volume*/(*area of base*) = 500 cu. ft/100 sq. ft = 5 feet. The answer is (D).

84. Point A is 10 miles West of Point B. Point B is 30 miles North of Point C. Point C is 20 miles East of Point D. What is the distance between points A and D?

(A) $10\sqrt{10}$ miles
(B) $10\sqrt{20}$ miles
(C) $20\sqrt{10}$ miles
(D) $30\sqrt{10}$ miles
(E) $30\sqrt{20}$ miles

Plotting the given points A, B, C, and D on a map according to given information yields the following figure:

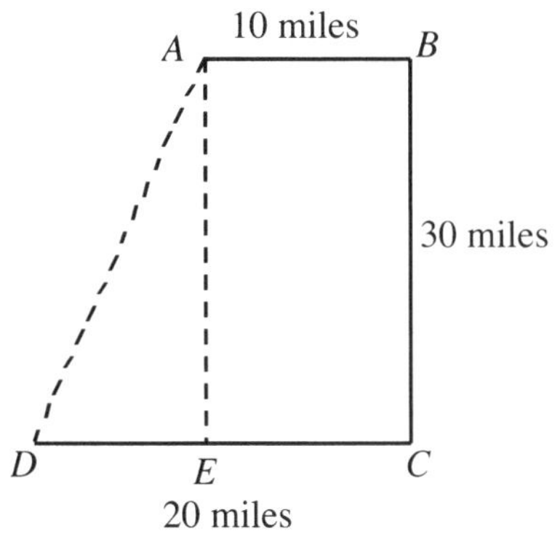

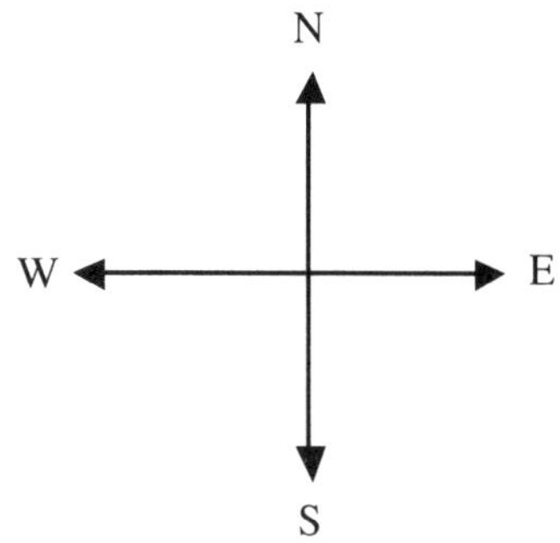

Let's drop a perpendicular from A to DC and name it AE. Also, let's join points A and D. Now, since AE is perpendicular to DC, $ABCE$ is a rectangle. Hence, opposite sides AB and EC are equal; and opposite sides AE and BC are equal and since AB equals 10 miles, EC equals 10 miles. Similarly, opposite sides AE and BC both equal 30 miles. Now, $DE = DC - EC = 20 - 10 = 10$ miles (from the figure). Applying The Pythagorean Theorem to the triangle AED yields

$$\begin{aligned} AD^2 &= AE^2 + DE^2 \\ &= 30^2 + 10^2 \\ &= 900 + 100 \\ &= 1000 \end{aligned}$$

Hence, $AD = \sqrt{1000} = 10\sqrt{10}$. The answer is (A).

85. Water is poured into an empty cylindrical tank at a constant rate. In 10 minutes, the height of the water increased by 7 feet. The radius of the tank is 10 feet. What is the rate at which the water is poured?

(A) 11/8 π cubic feet per minute.
(B) 11/3 π cubic feet per minute.
(C) 7/60 π cubic feet per minute.
(D) 11π cubic feet per minute.
(E) 70π cubic feet per minute.

The formula for the volume of a cylindrical tank is (*Area of the base*) × *height*. Since the base is circular (in a cylinder), the *Area of the base* = $\pi(radius)^2$. Also, The rate of filling the tank equals

The volume filled ÷ Time taken =

(*Area of the base* × *height*) ÷ *Time taken* =

[$\pi\ (radius)^2$ × height] ÷ *Time taken* =

$\pi(10\text{ feet})^2 \times 7\text{ feet} \div 10\text{ minutes} =$

70π cubic feet per minute

The answer is (E).

86. The length of a rectangle is increased by 25%. By what percentage should the width be decreased so that the area of the rectangle remains unchanged?

(A) 20
(B) 25
(C) 30
(D) 33.33
(E) 50

Let l and w be the initial length and width of the rectangle, respectively. Then by the formula *Area of Rectangle* = *Length* × *Width*, the area of the rectangle = lw.

When the length is increased by 25%, the new length is $l(1 + 25/100)$. Now, let x% be the percentage by which the width of the new rectangle is decreased so that the area is unchanged. Then the new width should equal $w(1 - x/100)$. The area now is $l(1 + 25/100) \times w(1 - x/100)$, which equals lw (area remained unchanged).

Thus, we have the equation

$l(1 + 25/100) \times w(1 - x/100) = lw$

$(1 + 25/100) \times (1 - x/100) = 1$ canceling l and w from both sides

$(125/100)(1 - x/100) = 1$

$(1 - x/100) = 100/125$

$-x/100 = 100/125 - 1$

$-x/100 = 100/125 - 125/125$

$-x/100 = -25/125$

$x = 100 \times 25/125 = 100 \times 1/5 = 20$

The answer is (A).

87. In the rectangular coordinate system shown, *ABCD* is a parallelogram. If the coordinates of the points *A*, *B*, *C*, and *D* are (0, 2), (*a*, *b*), (*a*, 2), and (0, 0), respectively, then $b =$

(A) 0
(B) 2
(C) 3
(D) 4
(E) 5

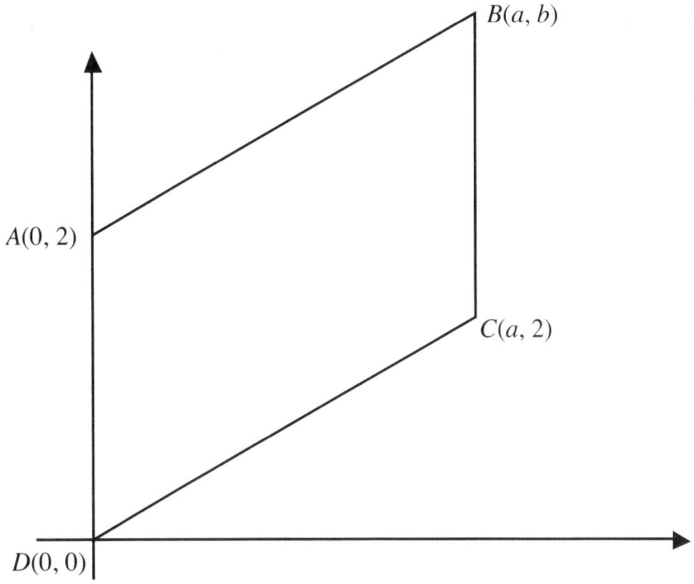

In the figure, points *A*(0, 2) and *D*(0, 0) have the same *x*-coordinate (which is 0). Hence, the two lines must be on the same vertical line in the coordinate system.

Similarly, the *x*-coordinates of points *B* and *C* are the same (both equal *a*). Hence, the points are on the same vertical line in the coordinate system.

Now, if *ABCD* is a parallelogram, then the opposite sides must be equal. Hence, *AD* must equal *BC*.

Since *AD* and *BC* are vertical lines, $AD = y$-coordinate difference of the points *A* and *D*, which equals $2 - 0 = 2$, and $BC =$ the *y*-coordinate difference of the points *B* and *C*, which equals $b - 2$. Equating *AD* and *BC* yields $b - 2 = 2$, or $b = 4$. The answer is (D).

88. In the figure, if $AB = 10$, what is the length of the side CD ?

(A) 5
(B) $5\sqrt{3}$
(C) $\frac{10}{\sqrt{3}}$
(D) 10
(E) $10\sqrt{3}$

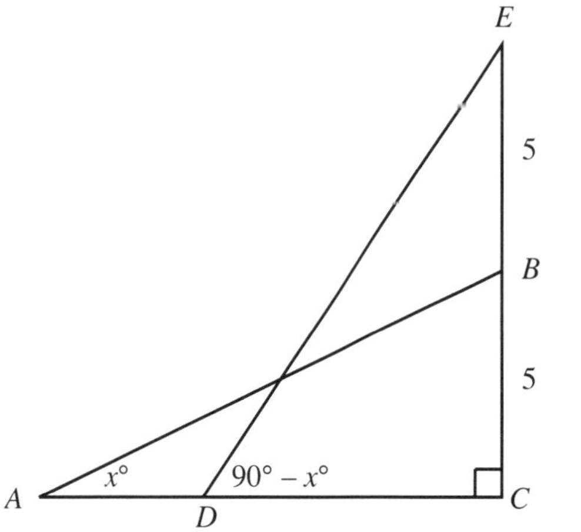

Applying The Pythagorean Theorem to the right triangle ABC yields

$$BC^2 + AC^2 = AB^2$$

$5^2 + AC^2 = 10^2$ given that $AB = 10$ and $BC = 5$ (from the figure)

$$AC^2 = 10^2 - 5^2 = 100 - 25 = 75$$

Square rooting yields $AC = \sqrt{75} = \sqrt{25 \cdot 3} = \sqrt{25} \times \sqrt{3} = 5\sqrt{3}$.

Hence, the sides opposite angles measuring $x°$ (A in ΔABC) and $90° - x°$ (B in ΔABC) are in the ratio $5 : 5\sqrt{3} = 1 : \sqrt{3}$.

Similarly, in ΔECD, the ratio of the sides opposite the angles E (measuring $x°$) and D (measuring $90° - x°$) must also be $1 : \sqrt{3}$.

Hence, we have

$$CD/EC = 1 : \sqrt{3}$$

$$CD/(5 + 5) = 1 : \sqrt{3}$$

$$CD = \frac{10}{\sqrt{3}}$$

The answer is (C).

89 In the figure, what is the area of ΔABC if $EC/CD = 3$?

(A) 12
(B) 24
(C) 81
(D) 121.5
(E) 143

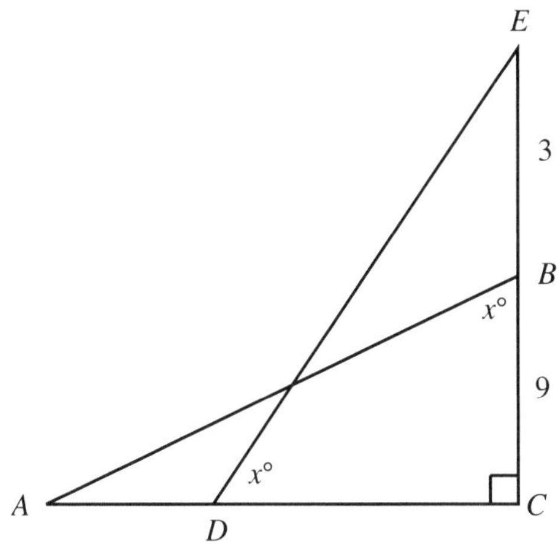

In right triangle EDC, $\angle C$ is right angled, $\angle D$ measures $x°$, and angle $\angle E$ measures $180° - (90° + x°) = 90° - x°$.

In right triangle ABC, $\angle C$ is right-angled, $\angle B$ measures $x°$, and angle $\angle A$ equals $180° - (90° + x°) = 90° - x°$.

Since corresponding angles in the two triangles are equal, both are similar triangles and the corresponding angle sides must be same. Hence, we have

$$AC/BC = EC/CD$$

$$AC/9 = 3$$

$$AC = 3 \cdot 9 = 27$$

Now, the area of $\Delta ABC = 1/2 \cdot base \cdot height = 1/2 \cdot AC \cdot BC = 1/2 \cdot 27 \cdot 9 = 121.5$.

The answer is (D).

90. *AB* and *CD* are chords of the circle, and *E* and *F* are the midpoints of the chords, respectively. The line *EF* passes through the center *O* of the circle. If $EF = 17$, then what is radius of the circle?

(A) 10
(B) 12
(C) 13
(D) 15
(E) 25

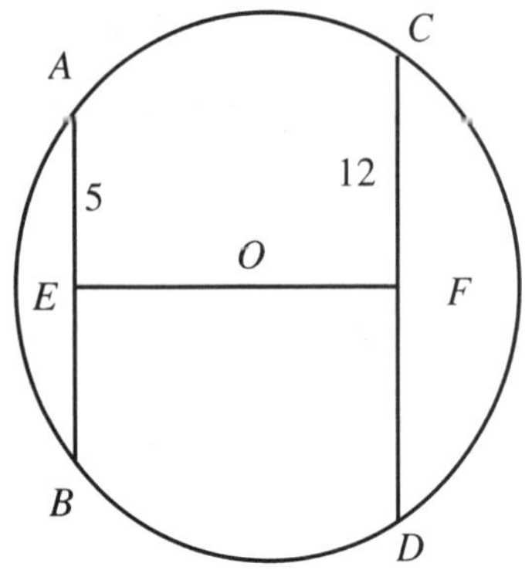

If a line joining the midpoints of two chords of a circle passes through the center of the circle, then it cuts both chords perpendicularly.

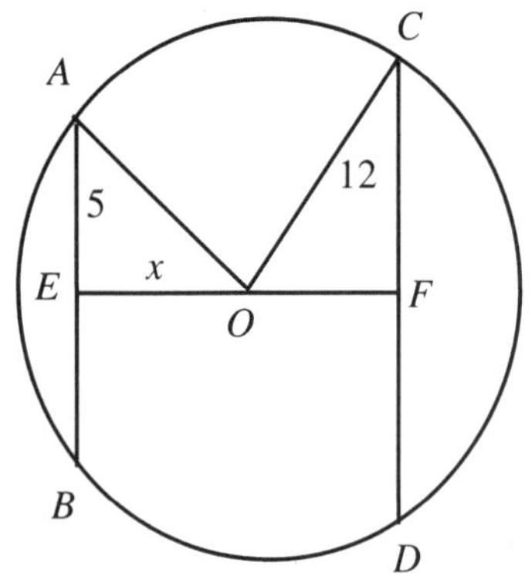

Let x be the length of the line segment from center of the circle to chord *AB* (at *E*). Since the length of *EF* is 17, $OF = 17 - x$.

Now, let r be the radius of the circle. Applying The Pythagorean Theorem to the right triangle *AEO* yields

$$AE^2 + EO^2 = AO^2$$
$$5^2 + x^2 = r^2 \qquad (1)$$

Also, applying The Pythagorean Theorem to ΔCFO yields

$$CF^2 + OF^2 = OC^2$$
$$12^2 + (17 - x)^2 = r^2 \qquad (2)$$

Equating the left-hand sides of equations (1) and (2), since their right-hand sides are the same, yields

$$x^2 + 5^2 = 12^2 + (17 - x)^2$$
$$x^2 + 5^2 = 12^2 + 17^2 + x^2 - 34x$$
$$34x = 12^2 + 17^2 - 5^2$$
$$34x = 144 + 289 - 25$$
$$34x = 408$$
$$x = 408/34$$
$$x = 12$$

Now, substituting this value of x in equation (1) yields

$$5^2 + 12^2 = r^2$$
$$25 + 144 = r^2$$
$$r = \sqrt{169} = 13$$

Hence, the radius is 13, and the answer is (C).

91. Column A — The perimeter of rectangle $ABCD$ is 5/2 times as long as the side AB. — Column B

Column A		Column B
Length of side AB	The perimeter of rectangle $ABCD$ is 5/2 times as long as the side AB.	Length of side BC

The ordering $ABCD$ indicates that AB and BC are adjacent sides of a rectangle with common vertex B (see figure below). Remember that the perimeter of a rectangle is equal to twice its length plus twice its width. Hence,

$$P = 2AB + 2BC$$

We are given that the perimeter is 5/2 times the length of side AB. Replacing the left-hand side of the equation with $\frac{5}{2}AB$ yields

$$\frac{5}{2}AB = 2AB + 2BC$$
$$5AB = 4AB + 4BC \quad \text{(by multiplying the equation by 2)}$$
$$AB = 4BC \quad \text{(by subtracting } 4AB \text{ from both sides)}$$

Thus, the length of side AB is four times as long as side BC. Hence, side AB is greater than side BC. The answer is (A).

92. A closed rectangular tank contains a certain amount of water. When the tank is placed on its 3 ft by 4 ft side, the height of the water in the tank is 5 ft. When the tank is placed on another side of dimensions 4 ft by 5 ft, what is the height, in feet, of the surface of the water above the ground?

(A) 2
(B) 3
(C) 4
(D) 5
(E) 6

When based on the 3 ft · 4 ft side, the height of water inside the rectangular tank is 5 ft. Hence, the volume of the water inside tank is *length* × *width* × *height* = 3 · 4 · 5 cu. ft.

When based on 4 ft · 5 ft side, let the height of water inside the rectangular tank be h ft. Then the volume of the water inside tank would be *length* × *width* × *height* = 4 · 5 · h cu. ft.

Equating the results for the volume of water, we have $3 \cdot 4 \cdot 5 = v = 4 \cdot 5 \cdot h$. Solving for h yields $h = (3 \cdot 4 \cdot 5)/(4 \cdot 5) = 3$ ft.

The answer is (B).

93. Column A | Column B

Column A	Column B
The clockwise angle made by the hour hand and the minute hand at 12:15pm	The counterclockwise angle made by the hour hand and the minute hand at 12:45pm

Draw sample pictures of the clock at 12:15pm and 12:45pm. The figures look like

From the figures, the clockwise angle between the hands at 12:15pm is less than 90°, and at 12:45pm the counterclockwise angle is more than 90°. Hence, Column B is greater. The answer is (B).

94. *AC*, a diagonal of the rectangle *ABCD*, measures 5 units. The area of the rectangle is 12 sq. units. What is the perimeter of the rectangle?

(A) 7
(B) 14
(C) 17
(D) 20
(E) 28

If l and w are the length and width of the rectangle, respectively, then we have

The perimeter = $2(l + w)$.
The length of a diagonal = $\sqrt{l^2 + w^2} = 5$ (given).
The area of the rectangle = $lw = 12$ (given).

Squaring both sides of the equation $\sqrt{l^2 + w^2} = 5$ yields $l^2 + w^2 = 5^2 = 25$.

Multiplying both sides of the equation $lw = 12$ by 2 yields $2lw = 24$.

Adding the equations $l^2 + w^2 = 25$ and $2lw = 24$ yields $l^2 + w^2 + 2lw = 49$.

Applying the Perfect Square Trinomial formula, $(a + b)^2 = a^2 + b^2 + 2ab$, to the left-hand side yields $(l + w)^2 = 7^2$. Square rooting yields $l + w = 7$ (positive since side lengths and their sum are positive). Hence, $2(l + w) = 2(7) = 14$. The answer is (B).

95. In the figure, ΔABC is inscribed in the circle. The triangle does not contain the center of the circle O. Which one of the following could be the value of x in degrees?

(A) 35
(B) 70
(C) 85
(D) 90
(E) 105

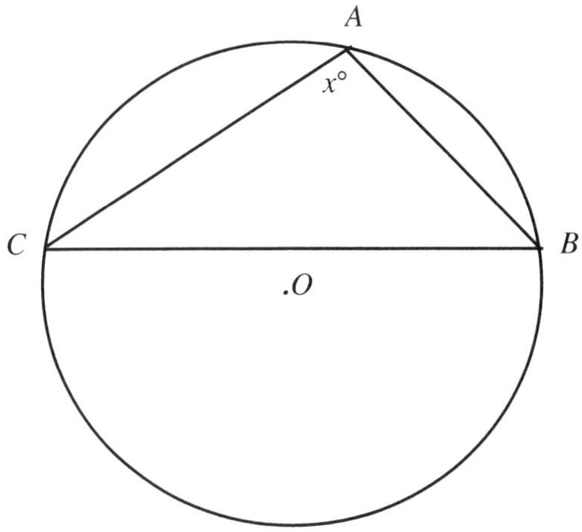

A chord makes an acute angle on the circle to the side containing the center of the circle and makes an obtuse angle to the other side. In the figure, BC is a chord and does not have a center to the side of point A. Hence, BC makes an obtuse angle at point A on the circle. Hence, $\angle A$, which equals $x°$, is obtuse and therefore is greater than 90°. Since 105 is the only obtuse angle offered, the answer is (E).

96. In the figure, $ABCD$ is a square, and BC is tangent to the circle with radius 3. P is the point of intersection of the line OC and the circle. If $PC = 2$, then what is the area of square $ABCD$?

(A) 9
(B) 13
(C) 16
(D) 18
(E) 25

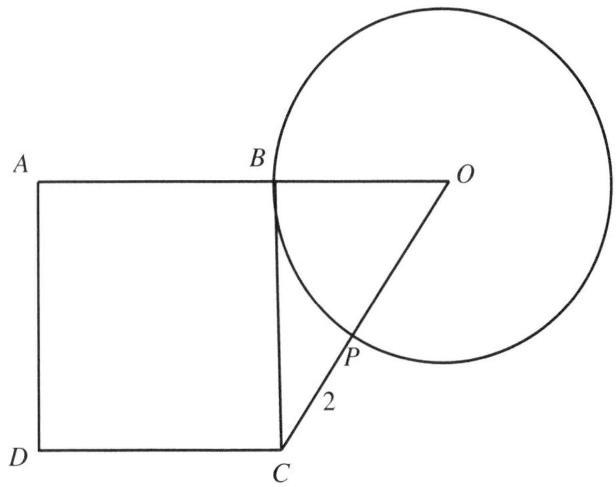

In the figure, since OB and OP are radii of the circle, both equal 3 units. Also, the length of line segment OC is $OP + PC = 3 + 2 = 5$. Now, since BC is tangent to the circle, $\angle OBC = 90$. Hence, triangle OBC is a right triangle. Applying The Pythagorean Theorem yields

$BC^2 + BO^2 = OC^2$

$BC^2 + 3^2 = 5^2$

$BC^2 = 5^2 - 3^2$

$BC^2 = 25 - 9$

$BC^2 = 16$

$BC = 4$

By the formula for the area of a square, the area of square *ABCD* is $side^2 = BC^2 = 4^2 = 16$. The answer is (C).

97. In the figure, *ABCD* is a square and *BCP* is an equilateral triangle. What is the measure of x ?

(A) 7.5
(B) 15
(C) 30
(D) 45
(E) 60

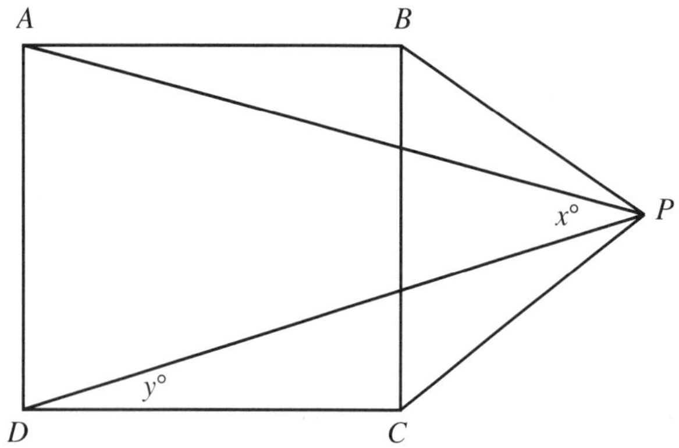

Through point *P*, draw a line *QP* parallel to the side *AB* of the square. Since *QP* cuts the figure symmetrically, it bisects $\angle BPC$. Hence, $\angle QPC$ = (Angle in equilateral triangle)/2 = 60°/2 = 30°. Now, $\angle QPD = \angle CDP$ (alternate interior angles are equal) = $y°$ (given). Since sides in a square are equal, $BC = CD$; and since sides in an equilateral triangle are equal, $PC = BC$. Hence, we have $CD = PC$ (= BC). Since angles opposite equal sides in a triangle are equal, in ΔCDP we have that $\angle DPC$ equals $\angle CDP$, which equals $y°$. Now, from the figure, $\angle QPD + \angle DPC = \angle QPC$ which equals 30° (we know from earlier work). Hence, $y° + y° = 30°$, or $y = 30/2 = 15$. Similarly, by symmetry across the line *QP*, $\angle APQ = \angle QPD = 15°$. Hence, $x = \angle APD = \angle APQ + \angle QPD = 15° + 15° = 30°$. The answer is (C).

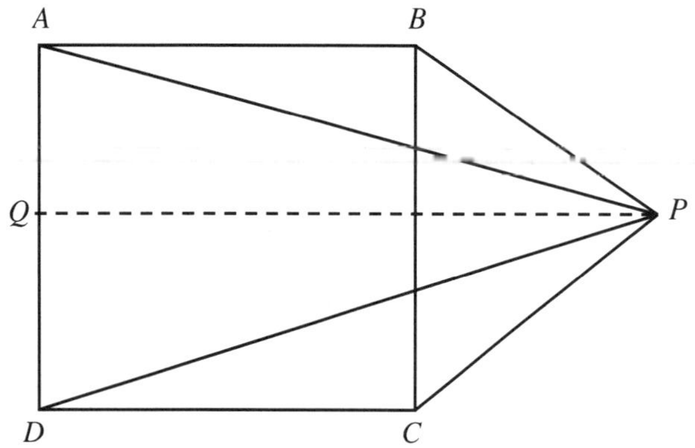

98. Column A | A circular park is enlarged uniformly such that it now occupies 21% more land. | Column B

The percentage increase in the radius of the park due to the enlargement | | The percentage increase in the area of the park due to the enlargement

Let the initial radius of the park be r, and let the radius after the enlargement be R. (Since the enlargement is uniform, the shape of the enlarged park is still circular.) By the formula for the area of a circle, the initial area of the park is πr^2, and the area after expansion is πR^2. Since the land occupied by the park is now 21% greater (given that the new area is 21% more), the new area is $(1 + 21/100)\pi r^2 = 1.21\pi r^2$. Equating this to πR^2 yields

$\pi R^2 = 1.21\pi r^2$

$R^2 = 1.21r^2$ by canceling π from both sides

$R = 1.1r$ by taking the square root of both sides

Now, Column A equals the percentage increase in the radius:

$$\frac{\text{Final radius} - \text{Initial radius}}{\text{Initial radius}} \cdot 100 =$$

$$\frac{1.1r - r}{r} \cdot 100 =$$

$$\frac{0.1r}{r} \cdot 100 =$$

$$\frac{1}{10} \cdot 100 =$$

10%

Column B equals the percentage increase in the area of the park due to the enlargement, which is same as the percentage increase in the area of the land, 21%. Hence, Column B is greater than Column A, and the answer is (B).

99. In the figure, $ABCD$ and $ABEC$ are parallelograms. The area of the quadrilateral $ABED$ is 6. What is the area of the parallelogram $ABCD$?

(A) 2
(B) 4
(C) 4.5
(D) 5
(E) 6

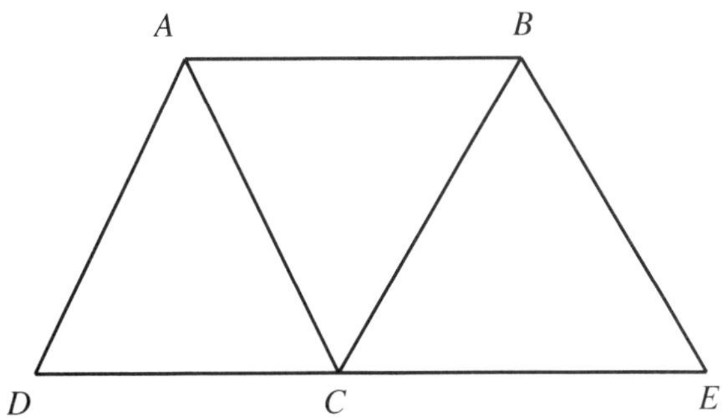

We know that a diagonal of a parallelogram divides the parallelogram into two triangles of equal area. Since AC is a diagonal of parallelogram $ABCD$, the area of ΔACD = the area of ΔABC; and since BC is a diagonal of parallelogram $ABEC$, the area of ΔCBE = the area of ΔABC. Hence, the areas of triangles ACD, ABC, and CBE, which form the total quadrilateral $ABED$, are equal. Since $ABCD$ forms only two triangles, ACD and ABC, of the three triangles, the area of the parallelogram equals two thirds of the area of the quadrilateral $ABED$. This equals $2/3 \times 6 = 4$. Hence, the answer is (B).

100. In the figure, lines l_1, l_2, and l_3 are parallel to one another. Line-segments AC and DF cut the three lines. If $AB = 3$, $BC = 4$, and $DE = 5$, then which one of the following equals DF ?

(A) 3/30
(B) 15/7
(C) 20/3
(D) 6
(E) 35/3

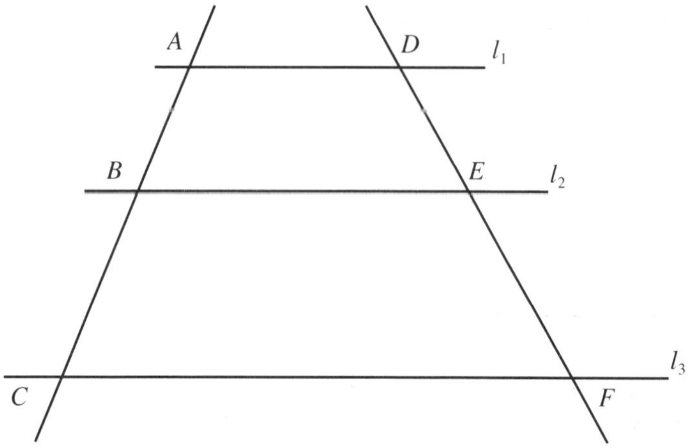

In the figure, AC and DF are transversals cutting the parallel lines l_1, l_2, and l_3. Let's move the line-segment DF horizontally until point D touches point A. The new figure looks like this:

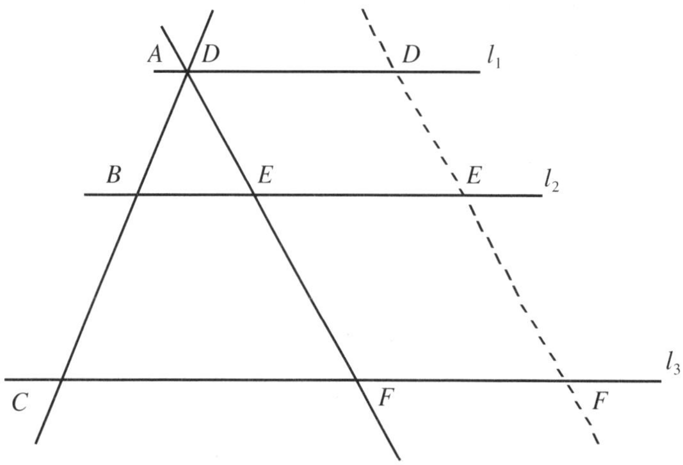

Now, in triangles ABE and ACF, $\angle B$ equals $\angle C$ and $\angle E$ equals $\angle F$ because corresponding angles of parallel lines (here l_2 and l_3) are equal. Also, $\angle A$ is a common angle of the two triangles. Hence, the three angles of triangle ABE equal the three corresponding angles of the triangle ACF. Hence, the two triangles are similar. Since the ratios of the corresponding sides of two similar triangles are equal, we have

$$\frac{AB}{AC} = \frac{AE}{AF}$$

$$\frac{AB}{AB + BC} = \frac{DE}{DF}$$

From the figure, $AC = AB + BC$. Also, from the new figure, point A is the same as point D. Hence, AE is the same as DE and AF is the same as DF.

$$\frac{3}{3+4} = \frac{5}{DF}$$

Substituting the given values

$$DF = \frac{35}{3}$$

By multiplying both sides by $7/3 \times DF$

The answer is (E).

101. In the figure, AB is parallel to CD. What is the value of x ?

(A) 36
(B) 45
(C) 60
(D) 75
(E) Cannot be determined

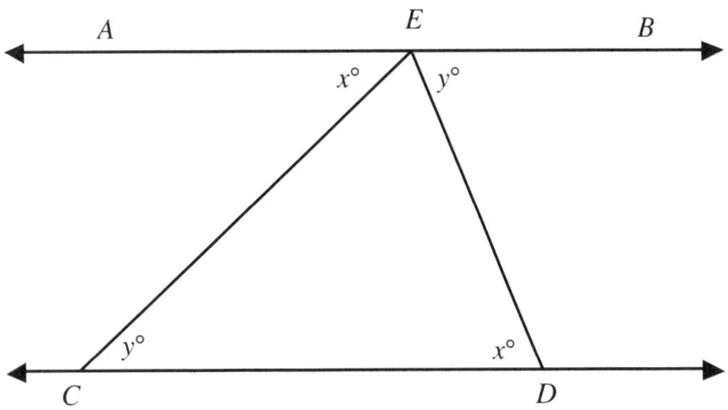

Lines AB and CD are parallel (given) and cut by transversal ED. Hence, the alternate interior angles x and y are equal. Since $x = y$, ΔECD is isosceles ($\angle C = \angle D$). Hence, angles x and y in ΔECD could each range between 0° and 90°. No unique value for x is derivable. Hence, the answer is (E).

102.	Column A	In the figure, P and Q are centers of the two circles of radii 3 and 4, respectively. A and B are the points at which a common tangent touches each circle.	Column B
	AB		PQ

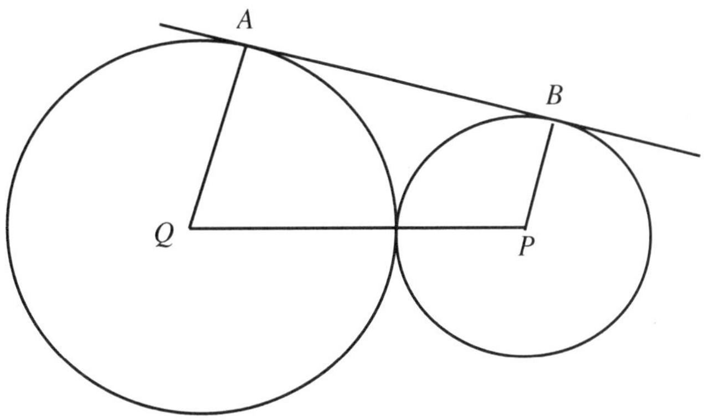

Since AB is tangent to both circles, $\angle BAQ = 90°$ and $\angle ABP = 90°$. Hence, AQ is parallel to BP. Now, draw a line through point P and parallel to AB as shown in the figure.

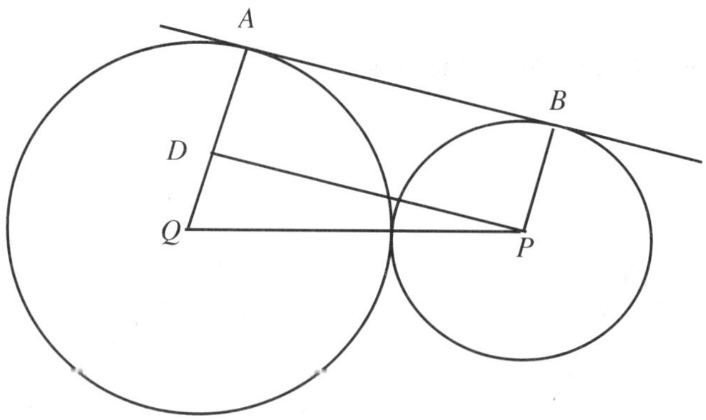

Then $ABPD$ must be a rectangle. Hence, $AB = DP$. Also, $\angle PDQ$ equals the corresponding angle, $\angle BAD$. So, both equal 90°. Since a right angle is the greatest angle in a triangle, the side opposite the angle is the longest. PQ is the side opposite the right angle $\angle PDQ$ in ΔPQD. Hence, PQ is greater than the other side DP. Hence, PQ is also greater than AB, which equals DP (we know). Hence, Column B is greater than Column A, and the answer is (B).

103. Column A | In the figure, ΔABC is inscribed in the circle. The triangle does not contain the center of the circle O. | Column B

x | | 90

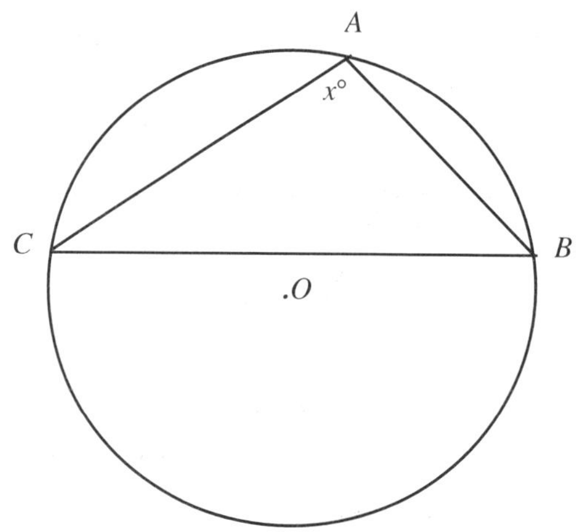

A chord makes an acute angle on the circle to the side containing the center of the circle and makes an obtuse angle to the other side. In the figure, BC is a chord and does not have a center to the side of point A. Hence, BC makes an obtuse angle at point A on the circle. Hence, $\angle A$, which equals $x°$, is obtuse and therefore is greater than 90°. Hence, Column A is greater than Column B, and the answer is (A).

104. Column A | | Column B

Measure of the largest angle (in degrees) of a triangle with sides of length 5, 6, and 7 | | 60

Since no two sides of the triangle in Column A are equal, no two angles of the triangle are equal. Since the sum of the three angles of the triangle is 180 degrees, at least one angle must be greater than 60 degrees. Otherwise, the angles would not sum to 180 degrees. Hence, Column A is greater than 60 (= Column B). The answer is (A).

➢ Very Hard

105. In the figure, if $y = 60$, then what is the value of z ?

(A) 20
(B) 30
(C) 55
(D) 75
(E) 90

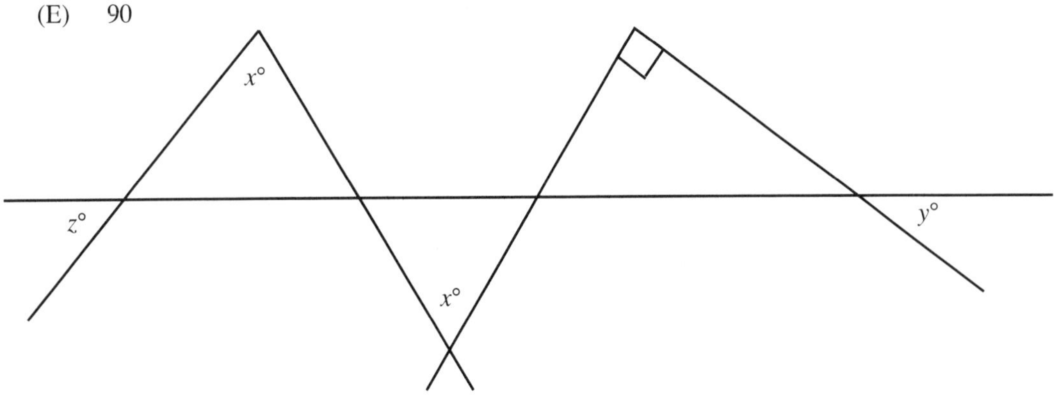

Let's name the vertices in the figure as shown below.

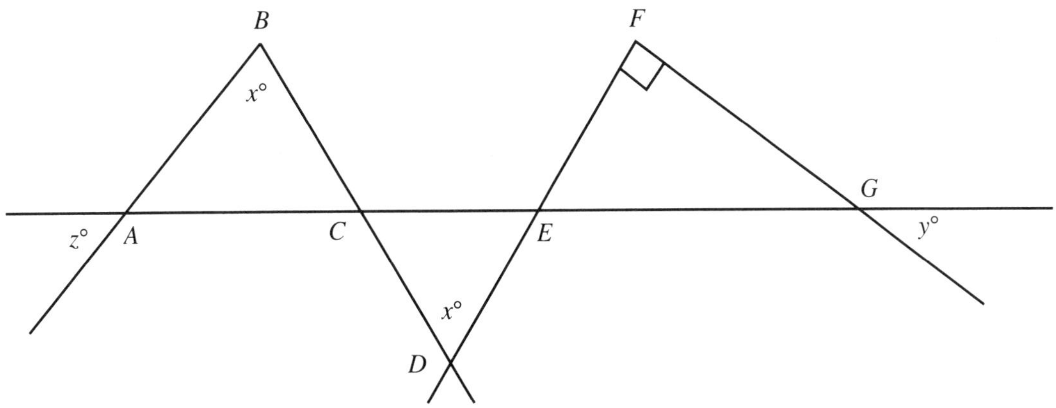

In the figure, $\angle FGE$ equals $y°$ (vertical angles). Summing the angles of ΔGEF to 180° yields

$$\angle GEF + \angle EFG + \angle FGE = 180$$
$$\angle GEF + 90 + y = 180$$
$$\angle GEF = 180 - (90 + y)$$
$$= 90 - y = 90 - 60 \quad \text{(since } y = 60\text{, given)}$$
$$= 30°$$

Now, since $\angle ABD = \angle BDF$ (both equal x, from the figure), lines AB and DF are parallel (alternate interior angles). Hence, $\angle CAB = \angle GEF$ (corresponding angles) = 30° (we know $\angle GEF$ = 30°). Now, $z = \angle CAB$ (vertical angles), and therefore $z = 30$. The answer is (B).

106. In the figure, *ABCD* is a rectangle and *AF* is parallel to *BE*. If $x = 5$, and $y = 10$, then what is the area of ΔAFD ?

(A) 2.5
(B) 5
(C) 12.5
(D) 50
(E) $50 + 5y$

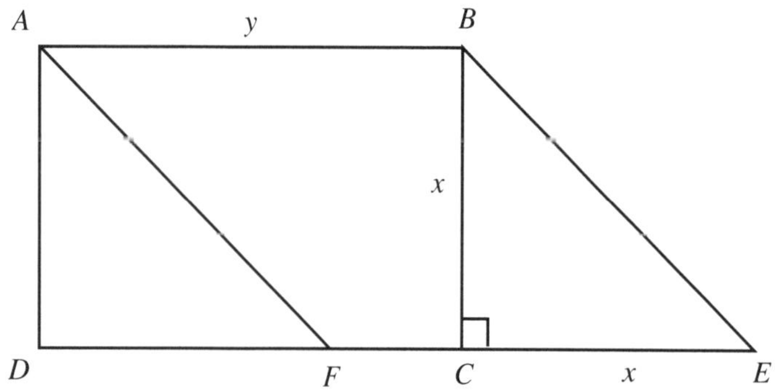

Since opposite sides of a rectangle are equal and parallel, *AD* is parallel to *BC* and $AD = BC$.

We are also given that *AF* is parallel to *BE*.

Further, points *D*, *C*, and *E* must be on same line because the angle made by points *D* and *E* at *C* is 180° ($\angle DCE = \angle DCB + \angle BCE$ = angle in a rectangle + a right angle [given] = 90° + 90° = 180°). So, *DC* and *CE* can be considered parallel.

Any three pairs of parallel lines (here *AF* and *BE*, *DF* and *CE*, and *AD* and *BC*) make two similar triangles (*AFD* and *BEC*). And if one pair of corresponding sides of two similar triangles are equal (here, $AD = BC$), then the triangles are congruent (equal).

The areas of congruent triangles are equal. So, area of ΔAFD = area of $\Delta BEC = 1/2 \cdot base \cdot height = 1/2 \cdot BC \cdot CE = 1/2 \cdot 5 \cdot 5 = 1/2 \cdot 25 = 12.5$. The answer is (C).

When Drawing a Geometric Figure or Checking a Given One, Be Sure to Include Drawings of Extreme Cases As Well As Ordinary Ones.

Example 1: In the figure to the right, what is the value of angle x?

(A) $x > 45°$
(B) $x < 45°$
(C) $x = 45°$
(D) $x \geq 45°$
(E) It cannot be determined from the information given

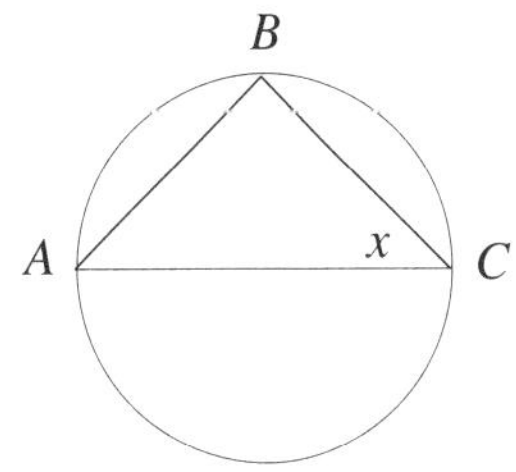

AC is a chord.
B is a point on the circle.

Although in the drawing AC looks to be a diameter, that cannot be assumed. All we know is that AC is a chord. Hence, numerous cases are possible, three of which are illustrated below:

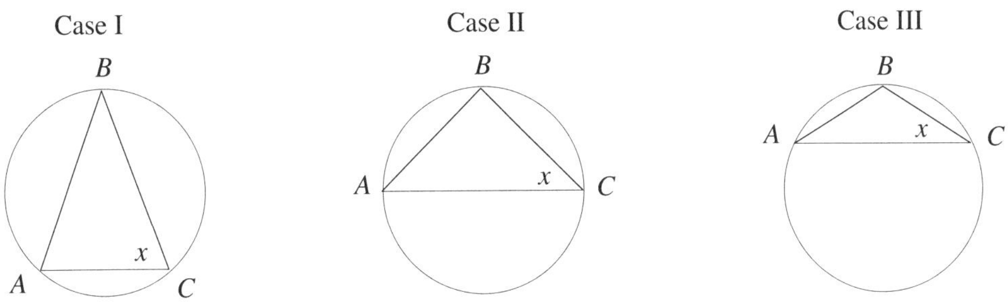

In Case I, x is greater than 45°; in Case II, x equals 45°; in Case III, x is less than 45°. Hence, the answer is (E).

Example 2: Three rays emanate from a common point and form three angles with measures p, q, r. Which one of the following is the measure of angle $q + r$?

(A) $q + r > 180°$
(B) $q + r < 180°$
(C) $q + r = 180°$
(D) $q + r \leq 180°$
(E) It cannot be determined from the information given

It is natural to make the drawing symmetric as follows:

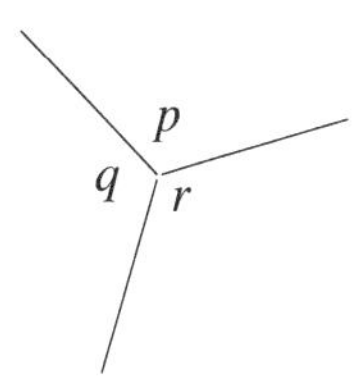

In this case, $p = q = r = 120°$, so $q + r = 240°$. However, there are other drawings possible. For example:

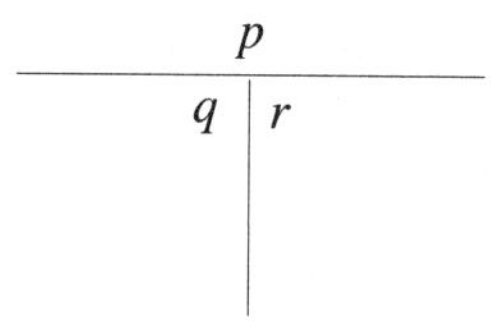

In this case, $q + r = 180°$ and therefore it cannot be determined from the information given. The answer is (E).

Problem Set J:

1. In triangle ABC, $AB = 5$ and $AC = 3$. Which one of the following is the measure of the length of side BC ?

   (A) $BC < 7$
   (B) $BC = 7$
   (C) $BC > 7$
   (D) $BC \le 7$
   (E) It cannot be determined from the information given

2. In the figure to the right, what is the area of ΔABC ?

   (A) 6
   (B) 7
   (C) 8
   (D) 9
   (E) It cannot be determined from the information given

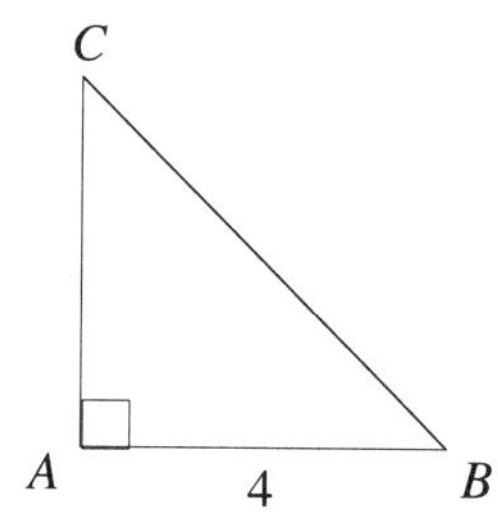

3. In the figure to the right, which one of the following is the measure of angle θ ?

   (A) $\theta < 45°$
   (B) $\theta > 45°$
   (C) $\theta = 45°$
   (D) $\theta \le 45°$
   (E) It cannot be determined from the information given

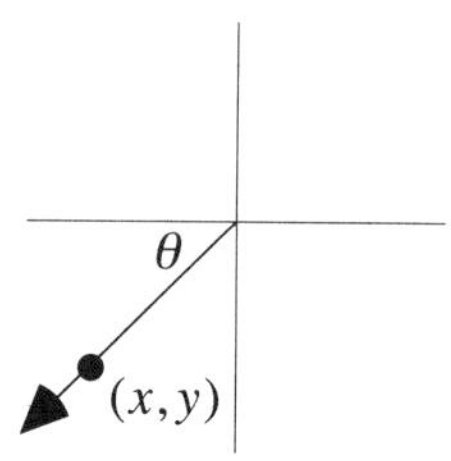

$x \ne y$

4. In isosceles triangle ABC, $CA = CB = 4$. Which one of the following is the area of triangle ABC ?

   (A) 7
   (B) 8
   (C) 9
   (D) 10
   (E) It cannot be determined from the information given

Answers and Solutions to Problem Set J

1. In triangle ABC, $AB = 5$ and $AC = 3$. Which one of the following is the measure of the length of side BC ?

   (A) $BC < 7$
   (B) $BC = 7$
   (C) $BC > 7$
   (D) $BC \leq 7$
   (E) It cannot be determined from the information given

The most natural drawing is the following:

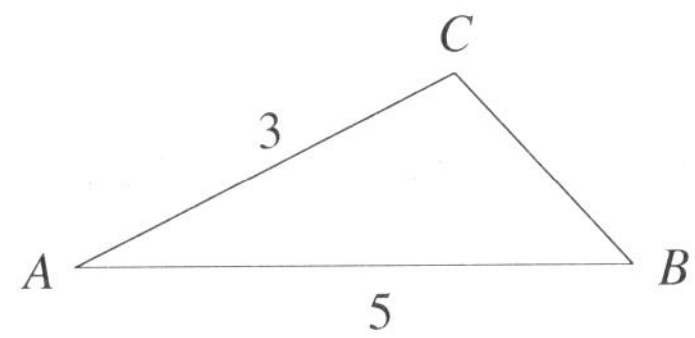

In this case, the length of side BC is less than 7. However, there is another drawing possible, as follows:

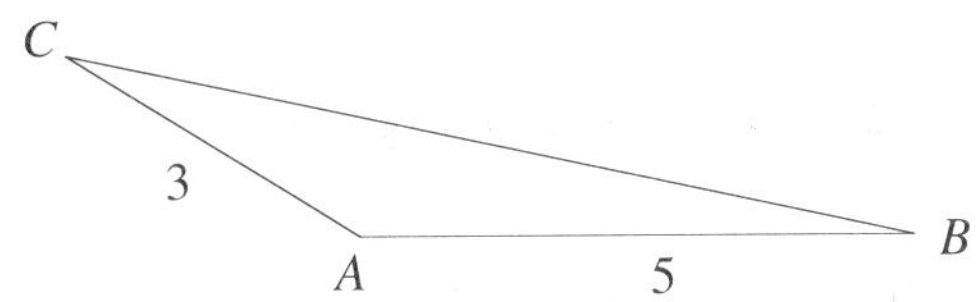

In this case, the length of side BC is greater than 7. Hence, there is not enough information to decide, and the answer is (E).

2. In the figure to the right, what is the area of ΔABC ?

   (A) 6
   (B) 7
   (C) 8
   (D) 9
   (E) It cannot be determined from the information given

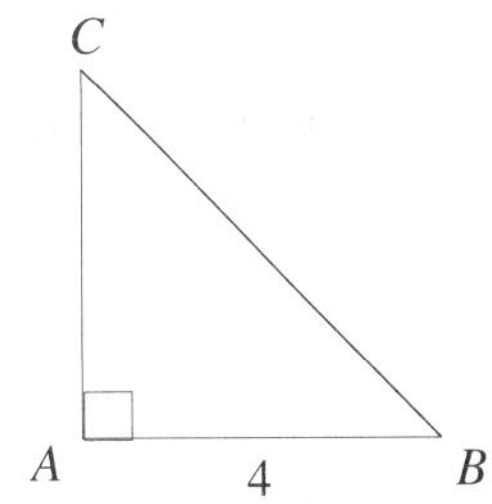

Although the drawing looks to be an isosceles triangle, that cannot be assumed. We are not given the length of side AC: it could be 4 units long or 100 units long, we don't know. Hence, the answer is (E).

3. In the figure to the right, which one of the following is the measure of angle θ ?

   (A) $\theta < 45^\circ$
   (B) $\theta > 45^\circ$
   (C) $\theta = 45^\circ$
   (D) $\theta \leq 45^\circ$
   (E) It cannot be determined from the information given

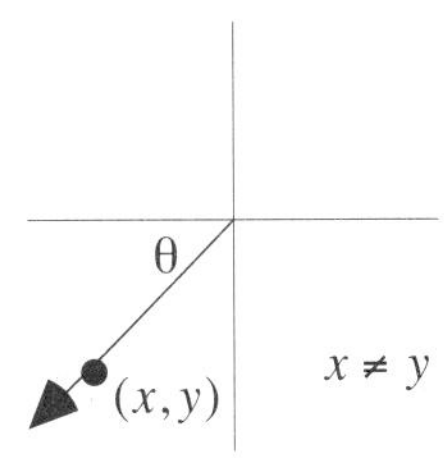

There are two possible drawings:

In Case I, $\theta < 45^\circ$. Whereas, in Case II, $\theta > 45^\circ$. This is a double case, and the answer therefore is (E).

In Case I, $\theta < 45°$. Whereas, in Case II, $\theta > 45°$. This is a double case, and the answer therefore is (E).

4. In isosceles triangle *ABC*, *CA* = *CB* = 4. Which one of the following is the area of triangle *ABC* ?

(A) 7
(B) 8
(C) 9
(D) 10
(E) It cannot be determined from the information given

There are many possible drawings for the triangle, two of which are listed below:

Case I

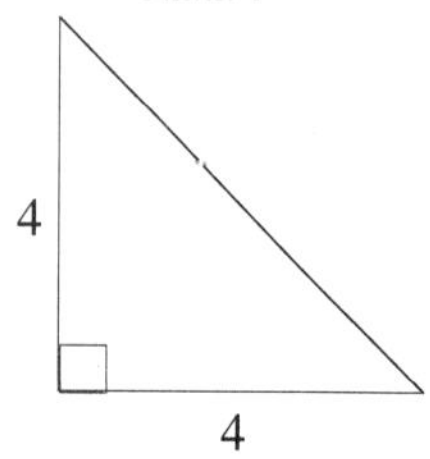

Case II

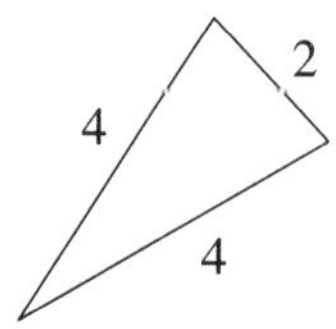

In Case I, the area is 8. In Case II, the area is $\sqrt{15}$ This is a double case and therefore the answer is (E).

Coordinate Geometry

On a number line, the numbers increase in size to the right and decrease to the left:

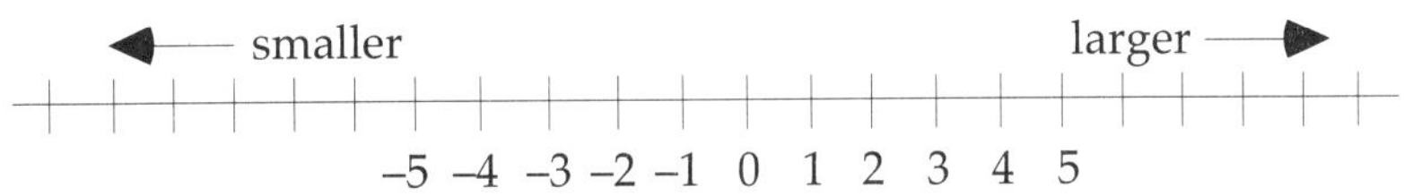

If we draw a line through the point 0 perpendicular to the number line, we will form a grid:

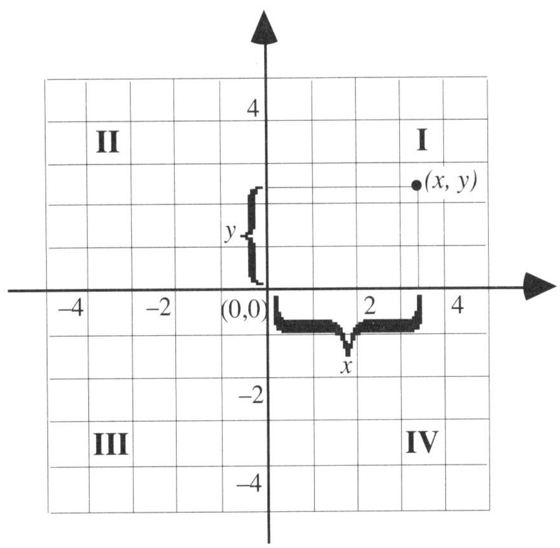

The thick horizontal line in the above diagram is called the x-axis, and the thick vertical line is called the y-axis. The point at which the axes meet, (0, 0), is called the origin. On the x-axis, positive numbers are to the right of the origin and increase in size to the right; further, negative numbers are to the left of the origin and decrease in size to the left. On the y-axis, positive numbers are above the origin and ascend in size; further, negative numbers are below the origin and descend in size. As shown in the diagram, the point represented by the ordered pair (x, y) is reached by moving x units along the x-axis from the origin and then moving y units vertically. In the ordered pair (x, y), x is called the *abscissa* and y is called the *ordinate*; collectively they are called coordinates. The x and y axes divide the plane into four quadrants, numbered I, II, III, and IV counterclockwise. Note, if $x \neq y$, then (x, y) and (y, x) represent different points on the coordinate system. The points (2, 3), (–3, 1), (–4, –4), and (4, –2) are plotted in the following coordinate system:

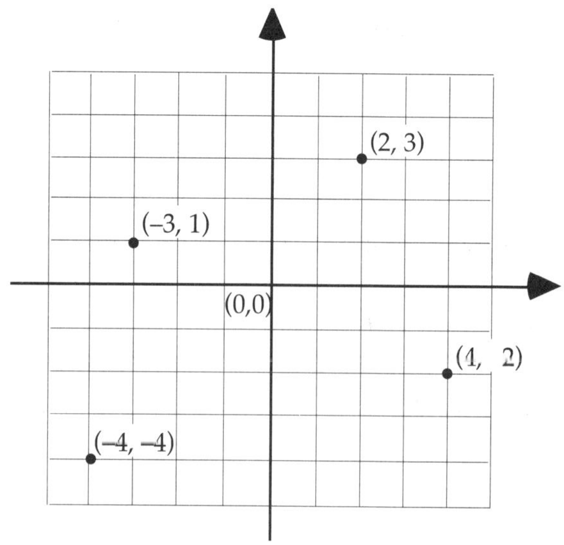

Example: If the point (a, b) is in Quadrant II and $|a| - |b| > 0$, then which one of the following is true?

(A) $a > 0$
(B) $b < 0$
(C) $a > b$
(D) $|a| < |b|$
(E) $a + b < 0$

We are given that the point (a, b) is in Quadrant II. In Quadrant II, the x-coordinate is negative and the y-coordinate is positive. Hence, $a < 0$ and $b > 0$. Reject choices (A) and (B), since they conflict with these inequalities.

Next, since a is negative and b is positive, $a < b$. Reject Choice (C) since it conflicts with this inequality.

Adding $|b|$ to both sides of the given equation $|a| - |b| > 0$ yields $|a| > |b|$. Reject Choice (D) since it conflicts with this inequality.

Since a is negative, $|a|$ equals $-a$; and since b is positive, $|b|$ equals b. So, $a + b = -|a| + |b| = -(|a| - |b|)$. Now, since $|a| - |b|$ is given to be positive, $-(|a| - |b|)$ must be negative. So, $a + b < 0$. The answer is (E).

Distance Formula:

The distance formula is derived by using the Pythagorean theorem. Notice in the figure below that the distance between the points (x, y) and (a, b) is the hypotenuse of a right triangle. The difference $y - b$ is the measure of the height of the triangle, and the difference $x - a$ is the length of base of the triangle. Applying the Pythagorean theorem yields

$$d^2 = (x - a)^2 + (y - b)^2$$

Taking the square root of both sides of this equation yields

$$d = \sqrt{(x - a)^2 + (y - b)^2}$$

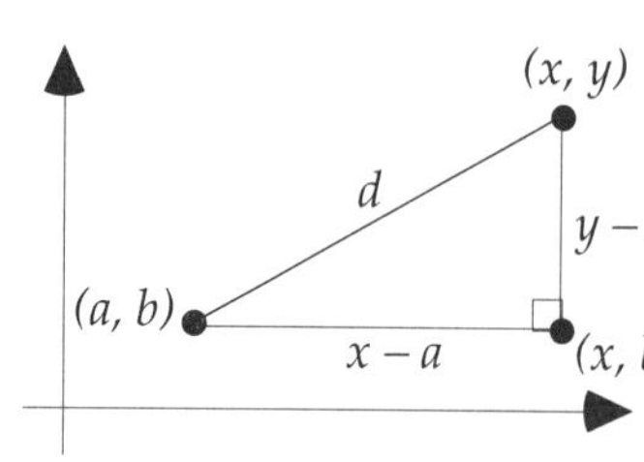

Example: If C is the midpoint of the points $A(-3, -4)$, and $B(-5, 6)$, then $AC =$

(A) 5
(B) $\sqrt{26}$
(C) $\sqrt{61}$
(D) 8
(E) 10

Using the distance formula to calculate the distance between A and B yields

$$AB = \sqrt{(-3-(-5))^2 + (-4-6)^2} = \sqrt{2^2 + (-10)^2} = \sqrt{4+100} = \sqrt{104} = 2\sqrt{26}$$

Since C is the midpoint of AB, $AC = AB/2 = \dfrac{2\sqrt{26}}{2} = \sqrt{26}$. The answer is (B).

Midpoint Formula:

The midpoint M between points (x, y) and (a, b) is given by

$$M = \left(\frac{x+a}{2}, \frac{y+b}{2}\right)$$

In other words, to find the midpoint, simply average the corresponding coordinates of the two points.

Example: If $(-3, -5)$ is the midpoint of the part of the line between the x and y axes, then what is the slope of the line?

(A) −5/3
(B) −3/5
(C) 3/5
(D) 5/3
(E) 10

We have that $(-3, -5)$ is the midpoint of the line between the x-intercept $(X, 0)$ and the y-intercept $(0, Y)$. The formula for the midpoint of two different points (x_1, y_1) and (x_2, y_2) is $((x_1 + x_2)/2, (y_1 + y_2)/2)$. Hence, the midpoint of $(X, 0)$ and $(0, Y)$ is $((X + 0)/2, (0 + Y)/2) = (X/2, Y/2)$. Equating this to the given midpoint yields $(X/2, Y/2) = (-3, -5)$. Equating corresponding coordinates yields $X/2 = -3$, or $X = -6$, and $Y/2 = -5$, or $Y = -10$. Hence, the slope of the line between $(X, 0)$, which equals $(-6, 0)$, and $(0, Y)$, which equals $(0, -10)$, is

$$\frac{y_1 - y_2}{x_1 - x_2} =$$
$$\frac{-10-0}{0-(-6)} =$$
$$\frac{-10}{6} =$$
$$-\frac{5}{3}$$

The answer is (A).

Slope Formula:

The slope of a line measures the inclination of the line. By definition, it is the ratio of the vertical change to the horizontal change (see figure below). The vertical change is called the *rise*, and the horizontal change is called the *run*. Thus, the slope is the *rise over the run*.

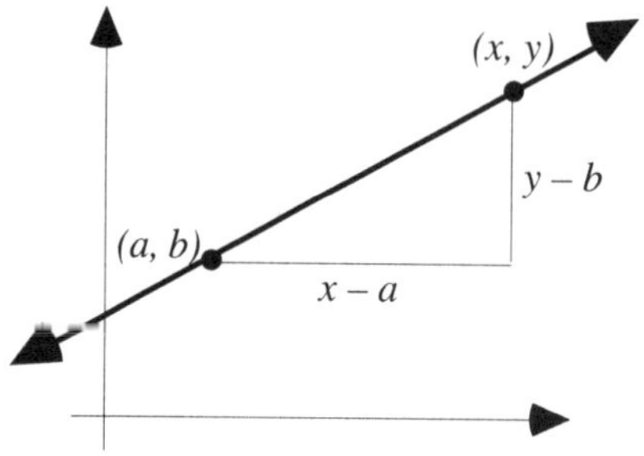

Forming the *rise over the run* in the above figure yields

$$m = \frac{y-b}{x-a}$$

Example: In the figure to the right, what is the slope of line passing through the two points?

(A) 1/4 (B) 1 (C) 1/2 (D) 3/2 (E) 2

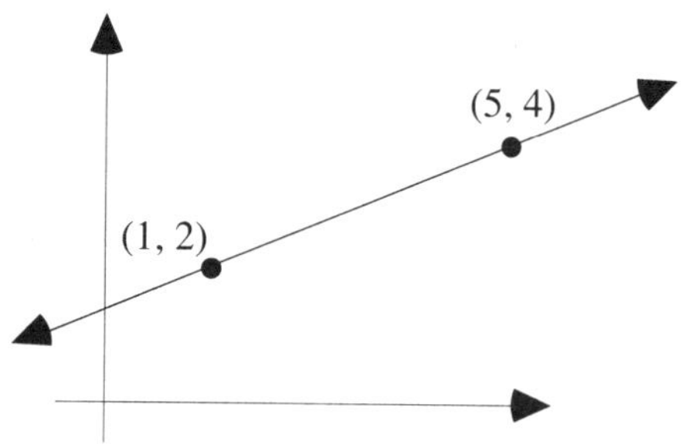

The slope formula yields $m = \frac{4-2}{5-1} = \frac{2}{4} = \frac{1}{2}$. The answer is (C).

Slope-Intercept Form:

Multiplying both sides of the equation $m = \frac{y-b}{x-a}$ by $x - a$ yields

$$y - b = m(x - a)$$

Now, if the line passes through the y-axis at $(0, b)$, then the equation becomes

$$y - b = m(x - 0)$$

or

$$y - b = mx$$

or

$$y = mx + b$$

This is called the slope-intercept form of the equation of a line, where m is the slope and b is the y-intercept. This form is convenient because it displays the two most important bits of information about a line: its slope and its y-intercept.

Example: In the figure to the right, the equation of the line is $y = \frac{9}{10}x + k$. Which one of the following must be true about line segments AO and BO?

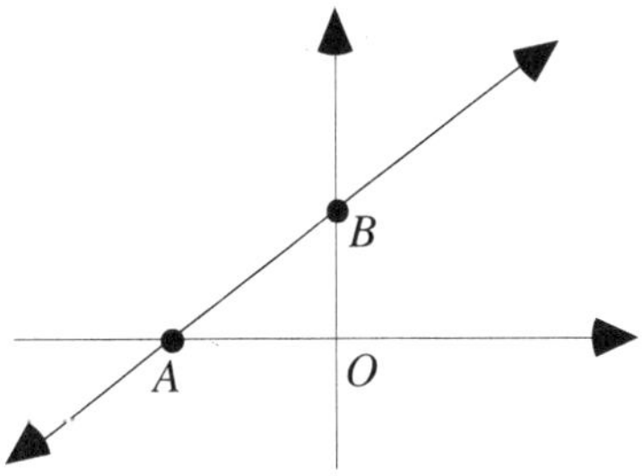

(A) $AO > BO$
(B) $AO < BO$
(C) $AO \leq BO$
(D) $AO = BO$
(E) $AO = BO/2$

Since $y = \frac{9}{10}x + k$ is in slope-intercept form, we know the slope of the line is $\frac{9}{10}$. Now, the ratio of BO to AO is the slope of the line (rise over run). Hence, $\frac{BO}{AO} = \frac{9}{10}$. Multiplying both sides of this equation by AO yields $BO = \frac{9}{10}AO$. In other words, BO is $\frac{9}{10}$ the length of AO. Hence, AO is longer. The answer is (A).

Intercepts:

The x-intercept is the point where the line crosses the x-axis. It is found by setting $y = 0$ and solving the resulting equation. The y-intercept is the point where the line crosses the y-axis. It is found by setting $x = 0$ and solving the resulting equation.

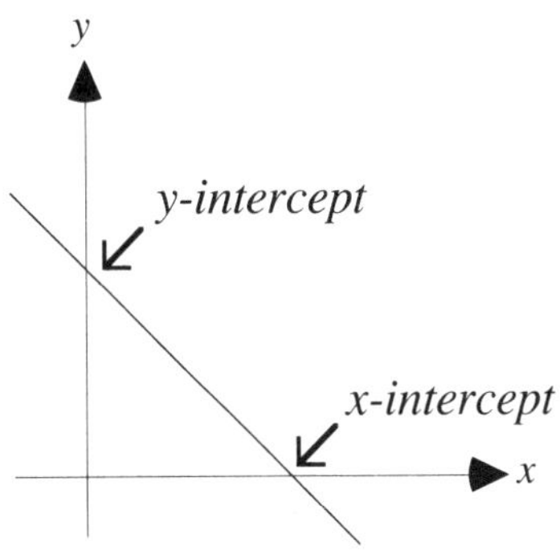

Example: Graph the equation $x - 2y = 4$.

Solution: To find the x-intercept, set $y = 0$. This yields $x - 2 \cdot 0 = 4$, or $x = 4$. So the x-intercept is (4, 0). To find the y-intercept, set $x = 0$. This yields $0 - 2y = 4$, or $y = -2$. So the y-intercept is (0, –2). Plotting these two points and connecting them with a straight line yields

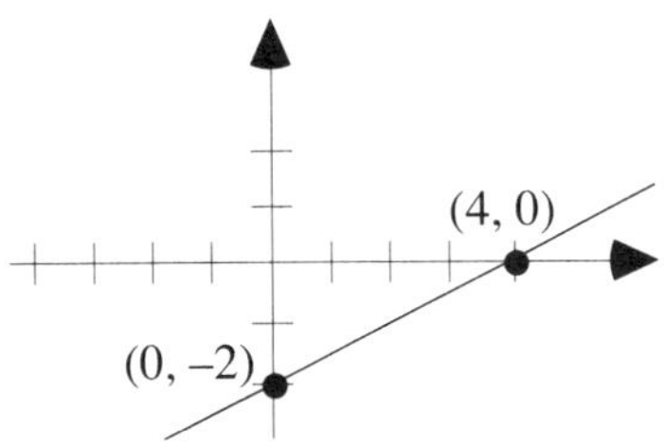

Areas and Perimeters:

Often, you will be given a geometric figure drawn on a coordinate system and will be asked to find its area or perimeter. In these problems, use the properties of the coordinate system to deduce the dimensions of the figure and then calculate the area or perimeter. For complicated figures, you may need to divide the figure into simpler forms, such as squares and triangles. A couple examples will illustrate:

Example: What is the area of the quadrilateral in the coordinate system to the right?

(A) 2
(B) 4
(C) 6
(D) 8
(E) 11

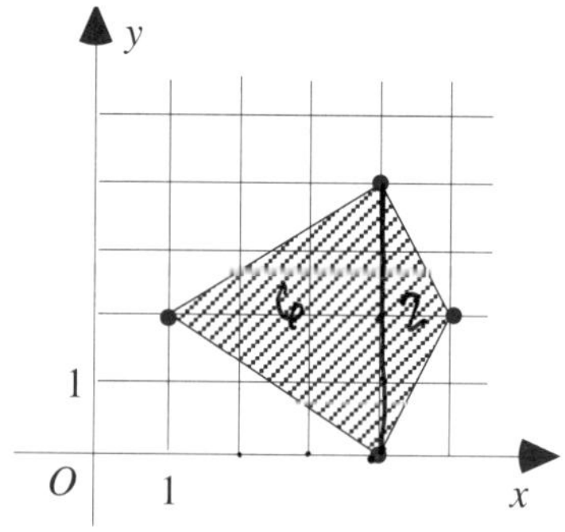

If the quadrilateral is divided horizontally through the line $y = 2$, two congruent triangles are formed. As the figure to the right shows, the top triangle has height 2 and base 4. Hence, its area is

$$A = \frac{1}{2}bh = \frac{1}{2}\cdot 4\cdot 2 = 4$$

The area of the bottom triangle is the same, so the area of the quadrilateral is 4 + 4 = 8. The answer is (D).

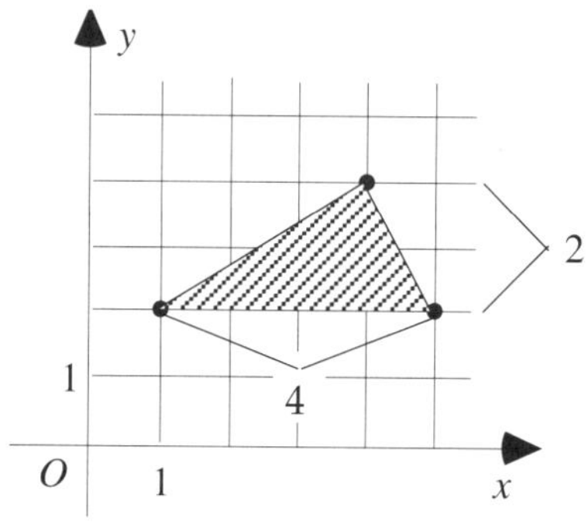

Example: What is the perimeter of Triangle ABC in the figure to the right?

(A) $5+\sqrt{5}+\sqrt{34}$
(B) $10+\sqrt{34}$
(C) $5+\sqrt{5}+\sqrt{28}$
(D) $2\sqrt{5}+\sqrt{34}$
(E) $\sqrt{5}+\sqrt{28}$

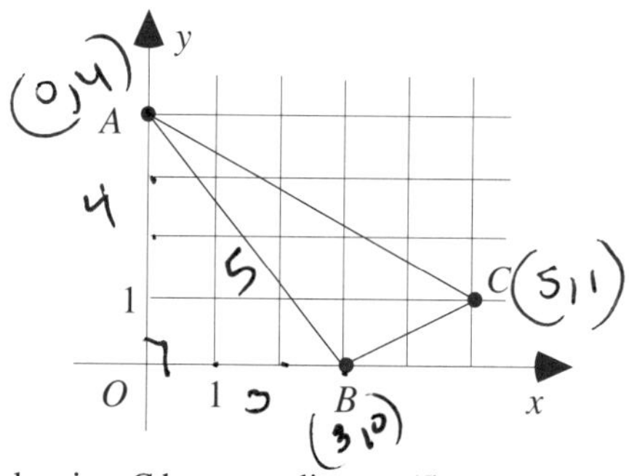

Point A has coordinates (0, 4), point B has coordinates (3, 0), and point C has coordinates (5, 1). Using the distance formula to calculate the distances between points A and B, A and C, and B and C yields

$$\overline{AB} = \sqrt{(0-3)^2+(4-0)^2} = \sqrt{9+16} = \sqrt{25} = 5$$
$$\overline{AC} = \sqrt{(0-5)^2+(4-1)^2} = \sqrt{25+9} = \sqrt{34}$$
$$\overline{BC} = \sqrt{(5-3)^2+(1-0)^2} = \sqrt{4+1} = \sqrt{5}$$

Adding these lengths gives the perimeter of Triangle ABC:

$$\overline{AB}+\overline{AC}+\overline{BC} = 5+\sqrt{34}+\sqrt{5}$$

The answer is (A).

Problem Set K:

➢ Easy

1. The slope of the line $2x + y = 3$ is NOT the same as the slope of which one of the following lines?

   (A) $2x + y = 5$
   (B) $x + y/2 = 3$
   (C) $x = -y/2 - 3$
   (D) $y = 7 - 2x$
   (E) $x + 2y = 9$

2. | Column A | | Column B |
   |---|---|---|
   | | C is the midpoint of points $A(-3, -4)$ and $B(-5, 6)$ | |
   | Slope of AC | | Slope of BC |

➢ Medium

3. | Column A | Column B |
   |---|---|
   | The slope of a line passing through (–3, –4) and the origin | The slope of a line passing through (–5, –6) and the origin |

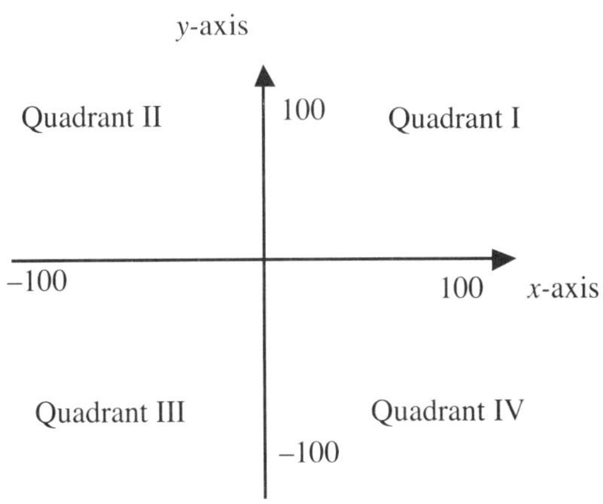

4. In the coordinate system shown, if (b, a) lies in Quadrant III, then in which quadrant can the point (a, b) lie?

   (A) I only
   (B) II only
   (C) III only
   (D) IV only
   (E) In any one of the four quadrants.

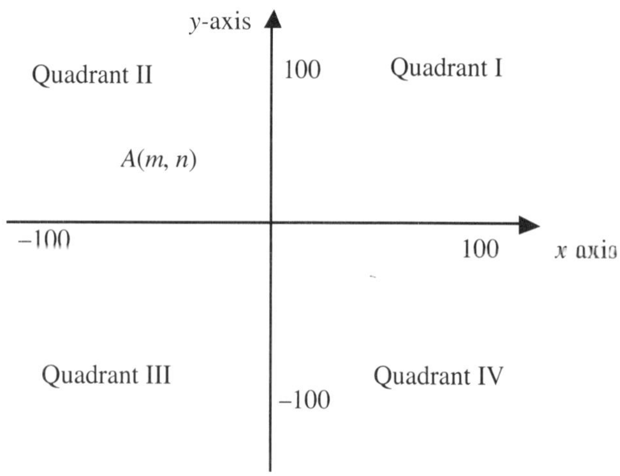

5. In the figure, the point $A(m, n)$ lies in Quadrant II as shown. In which region is the point $B(n, m)$?

   (A) Quadrant I
   (B) Quadrant II
   (C) Quadrant II
   (D) Quadrant IV
   (E) On the x-axis

6. Which of the following statements is true about the line segment with endpoints (–1, 1) and (1, –1)?

   (A) Crosses the x-axis only.
   (B) Crosses the y-axis only.
   (C) Crosses the y-axis on its positive side.
   (D) Passes through the origin (0, 0).
   (E) Crosses the x- and y-axes on their negative sides.

➢ **Hard**

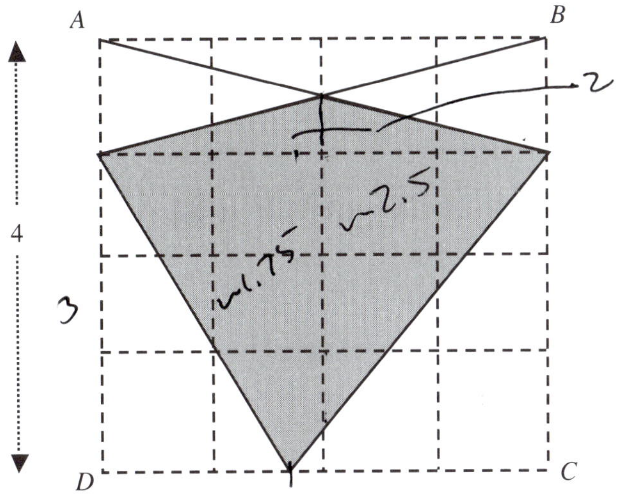

7. In the figure, the horizontal and vertical lines divide the square *ABCD* into 16 equal squares as shown. What is the area of the shaded region?

(A) 4
(B) 4.5
(C) 5
(D) 6.5
(E) 7

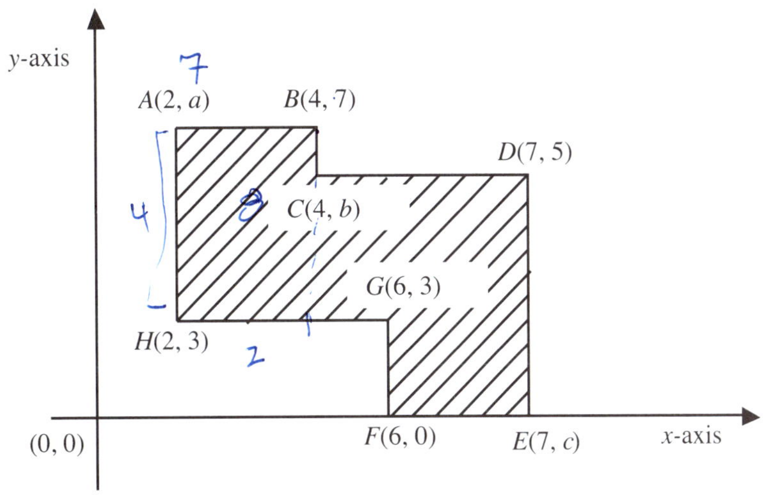

8. All the lines in the rectangular coordinate system shown in the figure are either horizontal or vertical with respect to the x-axis. What is the area of the figure *ABCDEFGH* ?

(A) 15
(B) 15.5
(C) 16
(D) 16.5
(E) 17

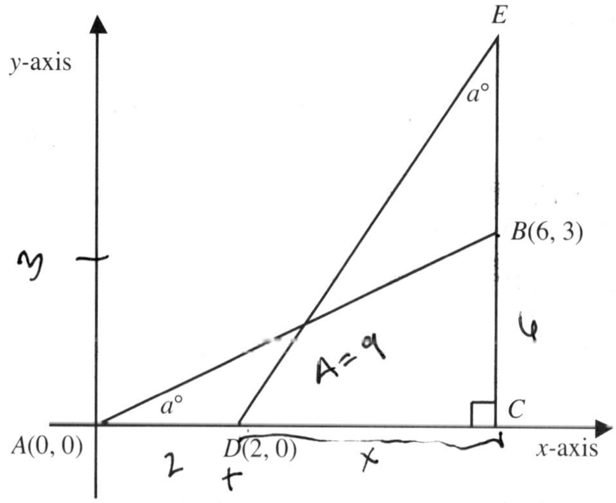

9. In the rectangular coordinate plane shown, what are the coordinates of point E ?

(A) (2, 0)
(B) (2, 3)
(C) (6, 2)
(D) (6, 6)
(E) (6, 8)

10.

Column A	In the two figures shown, line l represents the function f and line m represents the function g.	Column B
$f(10)$		$g(10)$

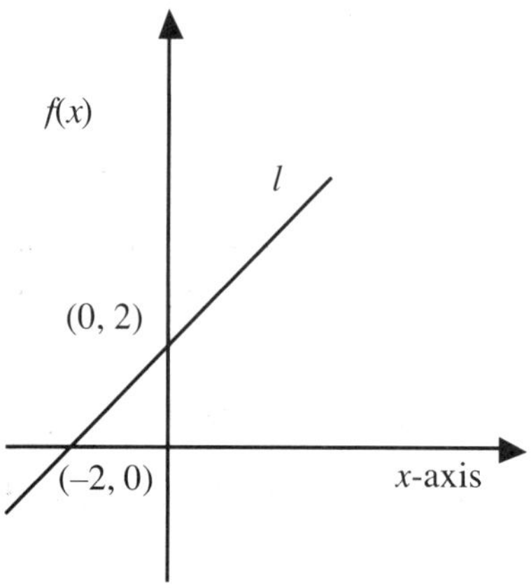

Fig. 1

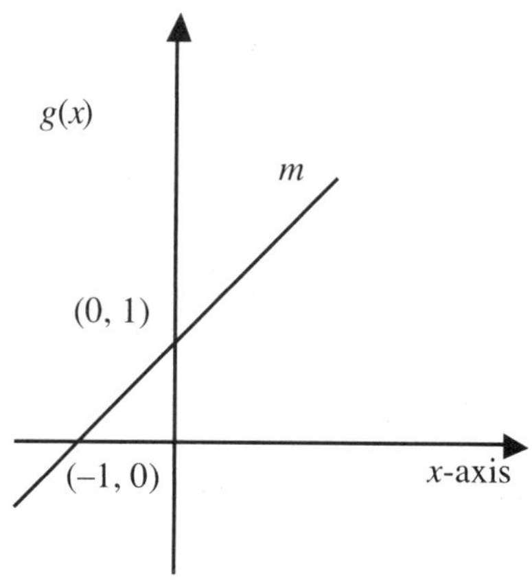

Fig. 2

Answers and Solutions to Problem Set K

➢ Easy

1. The slope of a line expressed as $y = mx + b$ is m. Expressing the given line $2x + y = 3$ in that format yields $y = -2x + 3$. Hence, the slope is -2, the coefficient of x. Let's express each line in the form $y = mx + b$ and pick the line whose slope is not -2.

Choice (A): $2x + y = 5$; $y = -2x + 5$, slope is -2. Reject.
Choice (B): $x + y/2 = 3$; $y = -2x + 6$, slope is -2. Reject.
Choice (C): $x = -y/2 - 3$; $y = -2x - 6$, slope is -2. Reject.
Choice (D): $y = 7 - 2x$; $y = -2x + 7$, slope is -2. Reject.
Choice (E): $x + 2y = 9$; $y = -\frac{1}{2}x + \frac{9}{2}$, slope is $-\frac{1}{2} \neq -2$. Accept the choice.

The answer is (E).

2. Any two points and their midpoint always lie on the same line. So, the slope of AC equals the slope of BC. Hence, the columns are equal, and the answer is (C).

➢ Medium

3. The formula for the slope of a line passing through two points (x_1, y_1) and (x_2, y_2) is $\frac{y_2 - y_1}{x_2 - x_1}$. Using this formula to calculate the slope in each column yields

Column A: Slope of the line through $(-3, -4)$ and the origin $(0, 0)$ is $\frac{0-(-4)}{0-(-3)} = \frac{4}{3}$.

Column B: Slope of the line through $(-5, -6)$ and the origin $(0, 0)$ is $\frac{0-(-6)}{0-(-5)} = \frac{6}{5}$.

Since $4/3 > 6/5$, Column A is greater than Column B. The answer is (A).

4. We are given that the point (b, a) lies in Quadrant III. In this quadrant, both x- and y-coordinates are negative. So, both b and a are negative. So, the point (a, b) also lies in the same quadrant (since both x- and y-coordinates are again negative). Hence, the answer is (C).

5. Since point $A(m, n)$ is in Quadrant II, the x-coordinate m is negative and the y-coordinate n is positive.

Hence, in point $B(n, m)$, the x-coordinate is positive, and the y-coordinate is negative. So, point B must be in Quadrant IV. The answer is (D).

6. Locating the points (–1, 1) and (1, –1) on the xy-plane gives

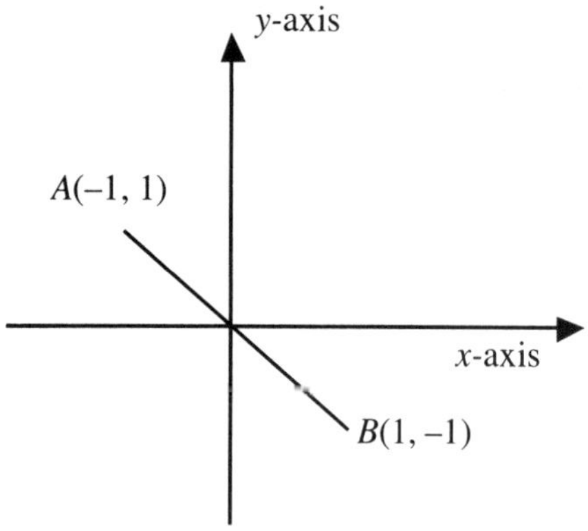

The midpoint of two points is given by

(Half the sum of the x-coordinates of the two points, Half the sum of the y-coordinates of the two points)

Hence, the midpoint of A and B is $\left(\frac{-1+1}{2}, \frac{1-1}{2}\right)$ = (0, 0).

Hence, the line-segment passes through the origin. The answer is (D).

> **Hard**

7. Since each side of the larger square measures 4 units and is divided by the four horizontal lines, the distance between any two adjacent horizontal lines must be 1 unit (= 4 units/4). Similarly, the larger square is divided into four vertical lines in the figure and any two adjacent lines are separated by 1 unit (= 4 units/4).

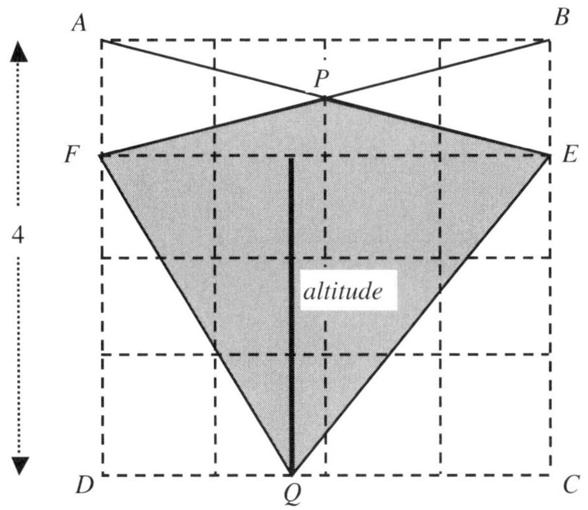

The shaded region can be divided into two sub-regions by the line *FE*, forming ΔFPE and ΔFQE.

Now, the line-segment *FE* is spread across the two opposite sides of the larger square and measures 4 units. The altitude to it, shown in the bottom triangle, is spread vertically across the three horizontal segments and therefore measures 3 units.

Hence, the area of the lower triangle = 1/2 × *base* × *height* = 1/2 × 4 × 3 = 6.

The shaded region above the line *FE* is one of the four sections formed by the two diagonals of the rectangle *ABEF* and therefore its area must equal one-fourth the area of rectangle *ABEF*. The area of rectangle *ABEF* is *length* × *width* = *FE* × *AF* = 4 × 1 = 4. Hence, the area of the shaded region above the line *FE* is 1/4 × 4 = 1.

Summing the areas of the two shaded regions yields the area of the total shaded region: 1 + 6 = 7. The answer is (E).

8. Drop a vertical line from C on to the side GH (to meet at, say, P), and draw a horizontal line from G on to the side DE (to meet at, say, Q). The resultant figure is as follows:

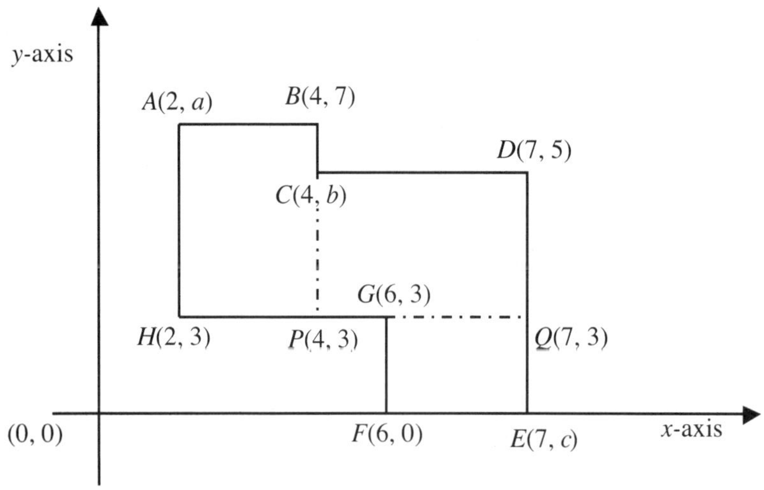

Since the point P is horizontal to the point H, its y-coordinate equals the y-coordinate of H, which is 3. Also, since P is vertical to the point C, its x-coordinate equals the x-coordinate of the point C, which is 4. Hence, the point P is (4, 3).

Similarly, the point Q is vertical to the point D and therefore its x-coordinate equals the x-coordinate of D, which is 7. Also, the point Q is horizontal to the point H and therefore takes its y-coordinate. Hence, the point Q is (7, 3).

Now, the shaded region in the given figure is the sum of the three rectangles $ABPH$, $CDQP$, and $GQEF$.

The area of the rectangle $ABPH$ is *length* $\cdot$ *width* =

$BP \cdot AB =$

(Difference in y-coordinates of B and P) $\cdot$ (Difference in x-coordinates of A and B) =

$(7 - 3)(4 - 2) =$

$4 \cdot 2 =$

8

The area of the rectangle $CDQP$ is *length* $\cdot$ *width* =

$CD \cdot DQ =$

(Difference in x-coordinates of C and D) $\cdot$ (Difference in y-coordinates of D and Q) =

$(7 - 4)(5 - 3) =$

$3 \cdot 2 =$

6

The area of the rectangle $GQEF$ is *length* $\cdot$ *width* =

$GQ \cdot QE =$

(Difference in x-coordinates of G and Q) $\cdot$ (Difference in y-coordinates of Q and E) =

$(7 - 6)(3 - 0) =$

$1 \cdot 3 =$

3

Hence, the total area is $8 + 6 + 3 = 17$ sq. units. The answer is (E).

9. From the figure, since EC is perpendicular to the x-axis, C is a point vertically below point B. Hence, both have the same x-coordinate.

The line AC is horizontal and therefore its length equals the x-coordinate difference of A and C, which equals $6 - 0 = 6$.

The line DC is horizontal and therefore its length equals the x-coordinate difference of D and C, which is $6 - 2 = 4$.

The length of the vertical line BC equals the y-coordinate difference of B and C, which is $3 - 0 = 3$.

The length of the vertical line EC equals the y-coordinate difference of E and C.

Now, in ΔABC, $\angle A = a°$, $\angle B = 90° - a°$, and $\angle C = 90°$. The sides opposite angles A and B are in the ratio $BC/AC = 3/6 = 1/2$.

Similarly, in ΔDEC, $\angle E = a°$, $\angle D = 90° - a°$, and $\angle C = 90°$ (So, ABC and DEC are similar triangles and their corresponding sides are proportional). Hence, the sides opposite angles E and D are in the ratio $DC/EC = 1/2$. Hence, we have

$$DC/EC = 1/2$$

$$EC = 2DC$$

$$EC = 2 \cdot 4 = 8$$

Hence, the y-coordinate of point E is 8, and the answer is (E).

10. If the two graphs in figures 1 and 2 were represented on a single coordinate plane, the figure would look like this:

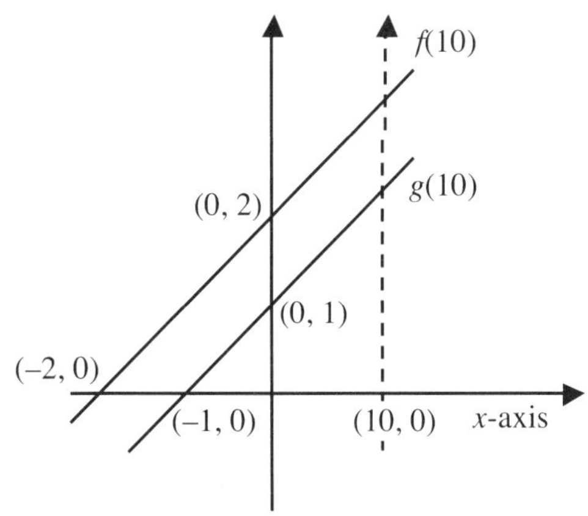

We know that the slope of a line through any two points (x_1, y_1) and (x_2, y_2) is given by $\dfrac{y_2 - y_1}{x_2 - x_1}$. Now, (–1, 0) and (0, 1) are two points on the line $g(x)$. Hence, the slope of the line is $\dfrac{1-0}{0-(-1)} = \dfrac{1}{1} = 1$. Similarly, (–2, 0) and (0, 2) are two points on the line $f(x)$. Hence, the slope of the line is $\dfrac{2-0}{0-(-2)} = \dfrac{2}{2} = 1$. Since the lines have same slope (1), they must be parallel.

Now, the point above the origin ($x = 0$) on the line $f(x)$ is (0, 2) and is 2 units above the origin. But, the corresponding point (above origin, i.e., above $x = 0$) on $g(x)$ is (0, 1) and is only 1 unit above the origin. This combined with the fact that the lines $f(x)$ and $g(x)$ are parallel shows that each point on $f(x)$ for a given value of x is above the corresponding point on $g(x)$. Hence, for any given value of x, $f(x)$ is greater than $g(x)$. Hence, $f(10)$ must be greater than $g(10)$. Hence, Column A > Column B, and the answer is (A).

Elimination Strategies

On hard problems, if you are asked to find the least (or greatest) number, then eliminate the least (or greatest) answer-choice.

This rule also applies to easy and medium problems. When people guess on these types of problems, they most often choose either the least or the greatest number. But if the least or the greatest number were the answer, most people would answer the problem correctly, and it therefore would not be a hard problem.

Example: What is the maximum number of points common to the intersection of a square and a triangle if no two sides coincide?

(A) 4
(B) 5
(C) 6
(D) 8
(E) 9

By the above rule, we eliminate answer-choice (E).

On hard problems, eliminate the answer-choice "not enough information."

When people cannot solve a problem, they most often choose the answer-choice "not enough information." But if this were the answer, then it would not be a "hard" problem.

Quantitative comparison problems are the lone exception to this rule. For often what makes a quantitative comparison problem hard is deciding whether there is enough information to make a decision.

On hard problems, eliminate answer-choices that <u>merely</u> repeat numbers from the problem.

Example: If the sum of x and 20 is 8 more than the difference of 10 and y, what is the value of $x + y$?

(A) –2
(B) 8
(C) 9
(D) 28
(E) not enough information

By the above rule, we eliminate choice (B) since it merely repeats the number 8 from the problem. By Strategy 2, we would also eliminate choice (E). **Caution:** If choice (B) contained more than the number 8, say, $8+\sqrt{2}$, then it would not be eliminated by the above rule.

On hard problems, eliminate answer-choices that can be derived from elementary operations.

Example: In the figure to the right, what is the perimeter of parallelogram ABCD?

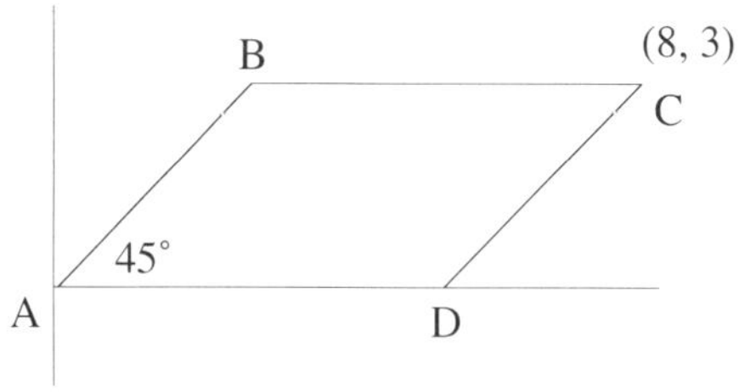

(A) 12
(B) $10+6\sqrt{2}$
(C) $20+\sqrt{2}$
(D) 24
(E) not enough information

Using the above rule, we eliminate choice (D) since $24 = 8 \cdot 3$. Further, using Strategy 2, eliminate choice (E). Note, 12 was offered as an answer-choice because some people will interpret the drawing as a rectangle tilted halfway on its side and therefore expect it to have one-half its original area.

After you have eliminated as many answer-choices as you can, choose from the more complicated or more unusual answer-choices remaining.

Example: Suppose you were offered the following answer-choices:

(A) $4+\sqrt{3}$
(B) $4+2\sqrt{3}$
(C) 8
(D) 10
(E) 12

Then you would choose either (A) or (B).

We have been discussing hard problems but have not mentioned how to identify a hard problem. Most of the time, we have an intuitive feel for whether a problem is hard or easy. But with tricky problems (problems that appear easy but are actually hard) our intuition can fail us.

On the test, your first question will be of medium difficulty. If you answer it correctly, the next question will be a little harder. If you again answer it correctly, the next question will be harder still, and so on. If your math skills are strong and you are not making any mistakes, you should reach the medium-hard or hard problems by about the fifth problem. Although this is not very precise, it can be quite helpful. Once you have passed the fifth question, you should be alert to subtleties in any seemingly simple problems.

Problem Set L:

1. What is the maximum number of 3x3 squares that can be formed from the squares in the 6x6 checker board to the right?

   (A) 4
   (B) 6
   (C) 12
   (D) 16
   (E) 24

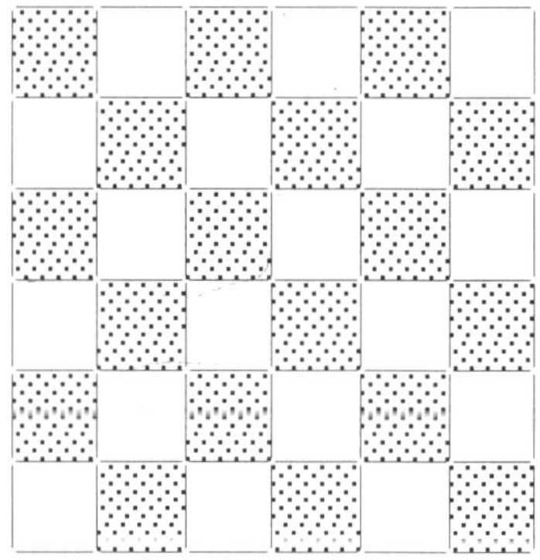

2. Let P stand for the product of the first 5 positive integers. What is the greatest possible value of m if $\frac{P}{10^m}$ is an integer?

   (A) 1
   (B) 2
   (C) 3
   (D) 5
   (E) 10

3. After being marked down 20 percent, a calculator sells for \$10. The original selling price was

   (A) \$20 (B) \$12.5 (C) \$12 (D) \$9 (E) \$7

4. The distance between cities A and B is 120 miles. A car travels from A to B at 60 miles per hour and returns from B to A along the same route at 40 miles per hour. What is the average speed for the round trip?

   (A) 48 (B) 50 (C) 52 (D) 56 (E) 58

5. If **w** is 10 percent less than **x,** and **y** is 30 percent less than **z**, then **wy** is what percent less than **xz**?

   (A) 10% (B) 20% (C) 37% (D) 40% (E) 100%

6. In the game of chess, the Knight can make any of the moves displayed in the diagram to the right. If a Knight is the only piece on the board, what is the greatest number of spaces from which not all 8 moves are possible?

   (A) 8
   (B) 24
   (C) 38
   (D) 48
   (E) 56

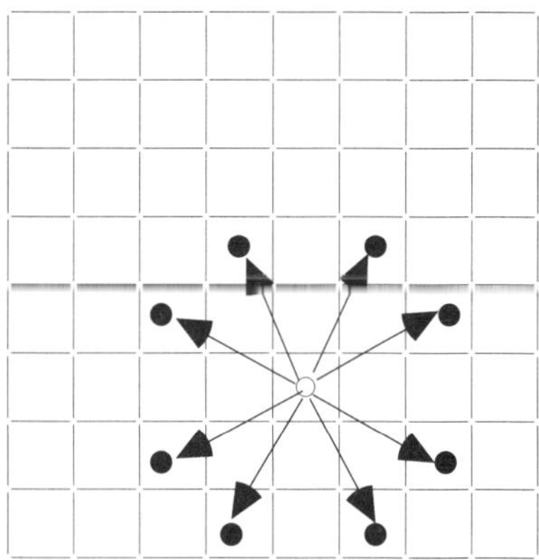

7. How many different ways can 3 cubes be painted if each cube is painted one color and only the 3 colors red, blue, and green are available? (Order is not considered, for example, green, green, blue is considered the same as green, blue, green.)

(A) 2
(B) 3
(C) 9
(D) 10
(E) 27

8. What is the greatest prime factor of $\left(2^4\right)^2 - 1$?

(A) 3
(B) 5
(C) 11
(D) 17
(E) 19

9. Suppose five circles, each 4 inches in diameter, are cut from a rectangular strip of paper 12 inches long. If the least amount of paper is to be wasted, what is the width of the paper strip?

(A) 5
(B) $4 + 2\sqrt{3}$
(C) 8
(D) $4\left(1 + \sqrt{3}\right)$
(E) not enough information

10. Let C and K be constants. If $x^2 + Kx + 5$ factors into $(x + 1)(x + C)$, the value of K is

(A) 0
(B) 5
(C) 6
(D) 8
(E) not enough information

Answers and Solutions to Problem Set L

1. What is the maximum number of 3x3 squares that can be formed from the squares in the 6x6 checker board to the right?

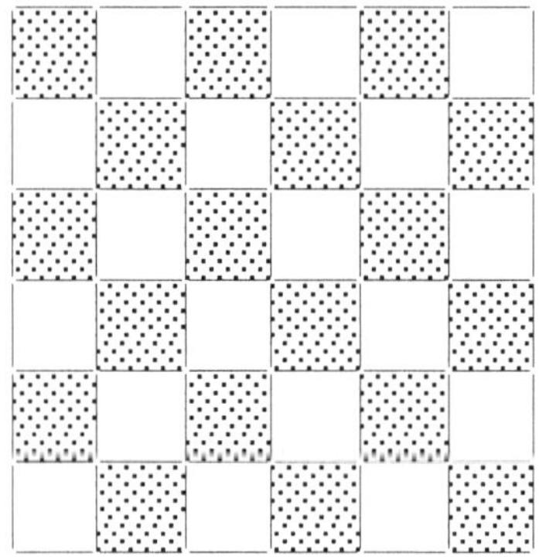

(A) 4
(B) 6
(C) 12
(D) 16
(E) 24

Clearly, there are more than four 3x3 squares in the checker board—eliminate (A). Next, eliminate (B) since it merely repeats a number from the problem. Further, eliminate (E) since it is the greatest. This leaves choices (C) and (D). If you count carefully, you will find sixteen 3x3 squares in the checker board. The answer is (D).

2. Let P stand for the product of the first 5 positive integers. What is the greatest possible value of m if $\frac{P}{10^m}$ is an integer?

(A) 1
(B) 2
(C) 3
(D) 5
(E) 10

Since we are to find the greatest value of m, we eliminate (E)—the greatest. Also, eliminate 5 because it is repeated from the problem. Now, since we are looking for the largest number, start with the greatest number remaining and work toward the smallest number. The first number that works will be the answer. To this end, let $m = 3$. Then $\frac{P}{10^m} = \frac{1 \cdot 2 \cdot 3 \cdot 4 \cdot 5}{10^3} = \frac{120}{1000} = \frac{3}{25}$. This is not an integer, so eliminate (C). Next, let $m = 2$. Then $\frac{P}{10^m} = \frac{1 \cdot 2 \cdot 3 \cdot 4 \cdot 5}{10^2} = \frac{120}{100} = \frac{6}{5}$. This still is not an integer, so eliminate (B). Hence, by process of elimination, the answer is (A).

3. After being marked down 20 percent, a calculator sells for \$10. The original selling price was

(A) \$20
(B) \$12.5
(C) \$12
(D) \$9
(E) \$7

Twenty dollars is too large. The discount was only 20 percent—eliminate (A). Both (D) and (E) are impossible since they are less than the selling price—eliminate. 12 is the eye-catcher: 20% of 10 is 2 and 10 + 2 = 12. This is too easy for a hard problem—eliminate. Thus, by process of elimination, the answer is (B).

4. The distance between cities A and B is 120 miles. A car travels from A to B at 60 miles per hour and returns from B to A along the same route at 40 miles per hour. What is the average speed for the round trip?

 (A) 48
 (B) 50
 (C) 52
 (D) 56
 (E) 58

We can eliminate 50 (the mere average of 40 and 60) since that would be too elementary. Now, the average must be closer to 40 than to 60 because the car travels for a longer time at 40 mph. But 48 is the only number given that is closer to 40 than to 60. The answer is (A).

It's instructive to also calculate the answer. $Average\ Speed = \dfrac{Total\ Distance}{Total\ Time}$. Now, a car traveling at 40 mph will cover 120 miles in 3 hours. And a car traveling at 60 mph will cover the same 120 miles in 2 hours. So the total traveling time is 5 hours. Hence, for the round trip, the average speed is $\dfrac{120+120}{5} = 48$.

5. If **w** is 10 percent less than **x,** and **y** is 30 percent less than **z**, then **wy** is what percent less than **xz**?

 (A) 10%
 (B) 20%
 (C) 37%
 (D) 40%
 (E) 100%

We eliminate (A) since it repeats the number 10 from the problem. We can also eliminate choices (B), (D), and (E) since they are derivable from elementary operations:

$$20 = 30 - 10$$

$$40 = 30 + 10$$

$$100 = 10 \cdot 10$$

This leaves choice (C) as the answer.

Let's also solve this problem directly. The clause

w is 10 percent less than x

translates into

$$w = x - .10x$$

Simplifying yields

1) $w = .9x$

Next, the clause

y is 30 percent less than **z**

translates into

$$y = z - .30z$$

Simplifying yields

2) $y = .7z$

Multiplying 1) and 2) gives

$$wy = (.9x)(.7z) = .63xz = xz - .37xz$$

Hence, **wy** is 37 percent less than **xz**. The answer is (C).

6. In the game of chess, the Knight can make any of the moves displayed in the diagram to the right. If a Knight is the only piece on the board, what is the greatest number of spaces from which not all 8 moves are possible?

(A) 8
(B) 24
(C) 38
(D) 48
(E) 56

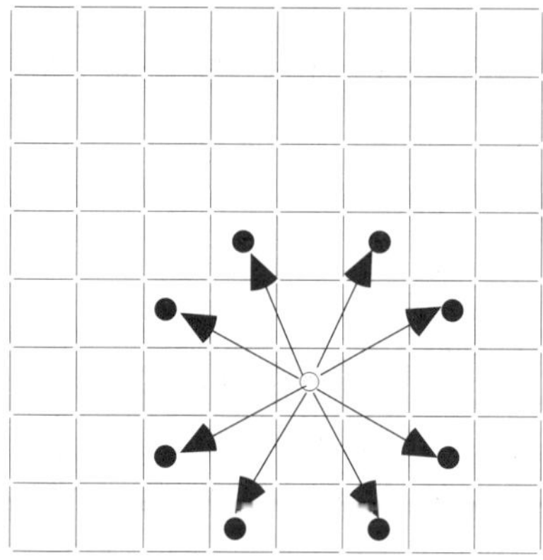

Since we are looking for the greatest number of spaces from which not all 8 moves are possible, we can eliminate the greatest number, 56. Now, clearly not all 8 moves are possible from the outer squares, and there are 28 outer squares—not 32. Also, not all 8 moves are possible from the next to outer squares, and there are 20 of them—not 24. All 8 moves are possible from the remaining squares. Hence, the answer is $28 + 20 = 48$. The answer is (D). Notice that 56, $(32 + 24)$, is given as an answer-choice to catch those who don't add carefully.

7. How many different ways can 3 cubes be painted if each cube is painted one color and only the 3 colors red, blue, and green are available? (Order is not considered, for example, green, green, blue is considered the same as green, blue, green.)

(A) 2
(B) 3
(C) 9
(D) 10
(E) 27

Clearly, there are more than 3 color combinations possible. This eliminates (A) and (B). We can also eliminate (C) and (E) because they are both multiples of 3, and that would be too ordinary, too easy, to be the answer. Hence, by process of elimination, the answer is (D).

Let's also solve this problem directly. The following list displays all 27 $(= 3 \cdot 3 \cdot 3)$ color combinations possible (without restriction):

RRR	BBB	GGG
RRB	BBR	GGR
RRG	BBG	GGB
RBR	BRB	GRG
RBB	BRR	GRR
RBG	BRG	GRB
RGR	BGB	GBG
RGB	BGR	GBR
RGG	BGG	GBB

If order is not considered, then there are 10 distinct color combinations in this list. You should count them.

8. What is the greatest prime factor of $\left(2^4\right)^2 - 1$?

(A) 3
(B) 5
(C) 11
(D) 17
(E) 19

$\left(2^4\right)^2 - 1 = (16)^2 - 1 = 256 - 1 = 255$. Since the question asks for the greatest prime factor, we eliminate 19, the greatest number. Now, we start with the next largest number and work our way up the list; the first number that divides into 255 evenly will be the answer. Dividing 17 into 255 gives

$$17\overline{)255} = 15$$

Hence, 17 is the largest prime factor of $\left(2^4\right)^2 - 1$. The answer is (D).

9. Suppose five circles, each 4 inches in diameter, are cut from a rectangular strip of paper 12 inches long. If the least amount of paper is to be wasted, what is the width of the paper strip?

(A) 5
(B) $4 + 2\sqrt{3}$
(C) 8
(D) $4\left(1 + \sqrt{3}\right)$
(E) not enough information

Since this is a hard problem, we can eliminate (E), "not enough information." And because it is too easily derived, we can eliminate (C), (8 = 4 + 4). Further, we can eliminate (A), 5, because answer-choices (B) and (D) form a more complicated set. At this stage we cannot apply any more elimination rules; so if we could not solve the problem, we would guess either (B) or (D).

Let's solve the problem directly. The drawing below shows the position of the circles so that the paper width is a minimum.

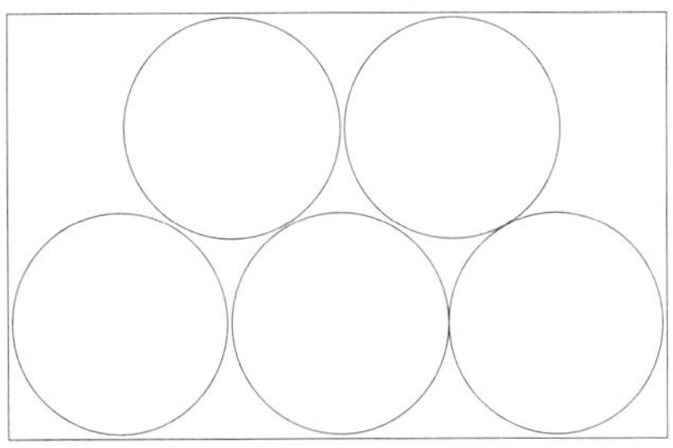

Now, take three of the circles in isolation, and connect the centers of these circles to form a triangle:

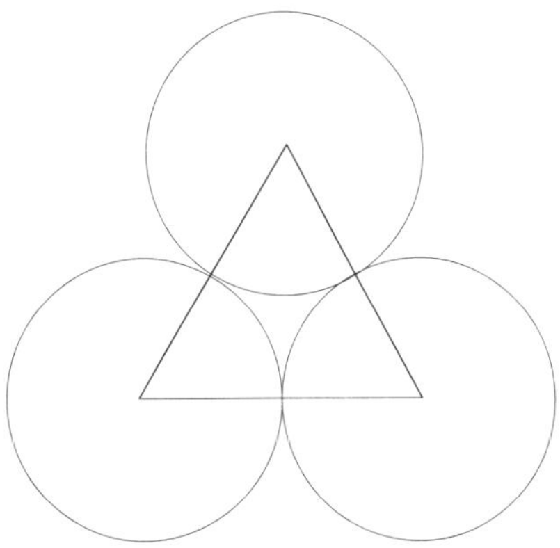

Since the triangle connects the centers of circles of diameter 4, the triangle is equilateral with sides of length 4.

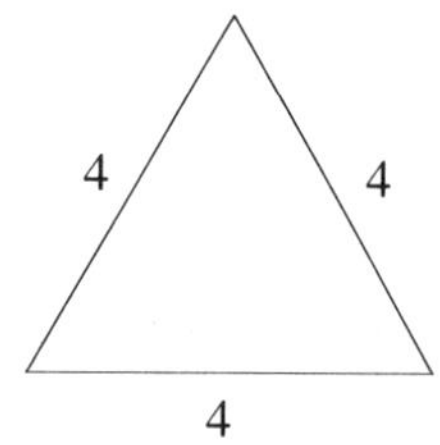

Drawing an altitude gives

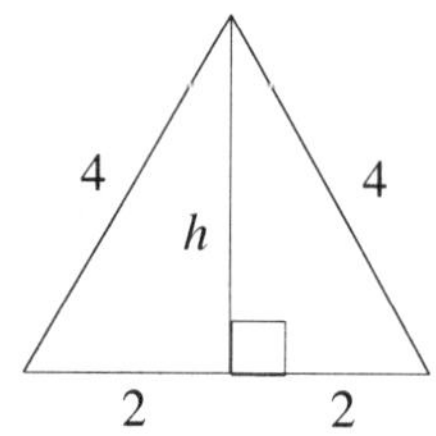

Applying the Pythagorean Theorem to either right triangle gives	$h^2 + 2^2 = 4^2$
Squaring yields	$h^2 + 4 = 16$
Subtracting 4 from both sides of this equation yields	$h^2 = 12$
Taking the square root of both sides yields	$h = \sqrt{12} = \sqrt{4 \cdot 3}$
Removing the perfect square 4 from the radical yields	$h = 2\sqrt{3}$
Summarizing gives	

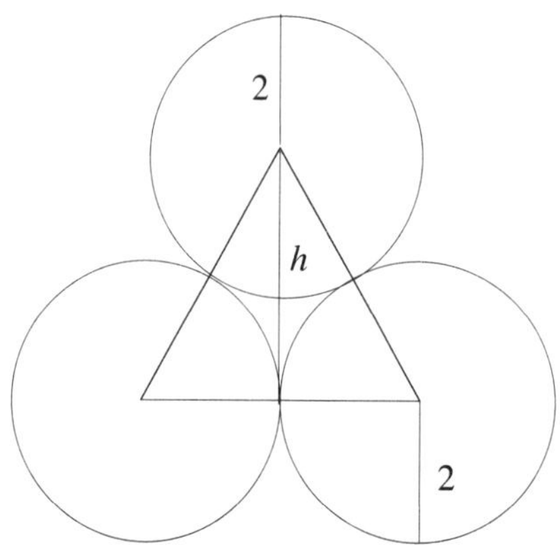

Adding to the height, $h = 2\sqrt{3}$, the distance above the triangle and the distance below the triangle to the edges of the paper strip gives

$$width = (2 + 2) + 2\sqrt{3} = 4 + 2\sqrt{3}$$

The answer is (B).

10. Let C and K be constants. If $x^2 + Kx + 5$ factors into $(x + 1)(x + C)$, the value of K is

(A) 0
(B) 5
(C) 6
(D) 8
(E) not enough information

Since the number 5 is merely repeated from the problem, we eliminate (B). Further, since this is a hard problem, we eliminate (E), "not enough information."

Now, since 5 is prime, its only factors are 1 and 5. So the constant C in the expression $(x + 1)(x + C)$ must be 5:

$$(x + 1)(x + 5)$$

Multiplying out this expression yields

$$(x + 1)(x + 5) = x^2 + 5x + x + 5$$

Combining like terms yields

$$(x + 1)(x + 5) = x^2 + 6x + 5$$

Hence, $K = 6$, and the answer is (C).

Inequalities

Inequalities are manipulated algebraically the same way as equations with one exception:

Note! **Multiplying or dividing both sides of an inequality by a negative number reverses the inequality. That is, if $x > y$ and $c < 0$, then $cx < cy$.**

Example: For which values of x is $4x + 3 > 6x - 8$?

As with equations, our goal is to isolate x on one side:

Subtracting $6x$ from both sides yields	$-2x + 3 > -8$
Subtracting 3 from both sides yields	$-2x > -11$
Dividing both sides by –2 and reversing the inequality yields	$x < 11/2$

Positive & Negative Numbers

A number greater than 0 is positive. On the number line, positive numbers are to the right of 0. A number less than 0 is negative. On the number line, negative numbers are to the left of 0. Zero is the only number that is neither positive nor negative; it divides the two sets of numbers. On the number line, numbers increase to the right and decrease to the left.

The expression $x > y$ means that x is greater than y. In other words, x is to the right of y on the number line:

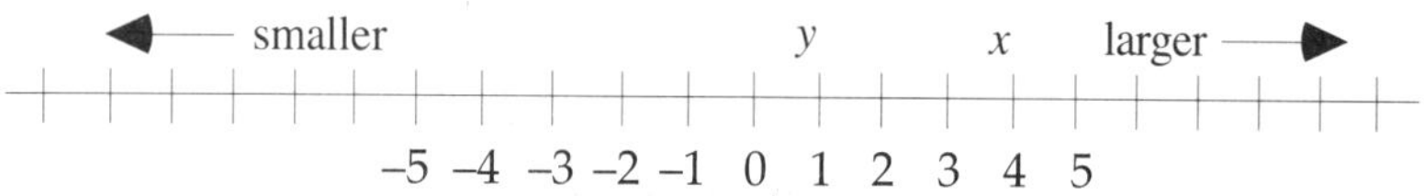

We usually have no trouble determining which of two numbers is larger when both are positive or one is positive and the other negative (e.g., 5 > 2 and 3.1 > –2). However, we sometimes hesitate when both numbers are negative (e.g., –2 > –4.5). When in doubt, think of the number line: if one number is to the right of the number, then it is larger. As the number line below illustrates, –2 is to the right of –4.5. Hence, –2 is larger than –4.5.

Miscellaneous Properties of Positive and Negative Numbers

1. The product (quotient) of positive numbers is positive.
2. The product (quotient) of a positive number and a negative number is negative.
3. The product (quotient) of an even number of negative numbers is positive.
4. The product (quotient) of an odd number of negative numbers is negative.
5. The sum of negative numbers is negative.
6. A number raised to an even exponent is greater than or equal to zero.

Example: If $xy^2z < 0$, then which one of the following statements must also be true?

I. $xz < 0$
II. $z < 0$
III. $xyz < 0$

(A) None (B) I only (C) III only (D) I and II (E) II and III

Since a number raised to an even exponent is greater than or equal to zero, we know that y^2 is positive (it cannot be zero because the product xy^2z would then be zero). Hence, we can divide both sides of the inequality $xy^2z < 0$ by y^2:

$$\frac{xy^2z}{y^2} < \frac{0}{y^2}$$

Simplifying yields $xz < 0$

Therefore, I is true, which eliminates (A), (C), and (E). Now, the following illustrates that $z < 0$ is not necessarily true:

$$-1 \cdot 2^2 \cdot 3 = -12 < 0$$

This eliminates (D). Hence, the answer is (B).

Absolute Value

The absolute value of a number is its distance on the number line from 0. Since distance is a positive number, absolute value of a number is positive. Two vertical bars denote the absolute value of a number: $|x|$. For example, $|3| = 3$ and $|-3| = 3$. This can be illustrated on the number line:

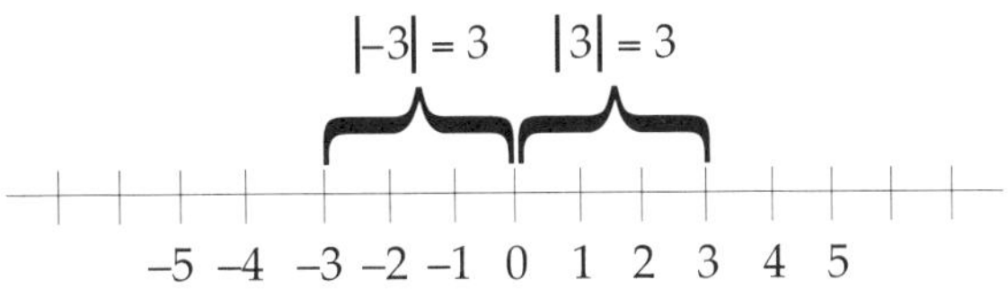

Students rarely struggle with the absolute value of numbers: if the number is negative, simply make it positive; and if it is already positive, leave it as is. For example, since –2.4 is negative, $|-24| = 2.4$ and since 5.01 is positive $|5.01| = 5.01$.

Further, students rarely struggle with the absolute value of positive variables: if the variable is positive, simply drop the absolute value symbol. For example, if $x > 0$, then $|x| = x$.

However, negative variables can cause students much consternation. If x is negative, then $|x| = -x$. This often confuses students because the absolute value is positive but the $-x$ appears to be negative. It is actually positive—it is the negative of a negative number, which is positive. To see this more clearly let $x = -k$, where k is a <u>positive</u> number. Then x is a negative number. So $|x| = -x = -(-k) = k$. Since k is positive so is $-x$. Another way to view this is $|x| = -x = (-1) \cdot x =$ (–1)(a negative number) = a positive number.

Example: If $x = -|x|$, then which one of the following statements could be true?

I. $x = 0$
II. $x < 0$
III. $x > 0$

(A) None (B) I only (C) III only (D) I and II (E) II and III

Statement I could be true because $-|0| = -(+0) = -(0) = 0$. Statement II could be true because the right side of the equation is always negative [$-|x|$ = –(a positive number) = a negative number]. Now, if one side of an equation is always negative, then the other side must always be negative, otherwise the opposite sides of the equation would not be equal. Since Statement III is the opposite of Statement II, it must be false. But let's show this explicitly: Suppose x were positive. Then $|x| = x$, and the equation $x = -|x|$ becomes $x = -x$. Dividing both sides of this equation by x yields $1 = -1$. This is contradiction. Hence, x cannot be positive. The answer is (D).

Higher Order Inequalities

These inequalities have variables whose exponents are greater than 1. For example, $x^2 + 4 < 2$ and $x^3 - 9 > 0$. The number line is often helpful in solving these types of inequalities.

Example: For which values of x is $x^2 > -6x - 5$?

First, replace the inequality symbol with an equal symbol: $x^2 = -6x - 5$

Adding $6x$ and 5 to both sides yields $x^2 + 6x + 5 = 0$

Factoring yields (see General Trinomials in the chapter Factoring) $(x + 5)(x + 1) = 0$

Setting each factor to 0 yields $x + 5 = 0$ and $x + 1 = 0$

Or $x = -5$ and $x = -1$

Now, the only numbers at which the expression can change sign are –5 and –1. So –5 and –1 divide the number line into three intervals. Let's set up a number line and choose test points in each interval:

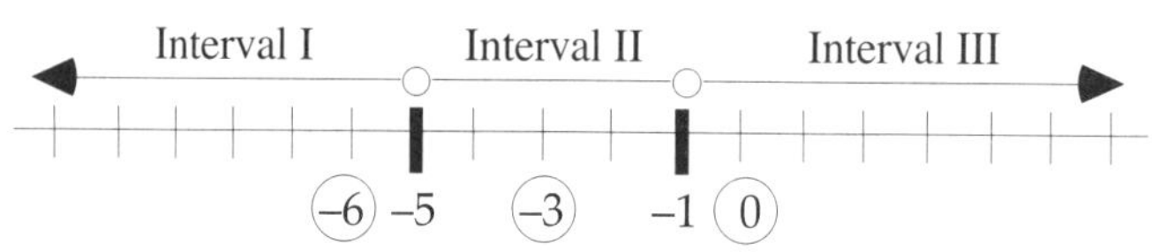

When $x = -6$, $x^2 > -6x - 5$ becomes $36 > 31$. This is true. Hence, all numbers in Interval I satisfy the inequality. That is, $x < -5$. When $x = -3$, $x^2 > -6x - 5$ becomes $9 > 13$. This is false. Hence, no numbers in Interval II satisfy the inequality. When $x = 0$, $x^2 > -6x - 5$ becomes $0 > -5$. This is true. Hence, all numbers in Interval III satisfy the inequality. That is, $x > -1$. The graph of the solution follows:

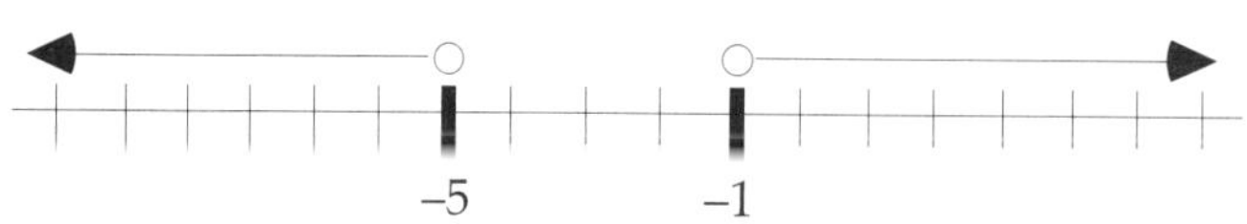

Note, if the original inequality had included the greater-than-or-equal symbol, $\geq$, the solution set would have included both –5 and –1. On the graph, this would have been indicated by filling in the circles above –5 and –1. The open circles indicate that –5 and –1 are not part of the solution.

Summary of steps for solving higher order inequalities:

1. Replace the inequality symbol with an equal symbol.
2. Move all terms to one side of the equation (usually the left side).
3. Factor the equation.
4. Set the factors equal to 0 to find zeros.
5. Choose test points on either side of the zeros.
6. If a test point satisfies the original inequality, then all numbers in that interval satisfy the inequality. Similarly, if a test point does not satisfy the inequality, then no numbers in that interval satisfy the inequality.

Transitive Property

If $x < y$ and $y < z$, then $x < z$

Example: If $\frac{1}{Q} > 1$, which of the following must be true?

(A) $1 < Q^2$ (B) $\frac{1}{Q^2} > 2$ (C) $1 > Q^2$ (D) $\frac{1}{Q^2} < 1$ (E) $Q < Q^2$

Since $\frac{1}{Q} > 1$ and $1 > 0$, we know from the transitive property that $\frac{1}{Q}$ is positive. Hence, Q is positive. Therefore, we can multiply both sides of $\frac{1}{Q} > 1$ by Q without reversing the inequality:

$$Q \cdot \frac{1}{Q} > 1 \cdot Q$$

Reducing yields $1 > Q$

Multiplying both sides again by Q yields $Q > Q^2$

Using the transitive property to combine the last two inequalities yields $1 > Q^2$

The answer is (C).

Like Inequalities Can Be Added

If $x < y$ and $w < z$, then $x + w < y + z$

Example: If $2 < x < 5$ and $3 < y < 5$, which of the following best describes $x - y$?

(A) $-3 < x - y < 2$
(B) $-3 < x - y < 5$
(C) $0 < x - y < 2$
(D) $3 < x - y < 5$
(E) $2 < x - y < 5$

Multiplying both sides of $3 < y < 5$ by -1 yields $-3 > -y > -5$. Now, we usually write the smaller number on the left side of the inequality. So $-3 > -y > -5$ becomes $-5 < -y < -3$. Add this inequality to the like inequality $2 < x < 5$:

$$\begin{array}{rr} & 2 < x < 5 \\ (+) & -5 < -y < -3 \\ \hline & -3 < x - y < 2 \end{array}$$

The answer is (A).

Problem Set M:

➢ **Easy**

1. If $ab > 0$, then which one of the following must be true?

(A) $a/b > 0$
(B) $a - b > 0$
(C) $a + b > 0$
(D) $b - a > 0$
(E) $a + b < 0$

2.

Column A		Column B
	$x/15 > y/25$	
$6y + 5x$		$10x + 3y$

➢ **Medium**

3.

Column A		Column B
	$3x + 7y > 7x + 3y$	
$x + 2y$		$y + 2x$

4.

Column A		Column B
	$1 < p < 3$	
p^2		$2p$

5. If $x + z > y + z$, then which of the following must be true?

(I) $x - z > y - z$
(II) $xz > yz$
(III) $x/z > y/z$

(A) I only
(B) II only
(C) III only
(D) I and II only
(E) II and III only

6. If $a = 3 + b$, which of the following is true?

(I) $a > b + 2.5$
(II) $a < b + 2.5$
(III) $a > 2 + b$

(A) I only
(B) II only
(C) III only
(D) I and II only
(E) I and III only

7. If $|x| + x = 4$, then which one of the following is odd?

(A) $x^2 + 3x$
(B) $x^2 + 3x + 2$
(C) $x^2 + 4x$
(D) $x^2 + 4x + 2$
(E) $x^2 + 4x + 3$

8. If $x < y < -1$, then which one of the following expressions is positive?

(A) $-x^2$
(B) y
(C) x^2y
(D) $\dfrac{x^2}{y^2}$
(E) $y - x^2$

9. If $x + 3$ is positive, then which one of the following must be positive?

(A) $x - 3$
(B) $(x - 3)(x - 4)$
(C) $(x - 3)(x + 3)$
(D) $(x - 3)(x + 4)$
(E) $(x + 3)(x + 6)$

10. If $(x + 1)^2 - 2x > 2(x + 1) + 2$, then x cannot equal which one of the following?

(A) -5
(B) -3
(C) 0
(D) 3
(E) 5

11. If $x^5 + x^2 < 0$, then which one of the following must be true?

(A) $x < -1$
(B) $x < 0$
(C) $x > 0$
(D) $x > 1$
(E) $x^4 < x^2$

12. In ΔPQR, $PQ = x$, $QR = x + 3$, and $PR = y$. If $x = y + 3$, then which one of the following is true?

(A) $\angle P < \angle Q < \angle R$
(B) $\angle Q < \angle R < \angle P$
(C) $\angle R < \angle P < \angle Q$
(D) $\angle P < \angle R < \angle Q$
(E) $\angle Q < \angle P < \angle R$

13. If $5 < x < 10$ and $y = x + 5$, what is the greatest possible integer value of $x + y$?

(A) 18
(B) 20
(C) 23
(D) 24
(E) 25

14.

Column A		Column B
$\dfrac{a-b}{a+b}$	$a > b > 0$	$\dfrac{a^2-b^2}{a^2+b^2}$

15. If $(x - y)^3 > (x - y)^2$, then which one of the following must be true?

(A) $x^3 < y^2$
(B) $x^5 < y^4$
(C) $x^3 > y^2$
(D) $x^5 > y^4$
(E) $x^3 > y^3$

16. If $x > 2$ and $x < 3$, then which of the following is positive?

(I) $(x - 2)(x - 3)$
(II) $(2 - x)(x - 3)$
(III) $(2 - x)(3 - x)$

(A) I only
(B) II only
(C) III only
(D) I and II only
(E) I and III only

17. If $x > y$ and $x < 0$, then which of the following must be true?

(I) $\frac{1}{x} < \frac{1}{y}$

(II) $\frac{1}{x-1} < \frac{1}{y-1}$

(III) $\frac{1}{x+1} < \frac{1}{y+1}$

(A) I only
(B) II only
(C) III only
(D) I and II only
(E) I and III only

18.

Column A	$a^2 + 7a < 0$	Column B
a		0

➢ **Hard**

19. Three workers A, B, and C are hired for 4 days. The daily wages of the three workers are as follows:

A's first day wage is \$4.
Each day, his wage increases by 2 dollars.

B's first day wage is \$3.
Each day, his wage increases by 2 dollars.

C's first day wage is \$1.
Each day, his wage increases by the prime numbers 2, 3, and 5 in that order.

Which one of the following is true about the wages earned by A, B, and C in the first 4 days?

(A) $A > B > C$
(B) $C > B > A$
(C) $A > C > B$
(D) $B > A > C$
(E) $C > A > B$

20. If $x^2 - y^2 = 16$ and $x + y > x - y$, then which one of the following could $x - y$ equal?

(A) 3
(B) 4
(C) 5
(D) 6
(E) 7

21. If it is true that $1/55 < x < 1/22$ and $1/33 < x < 1/11$, then which of the following numbers could x equal?

    I. 1/54
    II. 1/23
    III. 1/12

    (A) I only
    (B) II only
    (C) III only
    (D) I and II only
    (E) II and III only

22.

Column A		Column B
	$x/a > 4$ and $y/a < -6$ $a^2 = 9$ $ab^2 = -8$	
x		y

Answers and Solutions to Problem Set M

➢ Easy

1. The product ab is positive when both a and b are positive or when both a and b are negative; in either case, a/b is positive. Hence, choice (A) is always positive, and the answer is (A).

2. Multiplying the given inequality $x/15 > y/25$ by 75 yields $5x > 3y$.

Now, subtracting $3y$ and $5x$ from both columns yields

Column A		Column B
$3y$	$5x > 3y$	$5x$

Since we know that $5x > 3y$, Column B is greater than Column A and the answer is (B).

➢ Medium

3. We are given the inequality $3x + 7y > 7x + 3y$. Subtracting $3x + 3y$ from both sides of the inequality yields $4y > 4x$, and dividing both sides of this inequality by 4 yields $y > x$.

Column A		Column B
$x + 2y$	$y > x$	$y + 2x$

Now, subtracting $x + y$ from both columns yields

Column A		Column B
y	$y > x$	x

Since $y > x$, Column A is greater than Column B and the answer is (A).

4. If $p = 2$, then both columns equal 4. If $p \neq 2$, then the columns are unequal. This is a double case, and therefore the answer is (D).

5. Canceling z from both sides of the inequality $x + z > y + z$ yields $x > y$. Adding $-z$ to both sides yields $x - z > y - z$. Hence, I is true.

If z is negative, multiplying the inequality $x > y$ by z would flip the direction of the inequality resulting in the inequality $xz < yz$. Hence, II may not be true.

If z is negative, dividing the inequality $x > y$ by z would flip the direction of the inequality resulting in the inequality $x/z < y/z$. Hence, III may not be true.

The answer is (A) since we are asked for statements that MUST be true.

6. We are given that $a = 3 + b$. This equation indicates that a is 3 units larger than b, so a is greater than $b + 2.5$. Hence, I is true, and II is false.

Similarly, a is greater than $2 + b$. Hence, III is true.

Hence, the answer is (E): I and III are true.

7. We are given that $|x| + x = 4$. If x is negative or zero, then $|x|$ equals $-x$ and $|x| + x$ equals $-x + x = 0$. This conflicts with the given equation, so x is not negative nor equal to 0. Hence, x is positive and therefore $|x|$ equals x. Putting this in the given equation yields

$$|x| + x = 4$$
$$x + x = 4$$
$$2x = 4$$
$$x = 2$$

Now, select the answer-choice that results in an odd number when $x = 2$.

Choice (A): $x^2 + 3x = 2^2 + 3(2) = 4 + 6 = 10$, an even number. Reject.
Choice (B): $x^2 + 3x + 2 = 2^2 + 3(2) + 2 = 4 + 6 + 2 = 12$, an even number. Reject.
Choice (C): $x^2 + 4x = 2^2 + 4(2) = 4 + 8 = 12$, an even number. Reject.
Choice (D): $x^2 + 4x + 2 = 2^2 + 4(2) + 2 = 4 + 8 + 2 = 14$, an even number. Reject.
Choice (E): $x^2 + 4x + 3 = 2^2 + 4(2) + 3 = 4 + 8 + 3 = 15$, an odd number. Accept.

The answer is (E).

8. From the inequality $x < y < -1$, we have that x and y do not equal zero. Since the square of a nonzero number is positive, x^2 is positive.

Hence, $-x^2$ must be negative, and therefore choice (A) is negative.

Since we are given that $y < -1$, y is negative. Hence, choice (B) is negative.

Multiplying the inequality $y < -1$ by the positive value x^2 yields $x^2y < -x^2$, a negative value. Hence, choice (C) is also negative.

Dividing x^2, a positive value, by y^2 (also a positive value) yields a positive value. Hence, $\frac{x^2}{y^2}$ is positive, and choice (D) is positive.

Since y and $-x^2$ are negative, their sum $y + (-x^2) = y - x^2$ must also be negative. Hence, choice (E) is negative.

The only choice that is positive is (D), so the answer is (D).

9. We are given that $x + 3 > 0$.

Subtracting 3 from both sides yields $x > -3$.

In case x equals 2, $x - 3$ is negative. Hence, reject choice (A).

$(x - 3)(x - 4)$ is negative for the values of x between 3 and 4 [For example, when x equals 3.5, the expression is negative]. The known inequality $x > -3$ allows the values to be in this range. Hence, the expression can be negative. Reject choice (B).

$(x - 3)(x + 3)$ is negative for the values of x between 3 and -3 [For example, when x equals 0, the expression is negative]. The known inequality $x > -3$ allows the values to be in this range. Hence, the expression can be negative. Reject choice (C).

$(x - 3)(x + 4)$ is negative for the values of x between 3 and -4 [For example, when x equals 0, the expression is negative]. The known inequality $x > -3$ allows the values to be in this range. Hence, the expression can be negative. Reject choice (D).

$(x + 3)(x + 6)$ is negative only for the values of x between -3 and -6 [For example, when x equals -4, the expression is negative]. But the known inequality $x > -3$ does not allow the values to be in this range. Hence, the expression cannot be negative. Hence, accept choice (E).

10. We are given the inequality

$$(x+1)^2 - 2x > 2(x+1) + 2$$
$$x^2 + 2x + 1 - 2x > 2x + 2 + 2$$
$$x^2 + 1 > 2x + 4$$
$$x^2 - 2x + 1 > 4 \quad \text{by subtracting } 2x \text{ from both sides}$$
$$x^2 - 2(x)(1) + 1^2 > 4 \quad \text{expressing the left side in the form } a^2 - 2ab + b^2$$
$$(x-1)^2 > 4 \quad \text{by using the formula } a^2 - 2ab + b^2 = (a-b)^2$$

Square rooting both sides of the inequality yields two new inequalities: $x - 1 > 2$ or $x - 1 < -2$. Adding 1 to both sides of the solutions yields $x > 3$ and $x < -1$. Hence, x is either less than -1 or x is greater than 3. In either case, x does not equal 0. Hence, the answer is (C).

11. Subtracting x^2 from both sides of the given inequality yields $x^5 < -x^2$. Dividing the inequality by the positive value x^2 yields $x^3 < -1$. Taking the cube root of both sides yields $x < -1$. The answer is (A).

12. Subtracting 3 from both sides of the equation $x = y + 3$ by 3 yields $x - 3 = y$. Hence, $PR = x - 3$. Since $x - 3 < x < x + 3$, we have the inequality $PR < PQ < QR$ (since $PR = y = x - 3$, $PQ = x$, and $QR = x + 3$, given). Since the angle opposite the greater side is greater, in the triangle $\angle Q < \angle R < \angle P$. Hence, the answer is (B).

13. Adding x to both sides of the equation $y = x + 5$ yields $x + y = x + (x + 5)$, or $x + y = 2x + 5$. Hence, the greatest possible value of $x + y$ is the maximum possible value of $2x + 5$. Now, let's create this expression out of the given inequality $5 < x < 10$. Multiplying the inequality by 2 yields $10 < 2x < 20$. Adding 5 to each part of the inequality yields $10 + 5 < 2x + 5 < 20 + 5$, or $15 < 2x + 5 < 25$. So, $2x + 5$ is less than 25. The greatest possible integer value of $2x + 5$ is 24. Hence, the answer is (D).

14. Factoring Column B yields

Column A		Column B
$\dfrac{a-b}{a+b}$	$a > b > 0$	$\dfrac{(a-b)(a+b)}{a^2+b^2}$

Since $a > b$, $a - b > 0$. Hence, we can cancel $a - b$, a positive value, from both columns:

Column A		Column B
$\dfrac{1}{a+b}$	$a > b > 0$	$\dfrac{a+b}{a^2+b^2}$

Since $a > b > 0$, both a and b are positive. Hence, the Least Common Denominator of the fractions, $(a + b)(a^2 + b^2)$, is positive. So, we can multiply both columns by the LCD to clear the fractions:

Column A		Column B
$a^2 + b^2$	$a > b > 0$	$(a + b)(a + b)$

Performing the multiplication in Column B yields

Column A		Column B
$a^2 + b^2$	$a > b > 0$	$a^2 + b^2 + 2ab$

Subtracting $a^2 + b^2$ from both columns yields

Column A		Column B
0	$a > b > 0$	$2ab$

Since both a and b are positive, $2ab$ is positive. Hence, Column B is greater than Column A, and the answer is (B).

15. If $x - y$ equaled 0, the inequality $(x - y)^3 > (x - y)^2$ would not be valid. Hence, $x - y$ is nonzero. Since the square of a nonzero number is positive, $(x - y)^2$ is positive. Dividing both sides of the inequality $(x - y)^3 > (x - y)^2$ by the positive term $(x - y)^2$ yields $x - y > 1$. Adding y to both sides yields $x > y + 1$. This inequality says that x is at least one unit larger than y, so $x > y$. The relation can be carried to the cubes as $x^3 > y^3$, so the answer is (E). The remaining choices need *not* be true.

16. Combining the given inequalities $x > 2$ and $x < 3$ yields $2 < x < 3$. So, x lies between 2 and 3. Hence,

$x - 2$ is positive and $x - 3$ is negative. Hence, the product $(x - 2)(x - 3)$ is negative. I is false.

$2 - x$ is negative and $x - 3$ is negative. Hence, the product $(x - 2)(x - 3)$ is positive. II is true.

$2 - x$ is negative and $3 - x$ is positive. Hence, the product $(2 - x)(3 - x)$ is negative. III is false.

Hence, the answer is (B), II only is correct.

17. We are given the inequality $x > y$ and that x is negative. Since $x > y$, y must also be negative. Hence, xy, the product of two negative numbers, must be positive. Dividing the inequality by the positive expression xy yields $\frac{x}{xy} > \frac{y}{xy}$, or $\frac{1}{y} > \frac{1}{x}$. Rearranging yields $\frac{1}{x} < \frac{1}{y}$. Hence, I is true.

Since x is negative, $x - 1$ is also negative. Similarly, since y is negative, $y - 1$ is also negative. Hence, the product of the two, $(x - 1)(y - 1)$, must be positive. Subtracting –1 from both sides of the given inequality $x > y$ yields $x - 1 > y - 1$. Dividing the inequality by the positive value $(x - 1)(y - 1)$ yields $\frac{1}{y-1} > \frac{1}{x-1}$.

Rearranging the inequality yields $\frac{1}{x-1} < \frac{1}{y-1}$. Hence, II must be true.

Though x is negative, it is possible that $x + 1$ is positive while $y + 1$ is still negative. Here, $\frac{1}{x+1} < \frac{1}{y+1}$ is false because the left-hand side is positive while the right-hand side is negative. Hence, III need not be true.

Hence, the answer is (D), I and II must be true.

18. Factoring out the common factor a on the left-hand side of the inequality $a^2 + 7a < 0$ yields $a(a + 7) < 0$. The product of two numbers (here, a and $a + 7$) is negative when one is negative and the other is positive. Hence, we have two cases:

1) $a < 0$ and $a + 7 > 0$
2) $a > 0$ and $a + 7 < 0$

Case 2) is impossible since if a is positive then $a + 7$ cannot be negative.

Case 1) is valid for all values of a between 0 and –7. Any number in the range is less than 0. So, Column A is less than Column B, and the answer is (B).

➢ Hard

19. The payments to Worker A for the 4 days are the four integers 4, 6, 8, and 10. The sum of the payments is $4 + 6 + 8 + 10 = 28$.

The payments to Worker B for the 4 days are the four integers 3, 5, 7, and 9. The sum of the payments is $3 + 5 + 7 + 9 = 24$.

The payments to Worker C for the 4 days are 1, $1 + 2 = 3$, $3 + 3 = 6$, and $6 + 5 = 11$. The sum of the payments is $1 + 3 + 6 + 11 = 21$.

From the calculations, $A > B > C$. The answer is (A).

20. The choices given are positive. Multiplying both sides of the given inequality $x + y > x - y$ by the positive value $x - y$ yields

$$(x+y)(x-y) > (x-y)^2$$
$$x^2 - y^2 > (x-y)^2$$
$$16 > (x-y)^2$$
$$\sqrt{16} > x - y$$
$$4 > x - y$$

Since choice (A) is one such suitable choice, the answer is (A).

21. Combining the two given inequalities $1/55 < x < 1/22$ and $1/33 < x < 1/11$ yields $1/33 < x < 1/22$. Since among the three positive numbers 12, 23, and 54, the number 23 is the only one in the positive range between 22 and 33, only the number 1/23 lies between 1/33 and 1/22 and the numbers 1/54, and 1/12 do not. Hence, $x = 1/23$. The answer is (B), II only.

22. The possible solutions of the equation $a^2 = 9$ are $a = 3$ and $a = -3$.

Since the square of any nonzero number is positive, b^2 must be positive. We are given that $ab^2 = -8$, a negative number. Since the product of a and b^2 is negative, a must be negative. The negative solution of a is $a = -3$.

Substituting this value of a in the given inequality $x/a > 4$ yields

$$\frac{x}{-3} > 4$$

$x < -12$ by multiplying both sides by the negative number –3 and flipping the direction of the inequality

Hence, Column A is less than –12.

Now, replacing a with –3 in the inequality $y/a < -6$ yields

$$\frac{y}{-3} < -6$$

$y > 18$ by multiplying both sides by –3 and flipping the direction of the inequality

Hence, y is positive and therefore greater than x. Hence, Column B is greater, and the answer is (B).

Fractions & Decimals

Fractions

A fraction consists of two parts: a numerator and a denominator.

$$\frac{\textit{numerator}}{\textit{denominator}}$$

If the numerator is smaller than the denominator, the fraction is called *proper* and is less than one. For example: $\frac{1}{2}$, $\frac{4}{5}$, and $\frac{3}{\pi}$ are all proper fractions and therefore less than 1.

If the numerator is larger than the denominator, the fraction is called *improper* and is greater than 1. For example: $\frac{3}{2}$, $\frac{5}{4}$, and $\frac{\pi}{3}$ are all improper fractions and therefore greater than 1.

An improper fraction can be converted into a *mixed fraction* by dividing its denominator into its numerator. For example, since 2 divides into 7 three times with a remainder of 1, we get

$$\frac{7}{2} = 3\frac{1}{2}$$

To convert a mixed fraction into an improper fraction, multiply the denominator and the integer and then add the numerator. Then, write the result over the denominator. For example, $5\frac{2}{3} = \frac{3 \cdot 5 + 2}{3} = \frac{17}{3}$.

In a negative fraction, the negative symbol can be written on the top, in the middle, or on the bottom; however, when a negative symbol appears on the bottom, it is usually moved to the top or the middle: $\frac{5}{-3} = \frac{-5}{3} = -\frac{5}{3}$. If both terms in the denominator of a fraction are negative, the negative symbol is often factored out and moved to the top or middle of the fraction: $\frac{1}{-x-2} = \frac{1}{-(x+2)} = -\frac{1}{x+2}$ or $\frac{-1}{x+2}$.

To compare two fractions, cross-multiply. The larger number will be on the same side as the larger fraction.

Example: Which of the following fractions is larger?

$$\frac{9}{10} \qquad\qquad \frac{10}{11}$$

Cross-multiplying gives $9 \cdot 11$ versus $10 \cdot 10$, which reduces to 99 versus 100. Now, 100 is greater than 99. Hence, $\frac{10}{11}$ is greater than $\frac{9}{10}$.

Strategy

Always reduce a fraction to its lowest terms.

Example: If $x \neq -1$, then $\dfrac{2x^2 + 4x + 2}{(x+1)^2} =$

(A) 0 (B) 1 (C) 2 (D) 4 (E) 6

Factor out the 2 in the expression:

$$\frac{2\left(x^2 + 2x + 1\right)}{(x+1)^2}$$

Factor the quadratic expressions:

$$\frac{2(x+1)(x+1)}{(x+1)(x+1)}$$

Finally, canceling the $(x + 1)$'s gives 2. The answer is (C).

Strategy

To solve a fractional equation, multiply both sides by the LCD (lowest common denominator) to clear fractions.

Example: If $\dfrac{x+3}{x-3} = y$, what is the value of x in terms of y?

(A) $3 - y$ (B) $\dfrac{3}{y}$ (C) $\sqrt{y+12}$ (D) $\dfrac{-3y-3}{1-y}$ (E) $3y^2$

First, multiply both sides of the equation by $x - 3$: $(x-3)\dfrac{x+3}{x-3} = (x-3)y$

Cancel the $(x - 3)$'s on the left side of the equation: $x + 3 = (x - 3)y$

Distribute the y: $x + 3 = xy - 3y$

Subtract xy and 3 from both sides: $x - xy = -3y - 3$

Factor out the x on the left side of the equation: $x(1 - y) = -3y - 3$

Finally, divide both sides of the equation by $1 - y$: $x = \dfrac{-3y-3}{1-y}$

Hence, the answer is (D).

Note!

Complex Fractions: When dividing a fraction by a whole number (or vice versa), you must keep track of the main division bar:

$$\frac{a}{b/c} = a \cdot \frac{c}{b} = \frac{ac}{b}. \text{ But } \frac{a/b}{c} = \frac{a}{b} \cdot \frac{1}{c} = \frac{a}{bc}.$$

Example: $\dfrac{1 - \frac{1}{2}}{3} =$

(A) 6 (B) 3 (C) 1/3 (D) 1/6 (E) 1/8

Solution: $\dfrac{1-\frac{1}{2}}{3} = \dfrac{\frac{2}{2}-\frac{1}{2}}{3} = \dfrac{\frac{2-1}{2}}{3} = \dfrac{\frac{1}{2}}{3} = \dfrac{1}{2} \cdot \dfrac{1}{3} = \dfrac{1}{6}$. The answer is (D).

Example: If $z \neq 0$ and $yz \neq 1$, then $\dfrac{1}{y - \dfrac{1}{z}} =$

(A) $\dfrac{yz}{zy - 1}$ (B) $\dfrac{y - z}{z}$ (C) $\dfrac{yz - z}{z - 1}$ (D) $\dfrac{z}{zy - 1}$ (E) $\dfrac{y - z}{zy - 1}$

Solution: $\dfrac{1}{y - \dfrac{1}{z}} = \dfrac{1}{\dfrac{z}{z}y - \dfrac{1}{z}} = \dfrac{1}{\dfrac{zy - 1}{z}} = 1 \cdot \dfrac{z}{zy - 1} = \dfrac{z}{zy - 1}$. The answer is (D).

Multiplying fractions is routine: merely multiply the numerators and multiply the denominators: $\dfrac{a}{b} \cdot \dfrac{c}{d} = \dfrac{ac}{bd}$. For example, $\dfrac{1}{2} \cdot \dfrac{3}{4} = \dfrac{1 \cdot 3}{2 \cdot 4} = \dfrac{3}{8}$.

Two fractions can be added quickly by cross-multiplying: $\dfrac{a}{b} \pm \dfrac{c}{d} = \dfrac{ad \pm bc}{bd}$

Example: $\dfrac{1}{2} - \dfrac{3}{4} =$

(A) –5/4 (B) –2/3 (C) –1/4 (D) –1/2 (E) –2/3

Cross multiplying the expression $\dfrac{1}{2} - \dfrac{3}{4}$ yields $\dfrac{1 \cdot 4 - 2 \cdot 3}{2 \cdot 4} = \dfrac{4 - 6}{8} = \dfrac{-2}{8} = -\dfrac{1}{4}$. Hence, the answer is (C).

Example: Which of the following equals the average of x and $\dfrac{1}{x}$?

(A) $\dfrac{x + 2}{x}$ (B) $\dfrac{x^2 + 1}{2x}$ (C) $\dfrac{x + 1}{x^2}$ (D) $\dfrac{2x^2 + 1}{x}$ (E) $\dfrac{x + 1}{x}$

The average of x and $\dfrac{1}{x}$ is $\dfrac{x + \dfrac{1}{x}}{2} = \dfrac{\dfrac{x^2 + 1}{x}}{2} = \dfrac{x^2 + 1}{x} \cdot \dfrac{1}{2} = \dfrac{x^2 + 1}{2x}$. Thus, the answer is (B).

To add three or more fractions with different denominators, you need to form a common denominator of all the fractions.

For example, to add the fractions in the expression $\dfrac{1}{3} + \dfrac{1}{4} + \dfrac{1}{18}$, we have to change the denominator of each fraction into the common denominator 36 (note, 36 is a common denominator because 3, 4, and 18 all divide into it evenly). This is done by multiply the top and bottom of each fraction by an appropriate number (this does not change the value of the expression because any number divided by itself equals 1):

$$\frac{1}{3}\left(\frac{12}{12}\right) + \frac{1}{4}\left(\frac{9}{9}\right) + \frac{1}{18}\left(\frac{2}{2}\right) = \frac{12}{36} + \frac{9}{36} + \frac{2}{36} = \frac{12 + 9 + 2}{36} = \frac{23}{36}$$

You may remember from algebra that to find a common denominator of a set of fractions, you prime factor the denominators and then select each factor the greatest number of times it occurs in any of the factorizations. That is too cumbersome, however. A better way is to simply add the largest denominator to itself until all the other denominators divide into it evenly. In the above example, we just add 18 to itself to get the common denominator 36.

To find a common denominator of a set of fractions, simply add the largest denominator to itself until all the other denominators divide into it evenly.

Fractions often behave in unusual ways: Squaring a fraction makes it smaller, and taking the square root of a fraction makes it larger. (Caution: This is true only for proper fractions, that is, fractions between 0 and 1.)

Example: $\left(\frac{1}{3}\right)^2 = \frac{1}{9}$ and $\frac{1}{9}$ is less than $\frac{1}{3}$. Also $\sqrt{\frac{1}{4}} = \frac{1}{2}$ and $\frac{1}{2}$ is greater than $\frac{1}{4}$.

You can cancel only over multiplication, not over addition or subtraction.

For example, the c's in the expression $\frac{c+x}{c}$ cannot be canceled. However, the c's in the expression $\frac{cx+c}{c}$ can be canceled as follows: $\frac{cx+c}{c} = \frac{\not{c}(x+1)}{\not{c}} = x+1$.

Decimals

If a fraction's denominator is a power of 10, it can be written in a special form called a *decimal fraction*. Some common decimals are $\frac{1}{10} = .1, \frac{2}{100} = .02, \frac{3}{1000} = .003$. Notice that the number of decimal places corresponds to the number of zeros in the denominator of the fraction. Also, note that the value of the decimal place decreases to the right of the decimal point:

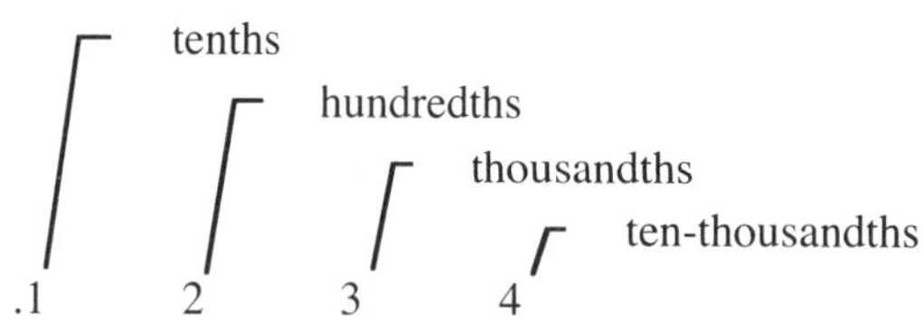

This decimal can be written in expanded form as follows:

$$.1234 = \frac{1}{10} + \frac{2}{100} + \frac{3}{1000} + \frac{4}{10000}$$

Sometimes a zero is placed before the decimal point to prevent misreading the decimal as a whole number. The zero has no affect on the value of the decimal. For example, .2 = 0.2.

Fractions can be converted to decimals by dividing the denominator into the numerator. For example, to convert $\frac{5}{8}$ to a decimal, divide 8 into 5 (note, a decimal point and as many zeros as necessary are added after the 5):

```
    .625
 8)5.000
   48
    20
    16
     40
     40
      0
```

The procedures for adding, subtracting, multiplying, and dividing decimals are the same as for whole numbers, except for a few small adjustments.

➤ **Adding and Subtracting Decimals:** To add or subtract decimals, merely align the decimal points and then add or subtract as you would with whole numbers.

$$\begin{array}{r} 1.369 \\ +\ \ 9.7 \\ \hline 11.069 \end{array} \qquad \begin{array}{r} 12.45 \\ -\ \ 6.367 \\ \hline 6.083 \end{array}$$

➤ **Multiplying Decimals:** Multiply decimals as you would with whole numbers. The answer will have as many decimal places as the sum of the number of decimal places in the numbers being multiplied.

$$\begin{array}{rl} 1.23 & \text{2 decimal places} \\ \times\ \ 2.4 & \text{1 decimal place} \\ \hline 492 & \\ 246 & \\ \hline 2.952 & \text{3 decimal places} \end{array}$$

➤ **Dividing Decimals:** Before dividing decimals, move the decimal point of the divisor all the way to the right and move the decimal point of the dividend the same number of spaces to the right (adding zeros if necessary). Then divide as you would with whole numbers.

$$.24\overline{).6} = \begin{array}{r} 2.5 \\ 24\overline{)60.0} \\ \underline{48} \\ 120 \\ \underline{120} \\ 0 \end{array}$$

Example: $\frac{1}{5}$ of .1 percent equals:

(A) 2 (B) .2 (C) .02 (D) .002 (E) .0002

Recall that percent means to divide by 100. So .1 percent equals $\frac{.1}{100} = .001$. To convert $\frac{1}{5}$ to a decimal, divide 5 into 1:

$$\begin{array}{r} .2 \\ 5\overline{)1.0} \\ \underline{10} \\ 0 \end{array}$$

In percent problems, "of" means multiplication. So multiplying .2 and .001 yields

$$\begin{array}{r} .001 \\ \times\ \ \ \ .2 \\ \hline .0002 \end{array}$$

Hence, the answer is (E). Note, you may be surprised to learn that the GRE would consider this to be a hard problem.

Example: The decimal .1 is how many times greater than the decimal $(.001)^3$?

(A) 10
(B) 10^2
(C) 10^5
(D) 10^8
(E) 10^{10}

Converting .001 to a fraction gives $\frac{1}{1000}$. This fraction, in turn, can be written as $\frac{1}{10^3}$, or 10^{-3}. Cubing this expression yields $(.001)^3 = \left(10^{-3}\right)^3 = 10^{-9}$. Now, dividing the larger number, .1, by the smaller number, $(.001)^3$, yields

$$\frac{.1}{(.001)^3} = \frac{10^{-1}}{10^{-9}} = 10^{-1-(-9)} = 10^{-1+9} = 10^8$$

Hence, .1 is 10^8 times as large as $(.001)^3$. The answer is (D).

Example: Let $x = .99$, $y = \sqrt{.99}$, and $z = (.99)^2$. Then which of the following is true?

(A) $x < z < y$
(B) $z < y < x$
(C) $z < x < y$
(D) $y < x < z$
(E) $y < z < x$

Converting .99 into a fraction gives $\frac{99}{100}$. Since $\frac{99}{100}$ is between 0 and 1, squaring it will make it smaller and taking its square root will make it larger. Hence, $(.99)^2 < .99 < \sqrt{.99}$. The answer is (C). Note, this property holds for all proper decimals (decimals between 0 and 1) just as it does for all proper fractions.

Problem Set N:

➤ **Easy**

1.

Column A	Column B
$\frac{2}{10}+\frac{3}{100}+\frac{4}{1000}+\frac{5}{10000}$	$\frac{4}{10}+\frac{3}{100}+\frac{2}{1000}$

2.

Column A	Column B
2/3 – 3/4	3/4 – 4/5

3.

Column A	Column B
$2 \times 10^1 + 3 \times 10^0 + 4 \times 10^{-1} + 5 \times 10^{-2}$	$1 \times 10^{-3} + 2 \times 10^{-2} + 3 \times 10^{-1} + 4 \times 10^0 + 5 \times 10^1$

4.

Column A	Column B
$\frac{30}{31}$ of $\frac{31}{32}$	$\frac{30}{31}$

5. Which one of the following is nearest to 0.313233 ?

(A) 3/10
(B) 31/100
(C) 313/1000
(D) 3132/10000
(E) 31323/10000

6. Column A $x = 1/y$ Column B

Column A	$x = 1/y$	Column B
$x + 1 + 1/x$		$y + 1 + 1/y$

7.

Column A	Column B
$1/2 + 1/4 + 1/8 + 1/16$	1

➢ Medium

8. 3/8 of a number is what fraction of 2 times the number?

(A) 3/16
(B) 3/8
(C) 1/2
(D) 4/6
(E) 3/4

9. If $p + q = 12$ and $pq = 35$, then $\frac{1}{p} + \frac{1}{q} =$

(A) 1/5
(B) 1/7
(C) 1/35
(D) 12/35
(E) 23/35

10.

Column A	$x = 1/y$	Column B
$\frac{x^2+1}{x}$		$\frac{y^2+1}{y}$

11. If x is not equal to 1 and $y = \frac{1}{x-1}$, then which one of the following cannot be the value of y ?

(A) 0
(B) 1
(C) 2
(D) 3
(E) 4

12. Which one of the following does the expression $\frac{2^x + 2^{x-1}}{2^{x+1} - 2^x}$ equal?

(A) 1
(B) 3/2
(C) 2
(D) 5/2
(E) 3

13. Column A — $x + 1/x$

$0 < x < y$

Column B — $y + 1/y$

14. Column A — Fraction of the cake eaten by Emily

Kate ate 1/3 of a cake, Fritz ate 1/5 of the remaining cake, and what was left was eaten by Emily.

Column B — 5/7

15. The product of two numbers x and y is twice the sum of the numbers. What is the sum of the reciprocals of x and y?

(A) 1/8
(B) 1/4
(C) 1/2
(D) 2
(E) 4

16. Column A — $\dfrac{x^2 + x + 2}{x}$

$x = 1/y$

Column B — $\dfrac{2y^2 + y + 1}{y}$

➢ Hard

17. There are 87 balls in a jar. Each ball is painted with at least one of two colors, red or green. It is observed that 2/7 of the balls that have red color also have green color, while 3/7 of the balls that have green color also have red color. What fraction of the balls in the jar have both red and green colors?

(A) 6/14
(B) 2/7
(C) 6/35
(D) 6/29
(E) 6/42

18. Which one of the following fractions is a result of the sum of an integer and its reciprocal?

(A) 15/8
(B) 17/5
(C) 36/7
(D) 37/5
(E) 65/8

19. Column A — $\dfrac{2 + x - x^2}{x}$

$x < 1/y$, and x and y are positive

Column B — $\dfrac{2y^2 + y - 1}{y}$

20. Which one of the following is closest to 1?

(A) $\frac{3}{3+0.3}$
(B) $\frac{3}{3+0.3^2}$
(C) $\frac{3}{3-0.3}$
(D) $\frac{3}{3-0.3^2}$
(E) $\frac{3}{3+0.33}$

21. Jane gave three-fifths of the amount of money she had to Jack. Jane now has 200 dollars. How much did she give to Jack?

(A) $80
(B) $120
(C) $200
(D) $300
(E) $500

➢ Very Hard

22. In a country, 60% of the male citizen and 70% of the female citizen are eligible to vote. 70% of male citizens eligible to vote voted, and 60% of female citizens eligible to vote voted. What fraction of the citizens voted during the election?

(A) 0.42
(B) 0.48
(C) 0.49
(D) 0.54
(E) 0.60

Answers and Solutions to Problem Set N

➢ Easy

1. First, cancel (subtract) the common term 3/100 from both columns:

$$\frac{2}{10}+\frac{4}{1000}+\frac{5}{10000} \qquad\qquad \frac{4}{10}+\frac{2}{1000}$$

Next, multiply both columns by 10000 to clear the fractions:

$$2000 + 40 + 5 \qquad\qquad 4000 + 20$$

Finally, add the numbers:

$$2045 \qquad\qquad 4020$$

The answer is (B).

Method II
Column A equals 2/10 + 3/100 + 4/1000 + 5/10000 = 0.2345.

Column B equals 4/10 + 3/100 + 2/1000 = 0.432.

Since 0.432 is greater than 0.2345, Column B is greater. The answer is (B).

2. The least common multiple of the denominators of all the fractions is 60. Multiplying both columns by 60 to clear the fractions yields

$$40 - 45 \qquad\qquad 45 - 48$$

Subtracting the numbers yields

$$-5 \qquad\qquad -3$$

Since $-5 < -3$, Column B is greater than Column A and the answer is (B).

3. The dominant term 10^1 appears in both columns, but has more weight (5) in Column B. Hence, Column B is greater. Let's still evaluate the expressions:

Column A = $2 \times 10^1 + 3 \times 10^0 + 4 \times 10^{-1} + 5 \times 10^{-2} = 20 + 3 + 0.4 + 0.05 = 23.45$.
Column B = $1 \times 10^{-3} + 2 \times 10^{-2} + 3 \times 10^{-1} + 4 \times 10^0 + 5 \times 10^1 = 0.001 + 0.02 + 0.3 + 4 + 50 = 54.321$.

Hence, Column B is greater than Column A. The answer is (B).

4. Column A: $\frac{30}{31}$ of $\frac{31}{32} = \frac{30}{31} \cdot \frac{31}{32} = \frac{30}{32}$. Now, $\frac{30}{32} < \frac{30}{31}$ because the fractions have the same numerators and the denominator of $\frac{30}{32}$ is larger than the denominator of $\frac{30}{31}$. Hence, Column B is larger than Column A, and the answer is (B).

5. The given decimal 0.313233 rounded to first, second, third, fourth and fifth digits after the decimal respectively equal 0.3 [= Choice (A)], 0.31 [= Choice (B)], 0.313 [= Choice (C)], 0.3132 [= Choice (D)], and 0.31323 [= Choice (E)]. Accuracy can be maintained by rounding the decimals only to later digits. So, choice (E) is the most accurate and hence the nearest. The answer is (E).

6. Substituting $1/y$ for x in Column A yields

$$\frac{1}{y}+1+\frac{1}{\frac{1}{y}}=$$

$$\frac{1}{y}+1+y=$$

Column B

The answer is (C).

7. Let's multiply both columns by 16 to clear the fractions. (Remember, this can only be done if the number you are multiplying by is positive.)

8 + 4 + 2 + 1	16
15	16

Hence, Column A is less than Column B, and the answer is (B).

➢ **Medium**

8. Let the number be x. Now, 3/8 of the number is $3x/8$, and 2 times the number is $2x$. Forming the fraction yields $\dfrac{\frac{3}{8}x}{2x}=\dfrac{\frac{3}{8}}{2}=\dfrac{3}{8}\cdot\dfrac{1}{2}=\dfrac{3}{16}$. The answer is (A).

9. Adding the fractions yields

$$\frac{1}{p}+\frac{1}{q}=\frac{p+q}{pq}=\frac{12}{35}$$ we are given that $p + q = 12$, and $pq = 35$

The answer is (D).

Method II:
Solving the given equation $pq = 35$ for q yields $q = 35/p$. Plugging this into the equation $p + q = 12$ yields

$p + 35/p = 12$
$p^2 + 35 = 12p$ by multiplying both sides by p
$p^2 - 12p + 35 = 0$ by subtracting $12p$ from both sides
$(p - 5)(p - 7) = 0$
$p - 5 = 0$ or $p - 7 = 0$
$p = 5$ or $p = 7$

When $p = 5$, the equation $p + q = 12$ shows that $q = 7$. Similarly, when $p = 7$, q equals 5. In either case, $\dfrac{1}{p}+\dfrac{1}{q}=\dfrac{1}{5}+\dfrac{1}{7}=\dfrac{7+5}{35}=\dfrac{12}{35}$. The answer is (D).

10. Substituting $1/y$ for x in Column A yields

$$\frac{\left(\frac{1}{y}\right)^2+1}{\frac{1}{y}}=$$

$$\frac{\frac{1}{y^2}+1}{\frac{1}{y}}=$$

$$\frac{\frac{1+y^2}{y^2}}{\frac{1}{y}}=$$

$$\frac{1+y^2}{y^2}\cdot\frac{y}{1}=$$

$$\frac{1+y^2}{y}=$$

Column B

The answer is (C).

11. Since the numerator of the fraction $\frac{1}{x-1}$ does not contain a variable, it can never equal 0. Hence, the fraction can never equal 0. The answer is (A).

12. The term 2^{x-1} equals $\frac{2^x}{2}$, and the term 2^{x+1} equals $2^x\cdot 2$. Hence, the given expression $\frac{2^x+2^{x-1}}{2^{x+1}-2^x}$ becomes

$$\frac{2^x+\frac{2^x}{2}}{2^x\cdot 2-2^x}=$$

$$\frac{2^x\left(1+\frac{1}{2}\right)}{2^x(2-1)}= \qquad \text{by factoring out } 2^x \text{ from both numerator and denominator}$$

$$\frac{\left(1+\frac{1}{2}\right)}{2-1}= \qquad \text{by canceling } 2^x \text{ from both numerator and denominator}$$

$$\frac{3/2}{1}=$$

$$\frac{3}{2}$$

The answer is (B).

13. If $x = 1$ and $y = 2$, then the columns become

1 + 1/1	2 + 1/2
2	2 1/2

In this case, Column B is greater.

If $x = 1/2$ and $y = 1$, the columns become

$\frac{1}{2}+\frac{1}{\frac{1}{2}}$	$1+\frac{1}{1}$
$2\frac{1}{2}$	2

In this case, Column A is greater.

Hence, we have a double case, and the answer is (D).

14. We are given that Kate ate 1/3 of the cake. So, the uneaten part of the cake is 1 – 1/3 = 2/3. Hence, regardless of how much Fritz ate, Emily could not have eaten more than 2/3 of the cake. Hence, Column A is less than 2/3; and since 2/3 < 5/7, Column B is larger. The answer is (B).

15. We are given that the product of x and y is twice the sum of x and y. Hence, we have $xy = 2(x + y)$.

Now, the sum of the reciprocals of x and y is

$$\frac{1}{x}+\frac{1}{y}=$$

$$\frac{y+x}{xy}=$$

$$\frac{x+y}{2(x+y)}=$$

$$\frac{1}{2}$$

The answer is (C).

16. Column A =

$$\frac{x^2+x+2}{x}=$$

$$\frac{x^2}{x}+\frac{x}{x}+\frac{2}{x}=$$

$$x+1+\frac{2}{x}$$

Now, substituting $1/y$ for x yields

$$\frac{1}{y}+1+\frac{2}{\frac{1}{y}}=$$

$$\frac{1}{y}+1+2y=$$

$$\frac{1+y+2y^2}{y}=$$

Column B

The answer is (C).

➢ Hard

17. Let T be the total number of balls, R the number of balls having red color, G the number having green color, and B the number having both colors.

So, the number of balls having only red is $R - B$, the number having only green is $G - B$, and the number having both is B. Now, the total number of balls is $T = (R - B) + (G - B) + B = R + G - B$.

We are given that 2/7 of the balls having red color have green also. This implies that $B = 2R/7$. Also, we are given that 3/7 of the green balls have red color. This implies that $B = 3G/7$. Solving for R and G in these two equations yields $R = 7B/2$ and $G = 7B/3$. Substituting this into the equation $T = R + G - B$ yields $T = 7B/2 + 7B/3 - B$. Solving for B yields $B = 6T/29$. Hence, 6/29 of all the balls in the jar have both colors. The answer is (D). Note that we did not use the information: "There are 87 balls." Sometimes, not all information in a problem is needed.

18. Let n be an integer. Then its reciprocal is $1/n$, and the sum of the two is $n+\frac{1}{n}=\frac{n^2+1}{n}$. The resultant fraction has a numerator 1 unit greater than the square of the denominator. Now, choose the answer-choice that has the fraction in this format.

Choice (A): Denominator is 8. Expected numerator is $8^2 + 1 = 65 \neq 15$, numerator in the choice. Hence, reject.
Choice (B): Denominator is 5. Expected numerator is $5^2 + 1 = 26 \neq 17$, numerator in the choice. Hence, reject.
Choice (C): Denominator is 7. Expected numerator is $7^2 + 1 = 50 \neq 36$, numerator in the choice. Hence, reject.
Choice (D): Denominator is 5. Expected numerator is $5^2 + 1 = 26 \neq 37$, numerator in the choice. Hence, reject.
Choice (E): Denominator is 8. Expected numerator is $8^2 + 1 = 65$, numerator in the choice. Correct.

The answer is (E).

19. Breaking up the fractions in both columns yields

Column A: $\dfrac{2+x-x^2}{x}=\dfrac{2}{x}+\dfrac{x}{x}-\dfrac{x^2}{x}=\dfrac{2}{x}+1-x$

Column B: $\dfrac{2y^2+y-1}{y}=\dfrac{2y^2}{y}+\dfrac{y}{y}-\dfrac{1}{y}=2y+1-\dfrac{1}{y}$

Now, multiplying both sides of the given inequality $x < 1/y$ by –1 and flipping the direction of the inequality yields

$-x > -1/y$ ...(1)

Since x and $1/y$ are both positive (since y is positive, so is its reciprocal $1/y$), we can safely invert both sides of the inequality $x < 1/y$ and flip the direction of the inequality to yield $1/x > y$. Multiplying both sides of this inequality by 2 yields

$2/x > 2y$ ...(2)

Now, adding inequalities (1) and (2) yields

$$\frac{2}{x}-x>2y-\frac{1}{y}$$

Adding 1 to both sides of this inequality yields

$$\frac{2}{x}+1-x>2y+1-\frac{1}{y}$$

Column A > Column B from the known results

Hence, the answer is (A).

20. Let's subtract 1 from each answer-choice. The answer-choice that has the lowest positive value should be closest to 1.

Choice (A): $\dfrac{3}{3+0.3}-1=\dfrac{3-(3+0.3)}{3+0.3}=\dfrac{-0.3}{3+0.3}=\dfrac{-0.3}{3.3}=\dfrac{-1}{11}$. Hence, Choice (A) is 1/11 units away from 1.

Choice (B): $\dfrac{3}{3+0.3^2}-1=\dfrac{3-(3+0.3^2)}{3+0.3^2}=\dfrac{-0.3^2}{3+0.3^2}=\dfrac{-0.09}{3+0.09}=\dfrac{-0.09}{3.09}=\dfrac{-9}{309}$. Hence, Choice (B) is 9/309 units away from 1. Since 9/309 is less than 1/11, Choice (B) is closer than Choice (A). Eliminate choice (A).

Choice (C): $\dfrac{3}{3-0.3}-1=\dfrac{3-(3-0.3)}{3-0.3}=\dfrac{0.3}{2.7}=\dfrac{1}{9}$. Hence, Choice (C) is 1/9 units away from 1. Clearly, this is greater than 9/309. Hence, Choice (B) is closer than Choice (C). Hence, eliminate choice (C).

Choice (D): $\dfrac{3}{3-0.3^2}-1=\dfrac{3-(3-0.3^2)}{3-0.3^2}=\dfrac{0.3^2}{3-0.3^2}=\dfrac{0.09}{2.91}=\dfrac{9}{291}$. Hence, Choice (D) is 9/291 units away from 1. This is greater than 9/309. Hence, Choice (B) is closer than Choice (D). Hence, eliminate choice (D).

Choice (E): $\dfrac{3}{3+0.33}-1=\dfrac{3-(3+0.33)}{3+0.33}=\dfrac{-0.33}{3.33}=\dfrac{-33}{333}$. So, Choice (E) is 33/333 units away from 1. This is greater than 9/309. Hence, Choice (B) is closer than Choice (E). Hence, eliminate choice (E).

The answer is (B).

21. Let the original amount of money Jane had be x. Since she gave 3/5 of her money to Jack, she now has 1 – 3/5 = 2/5 of the original amount. We are given that this 2/5 part equals 200 dollars. Hence, we have the equation $\frac{2}{5}x = 200$. Solving for x yields $x = 500$. Since she gave 3/5 of this amount to Jack, she gave him \$300 ($= \frac{3}{5} \cdot 500$). The answer is (D).

> **Very Hard**

22. Let the number of male and female citizens in the country be m and f, respectively.

Now, 60% of the male citizens are eligible to vote, and 60% of m is $60m/100$. 70% of female citizens are eligible to vote, and 70% of f is $70f/100$.

We are given that 70% of male citizens eligible to vote voted:

$$70\% \text{ of } 60m/100 \text{ is } \frac{70}{100} \times \frac{60m}{100} = \frac{70 \times 60m}{10{,}000} = 0.42m$$

We are also given that 60% of female citizens eligible to vote voted:

$$60\% \text{ of } 70f/100 \text{ is } \frac{60}{100} \times \frac{70f}{100} = \frac{60 \times 70f}{10{,}000} = 0.42f$$

So, out of the total $m + f$ citizens, the total number of voters who voted is

$$0.42m + 0.42f = 0.42(m + f)$$

Hence, the required fraction is

$$\frac{0.42(m + f)}{m + f} = 0.42$$

The answer is (A).

Equations

When simplifying algebraic expressions, we perform operations within parentheses first and then exponents and then multiplication and then division and then addition and lastly subtraction. This can be remembered by the mnemonic:

PEMDAS
Please **E**xcuse **M**y **D**ear **A**unt **S**ally

When solving equations, however, we apply the mnemonic in reverse order: **SADMEP**. This is often expressed as follows: inverse operations in inverse order. The goal in solving an equation is to isolate the variable on one side of the equal sign (usually the left side). This is done by identifying the main operation—addition, multiplication, etc.—and then performing the opposite operation.

Example: Solve the following equation for x: $2x + y = 5$

Solution: The main operation is addition (remember addition now comes before multiplication, SADMEP), so subtracting y from both sides yields

$$2x + y - y = 5 - y$$

Simplifying yields $\quad 2x = 5 - y$

The only operation remaining on the left side is multiplication. Undoing the multiplication by dividing both sides by 2 yields

$$\frac{2x}{2} = \frac{5-y}{2}$$

Canceling the 2 on the left side yields $\quad x = \dfrac{5-y}{2}$

Example: Solve the following equation for x: $3x - 4 = 2(x - 5)$

Solution: Here x appears on both sides of the equal sign, so let's move the x on the right side to the left side. But the x is trapped inside the parentheses. To release it, distribute the 2:

$$3x - 4 = 2x - 10$$

Now, subtracting $2x$ from both sides yields*

$$x - 4 = -10$$

Finally, adding 4 to both sides yields

$$x = -6$$

We often manipulate equations without thinking about what the equations actually say. The GRE likes to test this oversight. Equations are packed with information. Take for example the simple equation $3x + 2 = 5$. Since 5 is positive, the expression $3x + 2$ must be positive as well. An equation means that the terms on either side of the equal sign are equal in every way. Hence, any property one side of an equation has the

---

* Note, students often mistakenly add $2x$ to both sides of this equation because of the minus symbol between $2x$ and 10. But $2x$ is positive, so we subtract it. This can be seen more clearly by rewriting the right side of the equation as $-10 + 2x$.

other side will have as well. Following are some immediate deductions that can be made from simple equations.

Equation	**Deduction**
$y - x = 1$	$y > x$
$y^2 = x^2$	$y = \pm x$, or $\lvert y \rvert = \lvert x \rvert$. That is, x and y can differ only in sign.
$y^3 = x^3$	$y = x$
$y = x^2$	$y \geq 0$
$\frac{y}{x^2} = 1$	$y > 0$
$\frac{y}{x^3} = 2$	Both x and y are positive or both x and y are negative.
$x^2 + y^2 = 0$	$y = x = 0$
$3y = 4x$ and $x > 0$	$y > x$ and y is positive.
$3y = 4x$ and $x < 0$	$y < x$ and y is negative.
$y = \sqrt{x+2}$	$y \geq 0$ and $x \geq -2$
$y = 2x$	y is even
$y = 2x + 1$	y is odd
$yx = 0$	$y = 0$ or $x = 0$, or both

Note! **In Algebra, you solve an equation for, say, *y* by isolating *y* on one side of the equality symbol. On the GRE, however, you are often asked to solve for an entire term, say, 3 – *y* by isolating it on one side.**

Example: If $a + 3a$ is 4 less than $b + 3b$, then $a - b =$

(A) –4 (B) –1 (C) 1/5 (D) 1/3 (E) 2

Translating the sentence into an equation gives $a + 3a = b + 3b - 4$

Combining like terms gives $4a = 4b - 4$

Subtracting $4b$ from both sides gives $4a - 4b = -4$

Finally, dividing by 4 gives $a - b = -1$

Hence, the answer is (B).

Note! **Sometimes on the GRE, a system of 3 equations will be written as one long "triple" equation. For example, the three equations $x = y$, $y = z$, $x = z$, can be written more compactly as $x = y = z$.**

Example: If $w \neq 0$ and $w = 2x = \sqrt{2}y$, what is the value of $w - x$ in terms of y ?

(A) $2y$ (B) $\frac{\sqrt{2}}{2}y$ (C) $\sqrt{2y}$ (D) $\frac{4}{\sqrt{2}}y$ (E) y

The equation $w = 2x = \sqrt{2}y$ stands for three equations: $w = 2x$, $2x = \sqrt{2}y$, and $w = \sqrt{2}y$. From the last equation, we get $w = \sqrt{2}y$; and from the second equation, we get $x = \frac{\sqrt{2}}{2}y$. Hence, $w - x = \sqrt{2}y - \frac{\sqrt{2}}{2}y = \frac{2}{2}\sqrt{2}y - \frac{\sqrt{2}}{2}y = \frac{2\sqrt{2}y - \sqrt{2}y}{2} = \frac{\sqrt{2}y}{2}$. Hence, the answer is (B).

Note! **Often on the GRE, you can solve a system of two equations in two unknowns by merely adding or subtracting the equations—instead of solving for one of the variables and then substituting it into the other equation.**

Example: If p and q are positive, $p^2 + q^2 = 16$, and $p^2 - q^2 = 8$, then $q =$

(A) 2 (B) 4 (C) 8 (D) $2\sqrt{2}$ (E) $2\sqrt{6}$

Subtract the second equation from the first:

$$\begin{aligned} & p^2 + q^2 = 16 \\ (-)\ & p^2 - q^2 = 8 \\ & 2q^2 = 8 \end{aligned}$$

Dividing both sides of the equation by 2 gives $q^2 = 4$

Finally, taking the square root of both sides gives $q = \pm 2$

Hence, the answer is (A).

METHOD OF SUBSTITUTION (Four-Step Method)

Although on the GRE you can usually solve a system of two equations in two unknowns by merely adding or subtracting the equations, you still need to know a standard method for solving these types of systems.

The four-step method will be illustrated with the following system:

$$\begin{aligned} 2x + y &= 10 \\ 5x - 2y &= 7 \end{aligned}$$

1) *Solve one of the equations for one of the variables*:

   Solving the top equation for y yields $y = 10 - 2x$.

2) *Substitute the result from Step 1 into the other equation*:

   Substituting $y = 10 - 2x$ into the bottom equation yields $5x - 2(10 - 2x) = 7$.

3) *Solve the resulting equation*:

$$\begin{aligned} 5x - 2(10 - 2x) &= 7 \\ 5x - 20 + 4x &= 7 \\ 9x - 20 &= 7 \\ 9x &= 27 \\ x &= 3 \end{aligned}$$

4) *Substitute the result from Step 3 into the equation derived in Step 1*:

   Substituting $x = 3$ into $y = 10 - 2x$ yields $y = 10 - 2(3) = 10 - 6 = 4$.

Hence, the solution of the system of equations is the ordered pair (3, 4).

Problem Set O:

➢ Easy

1. If $x = y$ and $x + y = 10$, then $2x + y =$

(A) 3
(B) 15
(C) 18
(D) 24
(E) 30

2.

Column A		Column B
	$x + y = 1$	
x		y

3.

Column A		Column B
	$7x + 3y = 12$ $3x + 7y = 8$	
$x - y$		1

4. If $4p$ is equal to $6q$, then $2p - 3q$ equals which one of the following?

(A) 0
(B) 2
(C) 3
(D) 4
(E) 6

5. Which one of the following must equal $p + q$, if $x - y = p$ and $2x + 3y = q$?

(A) $x + y$
(B) $3x - 2y$
(C) $2x - 3y$
(D) $2x + 3y$
(E) $3x + 2y$

6. If $l + t = 4$ and $l + 3t = 9$, then which one of the following equals $l + 2t$?

(A) 13/2
(B) 19/2
(C) 15/2
(D) 17/3
(E) 21/4

7. If $(x - y)(x + y) = 15$ and $x + y = 5$, then what is the value of x/y ?

(A) 3
(B) 4
(C) 5
(D) 10
(E) 15

8. If $x - 4y = 1$ and $y = x/2 + 1$, then what is the value of x ?

(A) –5
(B) –2
(C) 2
(D) 5
(E) 8

➢ Medium

9. If $x^2 - 4x + 3$ equals 0, then what is the value of $(x-2)^2$?

(A) 0
(B) 1
(C) 2
(D) 3
(E) 4

10. If $x = a + 2$ and $b = x + 1$, then which one of the following must be true?

(A) $a > b$
(B) $a < b$
(C) $a = b$
(D) $a = b^2$
(E) $a = b^3$

11. If p is the sum of q and r, then which one of the following must equal $q - r$?

(A) $p - r$
(B) $p + r$
(C) $p - 2r$
(D) $p + 2r$
(E) $2q - p$

12. The sum of two numbers is 13, and their product is 30. What is the sum of the squares of the two numbers?

(A) –229
(B) –109
(C) 139
(D) 109
(E) 229

13. If $p + q = 7$ and $pq = 12$, then what is the value of $\frac{1}{p^2} + \frac{1}{q^2}$?

(A) 1/6
(B) 25/144
(C) 49/144
(D) 7/12
(E) 73/144

14. If $2x + 3y = 11$ and $3x + 2y = 9$, then $x + y =$

(A) 4
(B) 7
(C) 8
(D) 9
(E) 11

15. If $2x = 2y + 1$, then which one of the following is true?

(A) $x > y$
(B) $x < y$
(C) $x = y$
(D) $x = y + 1$
(E) $y = x + 1/2$

16. If $3x + y = x + 2y$, then $2x - y =$

(A) 0
(B) 1
(C) 2
(D) 3
(E) 4

17. If $yz - zx = 3$ and $zx - xy = 4$, then $xy - yz =$

(A) –7
(B) 1
(C) 3
(D) 4
(E) 7

18. If $(x+5) \div \left(\frac{1}{x} + \frac{1}{5}\right) = 5$, then $x =$

(A) –5
(B) 1/2
(C) 1
(D) 5
(E) 10

19. If $(x+5)\left(\frac{1}{x} + \frac{1}{5}\right) = 4$, then $x =$

(A) 1/5
(B) 1/2
(C) 1
(D) 5
(E) 10

20.

Column A		Column B
	$x + y = 7$ $x^2 + y^2 = 25$	
$x + y^2$		$3 + 4^2$

21. If both expressions $x^2 - 3x + 2$ and $x^2 - 4x + 3$ equal 0, then what is the value of $(x - 3)^2$?

(A) 0
(B) 1
(C) 2
(D) 3
(E) 4

22. If $42.42 = k(14 + 7/50)$, then what is the value of k?

(A) 1
(B) 2
(C) 3
(D) 4
(E) 5

23. If $|2x - 4|$ is equal to 2 and $(x - 3)^2$ is equal to 4, then what is the value of x?

(A) 1
(B) 2
(C) 3
(D) 4
(E) 5

24. If $(a + 2)(a - 3)(a + 4) = 0$ and $a > 0$, then $a =$

(A) 1
(B) 2
(C) 3
(D) 4
(E) 5

➤ **Hard**

25. If $x + y = 7$ and $x^2 + y^2 = 25$, then which one of the following equals the value of $x^3 + y^3$?

(A) 7
(B) 25
(C) 35
(D) 65
(E) 91

26. A system of equations is as shown below

$x + l = 6$
$x - m = 5$
$x + p = 4$
$x - q = 3$

What is the value of $l + m + p + q$?

(A) 2
(B) 3
(C) 4
(D) 5
(E) 6

27. If $mn = 3$ and $1/m + 1/n = 4/3$, then what is the value of $0.1 + 0.1^{1/m} + 0.1^{1/n}$?

(A) $0.2 + 0.1^{1/3}$
(B) $0.1 + 0.1^{1/3} + 0.1^{1/2}$
(C) $0.1 + 0.1^{4/3} + 0.1^{1/2}$
(D) $0.1 + 0.1^{1/3} + 0.1^{3/2}$
(E) $0.1 + 0.1^{1/4} + 0.1^{1/2}$

28.

Column A		Column B
	$(x - 2y)(x + 2y) = 5$ $(2x - y)(2x + y) = 35$	
$2x^2 - y^2$		$x^2 - 2y^2$

29. a, b, and c are three different numbers. None of the numbers equals the average of the other two. If $\frac{x}{a+b-2c} = \frac{y}{b+c-2a} = \frac{z}{c+a-2b}$, then $x + y + z =$

(A) 0
(B) 3
(C) 4
(D) 5
(E) 6

30. If a, b, and c are three different numbers and $\frac{x}{b-c} = \frac{y}{c-a} = \frac{z}{a-b}$, then what is the value of $ax + by + cz$?

(A) 0
(B) 1
(C) 2
(D) 3
(E) 4

31. a, b, and c are three different numbers, none of which equals the average of the other two. If $\frac{x}{b-c} = \frac{y}{c-a} = \frac{z}{a-b}$, then $x + y + z =$

(A) 0
(B) 1/2
(C) 1/3
(D) 2/3
(E) 3/4

32. If $x + y = 750$, then which one of the following additional details will determine the value of x ?

(A) $x + 2y = d$
(B) $2x + 4y = 2d$
(C) $2x + 2y = 1500$
(D) $3x = 2250 - 3y$
(E) $2x + y = 15$

➢ **Very Hard**

33. If a, b, and c are not equal to 0 or 1 and if $a^x = b$, $b^y = c$, and $c^z = a$, then $xyz =$

(A) 0
(B) 1
(C) 2
(D) a
(E) abc

Answers and Solutions to Problem Set O

➢ Easy

1. Substituting $x = y$ into the equation $x + y = 10$ yields $x + x = 10$. Combining like terms yields $2x = 10$. Finally, dividing by 2 yields $x = 5$. Hence, $2x + y = 2(5) + 5 = 15$. The answer is (B).

2. If $x = y = 1/2$, then $x + y = 1/2 + 1/2 = 1$ and the columns are equal.

But if $x = 1$ and $y = 0$, then $x + y = 1 + 0 = 1$ and Column A is larger than Column B.

Hence, we have a double case, and the answer is (D).

3. We are given the two equations:

$$7x + 3y = 12$$
$$3x + 7y = 8$$

Subtracting the bottom equation from the top equation yields

$$(7x + 3y) - (3x + 7y) = 12 - 8$$
$$7x + 3y - 3x - 7y = 4$$
$$4x - 4y = 4$$
$$4(x - y) = 4$$
$$x - y = 1$$

Hence, Column A equals 1; and since Column B also equals 1, the answer is (C).

4. We are given that $4p = 6q$. Dividing both sides by 2 yields $2p = 3q$. Subtracting $3q$ from both sides yields $2p - 3q = 0$. The answer is (A).

5. We are given the equations $x - y = p$ and $2x + 3y = q$. Adding the two equations yields

$$(x - y) + (2x + 3y) = p + q$$
$$x - y + 2x + 3y = p + q$$
$$3x + 2y = p + q.$$

Hence, $p + q = 3x + 2y$. The answer is (E).

6. Adding the two given equations $l + t = 4$ and $l + 3t = 9$ yields

$$(l + t) + (l + 3t) = 4 + 9$$
$$2l + 4t = 13$$
$$l + 2t = 13/2 \qquad \text{by dividing both sides by 2}$$

The answer is (A).

7. We are given

$$(x - y)(x + y) = 15 \qquad \ldots \text{(A)}$$
$$x + y = 5 \qquad \ldots \text{(B)}$$

Substituting the bottom equation in the top one yields $(x - y)(5) = 15$, or $x - y = 15/5 = 3$.

Adding this equation to equation (B) yields

$$x + y + (x - y) = 5 + 3$$
$$2x = 8$$
$$x = 8/2 = 4$$

Substituting this result in equation (B) yields $4 + y = 5$, or $y = 1$.

Hence, $x/y = 4/1 = 4$.

The answer is (B).

8. We have the system of equations

$$x - 4y = 1$$
$$y = x/2 + 1$$

Substituting the bottom equation in to the top yields

$$x - 4(x/2 + 1) = 1$$
$$x - 2x - 4 = 1$$
$$-x = 4 + 1$$
$$x = -5$$

The answer is (A).

➢ **Medium**

9. We have $x^2 - 4x + 3 = 0$. By the Perfect Square Trinomial formula, $(x - 2)^2 = x^2 - 4x + 4$; and this equals $(x^2 - 4x + 3) + 1 = 0 + 1 = 1$. Hence, the answer is (B).

10. We have the equations $x = a + 2$ and $b = x + 1$. Substituting the first equation into the second equation yields $b = (a + 2) + 1 = a + 3$. This equation shows that b is 3 units greater than a. The answer is (B).

11. We are given that $p = q + r$. Now, let's create the expression $q - r$ by subtracting $2r$ from both sides of this equation:

$$p - 2r = q + r - 2r$$

or

$$p - 2r = q - r$$

The answer is (C).

12. Let the two numbers be x and y. Since their sum is 13, $x + y = 13$. Since their product is 30, $xy = 30$. Solving the equation $xy = 30$ for y yields $y = 30/x$. Plugging this into the equation $x + y = 13$ yields

$$x + 30/x = 13$$
$$x^2 + 30 = 13x \quad \text{by multiplying both sides of the equation by } x$$
$$x^2 - 13x + 30 = 0 \quad \text{by subtracting } 13x \text{ from both sides of the equation}$$
$$(x - 3)(x - 10) = 0$$
$$x = 3 \text{ or } x = 10$$

Now, if $x = 3$, then $y = 13 - x = 13 - 3 = 10$. Hence, $x^2 + y^2 = 3^2 + 10^2 = 9 + 100 = 109$. The answer is (D).

Method II:
$(x + y)^2 = x^2 + y^2 + 2xy$. Hence, $x^2 + y^2 = (x + y)^2 - 2xy = 13^2 - 2(30) = 169 - 60 = 109$.

13. Solving the equation $p + q = 7$ for q yields $q = 7 - p$. Plugging this into the equation $pq = 12$ yields

$$p(7 - p) = 12$$
$$7p - p^2 = 12$$
$$p^2 - 7p + 12 = 0$$
$$(p - 3)(p - 4) = 0$$
$$p - 3 = 0 \text{ or } p - 4 = 0$$
$$p = 3 \text{ or } p = 4$$

If $p = 3$, then $q = 7 - p = 7 - 3 = 4$. Plugging these values into the expression $\frac{1}{p^2} + \frac{1}{q^2}$ yields

$$\frac{1}{3^2} + \frac{1}{4^2} =$$
$$\frac{1}{9} + \frac{1}{16} =$$
$$\frac{25}{144}$$

The result is the same for the other solution $p = 4$ (and then $q = 7 - p = 7 - 4 = 3$). The answer is (B).

Method II:

$$\frac{1}{p^2}+\frac{1}{q^2}=\frac{(q^2+p^2)}{p^2q^2}=\frac{(p+q)^2-2pq}{(pq)^2}=\frac{(7)^2-2(12)}{12^2}=\frac{49-24}{144}=\frac{25}{144}$$. The answer is (B).

14. Adding the two equations $2x + 3y = 11$ and $3x + 2y = 9$ yields $5x + 5y = 20$, or $x + y = 20/5 = 4$. The answer is (A).

15. Dividing the equation $2x = 2y + 1$ by 2 yields $x = y + 1/2$. Reading the equation yields "x is 1/2 unit greater than y." Or more simply, $x > y$. The answer is (A).

16. Subtracting x and $2y$ from both sides of the given equation $3x + y = x + 2y$ yields

$$3x + y - x - 2y = x + 2y - x - 2y$$
$$2x - y = 0$$

The answer is (A).

17. Adding the two given equations yields

$$(yz - zx) + (zx - xy) = 3 + 4$$
$$yz - zx + zx - xy = 7$$
$$yz - xy = 7$$
$$xy - yz = -7 \qquad \text{multiplying both sides by } -1$$

The answer is (A).

18. We are given the equation

$$(x+5)\div\left(\frac{1}{x}+\frac{1}{5}\right)=5$$
$$(x+5)\div\left(\frac{x+5}{5x}\right)=5$$
$$(x+5)\cdot\left(\frac{5x}{x+5}\right)=5$$
$$5x=5$$
$$x=1$$

The answer is (C). Note: If you solved the equation without getting a common denominator you may have gotten –5 as a possible solution. But, –5 is not a solution. Why? *

19. We are given the equation

$$(x+5)\left(\frac{1}{x}+\frac{1}{5}\right)=4$$
$$(x+5)\left(\frac{x+5}{5x}\right)=4$$
$$(x+5)^2=4x(5)$$
$$x^2+5^2+2x(5)=4x(5) \qquad \text{since } (a+b)^2=a^2+b^2+2ab$$
$$x^2+5^2-2x(5)=0$$
$$(x-5)^2=0 \qquad \text{since } (a-b)^2=a^2+b^2-2ab$$
$$x-5=0 \qquad \text{by taking the square root of both sides}$$
$$x=5$$

The answer is (D).

---

* Because –5 is not in the domain of the original equation since it causes the denominator to be 0. When you solve an equation, you are only finding possible solutions. The "solutions" may not work when plugged back into the equation.

20. We have the system of equations

$$x + y = 7$$
$$x^2 + y^2 = 25$$

Solving the top equation for y yields $y = 7 - x$. Substituting this into the bottom equation yields

$$x^2 + (7 - x)^2 = 25$$
$$x^2 + 49 - 14x + x^2 = 25$$
$$2x^2 - 14x + 24 = 0$$
$$x^2 - 7x + 12 = 0$$
$$(x - 3)(x - 4) = 0$$
$$x - 3 = 0 \text{ or } x - 4 = 0$$
$$x = 3 \text{ or } x = 4$$

Now, if $x = 3$, then $y = 7 - 3 = 4$ and Column A equals $x + y^2 = 3 + 4^2$ (= Column B); and if $x = 4$, then $y = 7 - 4 = 3$ and Column A = $x + y^2 = 4 + 3^2$ (≠ Column B). Hence, we have a double case, and the answer is (D).

21. We have the equation $x^2 - 3x + 2 = 0$. Factoring the left side yields $(x - 1)(x - 2) = 0$. Setting each factor to 0 yields $x - 1 = 0$ and $x - 2 = 0$. Solving for x yields $x = 1$ or 2.

Also, we have the equation $x^2 - 4x + 3 = 0$. Factoring the left side yields $(x - 1)(x - 3) = 0$. Setting each factor to 0 yields $x - 1 = 0$ and $x - 3 = 0$. Solving for x yields $x = 1$ or 3.

The common solution of the two equations is $x = 1$. Hence, $(x - 3)^2 = (1 - 3)^2 = (-2)^2 = 4$. The answer is (E).

Method II:
Subtracting the equation $x^2 - 4x + 3 = 0$ from the equation $x^2 - 3x + 2 = 0$ yields

$$x^2 - 3x + 2 - (x^2 - 4x + 3) = 0$$

$$x^2 - 3x + 2 - x^2 + 4x - 3 = 0$$

$$x - 1 = 0$$

$$x = 1$$

Hence, $(x - 3)^2 = (1 - 3)^2 = (-2)^2 = 4$. The answer is (E).

22. The given equation is

$$42.42 = k(14 + 7/50)$$
$$42.42 = k(14 + 14/100)$$
$$42.42 = k(14 + 0.14)$$
$$42.42 = k(14.14)$$
$$42.42/14.14 = k$$
$$3 = k$$

The answer is (C).

23. We have that $|2x - 4| = 2$. Since $|2x - 4|$ is only the positive value of $2x - 4$, the expression $2x - 4$ could equal 2 or -2. If $2x - 4$ equals 2, x equals 3; and if $2x - 4$ equals -2, x equals 1. We also have that $(x - 3)^2$ is equal to 4. By square rooting, we have that $x - 3$ may equal 2 (Here, $x = 3 + 2 = 5$), or $x - 3$ equals -2 (Here, $x = 3 - 2 = 1$). The common solution is $x = 1$. Hence, the answer is (A).

24. We are given that $a > 0$ and $(a + 2)(a - 3)(a + 4) = 0$. Hence, the possible solutions are

$a + 2 = 0$; $a = -2$, a is not greater than 0, so reject.
$a - 3 = 0$; $a = 3$, a is greater than 0, so accept.
$a + 4 = 0$; $a = -4$, a is not greater than 0, so reject.

The answer is (C).

➤ Hard

25. We are given the system of equations:

$$x + y = 7$$
$$x^2 + y^2 = 25$$

Solving the top equation for y yields $y = 7 - x$. Substituting this into the bottom equation yields

$$x^2 + (7 - x)^2 = 25$$
$$x^2 + 49 - 14x + x^2 = 25$$
$$2x^2 - 14x + 24 = 0$$
$$x^2 - 7x + 12 = 0$$
$$(x - 3)(x - 4) = 0$$
$$x - 3 = 0 \text{ or } x - 4 = 0$$
$$x = 3 \text{ or } x = 4$$

If $x = 3$, then $y = 7 - 3 = 4$. If $x = 4$, then $y = 7 - 4 = 3$. In either case, $x^3 + y^3 = 3^3 + 4^3 = 27 + 64 = 91$. The answer is (E).

26. The given system of equations is

$$x + l = 6$$
$$x - m = 5$$
$$x + p = 4$$
$$x - q = 3$$

Subtracting the second equation from the first one yields

$$(x + l) - (x - m) = 6 - 5$$
$$l + m = 1 \qquad \ldots (1)$$

Subtracting the fourth equation from the third one yields

$$(x + p) - (x - q) = 4 - 3$$
$$p + q = 1 \qquad \ldots (2)$$

Adding equations (1) and (2) yields

$$(l + m) + (p + q) = 1 + 1 = 2.$$
$$l + m + p + q = 2$$

The answer is (A).

27. We are given the two equations $1/m + 1/n = 4/3$ and $mn = 3$. From the second equation, we have $n = 3/m$. Substituting this in the equation $1/m + 1/n = 4/3$ yields $1/m + m/3 = 4/3$. Multiplying the equation by $3m$ yields $m^2 - 4m + 3 = 0$. The two possible solutions of this equation are 1 and 3.

When $m = 1$, $n = 3/m = 3/1 = 3$ and the expression $0.1 + 0.1^{1/m} + 0.1^{1/n}$ equals $0.1 + 0.1^{1/1} + 0.1^{1/3}$; and when $m = 3$, $n = 3/m = 3/3 = 1$ and the expression $0.1 + 0.1^{1/m} + 0.1^{1/n}$ equals $0.1 + 0.1^{1/3} + 0.1^{1/1}$.

In either case, the expressions equal $0.1 + 0.1^{1/3} + 0.1^{1/1} = 0.2 + 0.1^{1/3}$. Hence, the answer is (A).

28. We are given the two equations:

$$(x - 2y)(x + 2y) = 5$$
$$(2x - y)(2x + y) = 35$$

Applying the Difference of Squares formula, $(a + b)(a - b) = a^2 - b^2$, to the left-hand sides of each equation yields

$$x^2 - (2y)^2 = 5$$
$$(2x)^2 - y^2 = 35$$

Simplifying these two equations yields

$$x^2 - 4y^2 = 5$$
$$4x^2 - y^2 = 35$$

Subtracting the bottom equation from the top one yields

$$(x^2 - 4y^2) - (4x^2 - y^2) = 5 - 35$$
$$-3x^2 - 3y^2 = -30$$
$$-3(x^2 + y^2) = -30$$
$$x^2 + y^2 = -30/-3 = 10$$

Now, Column A – Column B = $(2x^2 - y^2) - (x^2 - 2y^2) = x^2 + y^2 = 10$. Hence, Column A is 10 units greater than Column B. The answer is (A).

29. Let each part of the given equation $\dfrac{x}{a+b-2c} = \dfrac{y}{b+c-2a} = \dfrac{z}{c+a-2b}$ equal t. Then we have

$$\frac{x}{a+b-2c} = \frac{y}{b+c-2a} = \frac{z}{c+a-2b} = t$$

Simplifying, we get $x = t(a + b - 2c) = at + bt - 2ct$, $y = t(b + c - 2a) = bt + ct - 2at$, and $z = t(c + a - 2b) = ct + at - 2bt$.

Hence, $x + y + z = (at + bt - 2ct) + (bt + ct - 2at) + (ct + at - 2bt) = 0$. The answer is (A).

30. Let each expression in the equation equal k. Then we have $\dfrac{x}{b-c} = \dfrac{y}{c-a} = \dfrac{z}{a-b} = k$. This reduces to $x = (b - c)k$, $y = (c - a)k$, $z = (a - b)k$. Now, $ax + by + cz$ equals

$$a(b - c)k + b(c - a)k + c(a - b)k =$$
$$k(ab - ac + bc - ba + ca - cb) =$$
$$k \times 0 =$$
$$0$$

The answer is (A).

31. Let each expression in the equation $\dfrac{x}{b-c} = \dfrac{y}{c-a} = \dfrac{z}{a-b}$ equal k. Then we have

$$\frac{x}{b-c} = \frac{y}{c-a} = \frac{z}{a-b} = k$$

Simplifying yields

$$x = k(b - c) = kb - kc$$
$$y = k(c - a) = kc - ka$$
$$z = k(a - b) = ka - kb$$

Hence, $x + y + z = (kb - kc) + (kc - ka) + (ka - kb) = 0$. The answer is (A).

32. Solving the given equation for y yields $y = 750 - x$. Now, let's substitute this into each answer-choice. The one that returns a numeric value for x is the answer.

Choice (A): $x + 2y = d$; $x + 2(750 - x) = d$; $x + 1500 - 2x = d$; $x = 1500 - d$; Since d is unknown, the value of x cannot be calculated. Reject.

Choice (B): $2x + 4y = 2d$; $2x + 4(750 - x) = 2d$; $2x + 3000 - 4x = 2d$; $-2x + 3000 = 2d$; $x = 1500 - d$; Since d is unknown, the value of x cannot be calculated. Reject.

Choice (C): $2x + 2y = 1500$; $2x + 2(750 - x) = 1500$; $2x + 1500 - 2x = 1500$; $0 = 0$, a known fact. No derivation is possible from this. Hence, the value of x cannot be calculated. Reject.

Choice (D): $3x = 2250 - 3y$; $3x = 2250 - 3(750 - x) = 3x$; $3x = 3x$; A known result. Hence, no derivation is possible. Reject.

Choice (E): $2x + y = 15$; $2x + 750 - x = 15$; $x + 750 = 15$; $x = 15 - 750 = -735$. We have numeric value for x here. Hence, accept it.

The answer is (E).

➤ Very Hard

33. We are given three equations $a^x = b$, $b^y = c$, and $c^z = a$. From the first equation, we have $b = a^x$. Substituting this in the second equation gives $(a^x)^y = c$. We can replace a in this equation with c^z (according to the third equation $c^z = a$):

$$\left(\left(c^z\right)^x\right)^y = c$$

$c^{xyz} = c^1$ By multiplying the exponents and writing c as c^1

$xyz = 1$ By equating the exponents of c on both sides

The answer is (B).

Averages

Problems involving averages are very common on the GRE. They can be classified into four major categories as follows.

Note! **The average of N numbers is their sum divided by N, that is,** $average = \dfrac{sum}{N}$.

Example 1: What is the average of x, $2x$, and 6?

(A) $x/2$
(B) $2x$
(C) $\dfrac{x+2}{6}$
(D) $x+2$
(E) $\dfrac{x+2}{3}$

By the definition of an average, we get $\dfrac{x+2x+6}{3}=\dfrac{3x+6}{3}=\dfrac{3(x+2)}{3}=x+2$. Hence, the answer is (D).

Note! ***Weighted average:*** **The average between two sets of numbers is closer to the set with more numbers.**

Example 2: If on a test three people answered 90% of the questions correctly and two people answered 80% correctly, then the average for the group is not 85% but rather $\dfrac{3\cdot 90+2\cdot 80}{5}=\dfrac{430}{5}=86$. Here, 90 has a weight of 3—it occurs 3 times. Whereas 80 has a weight of 2—it occurs 2 times. So the average is closer to 90 than to 80 as we have just calculated.

Note! **Using an average to find a number.**

Sometimes you will be asked to find a number by using a given average. An example will illustrate.

Example 3: If the average of five numbers is -10, and the sum of three of the numbers is 16, then what is the average of the other two numbers?

(A) –33 (B) –1 (C) 5 (D) 20 (E) 25

Let the five numbers be a, b, c, d, e. Then their average is $\dfrac{a+b+c+d+e}{5}=-10$. Now three of the numbers have a sum of 16, say, $a + b + c = 16$. So substitute 16 for $a + b + c$ in the average above: $\dfrac{16+d+e}{5}=-10$. Solving this equation for $d + e$ gives $d + e = -66$. Finally, dividing by 2 (to form the average) gives $\dfrac{d+e}{2}=-33$. Hence, the answer is (A).

Note! $Average\ Speed = \dfrac{Total\ Distance}{Total\ Time}$

Although the formula for average speed is simple, few people solve these problems correctly because most fail to find both the total distance and the total time.

Example 4: In traveling from city A to city B, John drove for 1 hour at 50 mph and for 3 hours at 60 mph. What was his average speed for the whole trip?

(A) 50
(B) $53\frac{1}{2}$
(C) 55
(D) 56
(E) $57\frac{1}{2}$

The total distance is $1 \cdot 50 + 3 \cdot 60 = 230$. And the total time is 4 hours. Hence,

$$Average\ Speed = \frac{Total\ Distance}{Total\ Time} = \frac{230}{4} = 57\frac{1}{2}$$

The answer is (E). Note, the answer is not the mere average of 50 and 60. Rather the average is closer to 60 because he traveled longer at 60 mph (3 hrs) than at 50 mph (1 hr).

Problem Set P:

➢ **Easy**

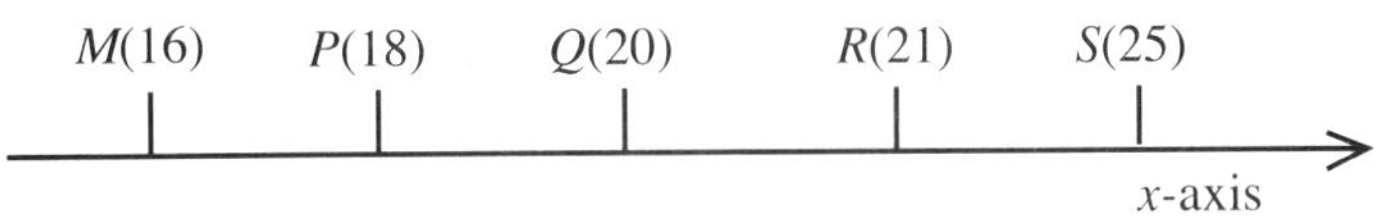

The figure is not drawn to scale.

1. Which one of the following points in the figure is the median of the points M, P, Q, R, and S ?

(A) M
(B) P
(C) Q
(D) R
(E) S

2.

Column A	Column B
The last digit in the average of the numbers 13 and 23	The last digit in the average of the numbers 113 and 123

3.

Column A		Column B
	The monthly rainfall (in inches) for the first eight months of 2008 was 2, 4, 4, 5, 7, 9, 10, 11.	
The mean monthly rainfall for thc 8 months		The median of the rainfall for thc 8 months

➢ Medium

4. A group of 30 employees of Cadre A has a mean age of 27. A different group of 70 employees of Cadre B has a mean age of 23. What is the mean age of the employees of the two groups together?

(A) 23
(B) 24.2
(C) 25
(D) 26.8
(E) 27

5. The difference between two angles of a triangle is 24°. The average of the same two angles is 54°. Which one of the following is the value of the greatest angle of the triangle?

(A) 45°
(B) 60°
(C) 66°
(D) 72°
(E) 78°

6. The average length of all the sides of a rectangle equals twice the width of the rectangle. If the area of the rectangle is 18, what is its perimeter?

(A) $6\sqrt{6}$
(B) $8\sqrt{6}$
(C) 24
(D) 32
(E) 48

7. In quadrilateral $ABCD$, $\angle A$ measures 20 degrees more than the average of the other three angles of the quadrilateral. Then $\angle A$ =

(A) 70°
(B) 85°
(C) 95°
(D) 105°
(E) 110°

8. The five numbers 1056, 1095, 1098, 1100, and 1126 are represented on a number line by the points A, B, C, D, and E, respectively, as shown in the figure. Which one of the following points represents the average of the five numbers?

(A) Point A
(B) Point B
(C) Point C
(D) Point D
(E) Point E

9. The arithmetic mean (average) of m and n is 50, and the arithmetic mean of p and q is 70. What is the arithmetic mean of m, n, p, and q?

(A) 55
(B) 65
(C) 60
(D) 120
(E) 130

10. Which one of the following numbers can be removed from the set $S = \{0, 2, 4, 5, 9\}$ without changing the average of set S?

(A) 0
(B) 2
(C) 4
(D) 5
(E) 9

> **Hard**

11. In a set of three numbers, the average of first two numbers is 2, the average of the last two numbers is 3, and the average of the first and the last numbers is 4. What is the average of three numbers?

(A) 2
(B) 2.5
(C) 3
(D) 3.5
(E) 4

12.

Column A		Column B
	The average ages of the players on team A and team B, are 20 and 30 years, respectively. The average age of the players of the teams together is 26.	
The number of players on team A		The number of players on team B

13. In 2007, the arithmetic mean of the annual incomes of Jack and Jill was \$3800. The arithmetic mean of the annual incomes of Jill and Jess was \$4800, and the arithmetic mean of the annual incomes of Jess and Jack was \$5800. What is the arithmetic mean of the incomes of the three?

(A) \$4000
(B) \$4200
(C) \$4400
(D) \$4800
(E) \$5000

14.

Column A		Column B
	Mike and Fritz ran a 30-mile Marathon. Mike ran 10 miles at 10 mph and the remaining 20 miles at 5 mph. Fritz ran one-third (by time) of the Marathon at 10 mph and the remaining two-thirds at 5 mph.	
Average speed of Mike		Average speed of Fritz

15.

Column A	Column B
The average of five consecutive integers starting from m	The average of six consecutive integers starting from m

16.

Column A		Column B
	The arithmetic mean (average) of the numbers a and b is 17. The geometric mean of the numbers a and b is 8.	
a		b

➢ Very Hard

17. 40% of the employees in a factory are workers. All the remaining employees are executives. The annual income of each worker is \$390. The annual income of each executive is \$420. What is the average annual income of all the employees in the factory together?

(A) 390
(B) 405
(C) 408
(D) 415
(E) 420

18.

Column A		Column B
	The average of five different positive numbers is 25. Δ is the decrease in the average when the smallest number among them is replaced by 0.	
Δ		5

Answers and Solutions to Problem Set P

➢ Easy

1. The definition of *median* is "When a set of numbers is arranged in order of size, the *median* is the middle number. If a set contains an even number of elements, then the median is the average of the two middle elements."

From the number line $M = 16$, $P = 18$, $Q = 20$, $R = 21$, and $S = 25$. The numbers arranged in order are 16, 18, 20, 21, and 25. The median is 20. Since $Q = 20$, the answer is (C).

2. The average of 13 and 23 is $(13 + 23)/2 = 36/2 = 18$, so the last digit is 8.

The average of 113 and 123 is $(113 + 123)/2 = 236/2 = 118$, so the last digit is also 8.

Hence, Column A equals Column B, and the answer is (C).

3. Column A: The mean rainfall for the 8 months is the sum of the eight rainfall measurements divided by 8:

$$\frac{2+4+4+5+7+9+10+11}{8} = 6.5$$

Column B: When a set of numbers is arranged in order of size, the *median* is the middle number. If a set contains an even number of elements, then the median is the average of the two middle elements. The average of 5 and 7 is 6, which is the median of the set. Hence, Column A is greater than Column B, and the answer is (A).

➢ Medium

4. Cadre A has 30 employees whose mean age is 27. Hence, the sum of their ages is $30 \times 27 = 810$. Cadre B has 70 employees whose mean age is 23. Hence, the sum of their ages is $23 \times 70 = 1610$. Now, the total sum of the ages of the 100 (= 30 + 70) employees is $810 + 1610 = 2420$. Hence, the average age is

The sum of the ages divided by the number of employees =

$$2420/100 =$$

$$24.2$$

The answer is (B).

5. Let a and b be the two angles in the question, with $a > b$. We are given that the difference between the angles is 24°, so $a - b = 24$. Since the average of the two angles is 54°, we have $(a + b)/2 = 54$. Solving for b in the first equation yields $b = a - 24$, and substituting this into the second equation yields

$$\frac{a+(a-24)}{2} = 54$$
$$\frac{2a-24}{2} = 54$$
$$2a - 24 = 54 \times 2$$
$$2a - 24 = 108$$
$$2a = 108 + 24$$
$$2a = 132$$
$$a = 66$$

Also, $b = a - 24 = 66 - 24 = 42$.

Now, let c be the third angle of the triangle. Since the sum of the angles in the triangle is 180°, $a + b + c = 180$. Plugging the previous results into the equation yields $66 + 42 + c = 180$. Solving for c yields $c = 72$. Hence, the greatest of the three angles a, b and c is c, which equals 72°. The answer is (D).

6. The perimeter of a rectangle is twice the sum of its length and width. Hence, if l and w are length and width, respectively, of the given rectangle, then the perimeter of the rectangle is $2(l + w)$. Also, the average side length of the rectangle is 1/4 times the sum. So, the average side length is $2(l + w)/4 = l/2 + w/2$.

Now, we are given that the average equals twice the width. Hence, we have $l/2 + w/2 = 2w$. Multiplying the equation by 2 yields $l + w = 4w$ and solving for l yields $l = 3w$.

Now, the area of the rectangle equals *length* × *width* = $l \times w = 18$ (given). Plugging $3w$ for l in the equation yields $3w \times w = 18$. Dividing the equation by 3 yields $w^2 = 6$, and square rooting both sides yields $w = \sqrt{6}$. Finally, the perimeter equals $2(l + w) = 2(3w + w) = 8w = 8\sqrt{6}$. The answer is (B).

7. Setting the angle sum of the quadrilateral to 360° yields $\angle A + \angle B + \angle C + \angle D = 360$. Subtracting $\angle A$ from both sides yields $\angle B + \angle C + \angle D = 360 - \angle A$. Forming the average of the three angles $\angle B$, $\angle C$, and $\angle D$ yields $(\angle B + \angle C + \angle D)/3$ and this equals $(360 - \angle A)/3$, since we know that $\angle B + \angle C + \angle D = 360 - \angle A$. Now, we are given that $\angle A$ measures 20 degrees more than the average of the other three angles. Hence, $\angle A = (360 - \angle A)/3 + 20$. Solving the equation for $\angle A$ yields $\angle A = 105$. The answer is (D).

8. The average of the five numbers 56, 95, 98, 100, and 126 is

$$\frac{56+95+98+100+126}{5} =$$
$$\frac{475}{5} =$$
$$95$$

Hence, the average of the five numbers 1056 (= 1000 + 56), 1095 (= 1000 + 95), 1098 (= 1000 + 98), 1100 (= 1000 + 100), and 1126 (= 1000 + 126) must be 1000 + 95 = 1095. The point that represents the number on the number line is point B. Hence, the answer is (B).

9. The arithmetic mean of m and n is 50. Hence, $(m + n)/2 = 50$. Multiplying the equation by 2 yields $m + n = 100$.

The arithmetic mean of p and q is 70. Hence, $(p + q)/2 = 70$. Multiplying the equation by 2 yields $p + q = 140$.

Now, the arithmetic mean of m, n, p, and q is

$$\frac{m+n+p+q}{4} =$$
$$\frac{(m+n)+(p+q)}{4} =$$
$$\frac{100+140}{4} =$$
$$\frac{240}{4} =$$
$$60$$

The answer is (C).

10. The average of the elements in the original set S is $(0 + 2 + 4 + 5 + 9)/5 = 20/5 = 4$. If we remove an element that equals the average, then the average of the new set will remain unchanged. The new set after removing 4 is $\{0, 2, 5, 9\}$. The average of the elements is $(0 + 2 + 5 + 9)/4 = 16/4 = 4$. The answer is (C).

➢ Hard

11. Let the three numbers be x, y, and z. We are given that

$$\frac{x+y}{2}=2$$

$$\frac{y+z}{2}=3$$

$$\frac{x+z}{2}=4$$

Summing the three equations yields

$$\frac{x+y}{2}+\frac{y+z}{2}+\frac{x+z}{2}=2+3+4$$

$$\frac{x}{2}+\frac{y}{2}+\frac{y}{2}+\frac{z}{2}+\frac{x}{2}+\frac{z}{2}=9$$

$$x+y+z=9$$

The average of the three numbers is $(x + y + z)/3 = 9/3 = 3$. The answer is (C).

12. Let the number of players on team A be a and the number of players on team B be b. Since the average age of the players on team A is 20, the sum of the ages of the players on the team is $20a$. Similarly, since the average age of the players on team B is b, the sum of the ages of the players on the team is $30b$.

Now, the average age of the players of the two teams together is

$$\frac{\text{Sum of the ages of the players on each team}}{\text{Total number of players on the teams}}=\frac{20a+30b}{a+b}$$

We are given that this average is 26. Hence, we have

$$\frac{20a+30b}{a+b}=26$$

$20a + 30b = 26a + 26b$ by multiplying both sides by $a + b$

$4b = 6a$ by subtracting $20a + 26b$ from both sides

$b/a = 6/4 = 3/2$

Since a and b are both positive (being the team strengths) and since b/a equals 3/2, which is greater than 1, b must be greater than a. Hence, Column B is greater than Column A, and the answer is (B).

13. Let a, b, and c be the annual incomes of Jack, Jill, and Jess, respectively.

Now, we are given that

The arithmetic mean of the annual incomes of Jack and Jill was \$3800. Hence, $(a + b)/2 = 3800$. Multiplying by 2 yields $a + b = 2 \times 3800 = 7600$.

The arithmetic mean of the annual incomes of Jill and Jess was \$4800. Hence, $(b + c)/2 = 4800$. Multiplying by 2 yields $b + c = 2 \times 4800 = 9600$.

The arithmetic mean of the annual incomes of Jess and Jack was \$5800. Hence, $(c + a)/2 = 5800$. Multiplying by 2 yields $c + a = 2 \times 5800 = 11{,}600$.

Summing these three equations yields

$$(a + b) + (b + c) + (c + a) = 7600 + 9600 + 11{,}600$$

$$2a + 2b + 2c = 28{,}800$$

$$a + b + c = 14{,}400$$

The average of the incomes of the three equals the sum of the incomes divided by 3:

$$(a + b + c)/3 =$$

$$14{,}400/3 =$$

$$4800$$

The answer is (D).

14. Mike ran 10 miles at 10 mph (for the *Time* = *Distance*/*Rate* = 10 miles/10 mph = 1 hour). He ran the remaining 20 miles at 5 mph (for the *Time* = *Distance*/*Rate* = 20 miles/5 mph = 4 hrs). The length of the Marathon track is 30 miles, and the total time taken to cover the track is 5 hours.

Now, let the time taken by Fritz to travel the 30-mile Marathon track be t hours. Then as given, Fritz ran at 10 mph for $t/3$ hours and at 5 mph for the rest of $2t/3$ hours. Hence, by formula, *Distance* = *Rate* × *time*, the distance covered is $(10 \text{ mph}) \cdot \frac{t}{3} + (5 \text{ mph}) \cdot \frac{2t}{3} = \left(\frac{10}{3} + \frac{10}{3}\right) \cdot t = \frac{20}{3} t = 30$ miles. Solving the equation for t yields $t = 90/20 = 4.5$ hours.

Since Fritz took less time to cover the Marathon than Mike, the average speed of Fritz is greater. Hence, Column B is greater than Column A, and the answer is (B).

15. Column A: The five consecutive integers starting from m are m, $m + 1$, $m + 2$, $m + 3$, and $m + 4$. The average of the five numbers equals

$$\frac{\text{The sum of the five numbers}}{5} =$$

$$\frac{m + (m+1) + (m+2) + (m+3) + (m+4)}{5} =$$

$$\frac{5m + 10}{5} =$$

$$m + 2$$

Column B: The six consecutive integers starting from m are m, $m + 1$, $m + 2$, $m + 3$, $m + 4$, and $m + 5$. The average of the six numbers equals

$$\frac{\text{The sum of the six numbers}}{6} =$$

$$\frac{m + (m+1) + (m+2) + (m+3) + (m+4) + (m+5)}{6} =$$

$$\frac{6m + 15}{6} =$$

$$m + \frac{5}{2} =$$

$$m + 2 + \frac{1}{2} =$$

$$(m + 2) + \frac{1}{2} =$$

$$\text{Column A} + \frac{1}{2}$$

Since Column B is 1/2 units greater than Column A, the answer is (B).

Method II
Choose any five consecutive integers, say, –2, –1, 0, 1 and 2. (We chose these particular numbers to make the calculation as easy as possible. But any five consecutive integers will do. For example, 1, 2, 3, 4, and 5.) Forming the average yields $\frac{-1+(-2)+0+1+2}{5}=\frac{0}{5}=0$. Now, add 3 to the set to form 6 consecutive integers: –2, –1, 0, 1, 2, and 3. Forming the average yields

$$\frac{-1+(-2)+0+1+2+3}{6}=$$

$$\frac{[-1+(-2)+0+1+2]+3}{6}=$$

$$\frac{[0]+3}{6}=$$ since the average of $-1+(-2)+0+1+2$ is zero, their sum must be zero

$$\frac{3}{6}=$$

$$\frac{1}{2}$$

Since 1/2 > 0, Column B is greater than Column A and the answer is (B).

16. The arithmetic mean of the numbers a and b is 17. Hence, $\frac{a+b}{2}=17$.

The geometric mean of the numbers a and b is 8. Hence, $\sqrt{ab}=8$.

Now, assume $a = b$. Then $\frac{a+b}{2}=\frac{a+a}{2}=a=17$. Hence, $\sqrt{ab}=\sqrt{aa}=a=17$. But, we are given that $\sqrt{ab}=8$. Hence, our assumption is wrong. So $a \neq b$.

Now, suppose the solution for the question is $a = s$ and $b = t$, and say s is greater than t. Then $(s + t)/2$ must equal 17 and $\sqrt{st}$ must equal 8. Here, Column A = $a > b$ = Column B.

Even $a = t$ and $b = s$ is possible (by substitution we get, $(t + s)/2 = 17$ and $\sqrt{st}=8$). Here, $a < b$. Hence, we have a double case, and the answer is (D).

➢ Very Hard

17. Let e be the number of employees.

We are given that 40% of the employees are workers. Now, 40% of e is $40/100 \times e = 0.4e$. Hence, the number of workers is $2e/5$.

All the remaining employees are executives, so the number of executives equals

(The number of Employees) – (The number of Workers) =

$e - 2e/5$ =

$3e/5$

The annual income of each worker is \$390. Hence, the total annual income of all the workers together is $2e/5 \times 390 = 156e$.

Also, the annual income of each executive is \$420. Hence, the total income of all the executives together is $3e/5 \times 420 = 252e$.

Hence, the total income of the employees is $156e + 252e = 408e$.

The average income of all the employees together equals

(The total income of all the employees) ÷ (The number of employees) =

$$408e/e =$$

$$408$$

The answer is (C).

18. Let $a, b, c, d,$ and e be the five positive numbers in the decreasing order of size such that e is the smallest number. We are given that the average of the five numbers is 25. Hence, we have the equation

$$\frac{a+b+c+d+e}{5} = 25$$

$a + b + c + d + e = 125$ ... (1) by multiplying by 5

The smallest number in a set is at least less than the average of the numbers in the set if at least one number is different. For example, the average of 1, 2, and 3 is 2, and the smallest number in the set 1 is less than the average 2. Hence, we have the inequality

$0 < e < 25$

$0 > -e > -25$ by multiplying both sides of the inequality by –1 and flipping the directions of the inequalities.

Adding this inequality to equation (1) yields

$0 + 125 > (a + b + c + d + e) + (-e) > 125 - 25$

$125 > a + b + c + d > 100$

$125 > a + b + c + d + 0 > 100$ by adding by 0

$25 > \dfrac{a+b+c+d+0}{5} = > 20$ by dividing the inequality by 5

$25 >$ The average of numbers a, b, c, d and $0 > 20$

Hence, Column A equals Δ =

(Average of the numbers a, b, c, d, and e) – (Average of the numbers a, b, c, and d) =

25 – (A number between 20 and 25) =

A number less than 5

Hence, Column A is less than Column B, and the answer is (B).

Ratio & Proportion

RATIO

A ratio is simply a fraction. The following notations all express the ratio of x to y: $x:y$, $x \div y$, or $\frac{x}{y}$. Writing two numbers as a ratio provides a convenient way to compare their sizes. For example, since $\frac{3}{\pi} < 1$, we know that 3 is less than π. A ratio compares two numbers. Just as you cannot compare apples and oranges, so to must the numbers you are comparing have the same units. For example, you cannot form the ratio of 2 feet to 4 yards because the two numbers are expressed in different units—feet vs. yards. It is quite common for the GRE to ask for the ratio of two numbers with different units. Before you form any ratio, make sure the two numbers are expressed in the same units.

Example 1: What is the ratio of 2 feet to 4 yards?

(A) 1 : 9 (B) 1 : 8 (C) 1 : 7 (D) 1 : 6 (E) 1 : 5

The ratio cannot be formed until the numbers are expressed in the same units. Let's turn the yards into feet. Since there are 3 feet in a yard, $4 \text{ yards} = 4 \times 3 \text{ feet} = 12 \text{ feet}$. Forming the ratio yields

$$\frac{2 \textit{ feet}}{12 \textit{ feet}} = \frac{1}{6} \textit{ or } 1:6$$

The answer is (D).

Note, taking the reciprocal of a fraction usually changes its size. For example, $\frac{3}{4} \neq \frac{4}{3}$. So order is important in a ratio: $3:4 \neq 4:3$.

PROPORTION

A proportion is simply an equality between two ratios (fractions). For example, the ratio of x to y is equal to the ratio of 3 to 2 is translated as

$$\frac{x}{y} = \frac{3}{2}$$

or in ratio notation,

$$x:y::3:2$$

Two variables are *directly proportional* if one is a constant multiple of the other:

$$y = kx$$

where k is a constant.

The above equation shows that as x increases (or decreases) so does y. This simple concept has numerous applications in mathematics. For example, in constant velocity problems, distance is directly proportional to time: $d = vt$, where v is a constant. Note, sometimes the word *directly* is suppressed.

Example 2: If the ratio of y to x is equal to 3 and the sum of y and x is 80, what is the value of y?

(A) –10 (B) –2 (C) 5 (D) 20 (E) 60

Translating *"the ratio of y to x is equal to 3"* into an equation yields

$$\frac{y}{x} = 3$$

Translating *"the sum of y and x is 80"* into an equation yields

$$y + x = 80$$

Solving the first equation for y gives $y = 3x$. Substituting this into the second equation yields

$$3x + x = 80$$
$$4x = 80$$
$$x = 20$$

Hence, $y = 3x = 3 \cdot 20 = 60$. The answer is (E).

In many word problems, as one quantity increases (decreases), another quantity also increases (decreases). This type of problem can be solved by setting up a *direct* proportion.

Example 3: If Biff can shape 3 surfboards in 50 minutes, how many surfboards can he shape in 5 hours?

(A) 16 (B) 17 (C) 18 (D) 19 (E) 20

As time increases so does the number of shaped surfboards. Hence, we set up a direct proportion. First, convert 5 hours into minutes: $5\ hours = 5 \times 60\ minutes = 300\ minutes$. Next, let x be the number of surfboards shaped in 5 hours. Finally, forming the proportion yields

$$\frac{3}{50} = \frac{x}{300}$$
$$\frac{3 \cdot 300}{50} = x$$
$$18 = x$$

The answer is (C).

Example 4: On a map, 1 inch represents 150 miles. What is the actual distance between two cities if they are $3\frac{1}{2}$ inches apart on the map?

(A) 225 (B) 300 (C) 450 (D) 525 (E) 600

As the distance on the map increases so does the actual distance. Hence, we set up a direct proportion. Let x be the actual distance between the cities. Forming the proportion yields

$$\frac{1\,in}{150\,mi} = \frac{3\,{}^1\!/_2\,in}{x\,mi}$$
$$x = 3\,{}^1\!/_2 \times 150$$
$$x = 525$$

The answer is (D).

Note, you need not worry about how you form the direct proportion so long as the order is the same on both sides of the equal sign. The proportion in Example 4 could have been written as $\frac{1\,in}{3\,{}^1\!/_2\,in} = \frac{150\,mi}{x\,mi}$. In this case, the order is inches to inches and miles to miles. However, the following is not a direct proportion because the order is not the same on both sides of the equal sign: $\frac{1\,in}{150\,mi} = \frac{x\,mi}{3\,{}^1\!/_2\,in}$. In this case, the order is inches to miles on the left side of the equal sign but miles to inches on the right side.

If one quantity increases (or decreases) while another quantity decreases (or increases), the quantities are said to be *inversely* proportional. The statement "y is inversely proportional to x" is written as

$$y = \frac{k}{x}$$

where k is a constant.

Multiplying both sides of $y = \frac{k}{x}$ by x yields

$$yx = k$$

Hence, in an inverse proportion, the product of the two quantities is constant. Therefore, instead of setting ratios equal, we set products equal.

In many word problems, as one quantity increases (decreases), another quantity decreases (increases). This type of problem can be solved by setting up a product of terms.

Example 5: If 7 workers can assemble a car in 8 hours, how long would it take 12 workers to assemble the same car?

(A) 3hrs (B) $3\frac{1}{2}$hrs (C) $4\frac{2}{3}$hrs (D) 5hrs (E) $6\frac{1}{3}$hrs

As the number of workers increases, the amount time required to assemble the car decreases. Hence, we set the products of the terms equal. Let x be the time it takes the 12 workers to assemble the car. Forming the equation yields

$$7 \cdot 8 = 12 \cdot x$$

$$\frac{56}{12} = x$$

$$4\frac{2}{3} = x$$

The answer is (C).

To summarize: if one quantity increases (decreases) as another quantity also increases (decreases), set ratios equal. If one quantity increases (decreases) as another quantity decreases (increases), set products equal.

The concept of proportion can be generalized to three or more ratios. A, B, and C are in the ratio 3:4:5 means $\frac{A}{B} = \frac{3}{4}$, $\frac{A}{C} = \frac{3}{5}$, and $\frac{B}{C} = \frac{4}{5}$.

Example 6: In the figure to the right, the angles A, B, C of the triangle are in the ratio 5:12:13. What is the measure of angle A?

(A) 15
(B) 27
(C) 30
(D) 34
(E) 40

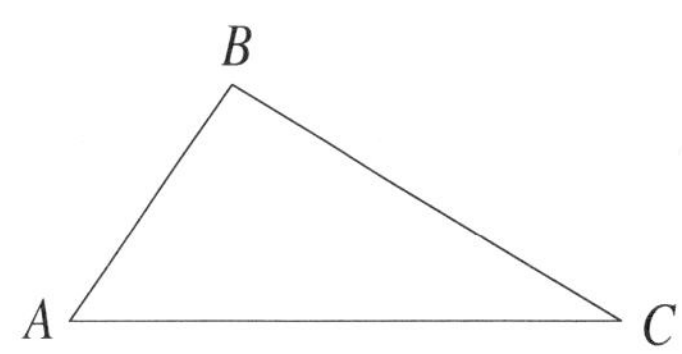

Since the angle sum of a triangle is 180°, $A + B + C = 180$. Forming two of the ratios yields

$$\frac{A}{B} = \frac{5}{12} \qquad \frac{A}{C} = \frac{5}{13}$$

Solving the first equation for B yields $B = \frac{12}{5}A$

Solving the second equation for C yields $C = \frac{13}{5}A$

Hence, $180 = A + B + C = A + \frac{12}{5}A + \frac{13}{5}A = 6A$. Therefore, $180 = 6A$, or $A = 30$. The answer is choice (C).

Problem Set Q

➢ Easy

1. At Stephen Stores, 3 pounds of cashews costs \$8. What is cost in cents of a bag weighing 9 ounces?

   (A) 30
   (B) 60
   (C) 90
   (D) 120
   (E) 150

➢ Medium

2. In the figure, what is the value of y if $x : y = 2 : 3$?

   (A) 16
   (B) 32
   (C) 48
   (D) 54
   (E) 72

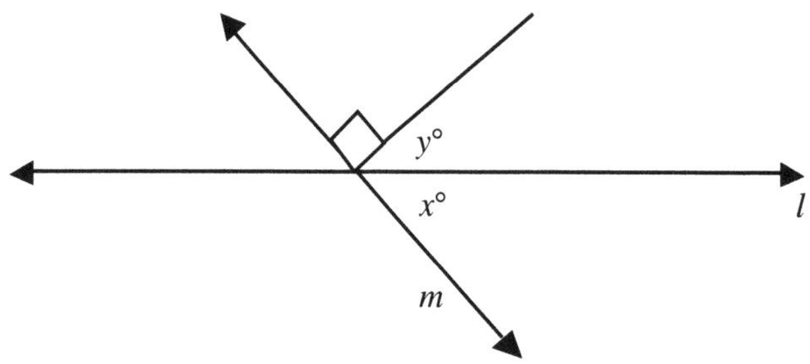

3. In the figure, $ABCD$ is a rectangle and points E, F, G and H are midpoints of its sides. What is the ratio of the area of the shaded region to the area of the un-shaded region in the rectangle?

   (A) 1 : 1
   (B) 1 : 2
   (C) 2 : 1
   (D) 1 : 3
   (E) 3 : 1

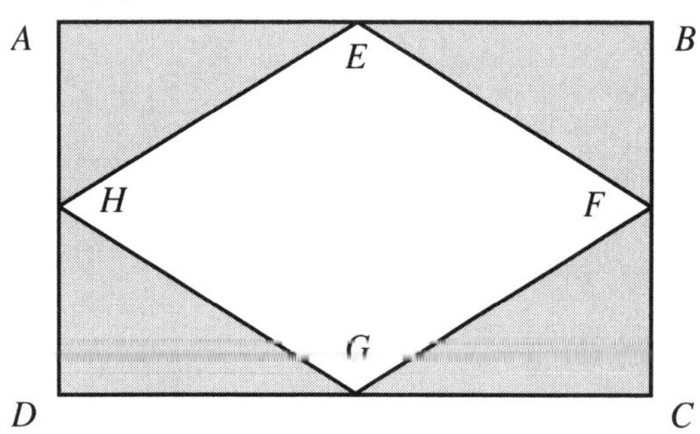

4. If r and s are two positive numbers and r is 25% greater than s, what is the value of the ratio r/s ?

   (A) 0.75
   (B) 0.8
   (C) 1
   (D) 1.2
   (E) 1.25

5. The ratio of the sum of the reciprocals of x and y to the product of the reciprocals of x and y is 1 : 3. What is sum of the numbers x and y?

(A) 1/3
(B) 1/2
(C) 1
(D) 2
(E) 4

6. The ratio of the number of chickens to the number of pigs to the number of horses on Richard's farm is 33 : 17 : 21. What fraction of the animals are either pigs or horses?

(A) 16/53
(B) 17/54
(C) 38/71
(D) 25/31
(E) 38/33

7.

Column A		Column B
	The savings from a person's income is the difference between his income and his expenditure. The ratio of the incomes of Marc and his boss are in the ratio 3 : 4. Their respective expenditures ratio is 1 : 2.	
The savings from income of Marc		The savings from income of Marc's Boss

8.

Column A		Column B
	A certain recipe requires 3/2 cups of sugar and makes 2-dozen cookies.	
The number of cups of sugar required for the same recipe to make 30 cookies		2

9. If a, b, and c are three different numbers and $ax : by : cz = 1 : 2 : -3$, then $ax + by + cz =$

(A) 0
(B) 1/2
(C) 3
(D) 6
(E) 9

10.

Column A		Column B
	$a : b = 2 : 3$. a is positive.	
$\frac{a+5}{b+5}$		1

➤ Hard

11. In Figure 1, $y = \sqrt{3}x$ and $z = 2x$. What is the ratio $p : q : r$ in Figure 2?

    (A) 1 : 2 : 3
    (B) $\sqrt{3}$: 1 : 2
    (C) 1 : $\sqrt{3/2}$: 1
    (D) 2 : $\sqrt{3}$: 1
    (E) 3 : 2 : 1

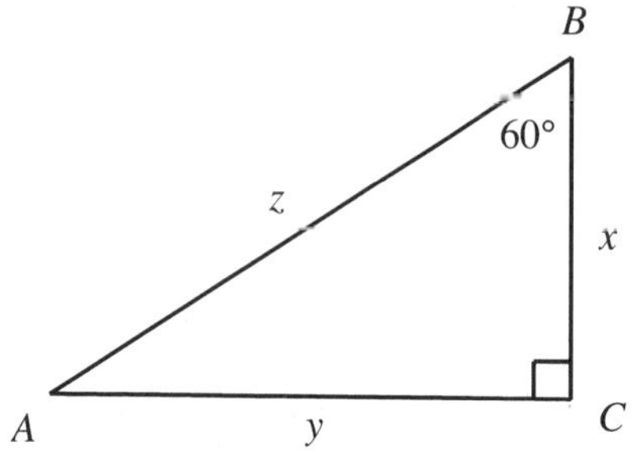

Figure 1

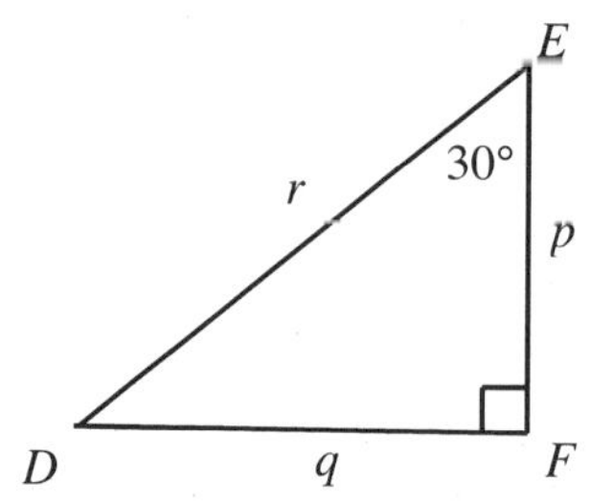

Figure 2

Note: The figures are not drawn to scale.

12. In the figure, the ratio of the area of parallelogram *ABCD* to the area of rectangle *AECF* is 5 : 3. What is the area of the rectangle *AECF* ?

    (A) 18
    (B) 24
    (C) 25
    (D) 50
    (E) 54

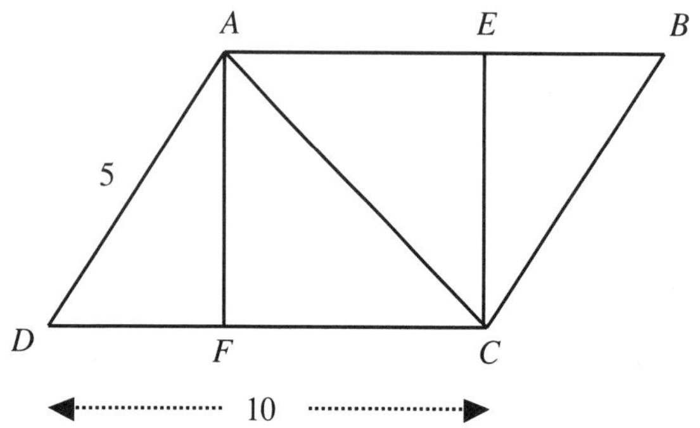

13. In a class, 10% of the girls have blue eyes, and 20% of the boys have blue eyes. If the ratio of girls to boys in the class is 3 : 4, then what is the fraction of the students in the class having blue eyes?

    (A) 11/70
    (B) 11/45
    (C) 14/45
    (D) 12/33
    (E) 23/49

14. The cost of production of a certain instrument is directly proportional to the number of units produced. The cost of production for 300 units is \$300. What is the cost of production for 270 units?

    (A) 270
    (B) 300
    (C) 325
    (D) 370
    (E) 395

15. Kelvin takes 3 minutes to inspect a car, and John takes 4 minutes to inspect a car. If they both start inspecting different cars at 8:30 AM, what would be the ratio of the number of cars inspected by Kelvin and John by 8:54 AM of the same day?

    (A) 1 : 3
    (B) 1 : 4
    (C) 3 : 4
    (D) 4 : 3
    (E) 7 : 4

16. Bank X pays a simple interest of \$80 on a principal of \$1,000 annually. Bank Y pays a simple interest of \$140 on a principal of \$1,000 annually. What is the ratio of the interest rates of Bank X to Bank Y ?

    (A) 5 : 8
    (B) 8 : 5
    (C) 14 : 8
    (D) 4 : 7
    (E) 5 : 7

17. If $x = 10a$, $y = 3b$, $z = 7c$, and $x : y : z = 10 : 3 : 7$, then $\dfrac{7x + 2y + 5z}{8a + b + 3c} =$

    (A) 111/12
    (B) 7/6
    (C) 8/15
    (D) 108/123
    (E) 135/202

18. The ratio of the angles in ΔABC is 2 : 3 : 4. Which one of the following triangles is similar to ΔABC ?

    (A) ΔDEF has angles in the ratio 4 : 3 : 2.
    (B) ΔPQR has angles in the ratio 1 : 2 : 3.
    (C) ΔLMN has angles in the ratio 1 : 1 : 1.
    (D) ΔSTW has sides in the ratio 1 : 1 : 1.
    (E) ΔXYZ has sides in the ratio 4 : 3 : 2.

19. In what proportion must rice at \$0.8 per pound be mixed with rice at \$0.9 per pound so that the mixture costs \$0.825 per pound?

    (A) 1 : 3
    (B) 1 : 2
    (C) 1 : 1
    (D) 2 : 1
    (E) 3 : 1

20.

Column A		Column B
	$x = a$, $y = 2b$, $z = 3c$. $x : y : z = 1 : 2 : 3$.	
$\dfrac{x + y + z}{a + b + c}$		2

21. A *perfect square* is a number that becomes an integer when square rooting it. A, B, and C are three positive integers. The ratio of the three numbers is 1 : 2 : 3, respectively. Which one of the following expressions must be a perfect square?

(A) $A + B + C$
(B) $A^2 + B^2 + C^2$
(C) $A^3 + B^3 + C^3$
(D) $3A^2 + B^2 + C^2$
(E) $3A^2 + 4B^2 + 4C^2$

22.

A precious stone was accidentally dropped and broke into 3 pieces of equal weight. The value of this type of stone is always proportional to the square of its weight.

Column A	Column B
The value of the 3 broken pieces together	The value of the original stone

23.

A spirit and water solution is sold in a market. The cost per liter of the solution is directly proportional to the part (fraction) of spirit (by volume) the solution has.

A solution of 1 liter of spirit and 1 liter of water costs x dollars.

A solution of 1 liter of spirit and 2 liters of water costs y dollars.

Column A	Column B

24.

a and b are positive.
$(a + 6) : (b + 6) = 5 : 6$

Column A	Column B
$\frac{a+10}{b+10}$	1

25.

Five years ago, in a zoo, the ratio of the number of cheetahs to the number of pandas was 1 : 3. The ratio is now 1 : 2.

Column A	Column B
The increase in the number of cheetahs in the zoo in the last five years	The increase in the number of pandas in the zoo in the last five years

26. Column A | In the figure, two circles with centers A and B touch a larger circle with center O internally. The ratio of the radii of Circle A to Circle B is 7 : 9. | Column B

OA | | OB

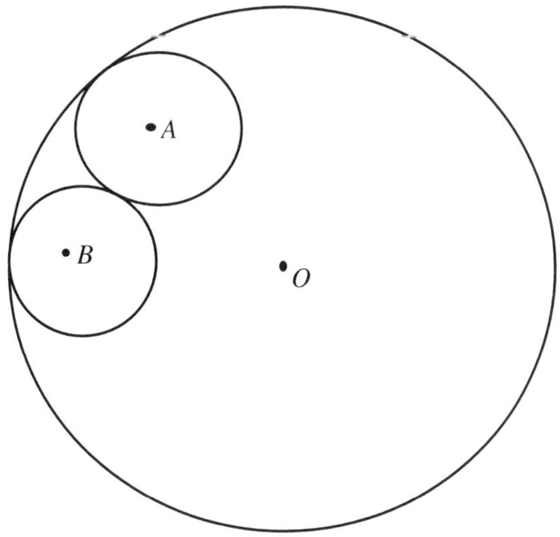

27. Column A | a and b are positive. $(a + 6) : (b + 6) = 5 : 6$ | Column B

b | | 1

28. Column A | The monthly incomes of A and B are in the ratio 3 : 2, and the monthly expenditures of the two are in the ratio 4 : 3.

The income of both is greater than the expenditures, and the *Savings* from income is defined as the Income – Expenditure. | Column B

Savings of A | | Savings of B

➢ **Very Hard**

29. Two alloys A and B are composed of two basic elements. The ratios of the compositions of the two basic elements in the two alloys are 5 : 3 and 1 : 2, respectively. A new alloy X is formed by mixing the two alloys A and B in the ratio 4 : 3. What is the ratio of the composition of the two basic elements in alloy X ?

(A) 1 : 1
(B) 2 : 3
(C) 5 : 2
(D) 4 : 3
(E) 7 : 9

30. Joseph bought two varieties of rice, costing 5 cents per ounce and 6 cents per ounce each, and mixed them in some ratio. Then he sold the mixture at 7 cents per ounce, making a profit of 20 percent. What was the ratio of the mixture?

(A) 1 : 10
(B) 1 : 5
(C) 2 : 7
(D) 3 : 8
(E) 5 : 7

31. The Savings of an employee equals Income minus Expenditure. If their Incomes ratio is 1 : 2 : 3 and their Expenses ratio is 3 : 2 : 1, then what is the order of the employees A, B, and C in the increasing order of the size of their savings?

(A) $A > B > C$
(B) $A > C > B$
(C) $B > A > C$
(D) $B > C > A$
(E) $C > B > A$

32. A cask initially contains pure alcohol up to the brim. The cask can be emptied by removing exactly 5 liters at a time . Each time this is done, the cask must be filled back to the brim with water. The capacity of the cask is 15 liters. When the cask is completely emptied and filled back to the brim two times, what is the ratio of alcohol to water in the cask?

(A) 1 : 2
(B) 2 : 1
(C) 1 : 1
(D) 4 : 5
(E) 5 : 4

Answers and Solutions to Problem Set Q

➢ Easy

1. This problem can be solved by setting up a proportion. Note that 1 pound has 16 ounces, so 3 pounds has 48 (= 3 × 16) ounces. Now, the proportion, in cents to ounces, is

$$\frac{800}{48} = \frac{\text{cents}}{9}$$

or

$$\text{cents} = 9 \cdot \frac{800}{48} = 150$$

The answer is (E).

➢ Medium

2. We know that the angle made by a line is 180°. Applying this to line m yields $x + y + 90 = 180$. Subtracting 90 from both sides of this equation yields $x + y = 90$. We are also given that $x : y = 2 : 3$. Hence, $x/y = 2/3$. Multiplying this equation by y yields $x = 2y/3$. Plugging this into the equation $x + y = 90$ yields

$$2y/3 + y = 90$$
$$5y/3 = 90$$
$$5y = 270$$
$$y = 54$$

The answer is (D).

3. Joining the midpoints of the opposite sides of the rectangle *ABCD* yields the following figure:

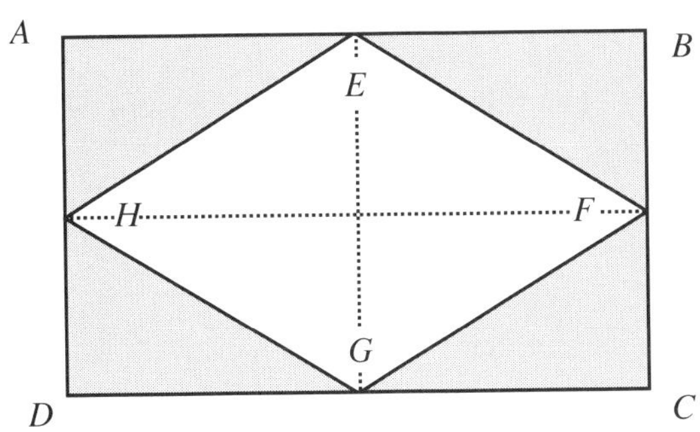

In the given figure, the bigger rectangle *ABCD* contains four small rectangles, each one divided by a diagonal. Since diagonals cut a rectangle into two triangles of equal area, in each of the small rectangles, the regions to either side (shaded and un-shaded) have equal area. Hence, even in the bigger rectangle, the area of the shaded and un-shaded regions are equal, so the required ratio is 1 : 1. The answer is (A).

4. We are given that r is 25% greater than s. Hence, $r = (1 + 25/100)s$. So, $r/s = 1 + 25/100 = 1.25$. The answer is (E).

5. We are given that the ratio of the sum of the reciprocals of x and y to the product of the reciprocals of x and y is 1 : 3. Writing this as an equation yields

$$\frac{\frac{1}{x}+\frac{1}{y}}{\frac{1}{x}\cdot\frac{1}{y}}=\frac{1}{3}$$

$$\frac{\frac{x+y}{xy}}{\frac{1}{xy}}=\frac{1}{3}$$

$$\frac{x+y}{xy}\cdot\frac{xy}{1}=\frac{1}{3}$$

$$x+y=\frac{1}{3}$$ by canceling xy from the numerator and denominator

The answer is (A).

6. Let the number of chickens, pigs and horses on Richard's farm be c, p, and h. Then forming the given ratio yields $c : p : h = 33 : 17 : 21$. Let c equal $33k$, p equal $17k$, and h equal $21k$, where k is a positive integer (such that $c : p : h = 33 : 17 : 21$). Then the total number of pigs and horses is $17k + 21k = 38k$; and the total number of pigs, horses and chickens is $17k + 21k + 33k = 71k$. Hence, the required fraction equals $38k/71k = 38/71$. The answer is (C).

7. Suppose Marc's income is \$300 and his boss's income is \$400 (incomes match the given ratio, 3 : 4). Also, suppose Marc's expenditure is \$100 and his boss's expenditure is \$200 (expenditures match the given ratio, 1 : 2). Then the savings from income of Marc is 300 – 100 = 200 and that of his boss is 400 – 200 = 200. In this case, Marc's savings equals his boss's savings.

Now, suppose the expenditures are different: Marc's expenditure is \$150 and his boss's expenditure is \$300 (expenditures match the given ratio, 1 : 2). In this case, Marc's savings is 300 – 150 = 150, and his boss's savings is 400 – 300 = 100 (savings are not equal). Hence, we have a double case, and the answer is (D).

8. This problem can be solved by setting up a proportion between the number of cookies and the number of cups of sugar required to make the corresponding number of cookies. Since there are 12 items in a dozen, 2-dozen cookies is 2 × 12 = 24 cookies. Since 3/2 cups are required to make the 24 cookies, we have the proportion

$$\frac{24 \text{ cookies}}{\frac{3}{2} \text{ cookies}}=\frac{30 \text{ cookies}}{x \text{ cups}}$$

$$24x = 30 \cdot 3/2 = 45$$ by Cross-multiplying

$$x = 45/24$$

Hence, Column A equals 45/24, which is less than 2 (= Column B). Hence, Column B is greater, and the answer is (B).

9. We are given that $ax : by : cz = 1 : 2 : -3$. Let $ax = t$, $by = 2t$, and $cz = -3t$ (such that $ax : by : cz = 1 : 2 : -3$). Then $ax + by + cz = t + 2t - 3t = 0$. The answer is (A).

10. Forming the ratio yields $a/b = 2/3$. Solving for b yields $b = 3a/2$. Since a is positive (given), this equation tells us that b is also positive. Hence, $b + 5$ is positive, and we can safely multiply both columns by $b + 5$. This yields

$a + 5$	$b + 5$

Subtracting 5 from both columns yields

a	b

Since both a and b are positive and b is one-and-a-half times larger than a ($b = 3a/2$), Column B is larger than Column A. The answer is (B).

➢ **Hard**

11. Angles $\angle B$ and $\angle C$ in triangle ABC equal 60° and 90°, respectively. Since the sum of the angles in a triangle is 180°, the third angle of the triangle, $\angle A$, must equal 180° – (60° + 90°) = 30°.

So, in the two triangles, ABC and EDF, we have $\angle A = \angle E = 30°$ and $\angle C = \angle F = 90°$ (showing that at least two corresponding angles are equal). So, the two triangles are similar. Hence, the ratios of the corresponding sides in the two triangles are equal. Hence, we have

$EF : DF : DE = AC : BC : AB$

$p : q : r = y : x : z$ after substitutions from the figure

$= \sqrt{3}x : x : 2x = \sqrt{3} : 1 : 2$ $y = \sqrt{3}x$ and $z = 2x$, given

The answer is (B).

12. The area of the parallelogram $ABCD$ equals *base* × *height* = $DC \times AF = 10AF$ (from the figure, DC = 10).

The area of rectangle $AECF$ equals *length* × *width* = $FC \times AF$.

We are given that the ratio of the two areas is 5 : 3. Forming the ratio gives

$$\frac{10AF}{FC \cdot AF} = \frac{5}{3}$$

$$\frac{10}{FC} = \frac{5}{3}$$

$$FC = 10 \cdot \frac{3}{5} = 6$$

From the figure, we have $DC = DF + FC$. Now, $DC = 10$ (from the figure) and $FC = 6$ (as we just calculated). So, $10 = DF + 6$. Solving this equation for DF yields $DF = 4$. Since F is one of the angles in the rectangle $AECF$, it is a right angle. Since $\angle AFC$ is right-angled, $\angle AFD$ must also measure 90°. So, ΔAFD is right-angled, and The Pythagorean Theorem yields

$$AF^2 + DF^2 = AD^2$$

$$AF^2 + 4^2 = 5^2$$

$$AF^2 = 5^2 - 4^2 = 25 - 16 = 9$$

$$AF = 3$$

Now, the area of the rectangle $AECF$ equals $FC \times AF = 6 \times 3 = 18$. The answer is (A).

13. Let the number of girls be x. Since the ratio of the girls to boys is 3 : 4, the number of boys equals $(4/3)x$. Hence, the number of students in the class equals $x + 4x/3 = 7x/3$. We are given that 10% of girls are blue eyed, and 10% of x is $10/100 \cdot x = x/10$. Also, 20% of the boys are blue eyed, and 20% of $4x/3$ is $(20/100)(4x/3) = 4x/15$.

Hence, the total number of blue-eyed students is $x/10 + 4x/15 = 11x/30$.

Hence, the required fraction is $\dfrac{\frac{11x}{30}}{\frac{7x}{3}} = \dfrac{11 \cdot 3}{30 \cdot 7} = \dfrac{11}{70}$. The answer is (A).

14. The cost of production is proportional to the number of units produced. Hence, we have the equation *The Cost of Production* = $k \times$ *Quantity*, where k is a constant.

We are given that 300 units cost 300 dollars. Putting this in the proportionality equation yields $300 = k \times 300$. Solving the equation for k yields $k = 300/300 = 1$. Hence, the *Cost of Production* of 270 units equals $k \times 270 = 1 \times 270 = 270$. The answer is (A).

15. Kelvin takes 3 minutes to inspect a car, and John takes 4 minutes to inspect a car. Hence, after t minutes, Kelvin inspects $t/3$ cars and John inspects $t/4$ cars. Hence, the ratio of the number of cars inspected by them is $t/3 : t/4 = 1/3 : 1/4 = 4 : 3$. The answer is (D).

16. The formula for simple interest rate (in percentage) equals $\dfrac{\text{Interest}}{\text{Principal}} \times 100$. From this formula, it is clear that the principal being the same (= \$1000 in the two given cases), the interest rates are directly proportional to the interests earned on them (\$80 in case of Bank A and \$140 in case of Bank B). Hence, the ratio of the interest rates is 80 : 140 = 4 : 7. The answer is (D).

17. We are given the equations $x = 10a$, $y = 3b$, $z = 7c$ and the proportion $x : y : z = 10 : 3 : 7$.

Substituting the first three equations into the last equation (ratio equation) yields $10a : 3b : 7c = 10 : 3 : 7$. Forming the resultant ratio yields $10a/10 = 3b/3 = 7c/7$. Simplifying the equation yields $a = b = c$. Hence, both a and b equal c. Substituting the result in the given equations $x = 10a$, $y = 3b$, $z = 7c$ yields $x = 10a = 10c$, $y = 3b = 3c$, and $z = 7c$. Now, we have

$$\frac{7x + 2y + 5z}{8a + b + 3c} =$$

$$\frac{7 \cdot 10c + 2 \cdot 3c + 5 \cdot 7c}{8c + c + 3c} = \qquad \text{because } x = 10c,\ y = 3c,\ z = 7c, \text{ and } a = b = c$$

$$\frac{111c}{12c} =$$

$$\frac{111}{12}$$

The answer is (A).

18. Two triangles are similar when their corresponding angle ratios or corresponding side ratios are the same.

Name the first triangle *ABC* and the second triangle *DEF*. Now, the angle ratio of the first triangle is $\angle A : \angle B : \angle C = 2 : 3 : 4$, and angle ratio for the second triangle is $\angle D : \angle E : \angle F = 4 : 3 : 2$, while the sum of the angles is 180 degrees.

The corresponding ordering ratio can be safely reversed and doing this for the second ratio equation yields $\angle F : \angle E : \angle D = 2 : 3 : 4$, while the sum of the angles is 180 degrees.

Since the angles in triangle *ABC* equal the corresponding angles in triangle *FED*, triangles *ABC* and *FED* are similar. Hence, the answer is (A).

19. Let 1 pound of the rice of the first type (\$0.8 per pound) be mixed with p pounds of the rice of the second type (\$0.9 per pound). Then the total cost of the $1 + p$ pounds of the rice is

$$(\$0.8 \text{ per pound} \times 1 \text{ pound}) + (\$0.9 \text{ per pound} \times p \text{ pounds}) =$$

$$0.8 + 0.9p$$

Hence, the cost of the mixture per pound is

$$\frac{\text{Cost}}{\text{Weight}} = \frac{0.8 + 0.9p}{1 + p}$$

If this equals \$0.825 per pound (given), then we have the equation

$$\frac{0.8 + 0.9p}{1 + p} = 0.825$$
$$0.8 + 0.9p = 0.825(1 + p)$$
$$0.8 + 0.9p = 0.825 + 0.825p$$
$$0.9p - 0.825p = 0.825 - 0.8$$
$$900p - 825p = 825 - 800$$
$$75p = 25$$
$$p = 25/75 = 1/3$$

Hence, the proportion of the two rice types is 1 : 1/3, which also equals 3 : 1. Hence, the answer is (E).

20. We are given the equations $x = a$, $y = 2b$, $z = 3c$, and the proportion $x : y : z = 1 : 2 : 3$. Substituting the first three equations into the last equation yields $a : 2b : 3c = 1 : 2 : 3$. Forming the resultant ratio yields $a/1 = 2b/2 = 3c/3$. Simplifying the equation yields $a = b = c$. Thus, we have that both a and b equal c. Hence, from the given equations, we have $x = a = c$, $y = 2b = 2c$, and $z = 3c$.

Now, Column A $= \dfrac{x + y + z}{a + b + c}$

$= \dfrac{c + 2c + 3c}{c + c + c}$ because $x = a$, $y = 2c$, $z = 3c$, and $a = b = c$

$= \dfrac{6c}{3c} = 2 =$ Column B.

Hence, the answer is (C).

21. Forming the given ratio yields

$A/1 = B/2 = C/3 = k$, for some integer
$A = k$, $B = 2k$, and $C = 3k$

Choice (A): $A + B + C = k + 2k + 3k = 6k$. This is a perfect square only when k is a product of 6 and a perfect square number. For example, when k is $6 \cdot 9^2$, $6k = 6^2 \cdot 9^2$, a perfect square. In all other cases (suppose $k = 2$, then $6k = 12$), it is not a perfect square. Hence, reject.

Choice (B): $A^2 + B^2 + C^2 = k^2 + (2k)^2 + (3k)^2 = k^2 + 4k^2 + 9k^2 = 14k^2$. This is surely not a perfect square. For example, suppose k equals 2. Then $14k^2 = 56$, which is not a perfect square. Hence, reject.

Choice (C): $A^3 + B^3 + C^3 = k^3 + (2k)^3 + (3k)^3 = k^3 + 8k^3 + 27k^3 = 36k^3$. This is a perfect square only when k^3 is perfect square. For example, suppose $k = 2$. Then $36k^3 = 288$, which is not a perfect square. Hence, reject.

Choice (D): $3A^2 + B^2 + C^2 = 3k^2 + (2k)^2 + (3k)^2 = 3k^2 + 4k^2 + 9k^2 = 16k^2 = 4^2k^2 = (4k)^2$. The square root of $(4k)^2$ is $4k$ and is an integer for any integer value of k. Hence, this expression must always result in a perfect square. Choose (D).

Choice (E): $3A^2 + 4B^2 + 4C^2 = 3k^2 + 4(2k)^2 + 4(3k)^2 = 3k^2 + 16k^2 + 36k^2 = 55k^2$. This is surely not a perfect square. Hence, reject.

The answer is (D).

22. Let x be the weight of the full stone. Then the weight of each of the three broken pieces is $x/3$.

Since we are given that the value of the stone is proportional to the square of its weight, we have that if kx^2 is the value of the full stone, then the value of each small stone should be $k\left(\frac{x}{3}\right)^2$, where k is the proportionality constant. Hence, the value of the three pieces together is

$$k\left(\frac{x}{3}\right)^2 + k\left(\frac{x}{3}\right)^2 + k\left(\frac{x}{3}\right)^2 =$$

$$k\left(\frac{x^2}{9}\right) + k\left(\frac{x^2}{9}\right) + k\left(\frac{x^2}{9}\right) =$$

$$\frac{3kx^2}{9} =$$

$$\frac{kx^2}{3}$$

Hence, Column A equals $kx^2/3$ and Column B equals kx^2. Hence, Column A is one third of Column B. The answer is (B).

23. Since the cost of *each* liter of the spirit water solution is proportionally to the part (fraction) of spirit the solution has, the cost per liter can be expressed as kf, where f is the fraction (part of) of pure spirit the solution has.

Now, m liters of the solution containing n liters of the spirit should cost $m \times$ cost of each liter $= m \times kf = m \times k(n/m) = kn$. Hence, the solution is only priced for the content of spirit the solution has (n here). Hence, the cost of the two samples given in the problem must be equal since both have exactly 1 liter of the spirit. Hence, x equals y and therefore Column A equals Column B. The answer is (C).

24. Since a and b are positive, $a + 6$ and $b + 6$ are positive. From the ratio $(a + 6) : (b + 6) = 5 : 6$, we get $\frac{a+6}{b+6} = \frac{5}{6}$. Since $5/6 < 1$, $a + 6 < b + 6$. Adding 4 to both sides yields $a + 10 < b + 10$. Since b is positive, $b + 10$ is positive. Dividing the inequality by $b + 10$ yields $\frac{a+10}{b+10} < 1$. Hence, Column A < Column B, and the answer is (B).

25. Let k and $3k$ be the number of cheetahs and pandas five years ago. Let d and $2d$ be the corresponding numbers now.

The increase in the number of cheetahs is $d - k$, and the increase in the number of pandas is $2d - 3k$.

Now, suppose the increase in the number of cheetahs equals the increase in the number of pandas. Then we have the equation $d - k = 2d - 3k$. Solving the equation for d yields $d = 2k$. Hence, we have a case here. Suppose $k = 3$. Then $d = 2k = 6$. The case supposes there were 3 cheetahs and 9 pandas 5 years ago, and now, there are 6 cheetahs and 12 pandas. Hence, the increase is the same.

Now, suppose the increase in the number of cheetahs is less than that of the pandas. Then we have the inequality $d - k < 2d - 3k$. Solving the inequality yields $d > 2k$. Hence, suppose $k = 3$ and $d = 7$. The case refers to when there are 3 cheetahs and 9 pandas five years ago and now there are 7 cheetahs and 14 pandas.

Now, suppose the increase in the number of cheetahs is greater than that of the pandas. Then we have the inequality $d - k > 2d - 3k$. Solving the inequality yields $d < 2k$. Hence, suppose $k = 3$ and $d = 5$. The case refers to when there are 3 cheetahs and 9 pandas five years ago, and now, there are 5 cheetahs and just 10 pandas. Here, the increase in cheetahs is greater than the increase in pandas.

Hence, we have a double case, and the answer is (D).

26. Since the circle with center O and the circle with center A touch each other internally, the distance between their centers equals

(Radius of larger circle) – (Radius of circle with center A) = Column A

Similarly, since the circle with center O and the circle with center B touch each other internally, the distance between their centers equals

(Radius of larger circle) – (Radius of circle with center B) = Column B

Since the ratio of the radii of the circles with center A and center B is given to be 7 : 9 and since $7 < 9$, (Radius of circle with center A) < (Radius of circle with center B). Multiplying the inequality by –1 and flipping direction yields

–(Radius of circle with center A) > –(Radius of circle with center B)

Adding "Radius of larger circle" to both sides yields

(Radius of larger circle) – (Radius of circle with center A) >
(Radius of larger circle) – (Radius of circle with center B)

Column A > Column B.

The answer is (A).

27. Forming the ratio yields $\frac{a+6}{b+6}=\frac{5}{6}$. Multiplying both sides of the equation by $6(b + 6)$ yields

$6(a + 6) = 5(b + 6)$

$6a + 36 = 5b + 30$

$6a = 5b - 6$

$a = 5b/6 - 1$

$0 < 5b/6 - 1$ since a is positive

$1 < 5b/6$

$6/5 < b$

$1.2 < b$

$1 < 1.2 < b$ since $1 < 1.2$

Column B < 1.2 < Column A

Hence, the answer is (A).

28. Let the income of A and B be $3s$ and $2s$, respectively, and let their expenditures be $4t$ and $3t$. Then since savings is defined as income minus expenditure, Column A = the saving of $A = 3s - 4t$, and Column B = saving of $B = 2s - 3t$.

Column A is greater than Column B when $3s - 4t > 2s - 3t$, or $s > t$.

Since money is positive, s and t are positive. Since income is greater than expenditure (given), the Income of $B = 2s$ > Expenditure of $B = 3t$. Hence, $s > 3t/2$. Clearly, $s > t$. Since we know that Column A > Column B when $s > t$, the answer is (A).

➢ Very Hard

29. The new alloy X is formed from the two alloys A and B in the ratio 4 : 3. Hence, 7 parts of the alloy contains 4 parts of alloy A and 3 parts of alloy B. Let $7x$ ounces of alloy X contain $4x$ ounces of alloy A and $3x$ ounces of alloy B.

Now, alloy A is formed of the two basic elements mentioned in the ratio 5 : 3. Hence, $4x$ ounces of alloy A contains $\frac{5}{5+3} \cdot 4x = \frac{5x}{2}$ ounces of first basic element and $\frac{3}{5+3} \cdot 4x = \frac{3x}{2}$ ounces of the second basic element.

Also, alloy B is formed of the two basic elements mentioned in the ratio 1 : 2. Hence, let the $3x$ ounces of alloy A contain $\frac{1}{1+2} \cdot 3x = x$ ounces of the first basic element and $\frac{2}{1+2} \cdot 3x = 2x$ ounces of the second basic element.

Then the total compositions of the two basic elements in the $7x$ ounces of alloy X would contain $5x/2$ ounces (from A) + x ounces (from B) = $7x/2$ ounces of first basic element, and $3/2\ x$ (from A) + $2x$ (from B) = $7x/2$ ounces of the second basic element. Hence, the composition of the two basic elements in alloy X is $7x/2 : 7x/2 = 1 : 1$. The answer is (A).

30. Let $1 : k$ be the ratio in which Joseph mixed the two types of rice. Then a sample of $(1 + k)$ ounces of the mixture should equal 1 ounce of rice of the first type, and k ounces of rice of the second type. The rice of the first type costs 5 cents an ounce and that of the second type costs 6 cents an ounce. Hence, it cost him

$$(1 \text{ ounce} \times 5 \text{ cents per ounce}) + (k \text{ ounces} \times 6 \text{ cents per ounce}) = 5 + 6k$$

Since he sold the mixture at 7 cents per ounce, he must have sold the net $1 + k$ ounces of the mixture at $7(1 + k)$.

Since he earned 20% profit doing this, $7(1 + k)$ must be 20% more than $5 + 6k$. Hence, we have the equation

$$7(1 + k) = (1 + 20/100)(5 + 6k)$$
$$7 + 7k = (120/100)(5 + 6k)$$
$$7 + 7k = (6/5)(5 + 6k)$$
$$7 + 7k = 6/5 \cdot 5 + 6/5 \cdot 6k$$
$$7 + 7k = 6 + 36k/5$$
$$1 = k/5$$
$$k = 5$$

Hence, the required ratio is $1 : k = 1 : 5$. The answer is (B).

31. We have that the incomes of A, B, and C are in the ratio 1 : 2 : 3. Let their incomes be i, $2i$, and $3i$, respectively. Also, their expenses ratio is 3 : 2 : 1. Hence, let their expenses be $3e$, $2e$, and e. Since the Saving = Income – Expenditure, the savings of the three employees A, B, and C is $i - 3e$, $2i - 2e$, and $3i - e$, respectively.

Now, the saving of C is greater the saving of B when $3i - e > 2i - 2e$, or $i + e > 0$ which surely is correct, since the income and expenditure, i and e, are both money and therefore positive.

Now, the saving of B is greater the saving of A when $2i - 2e > i - 3e$, or $i + e > 0$ which is surely correct, since the income and the expenditure, i and e, are both money and therefore positive.

Hence, the employees A, B, and C in the order of their savings is $C > B > A$. The answer is (E).

32. We have that the volume of the cask is 15 liters.

Emptying + Filling up (first time):
The volume of alcohol initially available is 15 liters. When 5 liters are removed, the cask has 10 liters of alcohol. When filled back to the brim (by 5 liters of water), the composition of the solution now becomes 10 liters of alcohol and 5 liters of water.

* *The cask now has* 10 *liters* of alcohol *and* 5 *liters of water.*

Now, the ratio of alcohol to water is 10 : 5 = 2 : 1.

Emptying (second time):
The next time 5 liters are removed, the removed solution is $\frac{2}{2+1} \times 5 = \frac{10}{3}$ liters of alcohol and 5/3 liters of water. Hence, the remaining solution is 10 – 10/3 = 20/3 liters of alcohol and 5 – 5/3 = 10/3 liters of water.

* *The cask now has* 20/3 *liters of alcohol and* 10/3 *liters of water.*

Filling up (second time):
When the cask is filled the second time (with water), the solution is now 20/3 liters of alcohol (already existing) and 10/3 + 5 = 25/3 liters of water.

* *The cask now has* 20/3 *liters of alcohol and* 25/3 *liters of water.*

Hence, the solution now has alcohol and water in the ratio $\frac{20}{3} : \frac{25}{3} = 4 : 5$. The answer is (D).

Exponents & Roots

EXPONENTS

Exponents afford a convenient way of expressing long products of the same number. The expression b^n is called a power and it stands for $b \times b \times b \times \cdots \times b$, where there are n factors of b. b is called the base, and n is called the exponent. By definition, $b^0 = 1$.*

There are six rules that govern the behavior of exponents:

Rule 1: $x^a \cdot x^b = x^{a+b}$ Example, $2^3 \cdot 2^2 = 2^{3+2} = 2^5 = 32$. Caution, $x^a + x^b \neq x^{a+b}$

Rule 2: $\left(x^a\right)^b = x^{ab}$ Example, $\left(2^3\right)^2 = 2^{3 \cdot 2} = 2^6 = 64$

Rule 3: $(xy)^a = x^a \cdot y^a$ Example, $(2y)^3 = 2^3 \cdot y^3 = 8y^3$

Rule 4: $\left(\frac{x}{y}\right)^a = \frac{x^a}{y^a}$ Example, $\left(\frac{x}{3}\right)^2 = \frac{x^2}{3^2} = \frac{x^2}{9}$

Rule 5: $\frac{x^a}{x^b} = x^{a-b}$, if $a > b$. Example, $\frac{2^6}{2^3} = 2^{6-3} = 2^3 = 8$

$\frac{x^a}{x^b} = \frac{1}{x^{b-a}}$, if $b > a$. Example, $\frac{2^3}{2^6} = \frac{1}{2^{6-3}} = \frac{1}{2^3} = \frac{1}{8}$

Rule 6: $x^{-a} = \frac{1}{x^a}$ Example, $z^{-3} = \frac{1}{z^3}$ Caution, a negative exponent does not make the number negative; it merely indicates that the base should be reciprocated. For example, $3^{-2} \neq -\frac{1}{3^2}$ or $-\frac{1}{9}$.

Problems involving these six rules are common on the GRE, and they are often listed as hard problems. However, the process of solving these problems is quite mechanical: simply apply the six rules until they can no longer be applied.

Example 1: If $x \neq 0$, $\frac{x\left(x^5\right)^2}{x^4} =$

(A) x^5 (B) x^6 (C) x^7 (D) x^8 (E) x^9

First, apply the rule $\left(x^a\right)^b = x^{ab}$ to the expression $\frac{x\left(x^5\right)^2}{x^4}$:

$$\frac{x \cdot x^{5 \cdot 2}}{x^4} = \frac{x \cdot x^{10}}{x^4}$$

Next, apply the rule $x^a \cdot x^b = x^{a+b}$:

---

* Any term raised to the zero power equals 1, no matter how complex the term is. For example, $\left(\frac{x+5\pi}{y}\right)^0 = 1$.

$$\frac{x \cdot x^{10}}{x^4} = \frac{x^{11}}{x^4}$$

Finally, apply the rule $\frac{x^a}{x^b} = x^{a-b}$:

$$\frac{x^{11}}{x^4} = x^{11-4} = x^7$$

The answer is (C).

Note: Typically, there are many ways of solving these types of problems. For this example, we could have begun with Rule 5, $\frac{x^a}{x^b} = \frac{1}{x^{b-a}}$:

$$\frac{x\left(x^5\right)^2}{x^4} = \frac{\left(x^5\right)^2}{x^{4-1}} = \frac{\left(x^5\right)^2}{x^3}$$

Then apply Rule 2, $\left(x^a\right)^b = x^{ab}$:

$$\frac{\left(x^5\right)^2}{x^3} = \frac{x^{10}}{x^3}$$

Finally, apply the other version of Rule 5, $\frac{x^a}{x^b} = x^{a-b}$:

$$\frac{x^{10}}{x^3} = x^7$$

Example 2:

Column A	Column B
$\frac{3 \cdot 3 \cdot 3 \cdot 3}{9 \cdot 9 \cdot 9 \cdot 9}$	$\left(\frac{1}{3}\right)^4$

Canceling the common factor 3 in Column A yields $\frac{1 \cdot 1 \cdot 1 \cdot 1}{3 \cdot 3 \cdot 3 \cdot 3}$, or $\frac{1}{3} \cdot \frac{1}{3} \cdot \frac{1}{3} \cdot \frac{1}{3}$. Now, by the definition of a power, $\frac{1}{3} \cdot \frac{1}{3} \cdot \frac{1}{3} \cdot \frac{1}{3} = \left(\frac{1}{3}\right)^4$ Hence, the columns are equal and the answer is (C).

Example 3:

Column A	Column B
$\frac{6^4}{3^2}$	$2^4 \cdot 3^2$

First, factor Column A: $\frac{(2 \cdot 3)^4}{3^2}$

Next, apply the rule $(xy)^a = x^a \cdot y^a$: $\frac{2^4 \cdot 3^4}{3^2}$

Finally, apply the rule $\frac{x^a}{x^b} = x^{a-b}$: $2^4 \cdot 3^2$

Hence, the columns are equal and the answer is (C).

ROOTS

The symbol $\sqrt[n]{b}$ is read the *n*th root of *b*, where *n* is called the index, *b* is called the base, and $\sqrt{\ }$ is called the radical. $\sqrt[n]{b}$ denotes that number which raised to the *n*th power yields *b*. In other words, *a* is the *n*th root of *b* if $a^n = b$. For example, $\sqrt{9} = 3$* because $3^2 = 9$, and $\sqrt[3]{-8} = -2$ because $(-2)^3 = -8$. Even roots occur in pairs: both a positive root and a negative root. For example, $\sqrt[4]{16} = 2$ since $2^4 = 16$, and $\sqrt[4]{16} = -2$ since $(-2)^4 = 16$. Odd roots occur alone and have the same sign as the base: $\sqrt[3]{-27} = -3$ since $(-3)^3 = -27$. If given an even root, you are to assume it is the positive root. However, if you introduce even roots by solving an equation, then you <u>must</u> consider both the positive and negative roots:

$$x^2 = 9$$
$$\sqrt{x^2} = \pm\sqrt{9}$$
$$x = \pm 3$$

Square roots and cube roots can be simplified by removing perfect squares and perfect cubes, respectively. For example,

$$\sqrt{8} = \sqrt{4 \cdot 2} = \sqrt{4}\sqrt{2} = 2\sqrt{2}$$
$$\sqrt[3]{54} = \sqrt[3]{27 \cdot 2} = \sqrt[3]{27}\sqrt[3]{2} = 3\sqrt[3]{2}$$

Radicals are often written with fractional exponents. The expression $\sqrt[n]{b}$ can be written as $b^{1/n}$. This can be generalized as follows:

$$b^{m/n} = \left(\sqrt[n]{b}\right)^m = \sqrt[n]{b^m}$$

Usually, the form $\left(\sqrt[n]{b}\right)^m$ is better when calculating because the part under the radical is smaller in this case. For example, . Using the form $\sqrt[n]{b^m}$ would be much harder in this case: $27^{2/3} = \sqrt[3]{27^2} = \sqrt[3]{729} = 9$. Most students know the value of $\sqrt[3]{27}$, but few know the value of $\sqrt[3]{729}$.

If *n* is even, then

$$\sqrt[n]{x^n} = |x|$$

For example, $\sqrt[4]{(-2)^4} = |-2| = 2$. With odd roots, the absolute value symbol is not needed. For example, $\sqrt[3]{(-2)^3} = \sqrt[3]{-8} = -2$.

To solve radical equations, just apply the rules of exponents to undo the radicals. For example, to solve the radical equation $x^{2/3} = 4$, we cube both sides to eliminate the cube root:

$$\left(x^{2/3}\right)^3 = 4^3$$
$$x^2 = 64$$
$$\sqrt{x^2} = \sqrt{64}$$
$$|x| = 8$$
$$x = \pm 8$$

Even roots of negative numbers do not appear on the GRE. For example, you will not see expressions of the form $\sqrt{-4}$; expressions of this type are called complex numbers.

---

* With square roots, the index is not written, $\sqrt[2]{9} = \sqrt{9}$.

The following rules are useful for manipulating roots:

$$\sqrt[n]{xy} = \sqrt[n]{x}\sqrt[n]{y} \qquad \text{For example, } \sqrt{3x} = \sqrt{3}\sqrt{x}.$$

$$\sqrt[n]{\frac{x}{y}} = \frac{\sqrt[n]{x}}{\sqrt[n]{y}} \qquad \text{For example, } \sqrt[3]{\frac{x}{8}} = \frac{\sqrt[3]{x}}{\sqrt[3]{8}} = \frac{\sqrt[3]{x}}{2}.$$

Caution: $\sqrt[n]{x+y} \neq \sqrt[n]{x} + \sqrt[n]{y}$. For example, $\sqrt{x+5} \neq \sqrt{x} + \sqrt{5}$. Also, $\sqrt{x^2 + y^2} \neq x + y$. This common mistake occurs because it is similar to the following valid property: $\sqrt{(x+y)^2} = x + y$ (If $x + y$ can be negative, then it must be written with the absolute value symbol: $|x + y|$). Note, in the valid formula, it's the whole term, $x + y$, that is squared, not the individual x and y.

To add two roots, both the index and the base must be the same. For example, $\sqrt[3]{2} + \sqrt[4]{2}$ cannot be added because the indices are different, nor can $\sqrt{2} + \sqrt{3}$ be added because the bases are different. However, $\sqrt[3]{2} + \sqrt[3]{2} = 2\sqrt[3]{2}$. In this case, the roots can be added because both the indices and bases are the same. Sometimes radicals with different bases can actually be added once they have been simplified to look alike. For example, $\sqrt{28} + \sqrt{7} = \sqrt{4 \cdot 7} + \sqrt{7} = \sqrt{4}\sqrt{7} + \sqrt{7} = 2\sqrt{7} + \sqrt{7} = 3\sqrt{7}$.

You need to know the approximations of the following roots: $\sqrt{2} \approx 1.4$ $\qquad \sqrt{3} \approx 1.7$ $\qquad \sqrt{5} \approx 2.2$

Example 4: Given the system $\begin{matrix} x^2 = 4 \\ y^3 = -8 \end{matrix}$, which of the following is NOT necessarily true?

(A) $y < 0$ (B) $x < 5$ (C) y is an integer (D) $x > y$ (E) $\frac{x}{y}$ is an integer

$y^3 = -8$ yields one cube root, $y = -2$. However, $x^2 = 4$ yields two square roots, $x = \pm 2$. Now, if $x = 2$, then $x > y$; but if $x = -2$, then $x = y$. Hence, choice (D) is not necessarily true. The answer is (D).

Example 5: If $x < 0$ and y is 5 more than the square of x, which one of the following expresses x in terms of y?

(A) $x = \sqrt{y-5}$ (B) $x = -\sqrt{y-5}$ (C) $x = \sqrt{y+5}$ (D) $x = \sqrt{y^2-5}$ (E) $x = -\sqrt{y^2-5}$

Translating the expression *"y is 5 more than the square of x"* into an equation yields:

$$y = x^2 + 5$$

$$y - 5 = x^2$$

$$\pm\sqrt{y-5} = x$$

Since we are given that $x < 0$, we take the negative root, $-\sqrt{y-5} = x$. The answer is (B).

RATIONALIZING

A fraction is not considered simplified until all the radicals have been removed from the denominator. If a denominator contains a single term with a square root, it can be rationalized by multiplying both the numerator and denominator by that square root. If the denominator contains square roots separated by a plus or minus sign, then multiply both the numerator and denominator by the conjugate, which is formed by merely changing the sign between the roots.

Example: Rationalize the fraction $\dfrac{2}{3\sqrt{5}}$.

Multiply top and bottom of the fraction by $\sqrt{5}$:

$$\frac{2}{3\sqrt{5}} \cdot \frac{\sqrt{5}}{\sqrt{5}} = \frac{2\sqrt{5}}{3 \cdot \sqrt{25}} = \frac{2\sqrt{5}}{3 \cdot 5} = \frac{2\sqrt{5}}{15}$$

Example: Rationalize the fraction $\dfrac{2}{3-\sqrt{5}}$.

Multiply top and bottom of the fraction by the conjugate $3+\sqrt{5}$:

$$\frac{2}{3-\sqrt{5}} \cdot \frac{3+\sqrt{5}}{3+\sqrt{5}} = \frac{2\left(3+\sqrt{5}\right)}{3^2 + 3\sqrt{5} - 3\sqrt{5} - \left(\sqrt{5}\right)^2} = \frac{2\left(3+\sqrt{5}\right)}{9-5} = \frac{2\left(3+\sqrt{5}\right)}{4} = \frac{3+\sqrt{5}}{2}$$

Problem Set R:

➢ **Easy**

1.

Column A		Column B
$x(x^2)^4$		$(x^3)^3$

➢ **Medium**

2.

Column A	$0 < x < y$	Column B
$7^{\frac{1}{x}-\frac{1}{y}}$		7^{x-y}

3. If $x = 10^{1.4}$, $y = 10^{0.7}$, and $x^z = y^3$, then what is the value of z ?

(A) 0.5
(B) 0.66
(C) 1.5
(D) 2
(E) 3

4. A perfect square is a positive integer which when square rooted results in an integer. If $N = 3^4 \cdot 5^3 \cdot 7$, then what is the biggest perfect square that is a factor of N ?

(A) 3^2
(B) 5^2
(C) 9^2
(D) $(9 \cdot 5)^2$
(E) $(3 \cdot 5 \cdot 7)^2$

5. If $p = \dfrac{\sqrt{3}-2}{\sqrt{2}+1}$, then which one of the following equals $p - 4$?

(A) $\sqrt{3} - 2$
(B) $\sqrt{3} + 2$
(C) 2
(D) $-2\sqrt{2} + \sqrt{6} - \sqrt{3} - 2$
(E) $-2\sqrt{2} + \sqrt{6} - \sqrt{3} + 2$

6. Column A: $\sqrt{12.5}+\sqrt{12.5}$ Column B: $\sqrt{25}$

➢ Hard

7. If $p = 216^{-1/3} + 243^{-2/5} + 256^{-1/4}$, then which one of the following is an integer?

(A) $p/19$
(B) $p/36$
(C) p
(D) $19/p$
(E) $36/p$

8. $\dfrac{4\left(\sqrt{6}+\sqrt{2}\right)}{\sqrt{6}-\sqrt{2}}-\dfrac{2+\sqrt{3}}{2-\sqrt{3}}=$

(A) 1
(B) $\sqrt{6}-\sqrt{2}$
(C) $\sqrt{6}+\sqrt{2}$
(D) 8
(E) 12

9. If $\sqrt[m]{27}=3^{3m}$ and $4^m > 1$, then what is the value of m ?

(A) –1
(B) –1/4
(C) 0
(D) 1/4
(E) 1

10. Column A: $\dfrac{\left(\left(\sqrt{7}\right)^x\right)^2}{\left(\sqrt{7}\right)^{11}}$ Column B: $\dfrac{7^x}{7^{11}}$

11. Column A: $\dfrac{\sqrt{3}+\sqrt{2}}{\sqrt{3}-\sqrt{2}}$ Column B: $\dfrac{\sqrt{5}+\sqrt{2}}{\sqrt{5}-\sqrt{2}}$

➢ Very Hard

12. In which one of the following choices must p be greater than q?

(A) $0.9^p = 0.9^q$
(B) $0.9^p = 0.9^{2q}$
(C) $0.9^p > 0.9^q$
(D) $9^p < 9^q$
(E) $9^p > 9^q$

Answers and Solutions to Problem Set R

➢ Easy

1. Column A: $x(x^2)^4 = x \cdot x^{2 \cdot 4} = x \cdot x^8 = x^{1+8} = x^9$.

Column B: $(x^3)^3 = x^{3 \cdot 3} = x^9$.

The answer is (C).

➢ Medium

2. From the inequality $0 < x < y$, x and y are positive and $x < y$. Hence, the sides of the inequality $x < y$ can be inverted and the direction safely flipped. Hence, $1/x > 1/y$.

Subtracting y from both sides of the inequality $x < y$ yields $x - y < 0$.

Finally, subtracting $1/y$ from both sides of the inequality $1/x > 1/y$ yields $1/x - 1/y > 0$.

Thus, Column A has 7 raised to a positive number, while Column B has 7 raised to a negative number. Hence, Column A is greater than 1, and Column B is less than 1. The answer is (A).

3. We are given that $x = 10^{1.4}$ and $y = 10^{0.7}$. Substituting these values in the given equation $x^z = y^3$ yields

$$(10^{1.4})^z = (10^{0.7})^3$$
$$10^{1.4z} = 10^{0.7 \cdot 3}$$
$$10^{1.4z} = 10^{2.1}$$
$$1.4z = 2.1 \qquad \text{since the bases are the same, the exponents must be equal}$$
$$z = 2.1/1.4 = 3/2$$

The answer is (C).

4. We are given that $N = 3^4 \cdot 5^3 \cdot 7$. Writing N as a product of perfect squares yields $(3^4 \cdot 5^2) \cdot 5 \cdot 7 = (3^2 \cdot 5)^2 \cdot 5 \cdot 7 = (9 \cdot 5)^2 \cdot 5 \cdot 7$. The product of the remaining numbers 5 and 7 cannot be written as a perfect square. Hence, $(9 \cdot 5)^2$ is the biggest perfect square factor of N. The answer is (D).

5. Since none of the answers are fractions, let's rationalize the given fraction by multiplying top and bottom by the conjugate of the bottom of the fraction:

$$p = \frac{\sqrt{3}-2}{\sqrt{2}+1} \cdot \frac{\sqrt{2}-1}{\sqrt{2}-1} \qquad \text{the conjugate of } \sqrt{2}+1 \text{ is } \sqrt{2}-1$$
$$= \frac{\sqrt{3}\sqrt{2}+\sqrt{3}(-1)+(-2)\sqrt{2}+(-2)(-1)}{\left(\sqrt{2}\right)^2 - 1^2}$$
$$= \frac{\sqrt{6}-\sqrt{3}-2\sqrt{2}+2}{2-1}$$
$$= \sqrt{6} - \sqrt{3} - 2\sqrt{2} + 2$$

Now, $p - 4 = (\sqrt{6} - \sqrt{3} - 2\sqrt{2} + 2) - 4 = \sqrt{6} - \sqrt{3} - 2\sqrt{2} - 2$. The answer is (D).

6. Since the columns are positive, we can square both columns without affecting the inequality relation between the columns. Doing this yields

Column A	Column B
$\left(\sqrt{12.5}+\sqrt{12.5}\right)^2 =$	$\left(\sqrt{25}\right)^2 =$
$12.5+12.5+2\sqrt{12.5}\sqrt{12.5} =$	25
$25+2\sqrt{12.5}\sqrt{12.5}$	

Subtracting 25 from both columns yields

Column A	Column B
$2\sqrt{12.5}\sqrt{12.5}$	0

Since $2\sqrt{12.5}\sqrt{12.5}$ is greater than 0, Column A is greater than Column B. The answer is (A).

Method II
Whatever the value of $\sqrt{12.5}$ is, it is greater than 3 since $\sqrt{12.5} > \sqrt{9} = 3$. Hence, $\sqrt{12.5}+\sqrt{12.5} > 3 + 3 = 6 > 5 =$ Column B.

➢ Hard

7. Simplifying the given equation yields

$$\begin{aligned} p &= 216^{-1/3} + 243^{-2/5} + 256^{-1/4} \\ &= (6^3)^{-1/3} + (3^5)^{-2/5} + (4^4)^{-1/4} && \text{because } 216 = 6^3,\ 243 = 3^5, \text{ and } 256 = 4^4 \\ &= 6^{3(-1/3)} + 3^{5(-2/5)} + 4^{4(-1/4)} \\ &= 6^{-1} + 3^{-2} + 4^{-1} \\ &= \frac{1}{6} + \frac{1}{9} + \frac{1}{4} \\ &= \frac{6+4+9}{36} \\ &= \frac{19}{36} \end{aligned}$$

Now,

Choice (A): $p/19 = (19/36)/19 = 1/36$, not an integer. Reject.
Choice (B): $p/36 = (19/36)/36 = 19/36^2$, not an integer. Reject.
Choice (C): $p = 19/36$, not an integer. Reject.
Choice (D): $19/p = 19/(19/36) = 19 \cdot 36/19 = 36$, an integer. Correct.
Choice (E): $36/p = 36/(19/36) = 36^2/19$, not an integer. Reject.

The answer is (D).

8. Let's rationalize both fractions by multiplying top and bottom of each fraction by the conjugate of its denominator:

$$\frac{\sqrt{6}+\sqrt{2}}{\sqrt{6}-\sqrt{2}}$$

$$=\frac{\sqrt{6}+\sqrt{2}}{\sqrt{6}-\sqrt{2}}\cdot\frac{\sqrt{6}+\sqrt{2}}{\sqrt{6}+\sqrt{2}}$$ the conjugate is $\sqrt{6}+\sqrt{2}$

$$=\frac{\left(\sqrt{6}+\sqrt{2}\right)^2}{\left(\sqrt{6}\right)^2-\left(\sqrt{2}\right)^2}$$ by the formula $(a+b)(a-b)=a^2-b^2$

$$=\frac{\left(\sqrt{6}\right)^2+\left(\sqrt{2}\right)^2+2\sqrt{6}\sqrt{2}}{6-2}$$

$$=\frac{6+2+4\sqrt{3}}{4}$$

$$=2+\sqrt{3}$$

$$\frac{4\left(\sqrt{6}+\sqrt{2}\right)}{\sqrt{6}-\sqrt{2}}=4\left(2+\sqrt{3}\right)=8+4\sqrt{3}$$

$$\frac{2+\sqrt{3}}{2-\sqrt{3}}$$

$$=\frac{2+\sqrt{3}}{2-\sqrt{3}}\cdot\frac{2+\sqrt{3}}{2+\sqrt{3}}$$ the conjugate is $2+\sqrt{3}$

$$=\frac{4+3+4\sqrt{3}}{4-3}$$

$$=7+4\sqrt{3}$$

Hence, $\frac{4\left(\sqrt{6}+\sqrt{2}\right)}{\sqrt{6}-\sqrt{2}}-\frac{2+\sqrt{3}}{2-\sqrt{3}}=\left(8+4\sqrt{3}\right)-\left(7+4\sqrt{3}\right)=8+4\sqrt{3}-7-4\sqrt{3}=1$. The answer is (A).

9. We have

$\sqrt[m]{27}=3^{3m}$

$\sqrt[m]{3^3}=3^{3m}$ By replacing 27 with 3^3

$(3^3)^{1/m}=3^{3m}$ Since by definition $\sqrt[m]{a}=a^{1/m}$

$3^{3/m}=3^{3m}$ Since $(x^a)^b$ equals x^{ab}

$3/m=3m$ By equating the powers on both sides

$m^2=1$ By multiplying both sides by $m/3$

$m=\pm 1$ By square rooting both sides

We have $4^m > 1$. If $m = -1$, then $4^m = 4^{-1} = 1/4 = 0.25$, which is not greater than 1. Hence, m must equal the other value 1. Here, $4^m = 4^1 = 4$, which is greater than 1. Hence, $m = 1$. The answer is (E).

10.

$$\frac{\left(\left(\sqrt{7}\right)^x\right)^2}{\left(\sqrt{7}\right)^{11}} = \qquad\qquad \frac{7^x}{7^{11}}$$

$$\frac{\left(\left(7^{\frac{1}{2}}\right)^x\right)^2}{\left(7^{\frac{1}{2}}\right)^{11}} = \qquad \text{since } \sqrt{7} = 7^{\frac{1}{2}}$$

$$\frac{7^{\frac{2x}{2}}}{7^{\frac{11}{2}}} =$$

$$\frac{7^x}{7^{\frac{11}{2}}} =$$

Canceling 7^x from both columns yields

$$\frac{1}{7^{\frac{11}{2}}} = \qquad\qquad \frac{1}{7^{11}}$$

$$\frac{1}{\left(7^{11}\right)^{\frac{1}{2}}} =$$

$$\frac{1}{\sqrt{7^{11}}} =$$

Multiplying both columns by $7^{11} \cdot \sqrt{7^{11}}$ yields

$$7^{11} \qquad\qquad \sqrt{7^{11}}$$

Now, clearly, Column A is greater than Column B, and the answer is (A).

11. Let's rationalize both fractions by multiplying top and bottom of each fraction by the conjugate of its denominator:

Column A equals

$$\frac{\sqrt{3}+\sqrt{2}}{\sqrt{3}-\sqrt{2}}$$

$$=\frac{\sqrt{3}+\sqrt{2}}{\sqrt{3}-\sqrt{2}}\cdot\frac{\sqrt{3}+\sqrt{2}}{\sqrt{3}+\sqrt{2}} \quad \text{the conjugate is } \sqrt{3}+\sqrt{2}$$

$$=\frac{\left(\sqrt{3}+\sqrt{2}\right)^2}{\left(\sqrt{3}\right)^2-\left(\sqrt{2}\right)^2} \quad \text{by the formula } (a+b)(a-b)=a^2-b^2$$

$$=\frac{\left(\sqrt{3}\right)^2+\left(\sqrt{2}\right)^2+2\sqrt{3}\sqrt{2}}{3-2}$$

$$=3+2+2\sqrt{6}$$

$$=5+2\sqrt{6}$$

Column B equals

$$\frac{\sqrt{5}+\sqrt{2}}{\sqrt{5}-\sqrt{2}}$$

$$=\frac{\sqrt{5}+\sqrt{2}}{\sqrt{5}-\sqrt{2}}\cdot\frac{\sqrt{5}+\sqrt{2}}{\sqrt{5}+\sqrt{2}} \quad \text{the conjugate is } \sqrt{5}+\sqrt{2}$$

$$=\frac{\left(\sqrt{5}+\sqrt{2}\right)^2}{\left(\sqrt{5}\right)^2-\left(\sqrt{2}\right)^2} \quad \text{by the formula } (a+b)(a-b)=a^2-b^2$$

$$=\frac{\left(\sqrt{5}\right)^2+\left(\sqrt{2}\right)^2+2\sqrt{5}\sqrt{2}}{5-2}$$

$$=\frac{5+2+2\sqrt{10}}{3}$$

$$=\frac{7}{3}+\frac{2\sqrt{10}}{3}$$

Now, we know that the first term in the Column A (5) is greater than first term in the Column B (7/3). Also, the second term in the Column A, $2\sqrt{6}$, equals $2\sqrt{6}$ ($2\sqrt{6}=\sqrt{2^2\times 6}=\sqrt{24}$) and is greater than $\frac{2}{3}\sqrt{10}$, which equals $\sqrt{4.44}$ ($\frac{2}{3}\sqrt{10}=\sqrt{\left(\frac{2}{3}\right)^2\times 10}=\sqrt{\frac{4}{9}\times 10}=\sqrt{\frac{40}{9}}=\sqrt{4.44}$). Hence, Column A, which equals $5+2\sqrt{6}$, is greater than Column B, which equals $\frac{7}{3}+\frac{2\sqrt{10}}{3}$. The answer is (A).

➢ Very Hard

12. Choice (A): $0.9^p = 0.9^q$. Equating exponents on both sides of the equation yields $p = q$. Hence, p is not greater than q. Reject.

Choice (B): $0.9^p = 0.9^{2q}$. Equating the exponents on both sides yields $p = 2q$. This is not sufficient information to determine whether p is greater than q. For example, in case both p and q are negative, $q > p$. In case both p and q are positive, $p > q$. Hence, reject the choice.

Cases for choices (C), (D), and (E):

A. $0.9^2 > 0.9^3$	$0.81 > 0.729$
B. $0.9^{1/2} < 0.9^{1/3}$	$0.94 < 0.96$
C. $0.9^{-2} < 0.9^{-3}$	$1.1 < 1.37$
D. $0.9^{-1/2} > 0.9^{-1/3}$	$1.06 > 1.04$
E. $9^2 < 9^3$	$81 < 729$
F. $9^{1/2} > 9^{1/3}$	$3 > 2.064$
G. $9^{-2} > 9^{-3}$	$1/81 > 1/729$
H. $9^{-1/2} < 9^{-1/3}$	$1/3 < 1/2.06$

Summary: If $a^x > a^y$, then

$x > y$ if $a > 1$
and
$x < y$ if $0 < a < 1$

Choice (C): $0.9^p > 0.9^q$:
Comparing the cases A and D against the inequality $0.9^p > 0.9^q$, we have that in A, p (= 2) is less than q (= 3). Hence, reject the choice. In case of D, p (= –1/2) is greater than q (= –1/3).

Choice (D): $9^p < 9^q$:
Comparing the case E against the inequality $9^p < 9^q$, we have that in E, p (= 2) is less than q (= 3). Hence, reject the choice.

Choice (E): $9^p > 9^q$:
The cases F and G match the inequality $9^p > 9^q$. In case of F, p (= 1/2) is greater than q (= 1/3) and also in case G, p (= –2) is greater than q (= –3). In either case, p is greater than q.

The answer is (E).

Factoring

To factor an algebraic expression is to rewrite it as a product of two or more expressions, called factors. In general, any expression on the GRE that can be factored should be factored, and any expression that can be unfactored (multiplied out) should be unfactored.

DISTRIBUTIVE RULE

The most basic type of factoring involves the distributive rule:

$$\boldsymbol{ax + ay = a(x + y)}$$

When this rule is applied from left to right, it is called factoring. When the rule is applied from right to left, it is called distributing.

For example, $3h + 3k = 3(h + k)$, and $5xy + 45x = 5xy + 9 \cdot 5x = 5x(y + 9)$. The distributive rule can be generalized to any number of terms. For three terms, it looks like $ax + ay + az = a(x + y + z)$. For example, $2x + 4y + 8 = 2x + 2 \cdot 2y + 2 \cdot 4 = 2(x + 2y + 4)$. For another example, $x^2y^2 + xy^3 + y^5 = y^2(x^2 + xy + y^3)$.

Example 1: If $x - y = 9$, then $\left(x - \frac{y}{3}\right) - \left(y - \frac{x}{3}\right) =$

(A) –4 (B) –3 (C) 0 (D) 12 (E) 27

$$\left(x - \frac{y}{3}\right) - \left(y - \frac{x}{3}\right) =$$

$x - \frac{y}{3} - y + \frac{x}{3} =$ by distributing the negative sign

$\frac{4}{3}x - \frac{4}{3}y =$ by combining the fractions

$\frac{4}{3}(x - y) =$ by factoring out the common factor $\frac{4}{3}$

$\frac{4}{3}(9) =$ since $x - y = 9$

$$12$$

The answer is (D).

Example 2:

Column A	Column B
$\dfrac{2^{20}-2^{19}}{2^{11}}$	2^8

$\dfrac{2^{20}-2^{19}}{2^{11}} = \dfrac{2^{19+1}-2^{19}}{2^{11}} =$

$\dfrac{2^{19}\cdot 2^1-2^{19}}{2^{11}} =$ by the rule $x^a \cdot x^b = x^{a+b}$

$\dfrac{2^{19}(2-1)}{2^{11}} =$ by the distributive property $ax + ay = a(x + y)$

$\dfrac{2^{19}}{2^{11}} =$

2^8 by the rule $\dfrac{x^a}{x^b} = x^{a-b}$

Hence, the columns are equal, and the answer is (C).

DIFFERENCE OF SQUARES

One of the most important formulas on the GRE is the difference of squares:

$$\boxed{x^2 - y^2 = (x+y)(x-y)}$$

Caution: a sum of squares, $x^2 + y^2$, does not factor.

Example 3: If $x \neq -2$, then $\dfrac{8x^2-32}{4x+8} =$

(A) $2(x-2)$ (B) $2(x-4)$ (C) $8(x+2)$ (D) $x-2$ (E) $x+4$

In most algebraic expressions involving multiplication or division, you won't actually multiply or divide, rather you will factor and cancel, as in this problem.

$\dfrac{8x^2-32}{4x+8} =$

$\dfrac{8(x^2-4)}{4(x+2)} =$ by the distributive property $ax + ay = a(x + y)$

$\dfrac{8(x+2)(x-2)}{4(x+2)} =$ by the difference of squares $x^2 - y^2 = (x+y)(x-y)$

$2(x-2)$ by canceling common factors

The answer is (A).

PERFECT SQUARE TRINOMIALS

Like the difference of squares formula, perfect square trinomial formulas are very common on the GRE.

$$x^2 + 2xy + y^2 = (x+y)^2$$
$$x^2 - 2xy + y^2 = (x-y)^2$$

For example, $x^2 + 6x + 9 = x^2 + 2(3x) + 3^2 = (x+3)^2$. Note, in a perfect square trinomial, the middle term is twice the product of the square roots of the outer terms.

Example 4: If $r^2 - 2rs + s^2 = 4$, then $(r-s)^6 =$

(A) −4 (B) 4 (C) 8 (D) 16 (E) 64

$r^2 - 2rs + s^2 = 4$

$(r-s)^2 = 4$ by the formula $x^2 - 2xy + y^2 = (x-y)^2$

$\left[(r-s)^2\right]^3 = 4^3$ by cubing both sides of the equation

$(r-s)^6 = 64$ by the rule $\left(x^a\right)^b = x^{ab}$

The answer is (E).

GENERAL TRINOMIALS

$$x^2 + (a+b)x + ab = (x+a)(x+b)$$

The expression $x^2 + (a+b)x + ab$ tells us that we need two numbers whose product is the last term and whose sum is the coefficient of the middle term. Consider the trinomial $x^2 + 5x + 6$. Now, two factors of 6 are 1 and 6, but $1 + 6 \neq 5$. However, 2 and 3 are also factors of 6, and $2 + 3 = 5$. Hence, $x^2 + 5x + 6 = (x+2)(x+3)$.

Example 5: Which of the following could be a solution of the equation $x^2 - 7x - 18 = 0$?

(A) −1 (B) 0 (C) 2 (D) 7 (E) 9

Now, both 2 and −9 are factors of 18, and $2 + (-9) = -7$. Hence, $x^2 - 7x - 18 = (x+2)(x-9) = 0$. Setting each factor equal to zero yields $x + 2 = 0$ and $x - 9 = 0$. Solving these equations yields $x = -2$ and 9. The answer is (E).

COMPLETE FACTORING

When factoring an expression, first check for a common factor, then check for a difference of squares, then for a perfect square trinomial, and then for a general trinomial.

Example 6: Factor the expression $2x^3 - 2x^2 - 12x$ completely.

Solution: First check for a common factor: $2x$ is common to each term. Factoring $2x$ out of each term yields $2x\left(x^2 - x - 6\right)$. Next, there is no difference of squares, and $x^2 - x - 6$ is not a perfect square trinomial since x does not equal twice the product of the square roots of x^2 and 6. Now, −3 and 2 are factors of −6 whose sum is −1. Hence, $2x\left(x^2 - x - 6\right)$ factors into $2x(x-3)(x+2)$.

Problem Set S:

➢ **Easy**

1. Column A — Number of zeros in n

   n equals $10^4 + (2 \times 10^2)$

   Column B — Number of zeros in n^2

2. If $b = a + c$ and $b = 3$, then $ab + bc =$

   (A) $\sqrt{3}$
   (B) 3
   (C) $3\sqrt{3}$
   (D) 9
   (E) 27

➢ **Medium**

3. Column A — $\dfrac{a^2 - b^2}{a - b}$

   $a = 49$, $b = 59$

   Column B — $\dfrac{a^2 - b^2}{a + b}$

4. Column A — $\dfrac{2x^2 - 72}{x - 6}$

   $x \neq 3$ and $x \neq 6$

   Column B — $\dfrac{2x^2 - 18}{x - 3}$

5. If $x - 3 = 10/x$ and $x > 0$, then what is the value of x ?

   (A) −2
   (B) −1
   (C) 3
   (D) 5
   (E) 10

6. If $x^2 - 4x + 3 = 0$, then what is the value of $(x - 2)^2$?

   (A) −1
   (B) 0
   (C) 1
   (D) 3
   (E) 4

Answers and Solutions to Problem Set S

➢ Easy

1. Since n equals $10^4 + (2 \times 10^2) = 10200$ and has 3 zeros, Column A equals 3.

By the Perfect Square Trinomial formula,

$$n^2 =$$
$$[10^4 + (2 \times 10^2)]^2 =$$
$$10^{4 \times 2} + (2 \times 10^2)^2 + 2(2 \times 10^2)10^4 =$$
$$10^8 + 4 \times 10^4 + 4 \times 10^6 =$$
$$104040000$$

There are 6 zero digits. So, Column B equals 6.

Column B is greater, and the answer is (B).

2. Factoring the common factor b from the expression $ab + bc$ yields $b(a + c) = b \cdot b$ [since $a + c = b$] = $b^2 = 3^2 = 9$. The answer is (D).

➢ Medium

3. Applying the Difference of Squares Formula $a^2 - b^2 = (a + b)(a - b)$ to both columns yields

$$\frac{(a+b)(a-b)}{a-b} \qquad \frac{(a+b)(a-b)}{a+b}$$

Reducing yields

$$a + b \qquad a - b$$

Since $a = 49$ and $b = 59$, Column A is greater than Column B. The answer is (A).

4. Start by factoring 2 from the numerators of each fraction:

$$\frac{2(x^2-36)}{x-6} \qquad \frac{2(x^2-9)}{x-3}$$

Next, apply the Difference of Squares Formula $a^2 - b^2 = (a + b)(a - b)$ to the expressions in both columns:

$$\frac{2(x+6)(x-6)}{x-6} \qquad \frac{2(x+3)(x-3)}{x-3}$$

Next, cancel the term $x - 6$ in Column A and the term $x - 3$ in Column B:

$$2(x + 6) \qquad 2(x + 3)$$

Next, distribute the 2 in each expression:

$$2x + 12 \qquad 2x + 6$$

Finally, cancel $2x$ from both columns:

$$12 \qquad 6$$

Hence, Column A is greater than Column B, and answer is (A).

5. We have the equation $x - 3 = 10/x$. Multiplying the equation by x yields $x^2 - 3x = 10$. Subtracting 10 from both sides yields $x^2 - 3x - 10 = 0$. Factoring the equation yields $(x - 5)(x + 2) = 0$. The possible solutions are 5 and –2. The only solution that also satisfies the given inequality $x > 0$ is $x = 5$. The answer is (D).

6. Adding 1 to both sides of the given equation $x^2 - 4x + 3 = 0$ yields $x^2 - 4x + 4 = 1$. Expanding $(x - 2)^2$ by the Difference of Squares formula $(a - b)^2 = a^2 - 2ab + b^2$ yields $x^2 - 4 \cdot x + 2^2 = x^2 - 4x + 4 = 1$. Hence, $(x - 2)^2 = 1$, and the answer is (C).

Algebraic Expressions

A mathematical expression that contains a variable is called an algebraic expression. Some examples of algebraic expressions are x^2, $3x - 2y$, $2z(y^3 - \frac{1}{z^2})$. Two algebraic expressions are called like terms if both the variable parts and the exponents are identical. That is, the only parts of the expressions that can differ are the coefficients. For example, $5y^3$ and $\frac{3}{2}y^3$ are like terms, as are $x + y^2$ and $-7(x + y^2)$. However, x^3 and y^3 are not like terms, nor are $x - y$ and $2 - y$.

ADDING & SUBTRACTING ALGEBRAIC EXPRESSIONS

Only like terms may be added or subtracted. To add or subtract like terms, merely add or subtract their coefficients:

$$x^2 + 3x^2 = (1+3)x^2 = 4x^2$$

$$2\sqrt{x} - 5\sqrt{x} = (2-5)\sqrt{x} = -3\sqrt{x}$$

$$.5\left(x + \frac{1}{y}\right)^2 + .2\left(x + \frac{1}{y}\right)^2 = (.5 + .2)\left(x + \frac{1}{y}\right)^2 = .7\left(x + \frac{1}{y}\right)^2$$

$$\left(3x^3 + 7x^2 + 2x + 4\right) + \left(2x^2 - 2x - 6\right) = 3x^3 + (7+2)x^2 + (2-2)x + (4-6) = 3x^3 + 9x^2 - 2$$

You may add or multiply algebraic expressions in any order. This is called the commutative property:

$$x + y = y + x$$

$$xy = yx$$

For example, $-2x + 5x = 5x + (-2x) = (5-2)x = 3x$ and $(x-y)(-3) = (-3)(x-y) = (-3)x - (-3)y = -3x + 3y$.

Caution: the commutative property does not apply to division or subtraction: $2 = 6 \div 3 \neq 3 \div 6 = \frac{1}{2}$ and $-1 = 2 - 3 \neq 3 - 2 = 1$.

When adding or multiplying algebraic expressions, you may regroup the terms. This is called the associative property.

$$x + (y + z) = (x + y) + z$$

$$x(yz) = (xy)z$$

Notice in these formulas that the variables have not been moved, only the way they are grouped has changed: on the left side of the formulas the last two variables are grouped together, and on the right side of the formulas the first two variables are grouped together.

For example, $(x-2x)+5x=(x+[-2x])+5x=x+(-2x+5x)=x+3x=4x$

and

$2(12x)=(2\cdot 12)x=24x$

The associative property doesn't apply to division or subtraction: $4=8\div 2=8\div(4\div 2)\neq(8\div 4)\div 2=2\div 2=1$

and

$-6=-3-3=(-1-2)-3\neq -1-(2-3)=-1-(-1)=-1+1=0.$

Notice in the first example that we changed the subtraction into negative addition: $(x-2x)=(x+[-2x])$. This allowed us to apply the associative property over addition.

PARENTHESES

When simplifying expressions with nested parentheses, work from the inner most parentheses out:

$$5x+(y-(2x-3x))=5x+(y-(-x))=5x+(y+x)=6x+y$$

Sometimes when an expression involves several pairs of parentheses, one or more pairs are written as brackets. This makes the expression easier to read:

$$\begin{gathered}2x(x-[y+2(x-y)])=\\2x(x-[y+2x-2y])=\\2x(x-[2x-y])=\\2x(x-2x+y)=\\2x(-x+y)=\\-2x^2+2xy\end{gathered}$$

ORDER OF OPERATIONS: (PEMDAS)

When simplifying algebraic expressions, perform operations within parentheses first and then exponents and then multiplication and then division and then addition and lastly subtraction. This can be remembered by the mnemonic:

PEMDAS
Please **E**xcuse **M**y **D**ear **A**unt **S**ally

This mnemonic isn't quite precise enough. Multiplication and division are actually tied in order of operation, as is the pair addition and subtraction. When multiplication and division, or addition and subtraction, appear at the same level in an expression, perform the operations from left to right. For example, $6\div 2\times 4=(6\div 2)\times 4=3\times 4=12$. To emphasize this left-to-right order, we can use parentheses in the mnemonic: **PE(MD)(AS)**.

Example 1: $2-\left(5-3^3[4\div 2+1]\right)=$

(A) −21 (B) 32 (C) 45 (D) 60 (E) 78

$2-\left(5-3^3[4\div 2+1]\right)=$

$2-\left(5-3^3[2+1]\right)=$ By performing the division within the innermost parentheses

$2-\left(5-3^3[3]\right)=$ By performing the addition within the innermost parentheses

$2-(5-27[3])=$ By performing the exponentiation

$2-(5-81)=$ By performing the multiplication within the parentheses

$2-(-76)=$ By performing the subtraction within the parentheses

$2+76=$ By multiplying the two negatives

78

The answer is (E).

FOIL MULTIPLICATION

You may recall from algebra that when multiplying two expressions you use the FOIL method: **F**irst, **O**uter, **I**nner, **L**ast:

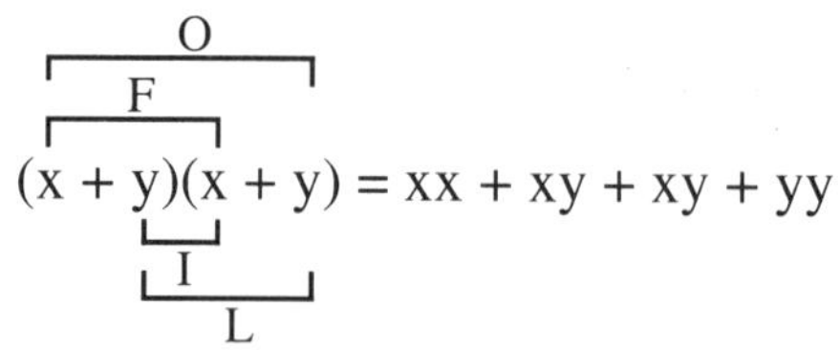

Simplifying the right side yields $(x+y)(x+y) = x^2 + 2xy + y^2$. For the product $(x-y)(x-y)$, we get $(x-y)(x-y) = x^2 - 2xy + y^2$. These types of products occur often, so it is worthwhile to memorize the formulas. Nevertheless, you should still learn the FOIL method of multiplying because the formulas do not apply in all cases.

Examples (FOIL):

$$(2-y)\left(x-y^2\right) = 2x - 2y^2 - xy + yy^2 = 2x - 2y^2 - xy + y^3$$

$$\left(\frac{1}{x} - y\right)\left(x - \frac{1}{y}\right) = \frac{1}{x}x - \frac{1}{x}\frac{1}{y} - xy + y\frac{1}{y} = 1 - \frac{1}{xy} - xy + 1 = 2 - \frac{1}{xy} - xy$$

$$\left(\frac{1}{2} - y\right)^2 = \left(\frac{1}{2} - y\right)\left(\frac{1}{2} - y\right) = \left(\frac{1}{2}\right)^2 - 2\left(\frac{1}{2}\right)y + y^2 = \frac{1}{4} - y + y^2$$

DIVISION OF ALGEBRAIC EXPRESSIONS

When dividing algebraic expressions, the following formula is useful:

$$\frac{x+y}{z} = \frac{x}{z} + \frac{y}{z}$$

This formula generalizes to any number of terms.

Examples:

$$\frac{x^2+y}{x} = \frac{x^2}{x} + \frac{y}{x} = x^{2-1} + \frac{y}{x} = x + \frac{y}{x}$$

$$\frac{x^2+2y-x^3}{x^2} = \frac{x^2}{x^2} + \frac{2y}{x^2} - \frac{x^3}{x^2} = x^{2-2} + \frac{2y}{x^2} - x^{3-2} = x^0 + \frac{2y}{x^2} - x = 1 + \frac{2y}{x^2} - x$$

When there is more than a single variable in the denomination, we usually factor the expression and then cancel, instead of using the above formula.

Example 2: $\dfrac{x^2-2x+1}{x-1} =$

(A) $x+1$ (B) $-x-1$ (C) $-x+1$ (D) $x-1$ (E) $x-2$

$\dfrac{x^2-2x+1}{x-1} = \dfrac{(x-1)(x-1)}{x-1} = x-1$. The answer is (D).

Problem Set T:

➢ **Easy**

1. If $x \neq 3$ and $x \neq 6$, then $\frac{2x^2 - 72}{x - 6} - \frac{2x^2 - 18}{x - 3} =$

(A) 3
(B) 6
(C) 9
(D) 12
(E) 15

➢ **Medium**

2.

Column A	Column B
$\frac{(2x-11)(2x+11)}{4}$	$(x-11)(x+11)$

3. If $|3x| \neq 2$, what is the value of $\frac{9x^2 - 4}{3x + 2} - \frac{9x^2 - 4}{3x - 2}$?

(A) –9
(B) –4
(C) 0
(D) 4
(E) 9

4. If $1/x + 1/y = 1/3$, then $\frac{xy}{x + y} =$

(A) 1/5
(B) 1/3
(C) 1
(D) 3
(E) 5

5. If $x = 2$ and $y = -1$, which one of the following expressions is greatest?

(A) $x + y$
(B) xy
(C) $-x + y$
(D) $x - y - 1$
(E) $-x - y$

6. If b is one-fourth of a, then what is the value of $\frac{a + b}{\sqrt{ab}}$?

(A) 1/5
(B) 1/3
(C) 1/2
(D) $1\frac{1}{2}$
(E) $2\frac{1}{2}$

7. If $\sqrt{3-2x}=1$, then what is the value of $(3-2x)+(3-2x)^2$?

(A) 0
(B) 1
(C) 2
(D) 3
(E) 4

8.

Column A		Column B
$(1111.0)^2-(999.0)^2$		$(1111.5)^2-(999.5)^2$

9.

Column A	$\lvert x\rvert \neq 1/2$	Column B
$\dfrac{4x^2-1}{2x+1}$		$\dfrac{4x^2-1}{2x-1}$

10.

Column A	$x=\dfrac{1}{1+\dfrac{1}{1+\dfrac{1}{2}}}$	Column B
x		1

Answers and Solutions to Problem Set T

➢ Easy

1. Start by factoring 2 from the numerators of each fraction:

$$\frac{2\left(x^2-36\right)}{x-6}-\frac{2\left(x^2-9\right)}{x-3}$$

Next, apply the Difference of Squares Formula $a^2-b^2=(a+b)(a-b)$ to both fractions in the expression:

$$\frac{2(x+6)(x-6)}{x-6}-\frac{2(x+3)(x-3)}{x-3}=$$

Next, cancel the term $x-6$ from the first fraction and $x-3$ from the second fraction:

$$2(x+6)-2(x+3)=2x+12-2x-6=6$$

Hence, the answer is (B).

➢ Medium

2. Applying the Difference of Squares formula $(a+b)(a-b)=a^2-b^2$ yields

Column A	Column B
$\frac{(2x)^2-11^2}{4}$	x^2-11^2

Column A	Column B
$\frac{4x^2-121}{4}$	x^2-121

Column A	Column B
$x^2-\frac{121}{4}$	x^2-121

Subtracting x^2 from both columns yields

Column A	Column B
$-121/4$	-121

Since $-121/4>-121$, Column A is greater than Column B and the answer is (A).

3. $\frac{9x^2-4}{3x+2}-\frac{9x^2-4}{3x-2}$

$=\left(9x^2-4\right)\left(\frac{1}{3x+2}-\frac{1}{3x-2}\right)$ by factoring out the common term $9x^2-4$

$=\left(9x^2-4\right)\frac{(3x-2)-(3x+2)}{(3x+2)(3x-2)}$

$=\left(9x^2-4\right)\frac{3x-2-3x-2}{(3x)^2-2^2}$

$=\left(9x^2-4\right)\frac{-4}{9x^2-4}$ Since $|3x|\neq 2$, $(3x)^2\neq 4$, and therefore $9x^2-4\neq 0$.

Hence, we can safely cancel $9x^2-4$ from numerator and denominator.

$=-4$

The answer is (B).

4. Multiplying the given equation $1/x + 1/y = 1/3$ by xy yields $y + x = xy/3$, or $x + y = xy/3$. Multiplying both sides of the equation $x + y = xy/3$ by $3/(x + y)$ yields $\frac{xy}{x+y} = 3$. The answer is (D).

5. Choice (A): $x + y = 2 + (-1) = 1$.
Choice (B): $xy = 2(-1) = -2$.
Choice (C): $-x + y = -2 + (-1) = -3$.
Choice (D): $x - y - 1 = 2 - (-1) - 1 = 2$.
Choice (E): $-x - y = -2 - (-1) = -1$.

The greatest result is Choice (D). The answer is (D).

6. We are given that b is 1/4 of a. Hence, we have the equation $b = a/4$. Multiplying both sides of this equation by $4/b$ yields $4 = a/b$.

Now,

$$\frac{a+b}{\sqrt{ab}} =$$

$$\frac{a}{\sqrt{ab}} + \frac{b}{\sqrt{ab}} =$$

$$\frac{\sqrt{a^2}}{\sqrt{ab}} + \frac{\sqrt{b^2}}{\sqrt{ab}} =$$

$$\sqrt{\frac{a^2}{ab}} + \sqrt{\frac{b^2}{ab}} =$$

$$\sqrt{\frac{a}{b}} + \sqrt{\frac{b}{a}} =$$

$$\sqrt{4} + \sqrt{\frac{1}{4}} =$$

$$2 + \frac{1}{2} =$$

$$2\frac{1}{2}$$

The answer is (E).

7. We have $\sqrt{3-2x} = 1$. Squaring both sides of the equation yields $(3 - 2x) = 1$. Squaring both sides of the equation again yields $(3 - 2x)^2 = 1$. Hence, $(3 - 2x) + (3 - 2x)^2 = 1 + 1 = 2$. The answer is (C).

8. Column B $= (1111.5)^2 - (999.5)^2 =$

$= (1111.5 - 999.5)(1111.5 + 999.5)$ — by the Difference of Squares formula $a^2 - b^2 = (a - b)(a + b)$

$= (1111 + 0.5 - 999 - 0.5)(1111 + 0.5 + 999 + 0.5)$
$= (1111 - 999)(1111 + 999 + 1)$
$= (1111 - 999)(1111 + 999) + (1111 - 999)$
$= 1111^2 - 999^2 + (1111 - 999)$ — by the Difference of Squares formula $(a - b)(a + b) = a^2 - b^2$

$=$ Column A $+ (1111 - 999)$
$=$ Column A + (a positive number)

From this equation, it is clear that Column B is greater than Column A, by $1111 - 999$, a positive number. Hence, the answer is (B).

9. Applying the Difference of Squares Formula $a^2 - b^2 = (a+b)(a-b)$ to both columns yields

$$\frac{(2x+1)(2x-1)}{2x+1} \qquad\qquad \frac{(2x+1)(2x-1)}{2x-1}$$

Reducing yields

$$2x - 1 \qquad\qquad 2x + 1$$

Subtracting $2x$ from both columns yields

$$-1 \qquad\qquad 1$$

The answer is (B).

10.

$$\begin{aligned} x &= \frac{1}{1+\frac{1}{1+\frac{1}{2}}} \\ &= \frac{1}{1+\frac{1}{\frac{3}{2}}} \\ &= \frac{1}{1+\frac{2}{3}} \\ &= \frac{1}{\frac{5}{3}} \\ &= \frac{3}{5} < 1 \end{aligned}$$

Hence, Column A is less than Column B, and the answer is (B).

Percents

Problems involving percent are common on the GRE. The word *percent* means "divided by one hundred." When you see the word "percent," or the symbol %, remember it means $\frac{1}{100}$. For example,

$$\begin{array}{c} 25 \text{ percent} \\ \downarrow \quad \downarrow \\ 25 \times \frac{1}{100} = \frac{1}{4} \end{array}$$

To convert a decimal into a percent, move the decimal point two places to the right. For example,

$$0.25 = 25\%$$
$$0.023 = 2.3\%$$
$$1.3 = 130\%$$

Conversely, to convert a percent into a decimal, move the decimal point two places to the left. For example,

$$47\% = .47$$
$$3.4\% = .034$$
$$175\% = 1.75$$

To convert a fraction into a percent, first change it into a decimal (by dividing the denominator [bottom] into the numerator [top]) and then move the decimal point two places to the right. For example,

$$\frac{7}{8} = 0.875 = 87.5\%$$

Conversely, to convert a percent into a fraction, first change it into a decimal and then change the decimal into a fraction. For example,

$$80\% = .80 = \frac{80}{100} = \frac{4}{5}$$

Following are the most common fractional equivalents of percents:

$$33\frac{1}{3}\% = \frac{1}{3} \qquad 20\% = \frac{1}{5}$$

$$66\frac{1}{3}\% = \frac{2}{3} \qquad 40\% = \frac{2}{5}$$

$$25\% = \frac{1}{4} \qquad 60\% = \frac{3}{5}$$

$$50\% = \frac{1}{2} \qquad 80\% = \frac{4}{5}$$

Note! **Percent problems often require you to translate a sentence into a mathematical equation.**

Example 1: What percent of 25 is 5?
(A) 10% (B) 20% (C) 30% (D) 35% (E) 40%

Translate the sentence into a mathematical equation as follows:

$$\begin{array}{cccccc} \text{What} & \text{percent} & \text{of} & 25 & \text{is} & 5 \\ \downarrow & \downarrow & \downarrow & \downarrow & \downarrow & \downarrow \\ x & \frac{1}{100} & \cdot & 25 & = & 5 \end{array}$$

$$\frac{25}{100}x = 5$$

$$\frac{1}{4}x = 5$$

$$x = 20$$

The answer is (B).

Example 2: 2 is 10% of what number
(A) 10 (B) 12 (C) 20 (D) 24 (E) 32

Translate the sentence into a mathematical equation as follows:

$$\begin{array}{cccccc} 2 & \text{is} & 10 & \% & \text{of} & \underline{\text{what number}} \\ \downarrow & \downarrow & \downarrow & \downarrow & \downarrow & \downarrow \\ 2 & = & 10 & \frac{1}{100} & \cdot & x \end{array}$$

$$2 = \frac{10}{100}x$$

$$2 = \frac{1}{10}x$$

$$20 = x$$

The answer is (C).

Example 3: What percent of a is $3a$?
(A) 100% (B) 150% (C) 200% (D) 300% (E) 350%

Translate the sentence into a mathematical equation as follows:

$$\begin{array}{cccccc} \text{What} & \text{percent} & \text{of} & a & \text{is} & 3a \\ \downarrow & \downarrow & \downarrow & \downarrow & \downarrow & \downarrow \\ x & \frac{1}{100} & \cdot & a & = & 3a \end{array}$$

$$\frac{x}{100} \cdot a = 3a$$

$$\frac{x}{100} = 3 \quad \text{(by canceling the } a\text{'s)}$$

$$x = 300$$

The answer is (D).

Example 4: If there are 15 boys and 25 girls in a class, what percent of the class is boys?

(A) 10%
(B) 15%
(C) 18%
(D) 25%
(E) 37.5%

The total number of students in the class is 15 + 25 = 40. Now, translate the main part of the sentence into a mathematical equation:

what	percent	of	the class	is	boys
↓	↓	↓	↓	↓	↓
x	$\frac{1}{100}$	$\cdot$	40	=	15

$$\frac{40}{100}x = 15$$
$$\frac{2}{5}x = 15$$
$$2x = 75$$
$$x = 37.5$$

The answer is (E).

Often you will need to find the percent of increase (or decrease). To find it, calculate the increase (or decrease) and divide it by the original amount:

$$\textbf{Percent of change: } \frac{\textit{Amount of change}}{\textit{Original amount}} \times 100\%$$

Example 5: The population of a town was 12,000 in 1980 and 16,000 in 1990. What was the percent increase in the population of the town during this period?

(A) $33\frac{1}{3}\%$
(B) 50%
(C) 75%
(D) 80%
(E) 120%

The population increased from 12,000 to 16,000. Hence, the change in population was 4,000. Now, translate the main part of the sentence into a mathematical equation:

Percent of change:

$$\frac{\textit{Amount of change}}{\textit{Original amount}} \times 100\% =$$
$$\frac{4000}{12000} \times 100\% =$$
$$\frac{1}{3} \times 100\% = \quad \text{(by canceling 4000)}$$
$$33\frac{1}{3}\%$$

The answer is (A).

Problem Set U:

➢ Easy

1. Column A | 3/11 of a number is 23. | Column B

3/5 of the number | | 60% of the number

2. Column A | | Column B

45% of 90 | | 90% of 45

➢ Medium

3. In January, the value of a stock increased by 25%; and in February, it decreased by 20%. In March, it increased by 50%; and in April, it decreased by 40%. If Jack invested $80 in the stock on January 1 and sold it at the end of April, what was the percentage change in the price of the stock?

(A) 0%
(B) 5%
(C) 10%
(D) 40%
(E) 50%

4. If b equals 10% of a and c equals 20% of b, then which one of the following equals 30% of c ?

(A) 0.0006% of a
(B) 0.006% of a
(C) 0.06% of a
(D) 0.6% of a
(E) 6% of a

5. Column A | Item A cost John $90 and item B cost him $100. | Column B

Percentage of profit that John earned by selling item A with a profit of $10 | | Percentage of profit that John earned by selling item B with a profit of $10

6. 8 is 4% of a, and 4 is 8% of b. c equals b/a. What is the value of c ?

(A) 1/32
(B) 1/4
(C) 1
(D) 4
(E) 32

7. Carlos & Co. generated revenue of $1,250 in 2006. This was 12.5% of its gross revenue. In 2007, the gross revenue grew by $2,500. What is the percentage increase in the revenue in 2007?

(A) 12.5%
(B) 20%
(C) 25%
(D) 50%
(E) 66.3%

8. If 80 percent of the number a is 80, then how much is 20 percent of the number a ?

(A) 20
(B) 40
(C) 50
(D) 60
(E) 80

9. In an acoustics class, 120 students are male and 100 students are female. 25% of the male students and 20% of the female students are engineering students. 20% of the male engineering students and 25% of the female engineering students passed the final exam. What percentage of engineering students passed the exam?

(A) 5%
(B) 10%
(C) 16%
(D) 22%
(E) 25%

10. If 9/100 of x is 9, then which one of the following is true?

(A) 25 percent of x is 25
(B) 1/4 of x is 0.25
(C) x is 120% of 80
(D) x is 9 percent of 90
(E) 90 percent of x is 9

11. If 50% of x equals the sum of y and 20, then what is the value of $x - 2y$?

(A) 20
(B) 40
(C) 60
(D) 80
(E) 100

12. If $\dfrac{x+y}{x-y} = \dfrac{4}{3}$ and $x \neq 0$, then what percentage (to the nearest integer) of $x + 3y$ is $x - 3y$?

(A) 20%
(B) 25%
(C) 30%
(D) 35%
(E) 40%

13.

Column A		Column B
	12 students from section A and 15 students from section B failed an Anthropology exam. Thus, equal percentage of attendees failed the exam from either section.	
Number of attendees for the exam from section A		Number of attendees for the exam from section B

14.

Column A		Column B
	500% of a equals $500b$. a and b are positive.	
a		b

15. Column A | $a/2$ is b% of 30, and a is c% of 50. b is positive. | Column B

Column A		Column B
b		c

16.

Column A	The annual exports of the company NeuStar increased by 25% last year. This year, it increased by 20%.	Column B
Increase in exports last year		Increase in exports in the current year

17.

Column A	Mr. Chang sold 100 oranges to customers for $300 and earned a profit.	Column B
The percentage of the profit that Chang got		The selling price of the oranges expressed as a percentage of the cost

➢ Hard

18. The selling price of 15 items equals the cost of 20 items. What is the percentage profit earned by the seller?

(A) 15
(B) 20
(C) 25
(D) 33.3
(E) 40

19. Each person in a group of 110 investors has investments in either equities or securities or both. Exactly 25% of the investors in equities have investments in securities, and exactly 40% of the investors in securities have investments in equities. How many have investments in equities?

(A) 65
(B) 80
(C) 120
(D) 135
(E) 150

20. The value of a share of stock was $30 on Sunday. The profile of the value in the following week was as follows: The value appreciated by $1.2 on Monday. It appreciated by $3.1 on Tuesday. It depreciated by $4 on Wednesday. It appreciated by $2 on Thursday and it depreciated by $0.2 on Friday. On Friday, the stock market closed for the weekend. By what percentage did the value of the share increase in the five days?

(A) 3.2%
(B) 4%
(C) 5.6%
(D) 7%
(E) 10%

21. The percentage of integers from 1 through 100 whose squares end with the digit 1 is $x\%$, and the percentage of integers from 1 through 200 whose squares end with the digit 1 is $y\%$. Which one of the following is true?

(A) $x = y$
(B) $x = 2y$
(C) $x = 4y$
(D) $y = 2x$
(E) $y = 4x$

22. The cost of painting a wall increases by a fixed percentage each year. In 1970, the cost was \$2,000; and in 1979, it was \$3,600. What was the cost of painting in 1988?

(A) \$1,111
(B) \$2,111
(C) \$3,600
(D) \$6240
(E) \$6480

23. The *list price* of a commodity is the price after a 20% discount on the retail price. The *festival discount price* on the commodity is the price after a 30% discount on the list price. Customers purchase commodities from stores at a festival discount price. What is the effective discount offered by the stores on the commodity on its retail price?

(A) 20%
(B) 30%
(C) 44%
(D) 50%
(E) 56%

24. The price of a car was m dollars. It then depreciated by $x\%$. Later, it appreciated by $y\%$ to n dollars. If there are no other changes in the price and if $y = \dfrac{x}{1 - \dfrac{x}{100}}$, then which one of the following must n equal?

(A) $3m/4$
(B) m
(C) $4m/3$
(D) $3m/2$
(E) $2m$

25.

Column A		Column B
	A stores sells Brand A items and Brand B items at 75% and 80%, respectively, off list prices. During festival season, discounts of 20% and 25%, on Brand A and Brand B, respectively, are offered. Spinach tins of Brand A and Brand B have the same list price.	
Cost of spinach tins of Brand A during festival season		Cost of spinach tins of Brand B during festival season

	Column A		Column B
26.	Increase in population in the first decade 1980-1990	The population of a country increases at a fixed percentage each year.	Increase in population in the second decade 1990–2000
27.	Sales tax paid by James	James purchased Medicine A for x dollars, which included a sales tax of 5%. Kate was charged 5% for sales tax on x dollars that Medicine B costs.	Sales tax paid by Kate
28.	The profit made by selling 100 apples	Miller sold apples at 125% of what it cost him.	The profit made by selling 200 apples at a further 10% discount
29.	The value of Fund A after 3 years	Each year, funds A and B grow by a particular percentage based on the following policy of the investment company: (1) The allowed percentages of growths on the two funds are 20% or 30%. (2) The growth percentages of the two funds are not the same in any year. (3) No fund will have the same percentage growth in any two consecutive years. Bob invested equal amounts in two funds *A* and *B*. In the first year, fund B was offered a growth of 30%.	The value of Fund B after 3 years
30.	The list price (in dollars) of the computer that John purchased	An off-season discount of 10% is being offered at a store for any purchase with list price above \$500. No other discounts are offered at the store. John purchased a computer from the store for \$459.	500
31.	$x - y$	Williams has x eggs. He sells y of them at a profit of 10 percent and the remaining eggs at a loss of 10 percent. He made a profit overall.	y

32. If $ab = cd$ and a is 25% greater than c, then by what percent is d greater than b?

(A) 16.66
(B) 20
(C) 25
(D) 30
(E) 33.33

➤ **Very Hard**

33. Selling 12 candies at a price of $10 yields a loss of a%. Selling 12 candies at a price of $12 yields a profit of a%. What is the value of a?

(A) 11/1100
(B) 11/100
(C) 100/11
(D) 10
(E) 11

34. The total income of Mr. Teng in the years 2003, 2004, and 2005 was $36,400. His income increased by 20% each year. What was his income in 2005?

(A) 5,600
(B) 8,800
(C) 10,000
(D) 12,000
(E) 14,400

Answers and Solutions to Problem Set U

➤ Easy

1. 60% of a number is 60/100 = 3/5 times the number. Hence, 60% of a number is 3/5 of the number. So, the columns are equal. The answer is (C). Note that we did not need the statement "3/11 of a number is 23" to solve the problem.

2. Column A: 45% of 90 = $\frac{45}{100} \times 90 = \frac{45 \times 90}{100}$.

Column B: 90% of 45 = $\frac{90}{100} \times 45 = \frac{90 \times 45}{100}$.

Since the columns are equal, the answer is (C).

➤ Medium

3. At the end of January, the value of the stock is \$80 + 25%(\$80) = \$80 + \$20 = \$100.

At the end of February, the value of the stock is \$100 – 20%(\$100) = \$100 – \$20 = \$80.

At the end of March, the value of the stock is \$80 + 50%(\$80) = \$80 + \$40 = \$120.

At the end of April, the value of the stock is \$120– 40%(\$120) = \$120 – \$48 = \$72.

Now, the percentage change in price is

$$\frac{\text{change in price}}{\text{original price}} = \frac{80-72}{80} = \frac{8}{80} = \frac{1}{10} = 10\%$$

The answer is (C).

4. $b = 10\%$ of $a = (10/100)a = 0.1a$.

$c = 20\%$ of $b = (20/100)b = 0.2b = (0.2)(0.1a)$

Now, 30% of $c = (30/100)c = 0.3c = (0.3)(0.2)(0.1a) = 0.006a = 0.6\%a$.

The answer is (D).

5. The formula for the profit percentage is $\frac{\text{Profit}}{\text{Cost}} \cdot 100$.

Hence, Column A equals 10/90 · 100 = 100/9% = 11.1%, and Column B equals 10/100 · 100 = 10%.

Hence, Column A is greater than Column B, and the answer is (A).

6. 4% of a is $4a/100$. Since this equals 8, we have $4a/100 = 8$. Solving for a yields $a = 8 \bullet \frac{100}{4} = 200$.

Also, 8% of b equals $8b/100$, and this equals 4. Hence, we have $\frac{8}{100} \bullet b = 4$. Solving for b yields $b = 50$.

Now, $c = b/a = 50/200 = 1/4$. The answer is (B).

7. We are given that Carlos & Co. generated revenue of \$1,250 in 2006 and that this was 12.5% of the gross revenue. Hence, if 1250 is 12.5% of the revenue, then 100% (gross revenue) is (100/12.5)(1250) = 10,000. Hence, the total revenue by end of 2007 is \$10,000. In 2006, revenue grew by \$2500. This is a growth of (2500/10000) × 100 = 25%. The answer is (C).

8. We have that 80 is 80 percent of a. Now, 80 percent of a is $80/100 \times a$. Equating the two yields $80/100 \times a = 80$. Solving the equation for a yields $a = 100/80 \times 80 = 100$. Now, 20 percent of a is $20/100 \times 100 = 20$. The answer is (A).

9. There are 100 female students in the class, and 20% of them are Engineering students. Now, 20% of 100 equals $20/100 \times 100 = 20$. Hence, the number of female engineering students in the class is 20.

Now, 25% of the female engineering students passed the final exam: 25% of 20 = $25/100 \times 20 = 5$. Hence, the number of female engineering students who passed is 5.

There are 120 male students in the class. And 25% of them are engineering students. Now, 25% of 120 equals $25/100 \times 120 = 1/4 \times 120 = 30$. Hence, the number of male engineering students is 30.

Now, 20% of the male engineering students passed the final exam: 20% of 30 = $20/100 \times 30 = 6$. Hence, the number of male engineering students who passed is 6.

Hence, the total number of Engineering students who passed is

(Female Engineering students who passed) + (Male Engineering students who passed) =

5 + 6 =

11

The total number of Engineering students in the class is

(Number of female engineering students) + (Number of male engineering students) =
30 + 20 =
50

Hence, the percentage of engineering students who passed is

$$\frac{\text{Total number of engineering students who passed}}{\text{Total number of engineering students}} \times 100 =$$

$$11/50 \times 100 =$$

22%

The answer is (D).

10. We are given that 9/100 of x is 9. Now, 9/100 of x can be expressed as 9% of x. Hence, 9% of x is 9. Hence, 25 percent of x must equal 25. The answer is (A).

11. 50% of x equals the sum of y and 20. Expressing this as an equation yields

$$(50/100)x = y + 20$$
$$x/2 = y + 20$$
$$x = 2y + 40$$
$$x - 2y = 40$$

The answer is (B).

12. Dividing both the numerator and the denominator of the given equation $\frac{x+y}{x-y} = \frac{4}{3}$ by y yields $\frac{\frac{x}{y}+1}{\frac{x}{y}-1} = \frac{4}{3}$. Cross-multiplying this equation yields $\frac{3x}{y} + 3 = \frac{4x}{y} - 4$. Solving for x/y yields $x/y = 7$.

Now, the percentage of $x + 3y$ the expression $x - 3y$ makes is $\frac{x-3y}{x+3y} \bullet 100$. Dividing both the numerator and the denominator of the expression by y yields

$$\frac{\frac{x}{y}-3}{\frac{x}{y}+3} \bullet 100 = \frac{7-3}{7+3} \bullet 100 = \frac{4}{10} \bullet 100 = 40\%$$

The answer is (E).

13. Given that an equal percent of attendees failed the exam in sections A and B. Let x be the percent. If a students took the exam from section A and b students took the exam from section B, then number of students who failed from the sections would be $a(x/100)$ and $b(x/100)$, respectively. Given that the two equal 12 and 15, respectively, we have $a(x/100) = 12$ and $b(x/100) = 15$. Since $12 < 15$, $a(x/100) < b(x/100)$. Canceling $x/100$ from both sides yields $a < b$. Hence, Column B > Column A, and the answer is (B).

14. We are given that 500% of a equals $500b$. Now, 500% of a is $\frac{500}{100}a = 5a$. Setting this equal to $500b$ yields $5a = 500b$. Dividing both sides of this equation by 5 yields $a = 100b$. Since both a and b are positive (given) and a is 100 times b, a is greater than b. Hence, Column A is greater than Column B, and the answer is (A).

15. We are given that $a/2$ is $b\%$ of 30. Now, $b\%$ of 30 is $\frac{30}{100}b$. Hence, $\frac{a}{2} = \frac{30}{100}b$. Solving for a yields $a = \frac{3}{5}b$. We are also given that a is $c\%$ of 50. Now, $c\%$ of 50 is $\frac{c}{100} \cdot 50 = \frac{c}{2}$. Hence, $a = c/2$. Plugging this into the equation $a = \frac{3}{5}b$ yields $\frac{c}{2} = \frac{3}{5}b$. Multiplying both sides by 2 yields $c = \frac{6}{5}b$. Since b is positive, c is also positive; and since $6/5 > 1$, $c > b$. Hence, the answer is (B).

16. Let x be the annual exports of the company before last year. It is given that the exports increased by 25% last year. The increase (Column A) equals $(25/100)x = x/4$, and the net exports equals $x + x/4 = 5x/4$. Now, exports increased by 20% this year. So, the increase (Column B) equals $(20/100)(5x/4) = x/4$. Hence, both columns equal $x/4$, and the answer is (C).

17. Let C be the cost of the oranges, and let S be the selling price of the oranges.

Column A: The percentage of the profit that Chang got is $\frac{S-C}{C} \cdot 100$.

Column B: The selling price expressed as a percentage of the cost is $\frac{S}{C} \cdot 100$.

Now,

$$\text{Column B} - \text{Column A} =$$
$$\frac{S}{C} \cdot 100 - \frac{S-C}{C} \cdot 100 =$$
$$\frac{S}{C} \cdot 100 - \left(\frac{S}{C} - 1\right) \cdot 100 =$$
$$\frac{S}{C} \cdot 100 - \frac{S}{C} \cdot 100 + 100 =$$
$$100$$

Hence, Column B is 100 units greater than Column A, and the answer is (B).

➢ Hard

18. Let c and s be the cost and the selling price, respectively, for the seller on each item.

We are given that the selling price of 15 items equals the cost of 20 items. Hence, we have $15s = 20c$, or $s = (20/15)c = 4c/3$. Now, the profit equals selling price – cost = $s - c = 4c/3 - c = c/3$. The percentage profit on each item is

$$\frac{\text{Profit}}{\text{Cost}} \bullet 100 = \frac{\frac{c}{3}}{c} \bullet 100 = \frac{100}{3} = 33.3\%$$

The answer is (D).

19. The investors can be categorized into three groups:

(1) Those who have investments in equities only.
(2) Those who have investments in securities only.
(3) Those who have investments in both equities and securities.

Let x, y, and z denote the number of people in the respective categories. Since the total number of investors is 110, we have

$$x + y + z = 110 \qquad (1)$$

Also,
The number of people with investments in equities is $x + z$ and
The number of people with investments in securities is $y + z$.

Since exactly 25% of the investors in equities have investments in securities, we have the equation

$25/100 \bullet (x + z) = z$
$25/100 \bullet x + 25/100 \bullet z = z$
$25/100 \bullet x = 75/100 \bullet z$
$x = 3z \qquad (2)$

Since exactly 40% of the investors in securities have investments in equities, we have the equation

$40/100 \bullet (y + z) = z$
$2/5 \bullet (y + z) = z$
$y + z = 5z/2$
$y = 3z/2 \qquad (3)$

Substituting equations (2) and (3) into equation (1) yields

$3z + 3z/2 + z = 110$
$11z/2 = 110$
$z = 110 \bullet 2/11 = 20$

Hence, the number of people with investments in equities is $x + z = 3z + z = 3 \bullet 20 + 20 = 60 + 20 = 80$. The answer is (B).

20. The initial price of the share is \$30.

After the \$1.2 appreciation on Monday, its price was 30 + 1.2 = \$31.2.
After the \$3.1 appreciation on Tuesday, its price was 31.2 + 3.1 = \$34.3.
After the \$4 depreciation on Wednesday, its price was 34.3 – 4 = \$30.3.
After the \$2 appreciation on Thursday, its price was 30.3 + 2 = \$32.3.
After the \$0.2 depreciation on Friday, its price was 32.3 – 0.2 = \$32.1.

The percentage increase in the price from the initial price is

$$(32.1 - 30)/30 \times 100 =$$
$$2.1/30 \times 100 =$$
$$2.1/3 \times 10 =$$
$$21/3 = 7$$

The answer is (D).

21. The square of an integer ends with the digit 1 only if the integer itself either ends with the digit 1 or with the digit 9. For example, $11^2 = 121$ and $19^2 = 361$. Now, there are ten integers ending with 1 from 1 through 100. The numbers are 1, 11, 21, …, 91. Also, there are ten integers ending with 9 from 1 through 100. They are 9, 19, 29, …, 99. Hence, the total number of integers from 1 through 100 whose squares end with the digit 1 is 20. The number 20 is 20/100 × 100 = 20% of 100. Hence, $x = 20$.

Similarly, there are twenty integers (1, 11, 21… 191) ending with 1, and twenty integers (9, 19, 29, …, 199) ending with 9. Hence, there are 20 + 20 = 40 integers ending with 1 or 9. Now, 40 is 40/200 × 100 = 20% of the total 200 integers from 1 through 200. So, y also equals 20. Since $x = y$, the answer is (A).

22. Since the cost of painting increases by a fixed percentage each year and it increased \$3,600/\$2,000 = 1.8 times in the 9-year period from 1970 to 1979, it must increase by the same number of times in the period 1979 to 1988. Hence, the amount becomes 1.8 × \$3,600 = \$6480 by 1988. The answer is (E).

23. Let r be the retail price. The list price is the price after a 20% discount on the retail price. Hence, it equals $r(1 - 20/100) = r(1 - 0.2) = 0.8r$.

The festival discount price is the price after a 30% discount on the list price. Hence, the festival discount price equals (list price)(1 – 30/100) = $(0.8r)(1 - 30/100) = (0.8r)(1 - 0.3) = (0.8r)(0.7) = 0.56r$.

Hence, the total discount offered is (Original Price – Price after discount)/Original Price × 100 = $(r - 0.56r)/ r \times 100 = 0.44 \times 100 = 44\%$.

The answer is (C).

24. After a depreciation of $x\%$ on the m dollars, the depreciated price of the car is $m(1 - x/100)$.

After an appreciation of $y\%$ on this price, the appreciated price, n, is $m(1 - x/100)(1 + y/100) = (m/100)(100 - x)(1 + y/100)$. Hence, $n = (m/100)(100 - x)(1 + y/100)$.

We are given that $y = \dfrac{x}{1-\dfrac{x}{100}} = \dfrac{x}{\dfrac{100-x}{100}} = \dfrac{100x}{100-x}$. Substituting this in the equation $n = (m/100)(100 - x)$ $(1 + y/100)$ yields

$$\frac{m}{100}(100-x)\left(1+\frac{\dfrac{100x}{100-x}}{100}\right)=$$

$$\frac{m}{100}(100-x)\left(1+\frac{x}{100-x}\right)=$$

$$\frac{m}{100}(100-x)\left(\frac{100-x+x}{100-x}\right)=$$

$$\frac{m}{100}(100-x)\left(\frac{100}{100-x}\right)=$$

$$m$$

Hence, $n = m$, and the answer is (B).

25. Since the spinach tins of either brand have the same list price, let each be x dollars. Now, 75% of x is $(75/100)x = 3x/4$, and 80% of x is $(80/100)x = 4x/5$. So, spinach tins of brands A and B are sold at $3x/4$ and $4x/5$, respectively. After festival discounts of 20% and 25% on the respective brand items, the cost of spinach tins drop to $\left(1-\dfrac{20}{100}\right)\left(\dfrac{3x}{4}\right)=\left(\dfrac{4}{5}\right)\left(\dfrac{3x}{4}\right)=\dfrac{3x}{5}$ and $\left(1-\dfrac{25}{100}\right)\left(\dfrac{4x}{5}\right)=\left(\dfrac{3}{4}\right)\left(\dfrac{4x}{5}\right)=\dfrac{3x}{5}$. Hence, the columns are equal, and the answer is (C).

26. The population of the country grows at a fixed percentage each year. Hence, just as we have a fixed percentage for each year, we will have a different and fixed percentage for each 10-year period (a decade). Let the later rate be $R\%$. Hence, if P is population in 1980, population in 1990 (after a decade) would grow by $PR/100$ (= Column A). So, population in 1990 would be Original Plus Increase = $P + PR/100$. Similarly, in the next decade (1990–2000) and at the same rate $R\%$ it would increase by $(P + PR/100)(R/100)$ (= Column B). Substituting the results in the columns yields

$PR/100$	$(P + PR/100)(R/100)$

Canceling $R/100$ from both sides yields

P	$P + PR/100$

Now, Column A = P and Column B = $P + PR/100$ = Column A + $PR/100$. Hence, Column B > Column A, and the answer is (B).

27. Let the original cost of Medicine A be a dollars. 5% tax on this equals $(5/100)a = a/20$. So, the total cost including sales tax is $a + a/20 = 21a/20$. This equals x dollars (what medicine A cost James). Hence, we have $21a/20 = x$. Solving for a yields $a = 20x/21$. Sales tax on A is $\dfrac{a}{20}=\dfrac{\dfrac{20x}{21}}{20}=\dfrac{x}{21}$ = Column A. Now, Kate was charged sales tax on Medicine B. The charge was 5% of the cost of the medicine (x). 5% of x is $(5/100)x = x/20$ = Column B. Since $1/20 > 1/21$, $x/20 > x/21$ and Column B > Column A. The answer is (B).

28. Let each apple cost Miller x dollars. Since he sold the apples at 125% of the cost, the profit made is

$$\begin{gathered}\text{Selling price} - \text{Cost} = \\ (125/100)x - x = \\ 5x/4 - x = \\ x/4\end{gathered}$$

The profit on 100 apples is $100 \cdot x/4 = 25x$. Hence, Column A equals $25x$.

Now, after a 10% discount on the selling price, Mr. Miller must be selling the apples at a price equal to

$$(\text{actual selling price}) \cdot \left(1 - \frac{\text{discount percent}}{100}\right) = \left(\frac{5x}{4}\right) \cdot \left(1 - \frac{10}{100}\right) = \left(\frac{5x}{4}\right) \cdot \left(\frac{9}{10}\right) = \frac{9x}{8}$$

Hence, the profit made on each apple equals Selling price – Cost = $9x/8 - x = x/8$. The profit on 200 apples is $200 \cdot x/8 = 25x$ = Column B.

Since the columns are equal, the answer is (C).

29. In the first year, fund B was given a growth of 30%. Hence, according to clauses (1) and (2), fund A must have grown by 20% (the other allowed growth percentage clause (1)).

In the second year, according to the clauses (1) and (3), the growth percentages of the two funds will swap between the only allowed values 30% and 20% (clause (1)). Hence, fund A grows by 30% and fund B grows by 20%.

In the third year, according to clauses (1) and (3), the growth percentages will again swap between the only two allowed values 20% and 30% (clause (1)). Hence, fund A grows by 20% and fund B grows by 30%.

Since fund B has a better growth percentage history than fund A in the first three years, the value of fund B will be greater than the value of fund A after three years. Hence, Column B is greater than Column A, and the answer is (B).

30. We do not know whether the $459 price that John paid for the computer was with the discount offer or without the discount offer.

If he did not get the discount offer, the list price of the computer should be $459 and John paid the exact amount for the computer. In this case, Column A (= 459) is less than Column B.

If the price corresponds to the price after the discount offer, then $459 should equal a 10% discount on the list price. Hence, if l represents the list price, then we have $\$459 = l(1 - 10/100) = l(1 - 1/10) = (9/10)l$. Solving the equation for l yields $l = (10/9)459 = 510$ dollars (a case when discount was offered because the list price is greater than $500). Hence, it is also possible that John got the 10% discount on the computer originally list priced at $510. Here, list price (= Column A) is greater than 500 (= Column B).

Hence, we have a double case, and the answer is (D).

31. Let a dollars be the cost of each egg to Williams. Hence, the net cost of the x eggs is ax dollars.

Now, the selling price of the eggs when selling at 10% profit is $a(1 + 10/100) = 11a/10$.

The selling price of the eggs when selling at 10% loss is $a(1 - 10/100) = 9a/10$.

We are given that Williams has x eggs and he sold y of them at 10 percent profit (at a selling price of $11a/10$) and the rest, $x - y$, at 10 percent loss (at a selling price of $9a/10$). Hence, the net selling price is $y(11a/10) + (x - y)(9a/10) = a(0.2y + 0.9x)$.

Since overall he made a profit, the net selling price must be greater than the net cost. Hence, we have the inequality $a(0.2y + 0.9x) > ax$.

Canceling a from both sides of the inequality yields $0.2y + 0.9x > x$. Subtracting $0.9x$ from both sides of the inequality yields $0.2y > 0.1x$. Multiplying both sides by 10 yields $2y > x$. Now, subtracting y from both sides yields $y > x - y$.

Hence, Column B is greater than Column A. The answer is (B).

32. Since a is 25% greater than c, we have $a = (1 + 0.25)c$. Substituting this in the given equation $ab = cd$ yields $[(1 + 0.25)c]b = cd$. Canceling c from both sides yields $(1 + 0.25)b = d$. This equation indicates that d is 25% greater than b. The answer is (C).

➤ Very Hard

33. Let c be the cost of each candy. Then the cost of 12 candies is $12c$. We are given that selling 12 candies at \$10 yields a loss of $a\%$. The formula for the loss percentage is $\frac{\text{cost - selling price}}{\text{cost}} \bullet 100$. Hence, $a = \frac{12c - 10}{12c} \bullet 100$. Let this be equation (1).

We are also given that selling 12 candies at \$12 yields a profit of $a\%$. The formula for profit percent is $\frac{\text{selling price - cost}}{\text{cost}} \bullet 100$. Hence, we have $\frac{12 - 12c}{12c} \bullet 100 = a\%$. Let this be equation (2).

Equating equations (1) and (2), we have

$$\frac{12 - 12c}{12c} \bullet 100 = \frac{12c - 10}{12c} \bullet 100$$

$$12 - 12c = 12c - 10 \qquad \text{by canceling } 12c \text{ and } 100 \text{ from both sides}$$

$$24c = 22$$

$$c = 22/24$$

From equation (1), we have

$$a = \frac{12c - 10}{12c} \bullet 100$$

$$= \frac{12 \cdot \frac{22}{24} - 10}{12 \cdot \frac{22}{24}} \bullet 100$$

$$= \frac{11 - 10}{11} \bullet 100$$

$$= \frac{100}{11}$$

The answer is (C).

34. Let p be the income of Mr. Teng in the year 2003.

We are given that his income increased by 20% each year. So, the income in the second year, 2004, must be $p(1 + 20/100) = p(1 + 0.2) = 1.2p$. The income in the third year, 2005, must be

$$1.2p(1 + 20/100) =$$

$$1.2p(1 + 0.2) =$$

$$1.2p(1.2) =$$

$$1.44p$$

Hence, the total income in the three years equals $p + 1.2p + 1.44p$. Since the total income is 36,400, we have the equation $p + 1.2p + 1.44p = 36{,}400$, or $3.64p = 36{,}400$, or $p = 36{,}400/3.64 = 10{,}000$. Hence, the income in the third year equals $1.44p = 1.44 \times 10{,}000 = 14{,}400$. The answer is (E).

Graphs

Questions involving graphs are common on the GRE. Rarely do these questions involve any significant calculating. Usually, the solution is merely a matter of interpreting the graph.

Questions 1-4 refer to the following graphs.

SALES AND EARNINGS OF CONSOLIDATED CONGLOMERATE

Note: Figure drawn to scale.

1. During which year was the company's earnings 10 percent of its sales?

(A) 85 (B) 86 (C) 87 (D) 88 (E) 90

Reading from the graph, we see that in 1985 the company's earnings were \$8 million and its sales were \$80 million. This gives

$$\frac{8}{10} = \frac{1}{10} = \frac{10}{100} = 10\%$$

The answer is (A).

2. During the years 1986 through 1988, what were the average earnings per year?

(A) 6 million (B) 7.5 million (C) 9 million (D) 10 million (E) 27 million

The graph yields the following information:

Year	Earnings
1986	\$5 million
1987	\$10 million
1988	\$12 million

Forming the average yields $\frac{5+10+12}{3} = \frac{27}{3} = 9$. The answer is (C).

3. In which year did sales increase by the greatest percentage over the previous year?

(A) 86 (B) 87 (C) 88 (D) 89 (E) 90

To find the percentage increase (or decrease), divide the numerical change by the original amount. This yields

Year	Percentage increase
86	$\frac{70-80}{80}=\frac{-10}{80}=\frac{-1}{8}=-12.5\%$
87	$\frac{50-70}{70}=\frac{-20}{70}=\frac{-2}{7}\approx -29\%$
88	$\frac{80-50}{50}=\frac{30}{50}=\frac{3}{5}=60\%$
89	$\frac{90-80}{80}=\frac{10}{80}=\frac{1}{8}=12.5\%$
90	$\frac{100-90}{90}=\frac{10}{90}=\frac{1}{9}\approx 11\%$

The largest number in the right-hand column, 60%, corresponds to the year 1988. The answer is (C).

4. If Consolidated Conglomerate's earnings are less than or equal to 10 percent of sales during a year, then the stockholders must take a dividend cut at the end of the year. In how many years did the stockholders of Consolidated Conglomerate suffer a dividend cut?

(A) None (B) One (C) Two (D) Three (E) Four

Calculating 10 percent of the sales for each year yields

Year	10% of Sales (millions)	Earnings (millions)
85	$.10 \times 80 = 8$	8
86	$.10 \times 70 = 7$	5
87	$.10 \times 50 = 5$	10
88	$.10 \times 80 = 8$	12
89	$.10 \times 90 = 9$	11
90	$.10 \times 100 = 10$	8

Comparing the right columns shows that earnings were 10 percent or less of sales in 1985, 1986, and 1990. The answer is (D).

Problem Set V:

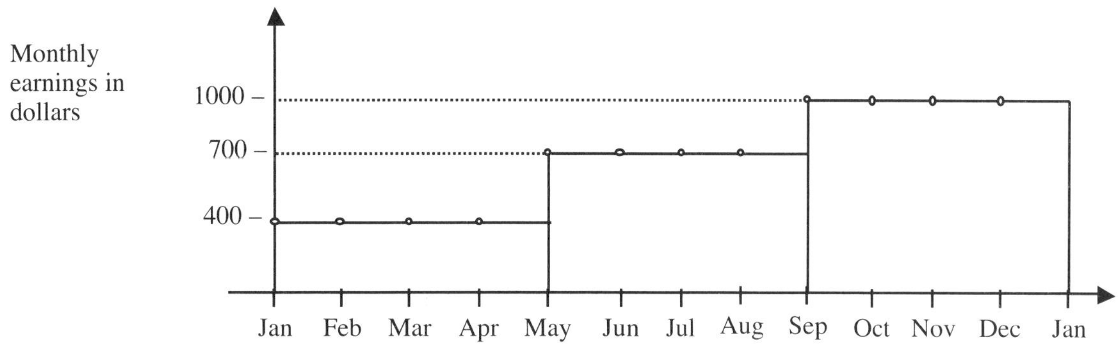

A's income profile during the year 2007

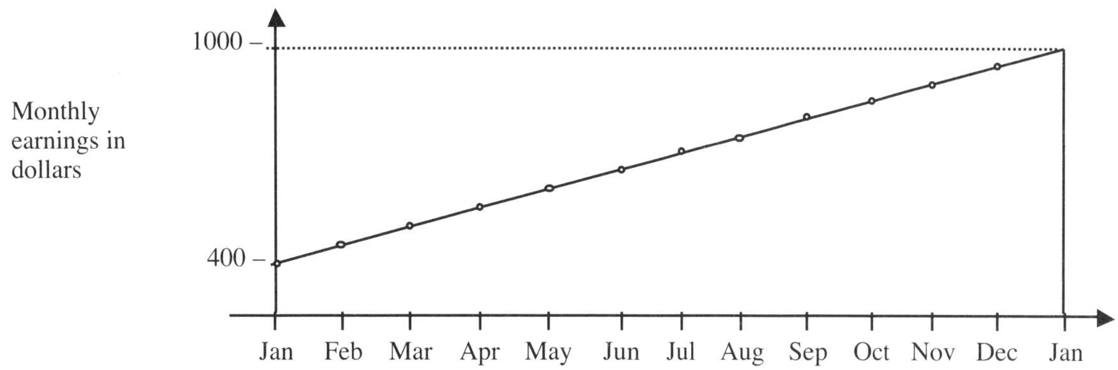

B's income profile during the year 2007

➢ **Medium**

1. *A* launched 3 products in the year 2007 and earns income from the sales of the products only. The top graph shows his monthly earnings for the year. *B*'s earnings consist of continuously growing salary, growing by same amount each month as shown in the figure. Which one of the following equals the total earnings of *A* and *B* in the year 2007?

   (A) 7500, 8100
   (B) 7850, 8300
   (C) 8150, 8400
   (D) 8400, 8100
   (E) 8400, 8700

Questions 2–4 refer to the following graph.

The graph below shows historical exchange rates between the Indian Rupee (INR) and the US Dollar (USD) between January 9 and February 8 of a particular year.

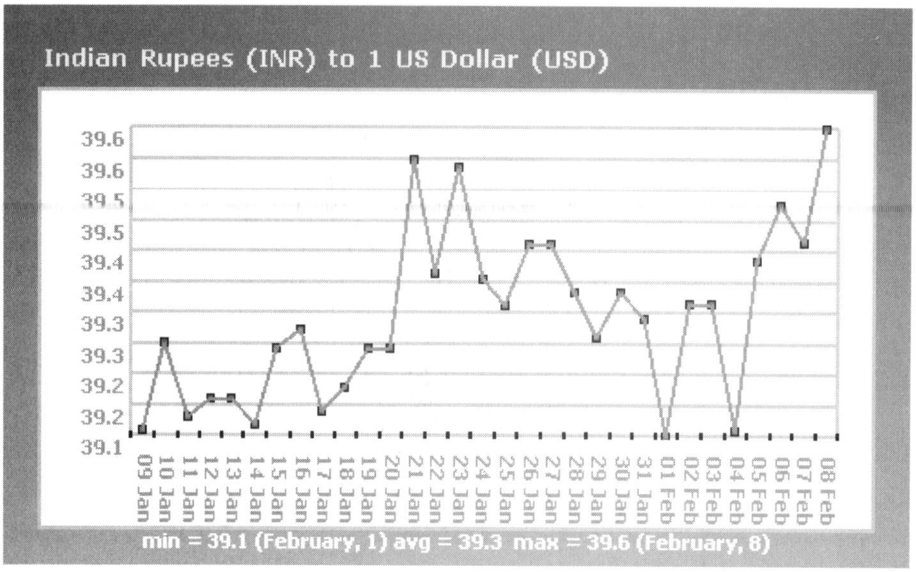

➤ **Easy**

2. On which day shown on the graph did the value of the US dollar increase against the Rupee by the greatest amount?

   (A) Jan. 10
   (B) Jan. 14
   (C) Jan. 21
   (D) Jan. 23
   (E) Feb. 4

➤ **Medium**

3. John had 100 dollars. The exchange rate converts the amount in US dollars to a number in Indian Rupees by directly multiplying by the value of the exchange rate. By what amount did John's $100 increase in terms of Indian Rupees from Jan. 9 to Feb. 8?

   (A) 5
   (B) 10
   (C) 15
   (D) 25
   (E) 50

➤ **Hard**

4. On February 8, the dollar value was approximately what percent of the dollar value on January 9?

   (A) 1.28
   (B) 12. 8
   (C) 101.28
   (D) 112. 8
   (E) 128

Questions 5–7 refer to the following graph.

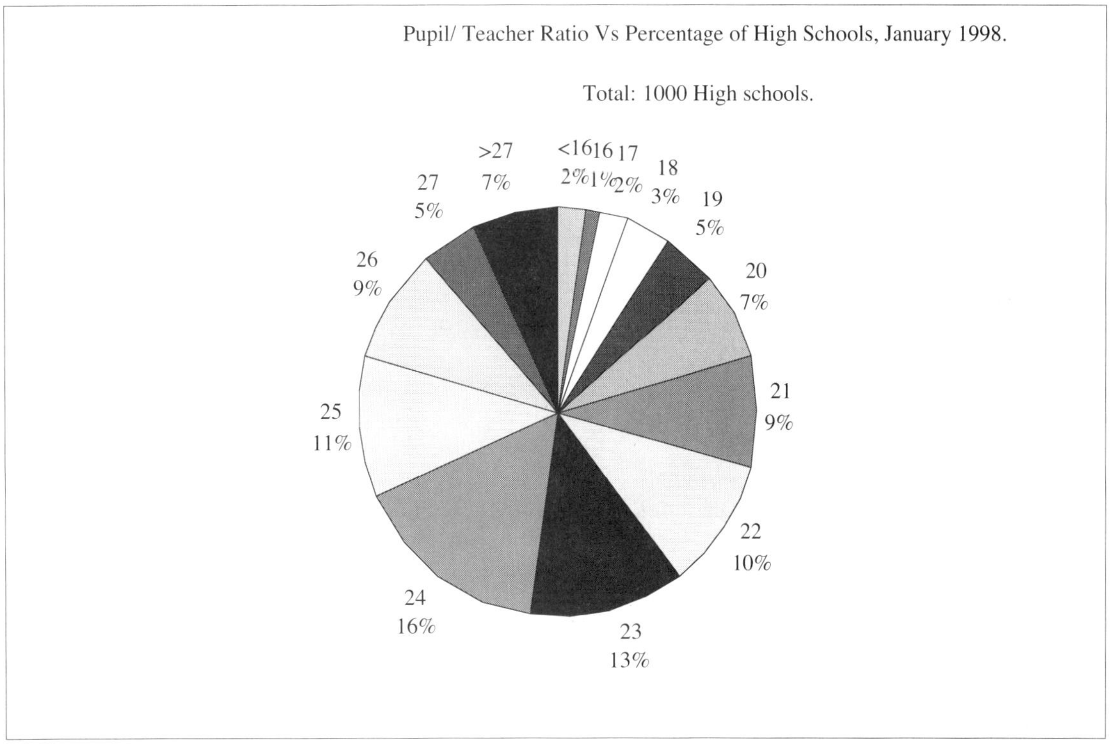

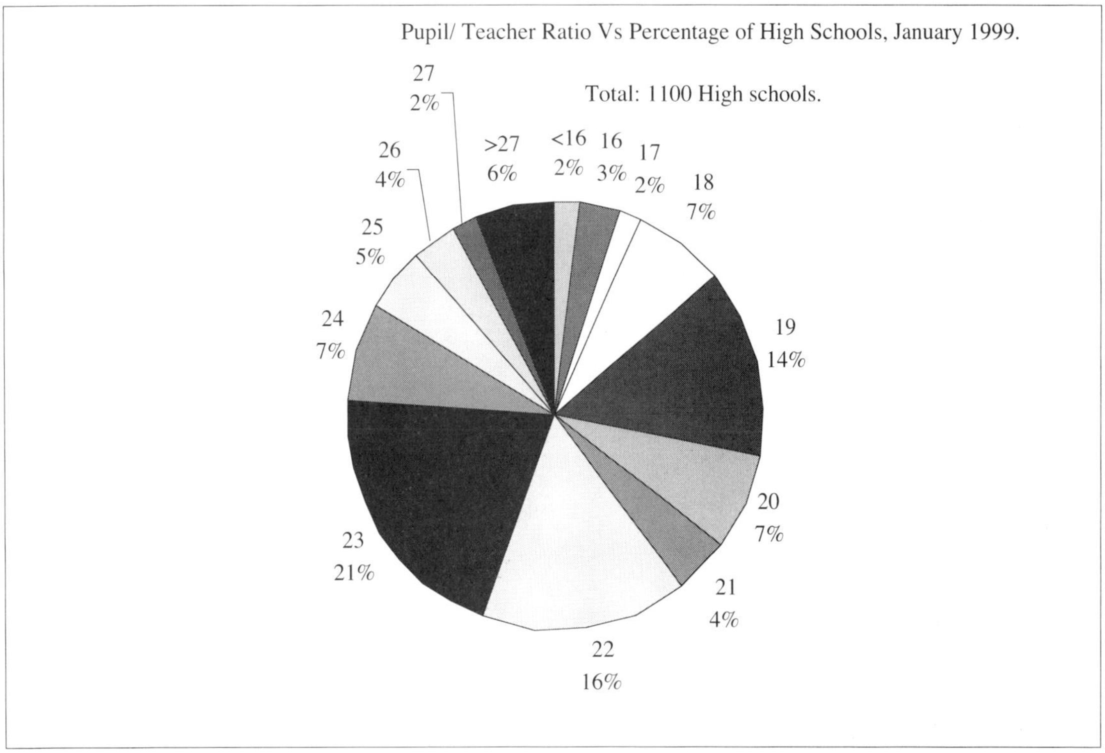

➢ **Medium**

5. By what percent did the number of schools with Pupil/Teacher ratio less than 16 increase in January 1999 over January 1998?

 (A) –2%
 (B) 0%
 (C) 2%
 (D) 10%
 (E) 12%

➢ **Hard**

6. In January 1998, what percent of high schools had a Pupil/Teacher ratio less than 23?

 (A) 25%
 (B) 39%
 (C) 50%
 (D) 60%
 (E) 75%

➢ **Hard**

7. If the areas of the sectors in the circle graphs are drawn in proportion to the percent shown, what is the measure, in degrees, of the sector representing the number of high schools with Pupil/Teacher ratio greater than 27 in 1999?

 (A) 21.6
 (B) 30
 (C) 45.7
 (D) 56.3
 (E) 72

Questions 8–10 refer to the following graph.

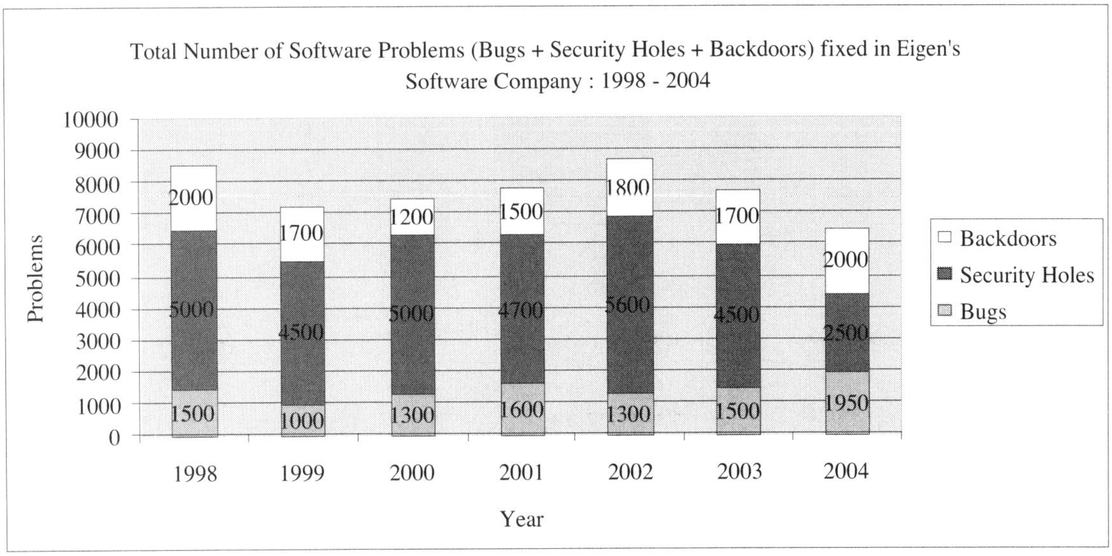

> **Easy**

8. What was the number of security holes fixed in 2003?

    (A) 1500
    (B) 1700
    (C) 4500
    (D) 6000
    (E) 6300

> **Medium**

9. For which year was the ratio of the Security holes to Bugs fixed by the software company the greatest?

    (A) 1998
    (B) 1999
    (C) 2000
    (D) 2001
    (E) 2002

> **Medium**

10. If the total number of software problems solved is a direct measure of the company's capability, then by approximately what percent did capability increase from 1999 to 2002?

    (A) 10%
    (B) 20%
    (C) 30%
    (D) 40%
    (E) 50%

Questions 11–15 refer to the following discussion.

The graphs below provide data on a common entrance examination conducted in different years.

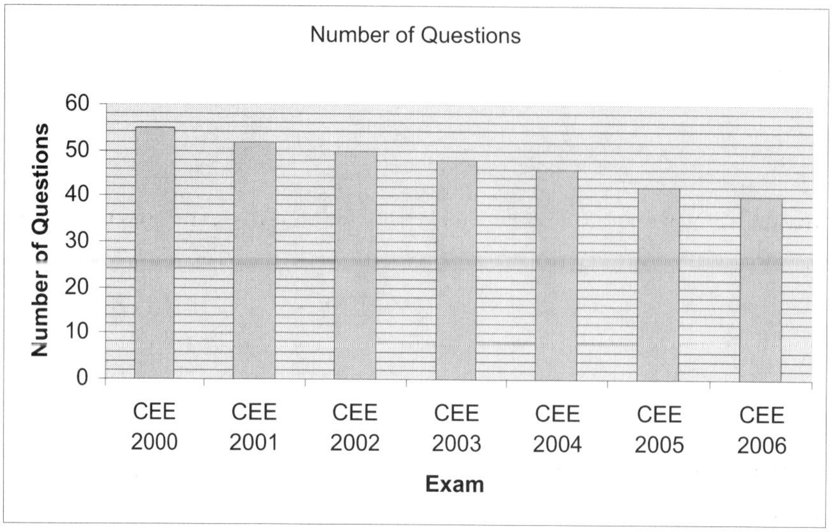

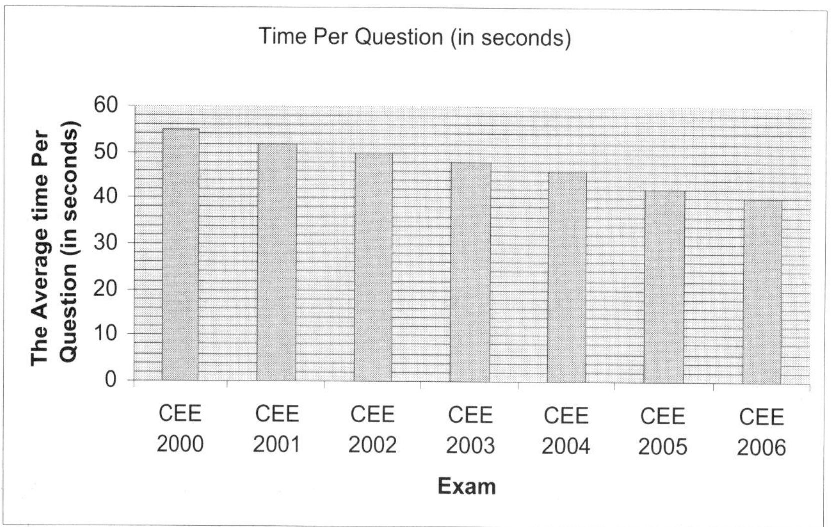

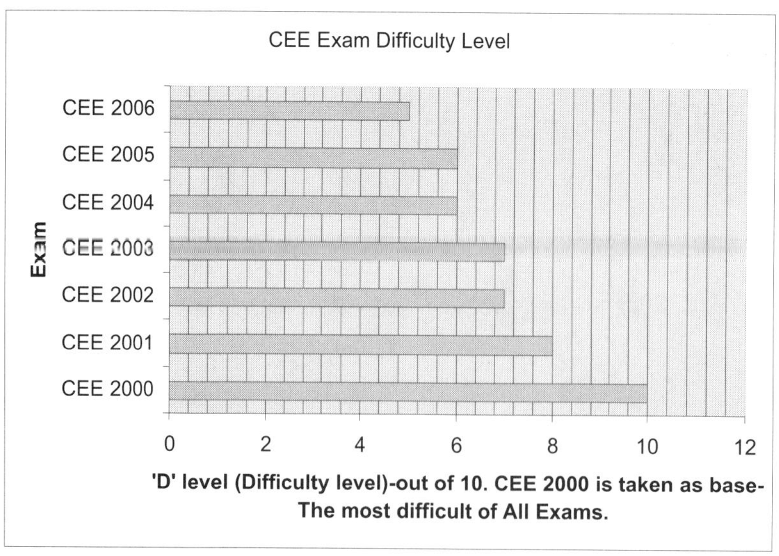

➢ **Easy**

11. In which year were the questions in the exam least difficult?

 (A) 2000
 (B) 2001
 (C) 2002
 (D) 2004
 (E) 2006

➢ **Medium**

12. By approximately what percent did the number of questions decrease from CEE 2000 to CEE 2006?

 (A) 11
 (B) 22
 (C) 27
 (D) 33
 (E) 37

➢ **Medium**

13. In which year were the test takers given the least time to answer all the questions?

 (A) 2000
 (B) 2001
 (C) 2002
 (D) 2004
 (E) 2006

➢ **Medium**

14. If the Pressure Factor for the examinees in an exam is defined as Difficulty level divided by Average Time (in minutes) given per question, then the Pressure Factor equals which one of the following in CEE 2006?

 (A) 7.5
 (B) 10
 (C) 12.5
 (D) 15
 (E) 17.5

➢ **Medium**

15. If the Stress Factor for the examinees in an exam is defined as the product of the Difficulty level and the Number of questions divided by the average time given per question, then the Stress Factor equals which one of the following in the exam CEE 2005?

 (A) 2 per second
 (B) 3 per second
 (C) 4 per second
 (D) 6 per second
 (E) 9 per second

Questions 16–19 refer to the following graph.

The table below provides the complete semantics of a Common Entrance Test (CET) conducted in different years.

Exam	Area	Questions	Marks per question	Total Duration (in minute)	Average time per question (in second)	Difficulty level 1 = Easy 2 = Average 3 = Difficult	Area wise Cut-off Scores	Overall cutoff mark as a percentage of maximum mark for the top five Institutes
CET – 1990	Quantitative	55	1	120	44	3	9	55
	Verbal	55	1			3	16	
	Analytical	55	1			2	12	
CET – 1991	Quantitative	50	1	120	48	2	14	65
	Verbal	50	1			1	19	
	Analytical	50	1			1	20	
CET – 1992	Quantitative	50	1	120	48	3	11	58
	Verbal	50	1			1	18	
	Analytical	50	1			3	14	
CET – 1993	Quantitative	50	1	120	48	3	10	60
	Verbal	50	1			1	18	
	Analytical	50	1			2	15	
CET – 1994	Quantitative	37	1	150	71	3	8	68
	Verbal	50	1			1	18	
	Analytical	39	1			3	9	

* The Difficulty Factor of the exam is the sum of the products of the number of questions of each type and the corresponding difficulty level. The Stress Factor is the Difficulty Factor divided by the Average Time Per Question.

➢ **Medium**

16. By approximately what percent did the number of questions decrease in CET 1994 over the previous year?

(A) 16%
(B) 19%
(C) 35%
(D) 40%
(E) 50%

➢ **Medium**

17. The Difficulty Factor is the greatest for which one of the following exams?

    (A) CET 1990
    (B) CET 1991
    (C) CET 1992
    (D) CET 1993
    (E) CET 1994

➢ **Medium**

18. Which one of the following exams has been marked as having the highest Stress Factor?

    (A) CET 1990
    (B) CET 1991
    (C) CET 1992
    (D) CET 1993
    (E) CET 1994

➢ **Hard**

19. Which one of the following statements can be inferred from the table?

    (I) As the Stress Factor increased, the cut off marks of the top five universities decreased
    (II) As the Difficulty Factor of the exam decreased, the cut off marks of the top five universities increased
    (III) As the Difficulty Factor increased, the Stress Factor increased

    (A) (I) only
    (B) (II) only
    (C) (III) only
    (D) (I), and (II)
    (E) (I), and (III)

Questions 20–23 refer to the following table.

2007 Composition of Maryland Employment by Industry (Annual Average by Place of Work).

Industry Groups and Totals	Number of Establishments	Employment	Percent of Total Employment
Federal Government	4,564	455,492	8.12%
State Government	849	1,121,712	19.99%
Local Government	345	96,972	1.73%
Total Government Sector	5,758	1,674,176	9.20%
Natural Resources and Mining	23,449	331,590	5.91%
Construction	749	6,836	0.30%
Manufacturing	19,335	188,420	3.36%
Service-Providing	14,283	136,334	2.43%
Utilities	121,238	1,041,777	31.04%
Wholesale Trade	2,320	9,711	0.17%
Retail Trade	11,342	94,997	1.69%
Transportation and Warehousing	18,593	299,648	5.34%
Information	3,998	65,765	1.17%
Financial Analysis	2,898	50,726	0.904082362
Professional and Business Services	14,828	344,565	6.14113352
Education and Health Services	36,384	347,821	6.19916475
Leisure and Hospitality	16,534	229,219	4.085337989
Other Services	13,733	87,309	1.556096024
Unclassified	1,802	1,878	0.03347133
Total Private Sector	301,486	3,936,596	70
Total Employment	307,244	5,610,772	100

➤ **Easy**

20. In 2007, how many industry groups consisted of more than 1 million employees?

(A) 0
(B) 1
(C) 2
(D) 3
(E) 4

➤ **Easy**

21. Which one of the following industry groups employs the maximum number of people?

(A) Utilities
(B) Information
(C) State Government
(D) Natural Resources and Mining
(E) Transportation and Warehousing

➢ **Hard**

22. Which one of the following industry groups employs more than 10 employees per establishment?

(A) Construction
(B) Manufacturing
(C) Wholesale Trade
(D) Retail Trade
(E) Transportation and Warehousing

➢ **Hard**

23. Which one of the following is a valid inference?

(I) The State Government can be inferred as employing the highest number of Employees per Establishment only because the Percentage Employment it provides is the highest. The number of Establishments is not important.
(II) The State Government can be inferred as employing the highest number of Employees per Establishment since it has the least number of organizations and offers the highest Employment.
(III) The State Government can be inferred as employing the highest number of Employees per Establishment since it has the least number of organizations and offers the highest Percentage of Employment.

(A) I only
(B) II only
(C) III only
(D) I and II
(E) II and III

Answers and Solutions to Problem Set V

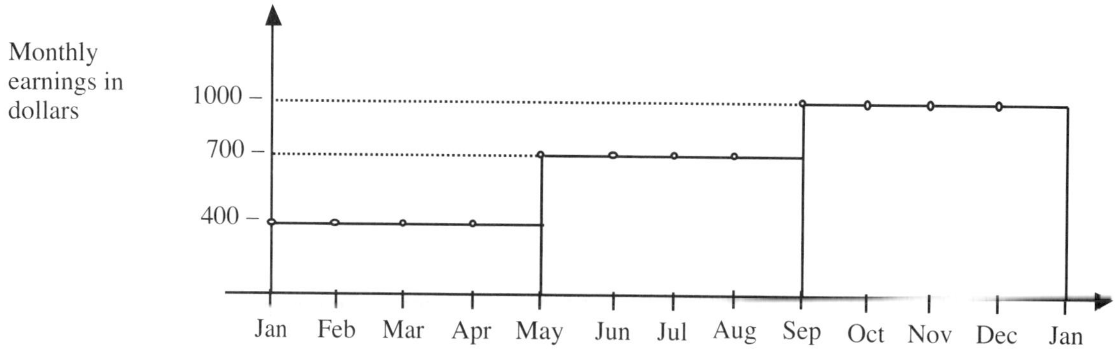

A's income profile during the year 2007

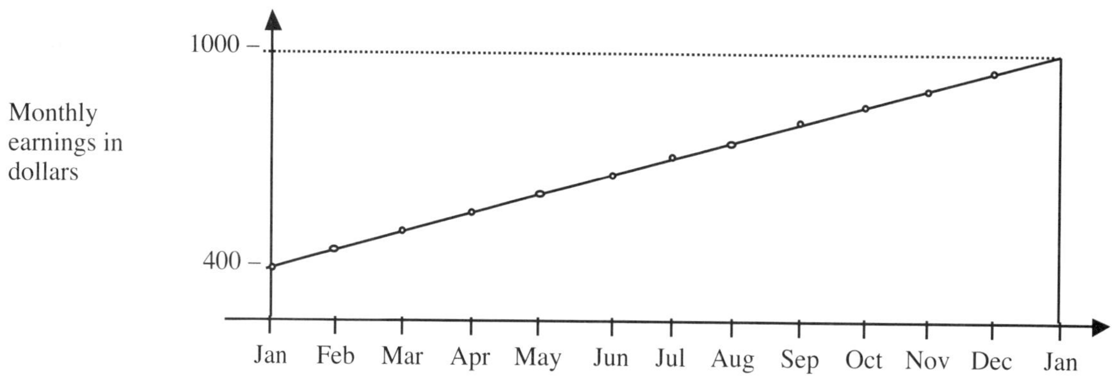

B's income profile during the year 2007

➢ **Medium**

1. *A* launched 3 products in the year 2007 and earns income from the sales of the products only. The top graph shows his monthly earnings for the year. *B*'s earnings consist of continuously growing salary, growing by same amount each month as shown in the figure. Which one of the following equals the total earnings of *A* and *B* in the year 2007?

   (A) 7500, 8100
   (B) 7850, 8300
   (C) 8150, 8400
   (D) 8400, 8100
   (E) 8400, 8700

From the figure, the monthly income of *A* for the first four months is \$400. Hence, the net earnings in the 4 months is $4 \times 400 = 1600$ dollars.

From the figure, the monthly income of *A* for the second four months is \$700. Hence, the net earnings in the 4 months is $4 \times 700 = 2800$ dollars.

From the figure, the monthly income of *A* for the last four months is \$1000. Hence, the net earnings in the 4 months is $4 \times 1000 = 4000$ dollars.

Hence, the total income in the year is $1600 + 2800 + 4000 = 8400$ dollars.

The monthly income of *B* grew regularly from 400 in January to 950 in December. Hence, the net income is $400 + 450 + 500 + 550 + 600 + 650 + 700 + 750 + 800 + 850 + 900 + 950 = 8100$.

Hence, the answer is (D).

Questions 2–4 refer to the following graph.

The graph below shows historical exchange rates between the Indian Rupee (INR) and the US Dollar (USD) between January 10 and February 8 of a particular year.

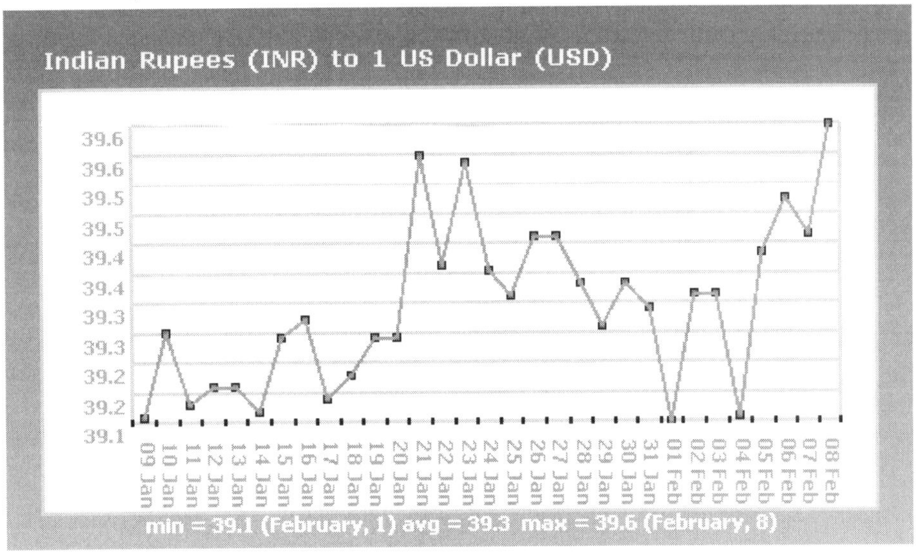

➢ **Easy**

2. On which day shown on the graph did the value of the US dollar increase against the Rupee by the greatest amount?

   (A) Jan. 10
   (B) Jan. 14
   (C) Jan. 21
   (D) Jan. 23
   (E) Feb. 4

Here, the scale of the x-axis is uniform. Hence, growth is greatest when the curve is steepest. The growth curve of the US dollar against the Indian Rupee is the steepest (increased by a bit more than six horizontal lines on the graph) on January 21. Hence, the answer is (C). On February 5th, the growth is the next greatest, growing by a bit less than 6 horizontal lines.

➢ **Medium**

3. John had 100 dollars. The exchange rate converts the amount in US dollars to a number in Indian Rupees by directly multiplying by the value of the exchange rate. By what amount did John's $100 increase in terms of Indian Rupees from Jan. 9 to Feb. 8?

   (A) 5
   (B) 10
   (C) 15
   (D) 25
   (E) 50

One dollar converted to 39.1 Rupees on Jan. 9. Hence, 100 dollars converts to 39.15 × 100 = 3915 Indian Rupees. On February 8, it converted to 39.65 Rupees. Hence, on that day, 100 dollars converted to 39.65 × 100 = 3965 Rupees. The increase in terms of Indian Rupees is 3965 – 3915 = 50. The answer is (E).

➢ **Hard**

4. On February 8, the dollar value was approximately what percent of the dollar value on January 9?

   (A) 1.28
   (B) 12. 8
   (C) 101.28
   (D) 112. 8
   (E) 128

On January 9, the dollar value was 39.15 Rupees, and on February 8 the dollar value was 39.65 Rupees. Hence, the dollar value on February 8th was 39.65/39.15 × 100 = (39.15 + 0.5)/39.15 × 100 =

100 + 0.5/39.15 × 100 = (100 + 1.28) = 101.28 percent of the value on January 9th. Hence, the answer is (C).

Questions 5–7 refer to the following graph.

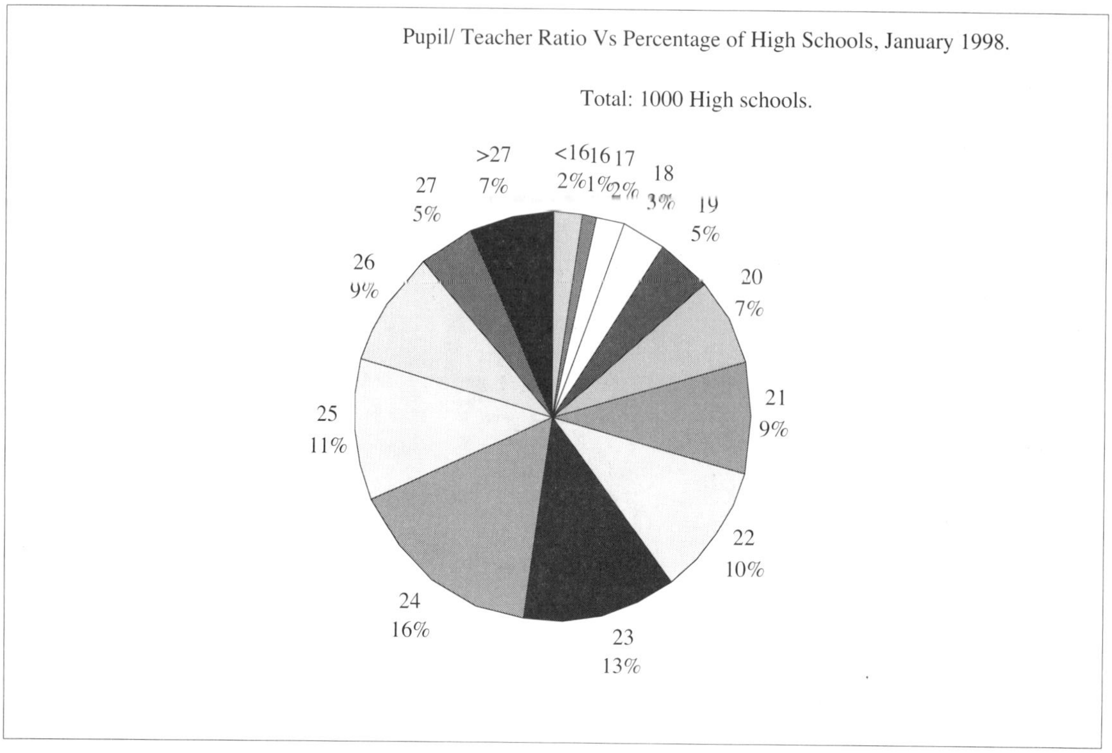

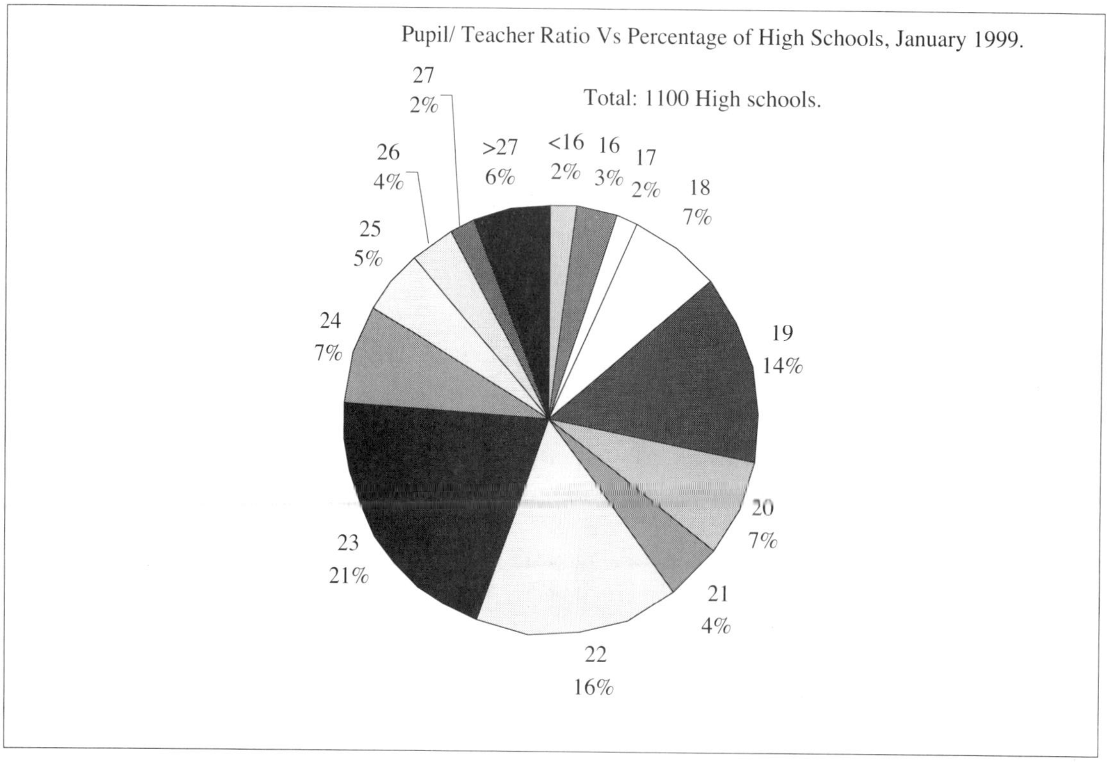

➤ **Medium**

5. By what percent did the number of schools with Pupil/Teacher ratio less than 16 increase in January 1999 over January 1998?

   (A) –2%
   (B) 0%
   (C) 2%
   (D) 10%
   (E) 12%

In January 1998, the Pupil/Teacher ratio is less than 16 in 2% of the schools. The number of schools in 1998 is 1000. Hence, 2% of 1000 is 2/100 × 1000 = 20. So, 20 schools have pupil/Teacher ratio less than 16.

In January 1999, the pupil/Teacher ratio is less than 16 in 2% of schools again. The number of schools in 1999 is 1100. Hence, 2% of 1100 is 2/100 × 1100 = 22. In 1999, there are 22 schools with the ratio less than 16.

The percentage increase equals (22 – 20)/20 × 100 = 2/20 × 100 = 10%. The answer is (D).

➤ **Hard**

6. In January 1998, what percent of high schools had a Pupil/Teacher ratio less than 23?

   (A) 25%
   (B) 39%
   (C) 50%
   (D) 60%
   (E) 75%

The number of schools having a ratio less than 23 is

> The number of schools having the Pupil/Teacher ratio less than 16
> \+ The number of schools having the Pupil/Teacher ratio equal to 16
> \+ The number of schools having the Pupil/Teacher ratio equal to 17
> \+ The number of schools having the Pupil/Teacher ratio equal to 18
> \+ The number of schools having the Pupil/Teacher ratio equal to 19
> \+ The number of schools having the Pupil/Teacher ratio equal to 20
> \+ The number of schools having the Pupil/Teacher ratio equal to 21
> \+ The number of schools having the Pupil/Teacher ratio equal to 22
> = 2% + 1% + 2% + 3% + 5% + 7% + 9% + 10%
> = 39%

The answer is (B).

Method II:
The number of schools having the ratio less than 23 equals

> 100%
> – (The number of schools having the Pupil Teacher ratio greater than 27
> \+ The number of schools having the Pupil/Teacher ratio equal to 27
> \+ The number of schools having the Pupil/Teacher ratio equal to 26
> \+ The number of schools having the Pupil/Teacher ratio equal to 25
> \+ The number of schools having the Pupil/Teacher ratio equal to 24
> \+ The number of schools having the Pupil/Teacher ratio equal to 23
>)
> = 100 – (7% + 5% + 9% + 11% + 16% + 13%) = 100 – 61% = 39%

The answer is (B).

7. If the areas of the sectors in the circle graphs are drawn in proportion to the percent shown, what is the measure, in degrees, of the sector representing the number of high schools with Pupil/Teacher ratio greater than 27 in 1999?

(A) 21.6
(B) 30
(C) 45.7
(D) 56.3
(E) 72

From the chart, in 1999, 6% of schools have a Pupil/Teacher ratio greater than 27. Hence, the fraction of the angle that the sector makes in the complete angle of the circle also equals 6% = 6/100. Since the complete angle is 360°, the part of the angle equals 6/100 × 360 = 21.6°. The answer is (A).

Questions 8–10 refer to the following graph.

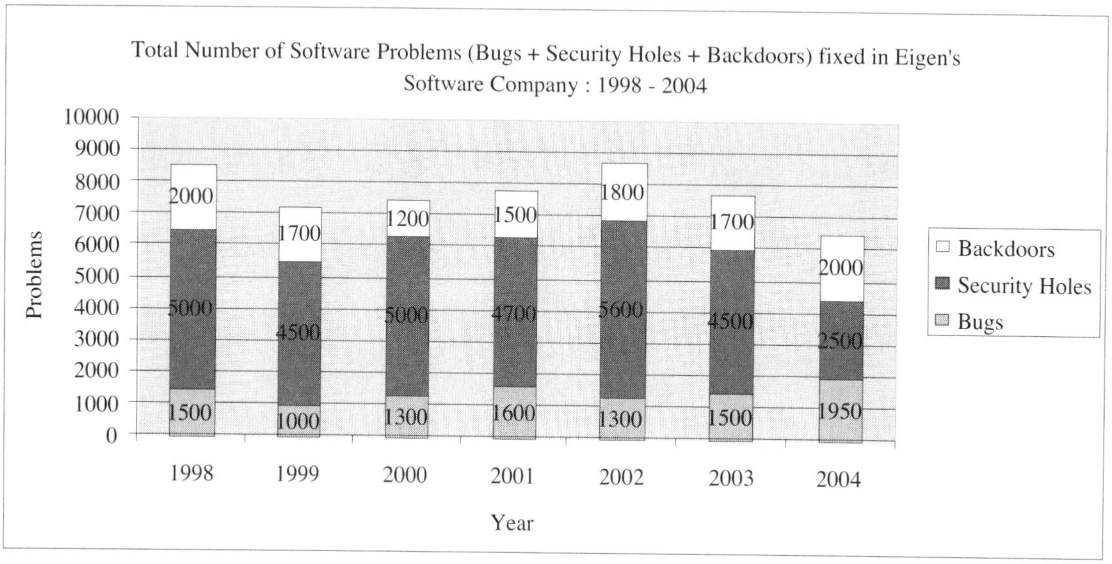

➢ **Easy**

8. What was the number of security holes fixed in 2003?

(A) 1500
(B) 1700
(C) 4500
(D) 6000
(E) 6300

From the graph, the number of security holes fixed in 2003 is 4500. The answer is (C).

➢ **Medium**

9. For which year was the ratio of the Security holes to Bugs fixed by the software company the greatest?

(A) 1998
(B) 1999
(C) 2000
(D) 2001
(E) 2002

Let's calculate the ratio and find the year in which the ratio is the greatest:

Choice (A): Year 1998. The number of security holes to bugs fixed is 5000/1500 = 10/3 = 3.33.

Choice (B): Year 1999. The number of security holes to bugs fixed is 4500/1000 = 9/2 = 4.5 > Choice (A). Reject choice (A).

Choice (C): Year 2000. The number of security holes to bugs fixed is 5000/1300 = 50/13 = 3.86 < Choice (B). Reject choice (C).

Choice (D): Year 2001. The number of security holes to bugs fixed is 4700/1600 = 47/16 = 2.9375 < Choice (B). Reject choice (D).

Choice (E): Year 2002. The number of security holes to bugs fixed is 5600/1300 = 56/13 = 4.3 < Choice (B). Reject choice (E).

The ratio is greatest in the year 1999. Hence, the answer is (B).

➢ **Medium**

10. If the total number of software problems solved is a direct measure of the company's capability, then by approximately what percent did capability increase from 1999 to 2002?

 (A) 10%
 (B) 20%
 (C) 30%
 (D) 40%
 (E) 50%

In 1999, the total number of software problems solved by Eigen's Software Company is Bugs + Security holes + Backdoors = 1000 + 4500 + 1700 = 7200.

In 2002, the total number of software problems solved by Eigen's Software Company is Bugs + Security holes + Backdoors = 1300 + 5600 + 1800 = 8700.

Hence, the percent increase in the number in the period is $\frac{8700 - 7200}{7200} \times 100 = 20.88\%$. The nearest answer is (B).

Questions 11–15 refer to the following discussion.

The graphs below provide data on a common entrance examination conducted in different years.

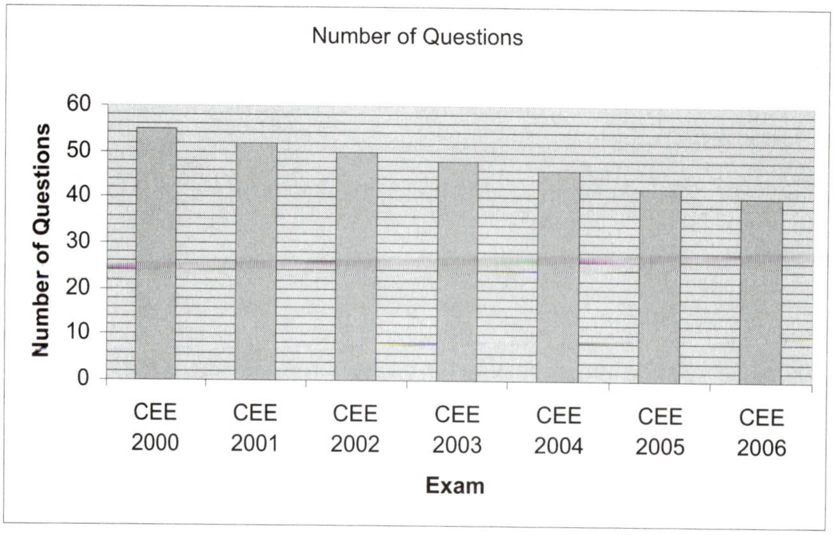

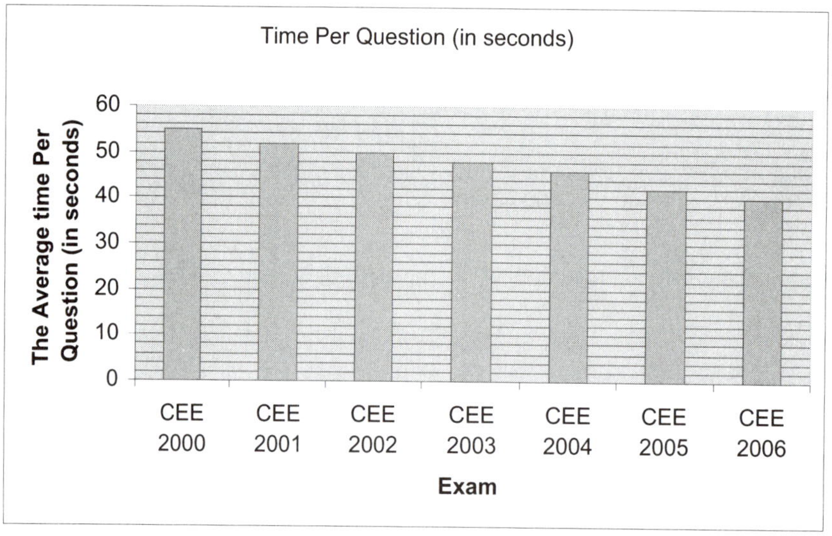

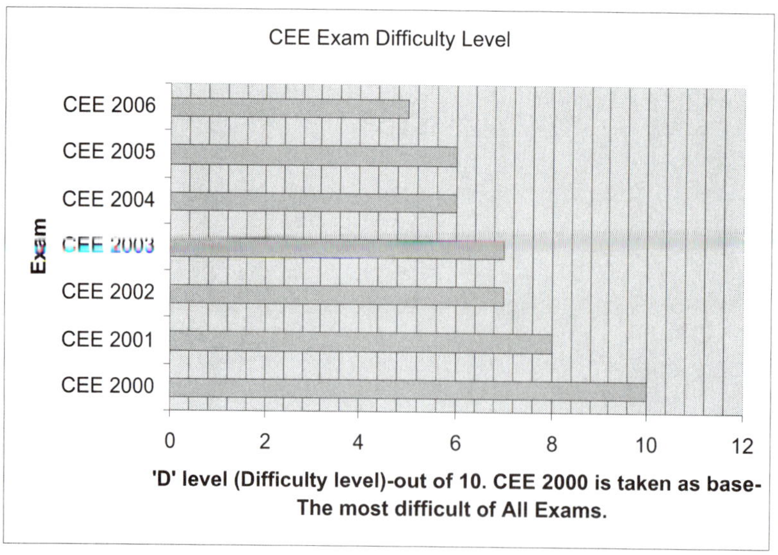

➤ **Easy**

11. Which year had the second most difficult exam?

    (A) 2000
    (B) 2001
    (C) 2002
    (D) 2004
    (E) 2006

Refer to the graph CEE Exam Difficulty Level. The graph starts at 0 for each exam and ends at 10 for the most difficult exam, CEE 2000. So, the difficulty actually increases with 'D-level' value. The second highest value corresponds to the Exam CEE 2001. The answer is (B).

➤ **Medium**

12. By approximately what percent did the number of questions decrease from CEE 2000 to CEE 2006?

    (A) 11
    (B) 22
    (C) 27
    (D) 33
    (E) 37

From the graph, the number of questions in CEE 2000 is 55. The number in CEE 2006 is 40. Hence, the percent drop is $\frac{55-40}{55} \times 100 = \frac{15}{55} \times 100 = \frac{3}{11} \times 100 = 27.27$. Since the nearest choice is (C), the answer is (C).

➤ **Medium**

13. In which year were the test takers given the least time to answer all the questions?

    (A) 2000
    (B) 2001
    (C) 2002
    (D) 2004
    (E) 2006

The time given can be evaluated as (Number of Questions) × (Time Per Question).

Both the number of questions and the time given per question are the least in 2006. Hence, their product should be minimum in that year.

Hence, the total time given is the least in 2006. The answer is (E).

➤ **Medium**

14. If the Pressure Factor for the examinees in an exam is defined as Difficulty level divided by Average Time (in minutes) given per question, then the Pressure Factor equals which one of the following in CEE 2006?

    (A) 7.5
    (B) 10
    (C) 12.5
    (D) 15
    (E) 17.5

The Pressure Factor in 2006 equals Difficulty level divided by Average Time given per question = 5/40 seconds or 5/(2/3 minutes) = 15/2 per minute. The answer is (A).

➤ **Medium**

15. If the Stress Factor for the examinees in an exam is defined as the product of the Difficulty level and the Number of questions divided by the average time given per question, then the Stress Factor equals which one of the following in the exam CEE 2005?

    (A) 2 per second
    (B) 3 per second
    (C) 4 per second
    (D) 6 per second
    (E) 9 per second

The Stress Factor equals

(The Difficulty level) × (Number of questions/Time given per question) =

$$6 \times \frac{42}{42 \text{ seconds per question}} =$$

6 per second

The answer is (D).

Questions 16–19 refer to the following graph.

The table below provides the complete semantics of a Common Entrance Test (CET) conducted in different years.

Exam	Area	Questions	Marks per question	Total Duration (in minute)	Average time per question (in second)	Difficulty level 1 = Easy 2 = Average 3 = Difficult	Area wise Cut-off Scores	Overall cutoff mark as a percentage of maximum mark for the top five Institutes
CET – 1990	Quantitative	55	1	120	44	3	9	55
	Verbal	55	1			3	16	
	Analytical	55	1			2	12	
CET – 1991	Quantitative	50	1	120	48	2	14	65
	Verbal	50	1			1	19	
	Analytical	50	1			1	20	
CET – 1992	Quantitative	50	1	120	48	3	11	58
	Verbal	50	1			1	18	
	Analytical	50	1			3	14	
CET – 1993	Quantitative	50	1	120	48	3	10	60
	Verbal	50	1			1	18	
	Analytical	50	1			2	15	
CET – 1994	Quantitative	37	1	150	71	3	8	68
	Verbal	50	1			1	18	
	Analytical	39	1			3	9	

*The Difficulty Factor of the exam is the sum of the products of the number of questions of each type and the corresponding difficulty level. The Stress Factor is the Difficulty Factor divided by the Average Time Per Question.

➢ **Medium**

16. By approximately what percent did the number of questions decrease in CET 1994 over the previous year?

    (A) 16%
    (B) 19%
    (C) 35%
    (D) 40%
    (E) 50%

CET 1993 asks 50 quantitative, 50 verbal, and 50 Analytical.

The total is 50 + 50 + 50 = 150.

CET 1994 asks 37 quantitative, 50 verbal, and 39 Analytical.

The total is 37 + 50 + 39 = 126.

The decrease percent is $\frac{150-126}{150} \times 100 = \frac{24}{150} \times 100 = 16\%$.

The answer is (A).

➢ **Medium**

17. The Difficulty Factor is the greatest for which one of the following exams?

    (A) CET 1990
    (B) CET 1991
    (C) CET 1992
    (D) CET 1993
    (E) CET 1994

The Difficulty Factor of the exam is the sum of the products of the number of questions of each type and the corresponding difficulty level.

Let's calculate the Difficulty Factor for each exam and pick the answer-choice that has greatest value:

Choice (A): In CET 1990, the Difficulty Factor is $(3 \times 55 + 3 \times 55 + 2 \times 55) = 165 + 165 + 110 = 440$.

Choice (B): In CET 1991, the Difficulty Factor is $(2 \times 50 + 1 \times 50 + 1 \times 50) = 100 + 50 + 50 = 200 <$ Choice (A). Reject the current choice.

Choice (C): In CET 1992, the Difficulty Factor is $(3 \times 50 + 1 \times 50 + 3 \times 50) = 350 <$ Choice (A). Reject the current choice.

Choice (D): In CET 1993, the Difficulty Factor is $(3 \times 50 + 1 \times 50 + 2 \times 50) = 300 <$ Choice (A). Reject the current choice.

Choice (E): In CET 1994, the Difficulty Factor is $(3 \times 37 + 1 \times 50 + 3 \times 39) = 111 + 50 + 117 = 278 <$ Choice (A). Reject the current choice.

The answer is (A).

➢ Medium

18. Which one of the following exams has been marked as having the highest Stress Factor?

    (A) CET 1990
    (B) CET 1991
    (C) CET 1992
    (D) CET 1993
    (E) CET 1994

The Difficulty Factor of the exam is the sum of the products of the number of questions of each type and the corresponding difficulty level.

Then

The Stress Factor = the Difficulty Factor divided by the average time per question.

Let's calculate the Stress Factor for each answer-choice and choose the one that has the highest value:

Choice (A): In CET 1990,
The Difficulty Factor is (3 × 55 + 3 × 55 + 2 × 55) = 165 + 165 + 110 = 440
The Stress Factor is 440/44 = 10.

Choice (B): In CET 1991,
Difficulty Factor is (2 × 50 + 1 × 50 + 1 × 50) = 100 + 50 + 50 = 200.
The Stress Factor is 200/48 < Choice (A). Reject.

Choice (C): In CET 1992,
Difficulty Factor is (3 × 50 + 1 × 50 + 3 × 50) = 350.
The Stress Factor is 350/48 < Choice (A). Reject.

Choice (D): In CET 1993,
Difficulty Factor is (3 × 50 + 1 × 50 + 2 × 50) = 300.
The Stress Factor is 300/48 < Choice (A). Reject.

Choice (E): In CET 1994,
Difficulty Factor is (3 × 37 + 1 × 50 + 3 × 39) = 111 + 50 + 117 = 278.
The Stress Factor is 278/71 < Choice (A). Reject.

Hence, the answer is (A).

➢ Hard

19. Which one of the following statements can be inferred from the table?

    (I) As the Stress Factor increased, the cut off marks of the top five universities decreased
    (II) As the Difficulty Factor of the exam decreased, the cut off marks of the top five universities increased
    (III) As the Difficulty Factor increased, the Stress Factor increased

    (A) (I) only
    (B) (II) only
    (C) (III) only
    (D) (I), and (II)
    (E) (I), and (III)

The increasing order of the Difficulty Factor is

CET 1991 (200) < CET 1994 (278) < CET 1993 (300) < CET 1992 (350) < CET 1990 (440).

The increasing order of the Stress Factor is

CET 1994 (278/71 = 3.92) < CET 1991 (200/48 = 4.16) < CET 1993(300/48 = 6.25) < CET 1992(350/48 = 7.29) < CET 1990 (440/44 = 10).

The decreasing order of the cut off marks is

CET 1994 (68) > CET 1991 (65) > CET 1993 (60) > CET 1992 (58) > CET 1990 (55).

The decreasing order of the cut off marks matches the increasing order of the Stress Factor.

The decreasing order of the cut off marks does *not* match the increasing order of the Difficulty Factor.

As the Difficulty Factor increased, the Stress Factor *did not* increase. Hence, III is false. The answer is (A), only I is true.

Questions 20–23 refer to the following table.

2007 Composition of Maryland Employment by Industry (Annual Average by Place of Work).

Industry Groups and Totals	Number of Establishments	Employment	Percent of Total Employment
Federal Government	4,564	455,492	8.12%
State Government	849	1,121,712	19.99%
Local Government	345	96,972	1.73%
Total Government Sector	5,758	1,674,176	9.20%
Natural Resources and Mining	23,449	331,590	5.91%
Construction	749	6,836	0.30%
Manufacturing	19,335	188,420	3.36%
Service-Providing	14,283	136,334	2.43%
Utilities	121,238	1,041,777	31.04%
Wholesale Trade	2,320	9,711	0.17%
Retail Trade	11,342	94,997	1.69%
Transportation and Warehousing	18,593	299,648	5.34%
Information	3,998	65,765	1.17%
Financial Analysis	2,898	50,726	0.904082362
Professional and Business Services	14,828	344,565	6.14113352
Education and Health Services	36,384	347,821	6.19916475
Leisure and Hospitality	16,534	229,219	4.085337989
Other Services	13,733	87,309	1.556096024
Unclassified	1,802	1,878	0.03347133
Total Private Sector	301,486	3,936,596	70
Total Employment	307,244	5,610,772	100

➢ **Easy**

20. In 2007, how many industry groups consisted of more than 1 million employees?

(A) 0
(B) 1
(C) 2
(D) 3
(E) 4

From the chart, the employment is greater than 1 million in the industry groups State Government and Utilities. Hence, the answer is 2, which is in choice (C).

➢ **Easy**

21. Which one of the following industry groups employs the maximum number of people?

(A) Utilities
(B) Information
(C) State Government
(D) Natural Resources and Mining
(E) Transportation and Warehousing

From the table, the State Government employs the maximum number. The number is 1,121,712. The answer is (C).

➢ **Hard**

22. Which one of the following industry groups employs more than 10 employees per establishment?

    (A) Construction
    (B) Manufacturing
    (C) Wholesale Trade
    (D) Retail Trade
    (E) Transportation and Warehousing

The correct choice is the industry that employs more than 10 employees per establishment in an average. Hence, the industry with the criterion: The Number of Establishments · 10 < the Number of Employees would be the correct choice.

Choice (A): Construction.
The number of establishments = 749.
The Number of Establishments × 10 = 7490.
The number of Employees = 6,836.
Here, The Number of Establishments × 10 is not less than The number of Employees.
Reject the choice.

Choice (B): Manufacturing.
The number of Establishments = 19,335.
The Number of Establishments × 10 = 193,350.
The number of Employees = 188,420.
Here, The Number of Establishments × 10 is not less than The number of Employees.
Reject the choice.

Choice (C): Wholesale Trade.
The number of Establishments = 2,320.
The Number of Establishments × 10 = 23,200.
The number of Employees = 9,711.
Here, The Number of Establishments × 10 is not less than The number of Employees.
Reject the choice.

Choice (D): Retail Trading.
The number of Establishments = 11,342.
The Number of Establishments × 10 = 113,420.
The number of Employees = 94,997.
Here, The Number of Establishments × 10 is not less than The number of Employees.
Reject the choice.

Choice (E): Transportation and Warehousing.
The number of Establishments = 18,593.
The Number of Establishments × 10 =.185,930.
The number of Employees = 299,648.
Here, The Number of Establishments × 10 is less than The number of Employees.
Accept.

The answer is (E).

➤ **Hard**

23. Which one of the following is a valid inference?

(I) The State Government can be inferred as employing the highest number of Employees per Establishment only because the Percentage Employment it provides is the highest. The number of Establishments is not important.

(II) The State Government can be inferred as employing the highest number of Employees per Establishment since it has the least number of organizations and offers the highest Employment.

(III) The State Government can be inferred as employing the highest number of Employees per Establishment since it has the least number of organizations and offers the highest Percentage of Employment.

(A) I only
(B) II only
(C) III only
(D) I and II
(E) II and III

The Employment per establishment is given as The Number of Employees/The Number of Establishments. The ratio is greatest when the numerator has the greatest positive value, and the denominator has the smallest positive value. Hence, II is true.

The highest employment can also be directly understood by the highest percentage employment. Hence, just as Statement II is true because of the highest employment, Statement III is also true because of the highest percentage employment.

Hence, II and III are correct and the answer is (E).

Word Problems

TRANSLATING WORDS INTO MATHEMATICAL SYMBOLS

Before we begin solving word problems, we need to be very comfortable with translating words into mathematical symbols. Following is a partial list of words and their mathematical equivalents.

Concept	Symbol	Words	Example	Translation
equality	=	is	2 plus 2 is 4	$2+2=4$
		equals	x minus 5 equals 2	$x-5=2$
		is the same as	multiplying x by 2 is the same as dividing x by 7	$2x=x/7$
addition	+	sum	the sum of y and π is 20	$y+\pi=20$
		plus	x plus y equals 5	$x+y=5$
		add	how many marbles must John add to collection P so that he has 13 marbles	$x+P=13$
		increase	a number is increased by 10%	$x+10\%x$
		more	the perimeter of the square is 3 more than the area	$P=3+A$
subtraction	–	minus	x minus y	$x-y$
		difference	the difference of x and y is 8	$\lvert x-y\rvert=8$
		subtracted	x subtracted from y	$y-x$ *
		less than	the circumference is 5 less than the area	$C=A-5$
multiplication	$\times$ *or* •	times	the acceleration is 5 times the velocity	$a=5v$
		product	the product of two consecutive integers	$x(x+1)$
		of	x is 125% of y	$x=125\%y$
division	÷	quotient	the quotient of x and y is 9	$x\div y=9$
		divided	if x is divided by y, the result is 4	$x\div y=4$

Although exact steps for solving word problems cannot be given, the following guidelines will help:

(1) First, choose a variable to stand for the least unknown quantity, and then try to write the other unknown quantities in terms of that variable.

For example, suppose we are given that Sue's age is 5 years less than twice Jane's and the sum of their ages is 16. Then Jane's age would be the least unknown, and we let $x = \textit{Jane's age}$. Expressing Sue's age in terms of x gives $\textit{Sue's age} = 2x - 5$.

(2) Second, write an equation that involves the expressions in Step 1. Most (though not all) word problems pivot on the fact that two quantities in the problem are equal. Deciding which two quantities should be set equal is usually the hardest part in solving a word problem since it can require considerable ingenuity to discover which expressions are equal.

For the example above, we would get $(2x - 5) + x = 16$.

(3) Third, solve the equation in Step 2 and interpret the result.

For the example above, we would get by adding the x's: $3x - 5 = 16$

Then adding 5 to both sides gives $3x = 21$

Finally, dividing by 3 gives $x = 7$

Hence, Jane is 7 years old and Sue is $2x - 5 = 2\cdot 7 - 5 = 9$ years old.

---

* Notice that with "minus" and "difference" the terms are subtracted in the same order as they are written, from left to right (x minus y —> $x - y$). However, with "subtracted" and "less than," the order of subtraction is reversed (x subtracted from y —> $y - x$). Many students translate "subtracted from" in the wrong order.

MOTION PROBLEMS

Virtually, all motion problems involve the formula *Distance* = *Rate* × *Time*, or

$$D = R \times T$$

Overtake: In this type of problem, one person catches up with or overtakes another person. The key to these problems is that at the moment one person overtakes the other they have traveled the same distance.

Example: Scott starts jogging from point X to point Y. A half-hour later his friend Garrett who jogs 1 mile per hour slower than twice Scott's rate starts from the same point and follows the same path. If Garrett overtakes Scott in 2 hours, how many miles will Garrett have covered?

(A) $2\frac{1}{5}$ (B) $3\frac{1}{3}$ (C) 4 (D) 6 (E) $6\frac{2}{3}$

Following Guideline 1, we let r = *Scott's rate*. Then $2r - 1$ = *Garrett's rate*. Turning to Guideline 2, we look for two quantities that are equal to each other. When Garrett overtakes Scott, they will have traveled the same distance. Now, from the formula $D = R \times T$, Scott's distance is $D = r \times 2\frac{1}{2}$

and Garrett's distance is $D = (2r - 1)2 = 4r - 2$

Setting these expressions equal to each other gives $4r - 2 = r \times 2\frac{1}{2}$

Solving this equation for r gives $r = \frac{4}{3}$

Hence, Garrett will have traveled $D = 4r - 2 = 4\left(\frac{4}{3}\right) - 2 = 3\frac{1}{3}$ miles. The answer is (B).

Opposite Directions: In this type of problem, two people start at the same point and travel in opposite directions. The key to these problems is that the total distance traveled is the sum of the individual distances traveled.

Example: Two people start jogging at the same point and time but in opposite directions. If the rate of one jogger is 2 mph faster than the other and after 3 hours they are 30 miles apart, what is the rate of the faster jogger?

(A) 3 (B) 4 (C) 5 (D) 6 (E) 7

Let r be the rate of the slower jogger. Then the rate of the faster jogger is $r + 2$. Since they are jogging for 3 hours, the distance traveled by the slower jogger is $D = rt = 3r$, and the distance traveled by the faster jogger is $3(r + 2)$. Since they are 30 miles apart, adding the distances traveled gives

$$3r + 3(r + 2) = 30$$
$$3r + 3r + 6 = 30$$
$$6r + 6 = 30$$
$$6r = 24$$
$$r = 4$$

Hence, the rate of the faster jogger is $r + 2 = 4 + 2 = 6$. The answer is (D).

Round Trip: The key to these problems is that the distance going is the same as the distance returning.

Example: A cyclist travels 20 miles at a speed of 15 miles per hour. If he returns along the same path and the entire trip takes 2 hours, at what speed did he return?

(A) 15 mph (B) 20 mph (C) 22 mph (D) 30 mph (E) 34 mph

Solving the formula $D = R \times T$ for T yields $T = \frac{D}{R}$. For the first half of the trip, this yields $T = \frac{20}{15} = \frac{4}{3}$ hours. Since the entire trip takes 2 hours, the return trip takes $2 - \frac{4}{3}$ hours, or $\frac{2}{3}$ hours. Now, the return trip is also 20 miles, so solving the formula $D = R \times T$ for R yields $R = \frac{D}{T} = \frac{20}{2/3} = 20 \cdot \frac{3}{2} = 30$. The answer is (D).

Compass Headings: In this type of problem, typically two people are traveling in perpendicular directions. The key to these problems is often the Pythagorean Theorem.

Example: At 1 PM, Ship A leaves port heading due west at x miles per hour. Two hours later, Ship B is 100 miles due south of the same port and heading due north at y miles per hour. At 5 PM, how far apart are the ships?

(A) $\sqrt{(4x)^2 + (100 + 2y)^2}$
(B) $x + y$
(C) $\sqrt{x^2 + y^2}$
(D) $\sqrt{(4x)^2 + (2y)^2}$
(E) $\sqrt{(4x)^2 + (100 - 2y)^2}$

Since Ship A is traveling at x miles per hour, its distance traveled at 5 PM is $D = rt = 4x$. The distance traveled by Ship B is $D = rt = 2y$. This can be represented by the following diagram:

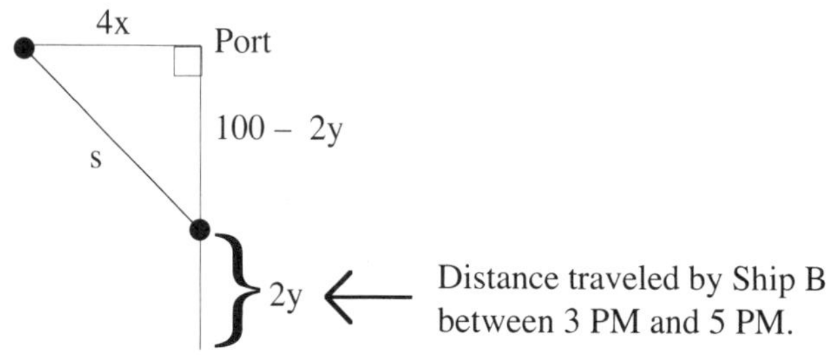

Applying the Pythagorean Theorem yields $s^2 = (4x)^2 + (100 - 2y)^2$. Taking the square root of this equation gives $s = \sqrt{(4x)^2 + (100 - 2y)^2}$. The answer is (E).

Circular Motion: In this type of problem, the key is often the arc length formula $S = R\theta$, where S is the arc length (or distance traveled), R is the radius of the circle, and θ is the angle.

Example: The figure to the right shows the path of a car moving around a circular racetrack. How many miles does the car travel in going from point A to point B ?

(A) $\pi/6$ (B) $\pi/3$ (C) π (D) 30 (E) 60

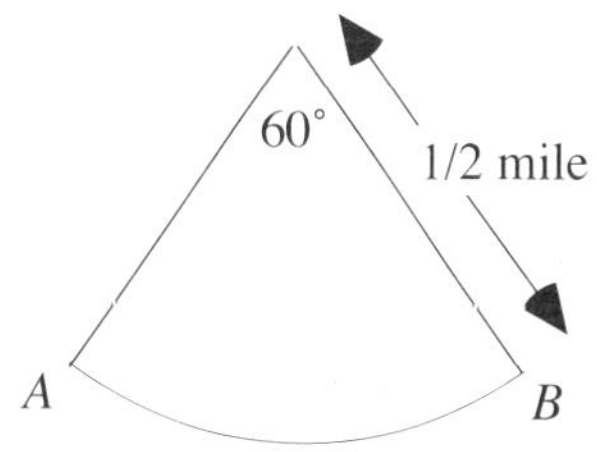

When calculating distance, degree measure must be converted to radian measure. To convert degree measure to radian measure, multiply by the conversion factor $\frac{\pi}{180}$. Multiplying 60° by $\frac{\pi}{180}$ yields $60 \cdot \frac{\pi}{180} = \frac{\pi}{3}$. Now, the length of arc traveled by the car in moving from point A to point B is S. Plugging this information into the formula $S = R\theta$ yields $S = \frac{1}{2} \cdot \frac{\pi}{3} = \frac{\pi}{6}$. The answer is (A).

Example: If a wheel is spinning at 1200 revolutions per minute, how many revolutions will it make in t seconds?

(A) $2t$ (B) $10t$ (C) $20t$ (D) $48t$ (E) $72t$

Since the question asks for the number of revolutions in t seconds, we need to find the number of revolutions per second and multiply that number by t. Since the wheel is spinning at 1200 revolutions per minute and there are 60 seconds in a minute, we get $\frac{1200 \text{ revolutions}}{60 \text{ seconds}} = 20 \text{ rev/sec}$. Hence, in t seconds, the wheel will make $20t$ revolutions. The answer is (C).

WORK PROBLEMS

The formula for work problems is *Work* = *Rate* × *Time*, or $W = R \times T$. The amount of work done is usually 1 unit. Hence, the formula becomes $1 = R \times T$. Solving this for R gives $R = \frac{1}{T}$.

Example: If Johnny can mow the lawn in 30 minutes and with the help of his brother, Bobby, they can mow the lawn 20 minutes, how long would it take Bobby working alone to mow the lawn?

(A) 1/2 hour (B) 3/4 hour (C) 1 hour (D) 3/2 hours (E) 2 hours

Let $r = 1/t$ be Bobby's rate. Now, the rate at which they work together is merely the sum of their rates:

Total Rate = *Johnny's Rate* + *Bobby's Rate*

$$\frac{1}{20} = \frac{1}{30} + \frac{1}{t}$$

$$\frac{1}{20} - \frac{1}{30} = \frac{1}{t}$$

$$\frac{30 - 20}{30 \cdot 20} = \frac{1}{t}$$

$$\frac{1}{60} = \frac{1}{t}$$

$$t = 60$$

Hence, working alone, Bobby can do the job in 1 hour. The answer is (C).

Example: A tank is being drained at a constant rate. If it takes 3 hours to drain $\frac{6}{7}$ of its capacity, how much longer will it take to drain the tank completely?

(A) 1/2 hour (B) 3/4 hour (C) 1 hour (D) 3/2 hours (E) 2 hours

Since 6/7 of the tank's capacity was drained in 3 hours, the formula $W = R \times T$ becomes $\frac{6}{7} = R \times 3$. Solving for R gives $R = 2/7$. Now, since 6/7 of the work has been completed, 1/7 of the work remains. Plugging this information into the formula $W = R \times T$ gives $\frac{1}{7} = \frac{2}{7} \times T$. Solving for T gives $T = 1/2$. The answer is (A).

MIXTURE PROBLEMS

The key to these problems is that the combined total of the concentrations in the two parts must be the same as the whole mixture.

Example: How many ounces of a solution that is 30 percent salt must be added to a 50-ounce solution that is 10 percent salt so that the resulting solution is 20 percent salt?

(A) 20 (B) 30 (C) 40 (D) 50 (E) 60

Let x be the ounces of the 30 percent solution. Then $30\%x$ is the amount of salt in that solution. The final solution will be $50 + x$ ounces, and its concentration of salt will be $20\%(50 + x)$. The original amount of salt in the solution is $10\% \cdot 50$. Now, the concentration of salt in the original solution plus the concentration of salt in the added solution must equal the concentration of salt in the resulting solution:

$$10\% \cdot 50 + 30\%x = 20\%(50 + x)$$

Multiply this equation by 100 to clear the percent symbol and then solving for x yields $x = 50$. The answer is (D).

COIN PROBLEMS

The key to these problems is to keep the quantity of coins distinct from the value of the coins. An example will illustrate.

Example: Laura has 20 coins consisting of quarters and dimes. If she has a total of \$3.05, how many dimes does she have?

(A) 3 (B) 7 (C) 10 (D) 13 (E) 16

Let D stand for the number of dimes, and let Q stand for the number of quarters. Since the total number of coins in 20, we get $D + Q = 20$, or $Q = 20 - D$. Now, each dime is worth 10¢, so the value of the dimes is $10D$. Similarly, the value of the quarters is $25Q = 25(20 - D)$. Summarizing this information in a table yields

	Dimes	Quarters	Total
Number	D	$20 - D$	20
Value	$10D$	$25(20 - D)$	305

Notice that the total value entry in the table was converted from \$3.05 to 305¢. Adding up the value of the dimes and the quarters yields the following equation:

$$10D + 25(20 - D) = 305$$
$$10D + 500 - 25D = 305$$
$$-15D = -195$$
$$D = 13$$

Hence, there are 13 dimes, and the answer is (D).

AGE PROBLEMS

Typically, in these problems, we start by letting x be a person's current age and then the person's age a years ago will be $x - a$ and the person's age a years in future will be $x + a$. An example will illustrate.

Example: John is 20 years older than Steve. In 10 years, Steve's age will be half that of John's. What is Steve's age?

(A) 2
(B) 8
(C) 10
(D) 20
(E) 25

Steve's age is the most unknown quantity. So we let x = Steve's age and then $x + 20$ is John's age. Ten years from now, Steve and John's ages will be $x + 10$ and $x + 30$, respectively. Summarizing this information in a table yields

	Age now	Age in 10 years
Steve	x	$x + 10$
John	$x + 20$	$x + 30$

Since "in 10 years, Steve's age will be half that of John's," we get

$$\frac{1}{2}(x+30) = x + 10$$
$$x + 30 = 2(x + 10)$$
$$x + 30 = 2x + 20$$
$$x = 10$$

Hence, Steve is 10 years old, and the answer is (C).

INTEREST PROBLEMS

These problems are based on the formula

$$\text{INTEREST} = \text{AMOUNT} \times \text{TIME} \times \text{RATE}$$

Often, the key to these problems is that the interest earned from one account plus the interest earned from another account equals the total interest earned:

Total Interest = (Interest from first account) + (Interest from second account)

An example will illustrate.

Example: A total of \$1200 is deposited in two savings accounts for one year, part at 5% and the remainder at 7%. If \$72 was earned in interest, how much was deposited at 5%?

(A) 410
(B) 520
(C) 600
(D) 650
(E) 760

Let x be the amount deposited at 5%. Then $1200 - x$ is the amount deposited at 7%. The interest on these investments is $.05x$ and $.07(1200 - x)$. Since the total interest is \$72, we get

$$.05x + .07(1200 - x) = 72$$
$$.05x + 84 - .07x = 72$$
$$-.02x + 84 = 72$$
$$-.02x = -12$$
$$x = 600$$

The answer is (C).

Problem Set W:

➢ Easy

1. Waugh jogged to a restaurant at x miles per hour, and jogged back home along the same route at y miles per hour. He took 30 minutes for the whole trip. If the restaurant is 2 miles from home, what is the average speed in miles per hour at which he jogged for the whole trip?

   (A) 0.13
   (B) 0.5
   (C) 2
   (D) 4
   (E) 8

2. A cyclist travels at 12 miles per hour. How many minutes will it take him to travel 24 miles?

   (A) 1
   (B) 2
   (C) 30
   (D) 60
   (E) 120

3.

Column A		Column B
	1 Pound = 16 Ounces	
Weight of 16,000 ounces of rice		Weight of 1,000 pounds of coal

➢ Medium

4.

Column A	Column B
Distance between point A and a point that is located 8 miles East of point P, if Point P is located 6 miles North of point A	Distance between point B and a point that is located 6 miles West of point Q, if Point Q is located 8 miles South of point B

5. A wheat bag weighs 5 pounds and 12 ounces. How much does the bag weigh in pounds?

   (A) 5 1/4
   (B) 5 1/2
   (C) 5 3/4
   (D) 6 1/4
   (E) 6 3/4

6. One ton has 2000 pounds, and one pound has 16 ounces. How many packets containing wheat weighing 16 pounds and 4 ounces each would totally fill a gunny bag of capacity 13 tons?

   (A) 1600
   (B) 1700
   (C) 2350
   (D) 2500
   (E) 8000

➢ Hard

7. Train X leaves New York at 10:00AM and travels East at a constant speed of x miles per hour. If another Train Y leaves New York at 11:30AM and travels East along the same tracks at speed $4x/3$, then at what time will Train Y catch Train X?

    (A) 2 PM of the same day
    (B) 3 PM of the same day
    (C) 3:30 PM of the same day
    (D) 4 PM of the same day
    (E) 8 PM of the same day

8. An old man distributed all the gold coins he had to his two sons into two different numbers such that the difference between the squares of the two numbers is 36 times the difference between the two numbers. How many coins did the old man have?

    (A) 24
    (B) 26
    (C) 30
    (D) 36
    (E) 40

9.

Column A		Column B
	Patrick purchased 80 pencils and sold them at a loss equal to the selling price of 20 pencils.	
Cost of 80 pencils		Selling price of 100 pencils

10. A man walks at a rate of 10 mph. After every ten miles, he rests for 6 minutes. How much time does he take to walk 50 miles?

    (A) 300
    (B) 318
    (C) 322
    (D) 324
    (E) 330

11. A project has three test cases. Three teams are formed to study the three different test cases. James is assigned to all three teams. Except for James, each researcher is assigned to exactly one team. If each team has exactly 6 members, then what is the exact number of researchers required?

    (A) 10
    (B) 12
    (C) 14
    (D) 15
    (E) 16

12. The combined salaries of three brothers is $90,000. Mr. Big earns twice what Mr. Small earns, and Mr. Middle earns 1 1/2 times what Mr. Small earns. What is the smallest salary of the three brothers?

    (A) 20,000
    (B) 22,000
    (C) 25,000
    (D) 30,000
    (E) 40,000

The next two questions refer to the discussion below:
Mike and Fritz ran a 30-mile Marathon. Mike ran 10 miles at 10 miles per hour and then ran at 5 miles per hour for the remaining 20 miles. Fritz ran the first one-third (by time) of the run at 10 miles per hour and the remaining two-thirds of the run at 5 miles per hour.

13. How much time in hours did Mike take to complete the Marathon?

(A) 3
(B) 3.5
(C) 4
(D) 4.5
(E) 5

14. How much time in hours did Fritz take to complete the Marathon?

(A) 3
(B) 3.5
(C) 4
(D) 4.5
(E) 5

15. A ship is sinking and 120 more tons of water would suffice to sink it. Water seeps in at a constant rate of 2 tons a minute while pumps remove it at a rate of 1.75 tons a minute. How much time in minutes has the ship to reach the shore before is sinks?

(A) 480
(B) 560
(C) 620
(D) 680
(E) 720

16. When the price of oranges is lowered by 40%, 4 more oranges can be purchased for \$12 than can be purchased for the original price. How many oranges can be purchased for 24 dollars at the original price?

(A) 8
(B) 12
(C) 16
(D) 20
(E) 24

17. John has \$42. He purchased fifty mangoes and thirty oranges with the whole amount. He then chose to return six mangoes for nine oranges as both quantities are equally priced. What is the price of each Mango?

(A) 0.4
(B) 0.45
(C) 0.5
(D) 0.55
(E) 0.6

18. In a market, a dozen eggs cost as much as a pound of rice, and a half-liter of kerosene costs as much as 8 eggs. If the cost of each pound of rice is \$0.33, then how many cents does a liter of kerosene cost? [One dollar has 100 cents.]

(A) 0.33
(B) 0.44
(C) 0.55
(D) 44
(E) 55

19. A father distributed his total wealth to his two sons. The elder son received 3/5 of the amount. The younger son received \$30,000. How much wealth did the father have?

(A) 15,000
(B) 45,000
(C) 60,000
(D) 75,000
(E) 89,000

20. Chelsea traveled from point A to point B and then from point B to point C. If she took 1 hour to complete the trip with an average speed of 50 mph, what is the total distance she traveled in miles?

(A) 20
(B) 30
(C) 50
(D) 70
(E) 90

21. A car traveled at 80 mph for the first half (by time) of a trip and at 40 mph for the second half of the trip. What is the average speed of the car during the entire trip?

(A) 20
(B) 40
(C) 50
(D) 60
(E) 80

22. Mr. Smith's average annual income in each of the years 1966 and 1967 is x dollars. His average annual income in each of the years 1968, 1969, and 1970 is y dollars. What is his average annual income in the five continuous years 1966 through 1970?

(A) $2x/5 + 3y/5$
(B) $x/2 + y/2$
(C) $5(x + y)$
(D) $5x/2 + 5y/2$
(E) $3x/5 + 2y/5$

23. Hose A can fill a tank in 5 minutes, and Hose B can fill the same tank in 6 minutes. How many tanks would Hose B fill in the time Hose A fills 6 tanks?

(A) 3
(B) 4
(C) 5
(D) 5.5
(E) 6

24.

Column A		Column B
	12 trophies cost 60 dollars. If the cost of each trophy decreases by 1 dollar, $12 + x$ trophies cost 60 dollars. If the cost of each trophy increases by 1 dollar, $12 - y$ trophies cost 60 dollars.	
x		y

➢ Very Hard

25. The costs of equities of type A and type B (in dollars) are two different positive integers. If 4 equities of type A and 5 equities of type B together costs 27 dollars, what is the total cost of 2 equities of type A and 3 equities of type B in dollars?

(A) 15
(B) 24
(C) 35
(D) 42
(E) 55

26. How many coins of 0.5 dollars each and 0.7 dollars each together make exactly 4.6 dollars?

(A) 1, 6
(B) 2, 7
(C) 3, 5
(D) 4, 3
(E) 5, 3

27.

Column A		Column B
	A train takes 15 seconds to cross a bridge at 50 mph, and at the same speed takes 10 seconds to cross the same bridge when the train's length is halved.	
Length of the bridge		Original length of the train

28. One day, Mr. Richards started 30 minutes late from home and reached his office 50 minutes late, while driving 25% slower than his usual speed. How much time in minutes does Mr. Richards usually take to reach his office from home?

(A) 20
(B) 40
(C) 60
(D) 80
(E) 100

29. Katrina has a wheat business. She purchases wheat from a local wholesaler at a particular cost per pound. The price of the wheat at her stores is \$3 per pound. Her faulty spring balance reads 0.9 pounds for a pound. Also, in the festival season, she gives a 10% discount on the wheat. She found that she made neither a profit nor a loss in the festival season. At what price did Katrina purchase the wheat from the wholesaler?

(A) 2.43
(B) 2.5
(C) 2.7
(D) 3
(E) 3.3

30. According to the stock policy of a company, each employee in the technical division is given 15 shares of the company and each employee in the recruitment division is given 10 shares. Employees belonging to both communities get 25 shares each. There are 20 employees in the company, and each one belongs to at least one division. The cost of each share is \$10. If the technical division has 15 employees and the recruitment division has 10 employees, then what is the total cost of the shares given by the company?

(A) 2,250
(B) 2,650
(C) 3,120
(D) 3,180
(E) 3,250

31. A car traveled 65% of the way from Town A to Town B at an average speed of 65 mph. The car traveled at an average speed of v mph for the remaining part of the trip. The average speed for the entire trip was 50 mph. What is v in mph?

    (A) 65
    (B) 50
    (C) 45
    (D) 40
    (E) 35

32.

Column A		Column B
	A piece of string 35 cm long is cut into three smaller pieces along the length of the string. The length of the longest piece is three times the length of the shortest piece.	
The length of the medium-size piece		15

Answers and Solutions to Problem Set W

➢ Easy

1. Remember that *Average Speed* = *Net Distance* ÷ *Time Taken*. We are given that the time taken for the full trip is 30 minutes. Hence, we only need the distance traveled. We are given that the restaurant is 2 miles from home. Since Waugh jogs back along the same route, the net distance he traveled equals 2 + 2 = 4 miles. Hence, the Average Speed equals 4 miles ÷ 30 minutes = 4 miles ÷ 1/2 hour = 8 miles per hour. The answer is (E).

2. Since the answer is in minutes, we must convert the cyclist's speed (12 miles per hour) into miles per minute. Since there are 60 minutes in an hour, his speed is 12/60 = 1/5 miles per minute.

Remember that *Distance* = *Rate* × *Time*. Hence,

$$24 = \frac{1}{5} \times t$$

Solving for t yields $t = 5 \times 24 = 120$. The answer is (E). [If you forgot to convert hours to minutes, you may have mistakenly answered (B).]

3. Column B has 1,000 pounds of coal, and there are 16 ounces in 1 pound. So, Column B has 1,000 pounds = 1,000(16 ounces) = 16,000 ounces. Hence, each column weighs 16,000 ounces. The answer is (C).

➢ Medium

4. Column A:
First, place point *A* arbitrarily. Then locate point *P* 6 miles North of point *A*, and then locate a new point 8 miles East of *P*. Name the new point *M*. Now, Column A equals *AM*. The map drawn should look like this:

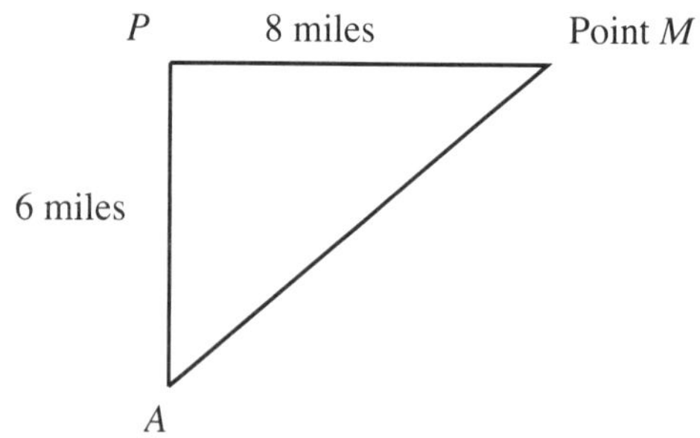

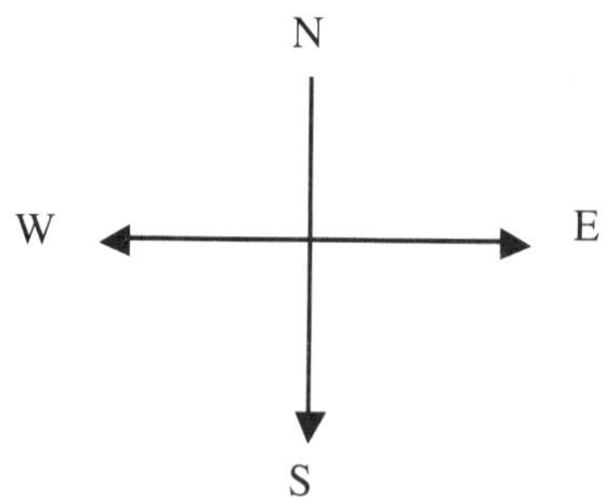

Since the angle between the standard directions East and North is 90°, the three points *A*, *P* and *M* form a right triangle, with right angle at *P*. So, *AM* is a hypotenuse. By The Pythagorean Theorem, the hypotenuse equals the square root of the sum of the squares of the other two sides. Hence,

$$\begin{aligned} AM &= \sqrt{AP^2 + PM^2} \\ &= \sqrt{6^2 + 8^2} \\ &= \sqrt{36 + 64} \\ &= \sqrt{100} \\ &= 10 \end{aligned}$$

Column B:
Similarly, place point *B* arbitrarily. Then locate point *Q* 8 miles South of it, and locate a new point 6 miles West of the point *Q*. Name the new point *N*. Now, Column B equals *BN*. The map should look like this:

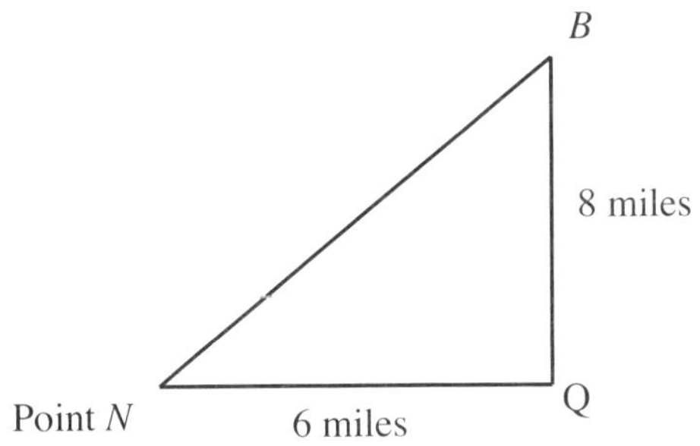

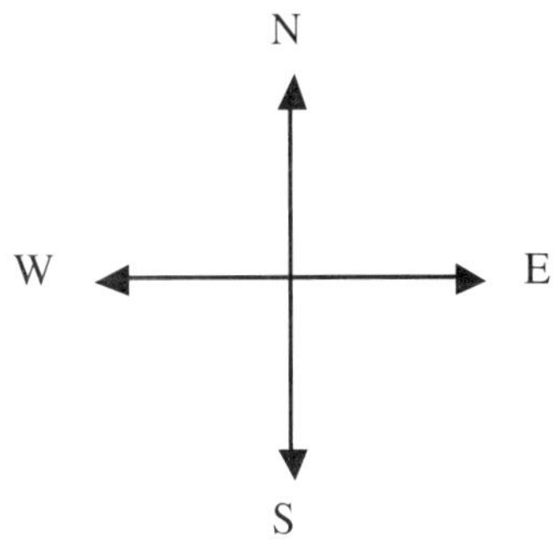

Again, since the angle between standard directions is 90°, we have a right triangle *BQN*, with right angle at *Q*, and, by The Pythagorean Theorem mentioned above, the hypotenuse *BN* equals

$$\sqrt{BQ^2 + QN^2}$$
$$= \sqrt{8^2 + 6^2}$$
$$= \sqrt{64 + 36}$$
$$= \sqrt{100}$$
$$= 10$$

Since both columns equal 10, the answer is (C).

5. There are 16 ounces in a pound. Hence, each ounce equals 1/16 pounds. Now, 12 ounces equals 12 × 1/16 = 3/4 pounds. Hence, 5 pounds + 12 ounces equals 5 3/4 pounds. The answer is (C).

6. One ton has 2000 pounds. The capacity of the gunny bag is 13 tons. Hence, its capacity in pounds would equal 13 × 2000 pounds.

One pound has 16 ounces. We are given the capacity of each packet is 16 pounds and 4 ounces. Converting it into pounds yields 16 pounds + 4/16 ounces = 16 1/4 pounds = (16 × 4 + 1)/4 = 65/4 pounds.

Hence, the number of packets required to fill the gunny bag equals

(Capacity of the gunny bag) ÷ (Capacity of the each packet) =
13 × 2000 pounds ÷ (65/4) pounds =
13 × 2000 × 4/65 =
2000 × 4/5 =
1600

The answer is (A).

➢ Hard

7. Train *X* started at 10:00AM. Let the time it has been traveling be *t*. Since Train *Y* started at 11:30AM, it has been traveling an hour and a half less. So, represent its time as $t - 1\ 1/2 = t - 3/2$.

Train *X* travels at speed *x* miles per hour, and Train *Y* travels at speed $4x/3$ miles per hour. By the formula *Distance = Speed · Time*, the respective distances they travel before meeting equals xt and $(4x/3)(t - 3/2)$. Since the trains started from the same point and traveled in the same direction, they will have traveled the same distance when they meet. Hence, we have

$xt = (4x/3)(t - 3/2)$
$t = (4/3)(t - 3/2)$ by canceling *x* from both sides
$t = 4t/3 - 2$ by distributing 4/3 on the right side
$t - 4t/3 = -2$
$-t/3 = -2$ by subtracting the expressions on the left side
$t = 6$ hours

Hence, Train *Y* will catch Train *X* at 4PM (10AM plus 6 hours is 4PM). The answer is (D).

8. Let x and y be the number of gold coins the two sons received. Since we are given that the difference between the squares of the two numbers is 36 times the difference between the two numbers, we have the equation

$x^2 - y^2 = 36(x - y)$

$(x - y)(x + y) = 36(x - y)$ by the Difference of Squares formula $a^2 - b^2 = (a - b)(a + b)$

$x + y = 36$ by canceling $(x - y)$ from both sides

Hence, the total number of gold coins the old man had is $x + y$ which equals 36. The answer is (D).

9. Let c be the cost of each pencil and s be the selling price of each pencil. Then the loss incurred by Patrick on each pencil is $c - s$. The net loss on 80 pencils is $80(c - s)$. Since we are given that the loss incurred on the 80 pencils equaled the selling price of 20 pencils (The selling price of 20 pencils is $20s$), we have the equation:

$80(c - s) = 20s$

$80c - 80s = 20s$

$80c = 100s$

Column A: Cost of 80 pencils equals $80c$.

Column B: Selling price of 100 pencils equals $100s$.

Column A = Column B.

Hence, the answer is (C).

10. Remember that *Time* = *Distance* ÷ *Speed*. Hence, the time taken by the man to walk 10 miles is 10 miles/10 mph = 1 hour.

Since the man walks 50 miles in five installments of 10 miles each, each installment should take him 1 hour. Hence, the total time for which he walked equals $5 \cdot 1$ hr = 5 hr = $5 \cdot 60$ = 300 mins.

Since he takes a break after each installment (until reaching the 50 mile point; one after 10 miles; one after 20 miles; one after 30 miles; final one after 40 miles. The 50th mile is his destination.), he takes four breaks; and since each break lasts 6 minutes, the net time spent in the breaks is $4 \cdot 6$ mins = 24 mins.

Hence, the total time taken to reach the destination is 300 + 24 = 324 mins. The answer is (D).

11. Since James is common to all three teams, he occupies one of six positions in each team. Since any member but James is with exactly one team, 5 different researchers are required for each team. Hence, apart from James, the number of researchers required is $5 \cdot 3 = 15$. Including James, there are $15 + 1 = 16$ researchers. The answer is (E).

12. Let s be the salary of Mr. Small. Since Mr. Big earns twice what Mr. Small earns, the salary of Mr. Big is $2s$; and since Mr. Middle earns 1 1/2 times what Mr. Small earns, the salary of Mr. Middle equals $(1\ 1/2)s = 3s/2$. Since $s < 3s/2$ and $s < 2s$, Mr. Small earns the smallest salary. Summing the salaries to 90,000 (given) yields

$2s + 3s/2 + s = 90,000$

$9s/2 = 90,000$

$s = 90,000 \cdot 2/9 = 20,000$

The answer is (A).

13. Mike ran 10 miles at 10 miles per hour (*Time* = *Distance*/*Rate* = 10 miles/10 miles per hour = 1 hour). He ran at 5 miles per hour for the remaining 20 miles (*Time* = *Distance*/*Rate* = 20 miles/5 miles per hour = 4 hrs). The net length of the Marathon track is 30 miles, and the total time taken to cover the track is 5 hours. Hence, the answer is (E).

14. Suppose Fritz took t hours to complete the 30-mile Marathon. Then as given, Fritz ran at 10 miles per hour for $t/3$ hours and 5 miles per hour for the remaining $2t/3$ hours. Now, by formula, *Distance* = *Rate* · *Time*, the net distance covered would be (10 miles per hour) · $t/3$ + (5 miles per hour) · $2t/3$ = $(10/3 + 10/3)t$ = 30 miles. Solving the equation for t yields $t = 90/20 = 4.5$ hours. The answer is (D).

15. We have that water enters the ship at 2 tons per minute and the pumps remove the water at 1.75 tons per minute. Hence, the effective rate at which water is entering the ship is 2 – 1.75 = 0.25 tons per minute. Since it takes an additional 120 tons of water to sink the ship, the time left is (120 tons)/(0.25 tons per minute) = 120/0.25 = 480 minutes. The answer is (A).

16. Let the original price of each orange be x dollars. Remember that *Quantity* = *Amount* ÷ *Rate*. Hence, we can purchase $12/x$ oranges for 12 dollars. After a 40% drop in price, the new price is $x(1 - 40/100) = 0.6x$ dollars per orange. Hence, we should be able to purchase $12/(0.6x) = 20/x$ oranges for the same 12 dollars. The excess number of oranges we get (for $12) from the lower price is $20/x - 12/x = (1/x)(20 - 12) = (1/x)(8) = 8/x = 4$ (given). Solving the equation $8/x = 4$ for x yields $x = 2$. Hence, the number of oranges that can be purchased for 24 dollars at original price x is 24/2 = 12. The answer is (B).

17. Since 6 mangoes are returnable for 9 oranges, if each mango costs m and each orange costs n, then $6m = 9n$, or $2m = 3n$. Solving for n yields, $n = 2m/3$. Now, since 50 mangoes and 30 oranges together cost 42 dollars,

$$\begin{gathered}50m + 30n = 42\\ 50m + 30(2m/3) = 42\\ m(50 + 30 \cdot 2/3) = 42\\ m(50 + 20) = 42\\ 70m = 42\\ m = 42/70 = 6/10 = 0.6\end{gathered}$$

The answer is (E).

18. One pound of rice costs 0.33 dollars. A dozen eggs cost as much as one pound of rice, and a dozen has 12 items. Hence, 12 eggs cost 0.33 dollars.

Now, since half a liter of kerosene costs as much as 8 eggs, one liter must cost 2 times the cost of 8 eggs, which equals the cost of 16 eggs.

Now, suppose 16 eggs cost x dollars. We know that 12 eggs cost 0.33 dollars. So, forming the proportion yields

$$\frac{0.33 \text{ dollars}}{12 \text{ eggs}} = \frac{x \text{ dollars}}{16 \text{ eggs}}$$

$$\begin{aligned}x &= 16 \times \frac{0.33}{12} = 4 \times \frac{0.33}{3} = 4 \times 0.11\\ &= 0.44 \text{ dollars} = 0.44\ (100 \text{ cents}) && \text{since one dollar has 100 cents}\\ &= 44 \text{ cents}\end{aligned}$$

The answer is (D).

19. Suppose x and y are the amounts received in dollars by the elder and the younger son, respectively. Then the amount the father had is $x + y$.

The elder son received 3/5 of the amount. Expressing this as an equation yields

$$x = (3/5)(x + y)$$
$$x = (3/5)x + (3/5)y$$
$$(2/5)x = (3/5)y$$
$$x = (3/2)y$$

Hence, $x + y$, the amount father had, equals $3y/2 + y = 5y/2 = (5/2)(30{,}000)$ [Given that the younger son received 30,000 dollars] = 75,000, and the answer is (D).

20. We have that her average speed is 50 mph. The formula for the *Average Speed* is *Distance Traveled* ÷ *Time Taken*. Hence, we have the equation 50 mph = *Distance Traveled* ÷ 1 hour. Solving this equation yields *Distance Traveled* = 50 miles. Hence, the answer is (C).

21. Let t be the entire time of the trip.

We have that the car traveled at 80 mph for $t/2$ hours and at 40 mph for the remaining $t/2$ hours. Remember that *Distance* = *Speed* × *Time*. Hence, the net distance traveled during the two periods equals $80 \times t/2 + 40 \times t/2$. Now, remember that

$$\textit{Average Speed} =$$
$$\frac{\textit{Net Distance}}{\textit{Time Taken}} =$$
$$\frac{80 \times \frac{t}{2} + 40 \times \frac{t}{2}}{t} =$$
$$80 \times \frac{1}{2} + 40 \times \frac{1}{2} =$$
$$40 + 20 =$$
$$60$$

The answer is (D).

22. Since Mr. Smith's average annual income in each of the two years 1966 and 1967 is x dollars, his total income in the two years is $2 \cdot x = 2x$.

Since Mr. Smith's average annual income in each of the next three years 1968 through 1970 is y dollars, his total income in the three years is $3 \cdot y = 3y$.

Hence, the net income in the five continuous years is $2x + 3y$.

Hence, the average income in the five years is

$$\text{(the net income)} \div 5 =$$
$$(2x + 3y)/5 =$$
$$2x/5 + 3y/5$$

The answer is (A).

23. Hose A takes 5 minutes to fill one tank. To fill 6 tanks, it takes $6 \cdot 5 = 30$ minutes. Hose B takes 6 minutes to fill one tank. Hence, in the 30 minutes, it would fill $30/6 = 5$ tanks. The answer is (C).

24. 12 trophies cost 60 dollars, so each trophy costs $60/12 = 5$ dollars.

If the price decreases by 1, the new price is $5 - 1 = 4$ dollars. Hence, 60 dollars can now buy 15 (= 60/4) trophies. Equating this to $12 + x$ yields $12 + x = 15$, or $x = 3$.

If the price increases by 1, the new price is $5 + 1 = 6$ dollars. Hence, 60 dollars can now buy 10 (= 60/6) trophies. Equating this to $12 - y$ yields $12 - y = 10$, or $y = 2$.

Now, Column A equals $x = 3$, and Column B equals $y = 2$. Hence, Column A > Column B, and the answer is (A).

➢ Very Hard

25. Let m and n be the costs of the equities of type A and type B, respectively. Since the costs are integers (given), m and n must be positive integers.

We have that 4 equities of type A and 5 equities of type B together cost 27 dollars. Hence, we have the equation $4m + 5n = 27$. Since m is a positive integer, $4m$ is a positive integer; and since n is a positive integer, $5n$ is a positive integer. Let $p = 4m$ and $q = 5n$. So, p is a multiple of 4 and q is a multiple of 5 and $p + q = 27$. Subtracting q from both sides yields $p = 27 - q$ [(a positive multiple of 4) equals 27 – (a positive multiple of 5)]. Let's seek such a solution for p and q:

If $q = 5$, $p = 27 - 5 = 22$, not a multiple of 4. Reject.

If $q = 10$, $p = 27 - 10 = 17$, not a multiple of 4. Reject.

If $q = 15$, $p = 27 - 15 = 12$, a multiple of 4. Acceptable. So, $n = p/4 = 3$ and $m = q/5 = 3$.

The following checks are not actually required since we already have an acceptable solution.

If $q = 20$, $p = 27 - 20 = 7$, not a multiple of 4. Reject.

If $q = 25$, $p = 27 - 25 = 2$, not a multiple of 4. Reject.

If $q \geq 30$, $p \leq 27 - 30 = -3$, not positive. Reject.

Hence, the cost of 2 equities of type A and 3 equities of type B is $2m + 3n = 2 \cdot 3 + 3 \cdot 3 = 15$. The answer is (A).

26. Let m coins of 0.5 dollars each and n coins of 0.7 dollars each add up to 4.6 dollars. Then, we have the equation $0.5m + 0.7n = 4.6$. Multiplying both sides by 10 to eliminate the decimals yields $5m + 7n = 46$. Since m is a positive integer, $5m$ is positive integer; and since n is a positive integer, $7n$ is a positive integer. Let $p = 5m$ and $q = 7n$. So, p is a multiple of 5 and q is a multiple of 7 and $p + q = 46$. Subtracting q from both sides yields $p = 46 - q$ [(a positive multiple of 5) equals 46 – (a positive multiple of 7)]. Let's seek such solution for p and q:

If $q = 7$, $p = 46 - 7 = 39$, not a multiple of 5. Reject.

If $q = 14$, $p = 46 - 14 = 32$, not a multiple of 5. Reject.

If $q = 21$, $p = 46 - 21 = 25$, a multiple of 5. Acceptable. So, $n = q/7 = 3$ and $m = p/5 = 5$.

The following checks are not actually required since we already have an acceptable solution.

If $q = 28$, $p = 46 - 28 = 18$, not a multiple of 5. Reject.

If $q = 35$, $p = 46 - 35 = 11$, not a multiple of 5. Reject.

If $q = 42$, $p = 46 - 42 = 4$, not a multiple of 5. Reject.

If $q \geq 49$, $p = 46 - 49 = -3$, not positive. Reject.

The answer is (E).

27. Let the length of the train be l and the length of the bridge be b. We are given that the train crosses the bridge in 15 seconds, and it crosses the same bridge in 10 seconds when its length is halved. In each case, the speed is the same. Hence, let's derive expressions for speed for each case and equate. The distance traveled by the train in crossing the bridge is

(*length of bridge*) + (*length of train*)

In the first case, distance traveled is $b + l$, and, in the second case, with train length halved, distance traveled is $b + l/2$. In the first case, time taken is 15 seconds, and in the second case, time taken is 10 seconds. By the formula *Speed* = *Distance*/*Time*, the speed in the first case is $(b + l)/15$ and in the second case is $(b + l/2)/10$. Since the speed in both cases is the same, we have

$$\frac{b+l}{15} = \frac{b+l/2}{10}$$

$$\frac{b+l}{3} = \frac{b+l/2}{2}$$

$$2(b + l) = 3(b + l/2)$$

$$2b + 2l = 3b + 3l/2$$

$$l/2 = b$$

$$l = 2b$$

Hence, the length of the train is twice the length of the bridge. So, Column B is greater, and the answer is (B).

28. If d is the distance to the office from home and if v is his usual speed, then the time taken is $t = d/v$ (by the formula *Time* = *Distance* ÷ *Speed*). Hence, we only need to find the ratio d/v.

Now, we are given that on a day after starting 30 minutes late and driving 25% slower [i.e., at speed $(1 - 25/100)v = 0.75$], he reached his office 50 minutes late.

Here, there are two delays: One delay is the "30 minute delay in starting from home" and the other is from driving slower than his regular speed. The total delay is 50 minutes. Hence, the delay caused by driving slower is 50 – 30 = 20.

Now, usually, Richards would have taken the time d/v to reach the office. But, on this day, he drove at a speed 25% slower [$Speed = v \cdot (1 - 25/100) = 0.75v$]. Hence, the time he would have taken for the trip equals $d/0.75v$, and the component of the delay from driving slower is

$$\frac{d}{0.75v} - \frac{d}{v} =$$

$$\frac{0.25d}{0.75v} =$$

$$\frac{d}{3v} = 20 \text{ minutes (given)}$$

Multiplying both sides by 3 yields $d/v = 60$. The answer is (C).

29. The cost of wheat at Katrina's store is \$3 per pound. After the 10% discount (festival season discount) the cost of the wheat would be, by the known formula the original price (1 – discount percent/100) = 3(1 – 10/100) = 3(90/100) = 2.7 dollars per pound.

Since her faulty balance was reading 0.9 pounds for a pound, she was unknowingly selling 1 pound in the name of 0.9 pounds This is equivalent to selling 1/0.9 = 10/9 pounds in the name of one pound. Hence, Katrina was trying to sell the wheat at 2.7 dollars per pound, but effectively she was selling the wheat at 2.7 dollars per (10/9 pounds) = 2.7 × 9/10 = 2.43 dollars a pound. Since she earned neither a profit nor a loss, she must have purchased the wheat at this same cost from the wholesaler. Hence, the answer is (A).

30. We have that the number of employees in the technical division is 15 and the number of employees in the recruitment division is 10. Each technical person is given 15 shares, and each recruitment person is given 10 shares. Hence, the net shares given equals $15 \cdot 15 + 10 \cdot 10 = 225 + 100 = 325$. Each share is worth 10 dollars, so the net worth of the shares is $325 \cdot 10 = 3{,}250$. The answer is (E).

Note: In the problem, employees belonging to the technical community get 15 shares, employees in the recruitment community get 10 shares, and the employees in both communities get 25 (= 15 + 10) shares. Effectively, independent treatment is given to the two communities. Hence, we did not give any special consideration to employees who are in both communities.

31. Let d be the distance between the towns A and B. 65% of this distance (= 65% of $d = 65/100 \times d = 0.65d$) was traveled at 65 mph and the remaining 100 – 65 = 35% of the distance was traveled at v mph. Now, Remember that *Time = Distance ÷ Rate*. Hence, the time taken by the car for the first 65% distance is $0.65d/65 = d/100$, and the time taken by the car for the last 35% distance is $0.35d/v$. Hence, the total time taken is $d/100 + 0.35d/v = d(1/100 + 0.35/v)$.

Now, remember that $\textit{Average Speed} = \dfrac{\textit{Total Distance Traveled}}{\textit{Total Time Taken}}$.

Hence, the average speed of the journey is $\dfrac{d}{d\left(\dfrac{1}{100}+\dfrac{0.35}{v}\right)}$. Equating this to the given value for the average speed yields

$$\frac{d}{d\left(\frac{1}{100}+\frac{0.35}{v}\right)} = 50$$

$$\frac{1}{\frac{1}{100}+\frac{0.35}{v}} = 50$$

$$1 = 50 \times \frac{1}{100} + 50 \times \frac{0.35}{v}$$

$$1 = \frac{1}{2} + 50 \times \frac{0.35}{v}$$

$$\frac{1}{2} = 50 \times \frac{0.35}{v}$$

$$v = 2 \times 50 \times 0.35 = 100 \times 0.35 = 35$$

The answer is (E).

32. The string is cut into three along its length. Let l be the length of the smallest piece. Then the length of the longest piece is $3l$, and the total length of the three pieces is 35 cm. The length of the longest and shortest pieces together is $l + 3l = 4l$. Hence, the length of the third piece (medium-size piece) must be $35 - 4l$. Arranging the lengths of the three pieces in increasing order of length yields the following inequality:

$l < 35 - 4l < 3l$	
$5l < 35 < 7l$	by adding $4l$ to each part of the inequality
$5l < 35$ and $35 < 7l$	by separating into two inequalities
$l < 7$ and $5 < l$	by dividing first inequality by 5 and the second inequality by 7
$5 < l < 7$	combining the two inequalities
$20 < 4l < 28$	multiplying each part by 4
$-20 > -4l > -28$	by multiplying the inequalities by -1 and flipping the directions of the inequalities
$35 - 20 > 35 - 4l > 35 - 28$	adding 35 to each part
$15 > 35 - 4l > 7$	
15 > The length of the medium-size piece > 7	
15 > Column A > 7	
Column B > Column A > 7	

Hence, the answer is (B).

Method II:

Had the length of the medium-size piece been greater than or equal to 15, the length of the longest-size piece would be greater than 15 and the length of the smallest piece, which equals 1/3 the length of the longest piece, would be greater than $15/3 = 5$. Hence, the net sum of the three lengths exceeds $15 + 15 + 5$ (= 35). Since this is impossible, our assumption that the length of the medium-sized piece is greater than or equal to 15 is false. Hence, it is less than 15 and therefore Column A is less than Column B. The answer is (B).

Sequences & Series

SEQUENCES

A sequence is an ordered list of numbers. The following is a sequence of odd numbers:

$$1, 3, 5, 7, \ldots$$

A term of a sequence is identified by its position in the sequence. In the above sequence, 1 is the first term, 3 is the second term, etc. The ellipsis symbol (. . .) indicates that the sequence continues forever.

Example 1: In sequence S, the 3rd term is 4, the 2nd term is three times the 1st, and the 3rd term is four times the 2nd. What is the 1st term in sequence S?

(A) 0 (B) 1/3 (C) 1 (D) 3/2 (E) 4

We know *"the 3rd term of S is 4,"* and that *"the 3rd term is four times the 2nd."* This is equivalent to saying the 2nd term is 1/4 the 3rd term: $\frac{1}{4} \cdot 4 = 1$. Further, we know *"the 2nd term is three times the 1st."* This is equivalent to saying the 1st term is $\frac{1}{3}$ the 2nd term: $\frac{1}{3} \cdot 1 = \frac{1}{3}$. Hence, the first term of the sequence is fully determined:

$$\frac{1}{3}, 1, 4$$

The answer is (B).

Example 2: Except for the first two numbers, every number in the sequence –1, 3, –3, . . . is the product of the two immediately preceding numbers. How many numbers of this sequence are odd?

(A) one (B) two (C) three (D) four (E) more than four

Since *"every number in the sequence –1, 3, –3, . . . is the product of the two immediately preceding numbers,"* the forth term of the sequence is $-9 = 3(-3)$. The first 6 terms of this sequence are

$$-1, 3, -3, -9, 27, -243, \ldots$$

At least six numbers in this sequence are odd: –1, 3, –3, –9, 27, –243. The answer is (E).

Arithmetic Progressions

An arithmetic progression is a sequence in which the difference between any two consecutive terms is the same. This is the same as saying: each term exceeds the previous term by a fixed amount. For example, 0, 6, 12, 18, . . . is an arithmetic progression in which the common difference is 6. The sequence 8, 4, 0, –4, . . . is arithmetic with a common difference of –4.

Example 3: The seventh number in a sequence of numbers is 31 and each number after the first number in the sequence is 4 less than the number immediately preceding it. What is the fourth number in the sequence?

(A) 15
(B) 19
(C) 35
(D) 43
(E) 51

Since each number *"in the sequence is 4 less than the number immediately preceding it,"* the sixth term is 31 + 4 = 35; the fifth number in the sequence is 35 + 4 = 39; and the fourth number in the sequence is 39 + 4 = 43. The answer is (D). Following is the sequence written out:

55, 51, 47, 43, 39, 35, 31, 27, 23, 19, 15, 11, . . .

Advanced concepts: (Sequence Formulas)

Students with strong backgrounds in mathematics may prefer to solve sequence problems by using formulas. Note, none of the formulas in this section are necessary to answer questions about sequences on the GRE.

Since each term of an arithmetic progression *"exceeds the previous term by a fixed amount,"* we get the following:

first term	$a + 0d$	where a is the first term and d is the common difference
second term	$a + 1d$	
third term	$a + 2d$	
fourth term	$a + 3d$	
	. . .	
nth term	$a + (n - 1)d$	This formula generates the nth term

The sum of the first n terms of an arithmetic sequence is

$$\frac{n}{2}\left[2a+(n-1)d\right]$$

Geometric Progressions

A geometric progression is a sequence in which the ratio of any two consecutive terms is the same. Thus, each term is generated by multiplying the preceding term by a fixed number. For example, –3, 6, –12, 24, . . . is a geometric progression in which the common ratio is –2. The sequence 32, 16, 8, 4, . . . is geometric with common ratio 1/2.

Example 4: What is the sixth term of the sequence 90, –30, 10, –10/3, . . . ?

(A) 1/3
(B) 0
(C) –10/27
(D) 3
(E) –100/3

Since the common ratio between any two consecutive terms is $-\frac{1}{3}$, the fifth term is $\frac{10}{9}=\left(-\frac{1}{3}\right)\cdot\left(-\frac{10}{3}\right)$. Hence, the sixth number in the sequence is $-\frac{10}{27}=\left(-\frac{1}{3}\right)\cdot\left(\frac{10}{9}\right)$. The answer is (C).

Advanced concepts: (Sequence Formulas)

Note, none of the formulas in this section are necessary to answer questions about sequences on the GRE.

Since each term of a geometric progression *"is generated by multiplying the preceding term by a fixed number,"* we get the following:

first term	a	
second term	ar^1	where r is the common ratio
third term	ar^2	
fourth term	ar^3	
	...	
nth term	$a_n = ar^{n-1}$	This formula generates the nth term

The sum of the first n terms of an geometric sequence is

$$\frac{a\left(1-r^n\right)}{1-r}$$

SERIES

A series is simply the sum of the terms of a sequence. The following is a series of even numbers formed from the sequence 2, 4, 6, 8, . . . :

$$2 + 4 + 6 + 8 + \cdots$$

A term of a series is identified by its position in the series. In the above series, 2 is the first term, 4 is the second term, etc. The ellipsis symbol (. . .) indicates that the series continues forever.

Example 5: The sum of the squares of the first n positive integers $1^2 + 2^2 + 3^2 + \ldots + n^2$ is $\frac{n(n+1)(2n+1)}{6}$. What is the sum of the squares of the first 9 positive integers?

(A) 90 (B) 125 (C) 200 (D) 285 (E) 682

We are given a formula for the sum of the squares of the first n positive integers. Plugging $n = 9$ into this formula yields

$$\frac{n(n+1)(2n+1)}{6} = \frac{9(9+1)(2\cdot 9+1)}{6} = \frac{9(10)(19)}{6} = 285$$

The answer is (D).

Example 6: For all integers $x > 1$, $\langle x\rangle = 2x + (2x-1) + (2x-2) + \ldots + 2 + 1$. What is the value of $\langle 3\rangle\cdot\langle 2\rangle$?

(A) 60 (B) 116 (C) 210 (D) 263 (E) 478

$\langle 3\rangle = 2(3) + (2\cdot 3-1) + (2\cdot 3-2) + (2\cdot 3-3) + (2\cdot 3-4) + (2\cdot 3-5) = 6+5+4+3+2+1 = 21$

$\langle 2\rangle = 2(2) + (2\cdot 2-1) + (2\cdot 2-2) + (2\cdot 2-3) = 4+3+2+1 = 10$

Hence, $\langle 3\rangle\cdot\langle 2\rangle = 21\cdot 10 = 210$, and the answer is (C).

Problem Set X:

➤ Medium

1. In a sequence, the nth term a_n is defined by the rule $(a_{n-1} - 3)^2$, $a_1 = 1$. What is the value of a_4 ?

   (A) 1
   (B) 4
   (C) 9
   (D) 16
   (E) 25

2. If the nth term in a sequence of numbers $a_0, a_1, a_2, \ldots, a_n$ is defined to equal $2n + 1$, then what is the numerical difference between the 5th and 6th terms in the sequence?

   (A) 1
   (B) 2
   (C) 4
   (D) 5
   (E) 6

3. A sequence of numbers $a_1, a_2, a_3, \ldots, a_n$ is generated by the rule $a_{n+1} = 2a_n$. If $a_7 - a_6 = 96$, then what is the value of a_7 ?

   (A) 48
   (B) 96
   (C) 98
   (D) 192
   (E) 198

4. A sequence of positive integers $a_1, a_2, a_3, \ldots, a_n$ is given by the rule $a_{n+1} = 2a_n + 1$. The only even number in the sequence is 38. What is the value of a_2 ?

   (A) 11
   (B) 25
   (C) 38
   (D) 45
   (E) 77

5. The sum of the first n terms of an arithmetic series whose nth term is n can be calculated by the formula $n(n + 1)/2$. Which one of the following equals the sum of the first eight terms in a series whose nth term is $2n$?

   (A) 24
   (B) 48
   (C) 56
   (D) 72
   (E) 96

6. The sum of the first n terms of a series is 31, and the sum of the first $n - 1$ terms of the series is 20. What is the value of nth term in the series?

   (A) 9
   (B) 11
   (C) 20
   (D) 31
   (E) 51

7. In the sequence a_n, the nth term is defined as $(a_{n-1} - 1)^2$. If $a_1 = 4$, then what is the value of a_2 ?

(A) 2
(B) 3
(C) 4
(D) 5
(E) 9

➤ Hard

8. A worker is hired for 7 days. Each day, he is paid 10 dollars more than what he is paid for the preceding day of work. The total amount he was paid in the first 4 days of work equaled the total amount he was paid in the last 3 days. What was his starting pay?

(A) 90
(B) 138
(C) 153
(D) 160
(E) 163

9. A sequence of numbers is represented as $a_1, a_2, a_3, \ldots, a_n$. Each number in the sequence (except the first and the last) is the mean of the two adjacent numbers in the sequence. If $a_1 = 1$ and $a_5 = 3$, what is the value of a_3 ?

(A) 1/2
(B) 1
(C) 3/2
(D) 2
(E) 5/2

10. A series has three numbers a, ar, and ar^2. In the series, the first term is twice the second term. What is the ratio of the sum of the first two terms to the sum of the last two terms in the series?

(A) 1 : 1
(B) 1 : 2
(C) 1 : 4
(D) 2 : 1
(E) 4 : 1

11. The sequence of numbers a, ar, ar^2, and ar^3 are in geometric progression. The sum of the first four terms in the series is 5 times the sum of first two terms and $r \neq -1$ and $a \neq 0$. How many times larger is the fourth term than the second term?

(A) 1
(B) 2
(C) 4
(D) 5
(E) 6

12. In the sequence a_n, the nth term is defined as $(a_{n-1} - 1)^2$. If $a_3 = 64$, then what is the value of a_2?

(A) 2
(B) 3
(C) 4
(D) 5
(E) 9

Answers and Solutions to Problem Set X

➢ Medium

1. The rule for the terms in the sequence is given as $a_n = (a_{n-1} - 3)^2$.

Substituting $n = 2$ in the rule yields

$$a_2 = (a_{2-1} - 3)^2 = (a_1 - 3)^2 = (1 - 3)^2 = (-2)^2 = 4$$

Substituting $n = 3$ in the rule yields

$$a_3 = (a_{3-1} - 3)^2 = (a_2 - 3)^2 = (4 - 3)^2 = 1^2 = 1$$

Substituting $n = 4$ in the rule yields

$$a_4 = (a_{4-1} - 3)^2 = (a_3 - 3)^2 = (1 - 3)^2 = (-2)^2 = 4$$

Hence, the answer is (B).

2. We have the rule $a_n = 2n + 1$. By this rule,

$$a_5 = 2(5) + 1 = 11$$
$$a_6 = 2(6) + 1 = 13$$

Forming the difference $a_6 - a_5$ yields

$$a_6 - a_5 = 13 - 11 = 2$$

The answer is (B).

3. Putting $n = 6$ in the given rule $a_{n+1} = 2a_n$ yields $a_{6+1} = 2a_6$, or $a_7 = 2a_6$. Since we are given that $a_7 - a_6 = 96$, we have $2a_6 - a_6 = 96$, or $a_6 = 96$. Hence, $a_7 = 2a_6 = 2 \cdot 96 = 192$. The answer is (D).

4. 2(an integer) + 1 is always odd. The rule $a_{n+1} = 2a_n + 1$ indicates that each term in the series, except possibly the first one, must be odd. The first term may be even. Hence, assign the even number 38 to the only possible even term in the sequence. By the rule $a_{n+1} = 2a_n + 1$, we have $a_2 = 2a_1 + 1 = 2(38) + 1 = 77$. The answer is (E).

5. The sum of the first n terms of an arithmetic series whose nth term is n is $n(n + 1)/2$. Hence, we have

$$1 + 2 + 3 + \dots + n = n(n + 1)/2$$

Multiplying each side by 2 yields

$$2 + 4 + 6 + \dots + 2n = 2n(n + 1)/2 = n(n + 1)$$

Hence, the sum to 8 terms equals $n(n + 1) = 8(8 + 1) - 8(9) = 72$. The answer is (D).

6. (The sum of the first n terms of a series) = (The sum of the first $n - 1$ terms) + (The nth term).

Substituting the given values in the equation yields 31 = 20 + nth term. Hence, the nth term is $31 - 20 = 11$. The answer is (B).

7. Replacing n with 2 in the given formula $a_n = (a_{n-1} - 1)^2$ yields $a_2 = (a_{2-1} - 1)^2 = (a_1 - 1)^2$. We are given that $a_1 = 4$. Putting this in the formula $a_2 = (a_1 - 1)^2$ yields $a_2 = (4 - 1)^2 = 3^2 = 9$. The answer is (E).

➤ Hard

8. This problem can be solved with a series. Let the payments for the 7 continuous days be $a_1, a_2, a_3, \ldots, a_7$. Since each day's pay was 10 dollars more than the previous day's pay, the rule for the series is $a_{n+1} = a_n + 10$.

By the rule, let the payments for each day be listed as

$$a_1$$
$$a_2 = a_1 + 10$$
$$a_3 = a_2 + 10 = (a_1 + 10) + 10 = a_1 + 20$$
$$a_4 = a_3 + 10 = (a_1 + 20) + 10 = a_1 + 30$$
$$a_5 = a_4 + 10 = (a_1 + 30) + 10 = a_1 + 40$$
$$a_6 = a_5 + 10 = (a_1 + 40) + 10 = a_1 + 50$$
$$a_7 = a_6 + 10 = (a_1 + 50) + 10 = a_1 + 60$$

We are given that the net pay for the first 4 days equals the net pay for the last 3 days.

The net pay for first 4 days is $a_1 + (a_1 + 10) + (a_1 + 20) + (a_1 + 30) = 4a_1 + 10(1 + 2 + 3)$.

The net pay for last (next) 3 days is $(a_1 + 40) + (a_1 + 50) + (a_1 + 60) = 3a_1 + 10(4 + 5 + 6)$.

Equating the two yields

$$4a_1 + 10(1 + 2 + 3) = 3a_1 + 10(4 + 5 + 6)$$

$$a_1 = 10(4 + 5 + 6 - 1 - 2 - 3) = 90$$

The answer is (A).

9. Since each number in the sequence (except the first and the last) is the mean of the adjacent two numbers in the sequence, we have

$$a_2 = (a_1 + a_3)/2$$
$$a_3 = (a_2 + a_4)/2$$
$$a_4 = (a_3 + a_5)/2$$

Substituting the given values $a_1 = 1$ and $a_5 = 3$ yields

$$a_2 = (1 + a_3)/2$$
$$a_3 = (a_2 + a_4)/2$$
$$a_4 = (a_3 + 3)/2$$

Substituting the top and the bottom equations into the middle one yields

$$a_3 = \frac{a_2 + a_4}{2}$$

$$a_3 = \frac{\frac{1 + a_3}{2} + \frac{a_3 + 3}{2}}{2}$$

$$a_3 = \frac{\frac{1}{2} + \frac{a_3}{2} + \frac{a_3}{2} + \frac{3}{2}}{2}$$

$$a_3 = \frac{2 + a_3}{2} = 1 + \frac{a_3}{2}$$

Subtracting $a_3/2$ from both sides yields $a_3/2 = 1$, or $a_3 = 2$. The answer is (D).

10. Since "the first term in the series is twice the second term," we have $a = 2(ar)$. Canceling a from both sides of the equation yields $1 = 2r$. Hence, $r = 1/2$.

Hence, the three numbers a, ar, and ar^2 become a, $a(1/2)$, and $a(1/2)^2$, or a, $a/2$, and $a/4$.

The sum of first two terms is $a + a/2$ and the sum of the last two terms is $a/2 + a/4$. Forming their ratio yields

$$\frac{a+\frac{a}{2}}{\frac{a}{2}+\frac{a}{4}} =$$

$$\frac{\frac{2a+a}{2}}{\frac{2a+a}{4}} =$$

$$\frac{\frac{3a}{2}}{\frac{3a}{4}} =$$

$$\left(\frac{3a}{2}\right)\left(\frac{4}{3a}\right) =$$

$$2 =$$

$$\frac{2}{1} \text{ or } 2:1$$

The answer is (D).

11. In the given progression, the sum of first two terms is $a + ar$, and the sum of first four terms is $a + ar + ar^2 + ar^3$. Since "the sum of the first four terms in the series is 5 times the sum of the first two terms," we have

$$a + ar + ar^2 + ar^3 = 5(a + ar)$$

$$\frac{a + ar + ar^2 + ar^3}{a + ar} = 5 \quad \text{by dividing both sides by } a + ar$$

$$\frac{(a + ar) + r^2(a + ar)}{a + ar} = 5$$

$$\frac{(a + ar)(1 + r^2)}{a + ar} = 5 \quad \text{by factoring out the common term } a + ar$$

$$1 + r^2 = 5 \quad \text{by canceling } a + ar \text{ from both numerator and denominator}$$

$$r^2 = 5 - 1 = 4.$$

Now, the fourth term is $ar^3/ar = r^2 = 4$ times the second term. Hence, the answer is (C).

12. Replacing n with 3 in the formula $a_n = (a_{n-1} - 1)^2$ yields $a_3 = (a_{3-1} - 1)^2 = (a_2 - 1)^2$. We are given that $a_3 = 64$. Putting this in the formula $a_3 = (a_2 - 1)^2$ yields

$$64 = (a_2 - 1)^2$$

$$a_2 - 1 = \pm 8$$

$$a_2 = -7 \text{ or } 9$$

Since, we know that a_2 is the result of the square of number $[a_2 = (a_1 - 1)^2]$, it cannot be negative. Hence, pick the positive value 9 for a_2. The answer is (E).

Counting

Counting may have been one of humankind's first thought processes; nevertheless, counting can be deceptively hard. In part, because we often forget some of the principles of counting, but also because counting can be inherently difficult.

When counting elements that are in overlapping sets, the total number will equal the number in one group plus the number in the other group minus the number common to both groups. Venn diagrams are very helpful with these problems.

Example 1: If in a certain school 20 students are taking math and 10 are taking history and 7 are taking both, how many students are taking either math or history?

(A) 20 (B) 22 (C) 23 (D) 25 (E) 29

Solution:

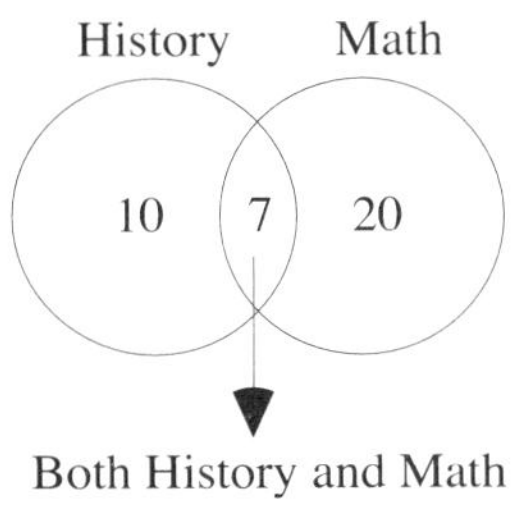

By the principle stated above, we add 10 and 20 and then subtract 7 from the result. Thus, there are $(10 + 20) - 7 = 23$ students. The answer is (C).

The number of integers between two integers <u>inclusive</u> is one more than their difference.

Example 2: How many integers are there between 49 and 101, inclusive?

(A) 50 (B) 51 (C) 52 (D) 53 (E) 54

By the principle stated above, the number of integers between 49 and 101 inclusive is $(101 - 49) + 1 = 53$. The answer is (D). To see this more clearly, choose smaller numbers, say, 9 and 11. The difference between 9 and 11 is 2. But there are three numbers between them inclusive—9, 10, and 11—one more than their difference.

***Fundamental Principle of Counting*: If an event occurs m times, and each of the m events is followed by a second event which occurs k times, then the first event follows the second event $m \cdot k$ times.**

The following diagram illustrates the fundamental principle of counting for an event that occurs 3 times with each occurrence being followed by a second event that occurs 2 times for a total of $3 \cdot 2 = 6$ events:

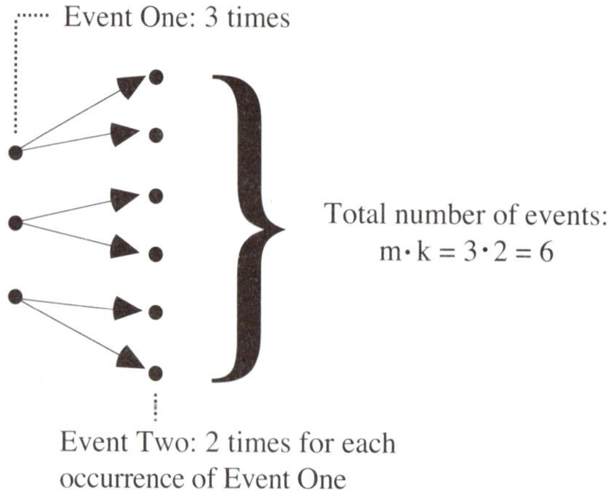

Example 3: A drum contains 3 to 5 jars each of which contains 30 to 40 marbles. If 10 percent of the marbles are flawed, what is the greatest possible number of flawed marbles in the drum?

(A) 51 (B) 40 (C) 30 (D) 20 (E) 12

There is at most 5 jars each of which contains at most 40 marbles; so by the fundamental counting principle, there is at most $5 \cdot 40 = 200$ marbles in the drum. Since 10 percent of the marbles are flawed, there is at most $20 = 10\% \cdot 200$ flawed marbles. The answer is (D).

MISCELLANEOUS COUNTING PROBLEMS

Example 4: In a legislative body of 200 people, the number of Democrats is 50 less than 4 times the number of Republicans. If one fifth of the legislators are neither Republican nor Democrat, how many of the legislators are Republicans?

(A) 42 (B) 50 (C) 71 (D) 95 (E) 124

Let D be the number of Democrats and let R be the number of Republicans. "One fifth of the legislators are neither Republican nor Democrat," so there are $\frac{1}{5} \cdot 200 = 40$ legislators who are neither Republican nor Democrat. Hence, there are $200 - 40 = 160$ Democrats and Republicans, or $D + R = 160$. Translating the clause "the number of Democrats is 50 less than 4 times the number of Republicans" into an equation yields $D = 4R - 50$. Plugging this into the equation $D + R = 160$ yields

$$4R - 50 + R = 160$$
$$5R - 50 = 160$$
$$5R = 210$$
$$R = 42$$

The answer is (A).

Example 5: Speed bumps are being placed at 20 foot intervals along a road 1015 feet long. If the first speed bump is placed at one end of the road, how many speed bumps are needed?

(A) 49 (B) 50 (C) 51 (D) 52 (E) 53

Since the road is 1015 feet long and the speed bumps are 20 feet apart, there are $\frac{1015}{20} = 50.75$, or 50 full sections in the road. If we ignore the first speed bump and associate the speed bump at the end of each section with that section, then there are 50 speed bumps (one for each of the fifty full sections). Counting the first speed bump gives a total of 51 speed bumps. The answer is (C).

SETS

A *set* is a collection of objects, and the objects are called *elements* of the set. You may be asked to form the *union* of two sets, which contains all the objects from either set. You may also be asked to form the *intersection* of two sets, which contains only the objects that are in both sets. For example, if Set $A = \{1, 2, 5\}$ and Set B = $\{5, 10, 21\}$, then the union of sets A and B would be $\{1, 2, 5, 10, 21\}$ and the intersection would be $\{5\}$.

Problem Set Y:

➢ **Easy**

1. Column A | In a jar, 60% of the marbles are red and the rest are green. | Column B

40% of the red marbles in the jar

60% of the green marbles in the jar

➢ **Medium**

2. In a zoo, each pigeon has 2 legs, and each rabbit has 4 legs. The head count of the two species together is 12, and the leg count is 32. How many pigeons and how many rabbits are there in the zoo?

(A) 4, 8
(B) 6, 6
(C) 6, 8
(D) 8, 4
(E) 8, 6

3. Column A | Column B

Fraction of numbers from 0 through 1000 that are divisible by both 7 and 10

Fraction of numbers from 0 through 1000 that are divisible by both 5 and 14

➢ **Hard**

4. For how many positive integers n is it true that the sum of $13/n$, $18/n$, and $29/n$ is an integer?

(A) 6
(B) 60
(C) Greatest common factor of 13, 18, and 29
(D) Least common multiple of 13, 18, and 29
(E) 12

5. For how many integers n between 5 and 20, inclusive, is the sum of $3n$, $9n$, and $11n$ greater than 200?

(A) 4
(B) 8
(C) 12
(D) 16
(E) 20

6. In a factory, there are workers, executives and clerks. 59% of the employees are workers, 460 are executives, and the remaining 360 employees are clerks. How many employees are there in the factory?

(A) 1500
(B) 2000
(C) 2500
(D) 3000
(E) 3500

7. In the town of Windsor, 250 families have at least one car while 60 families have at least two cars. How many families have exactly one car?

(A) 30
(B) 190
(C) 280
(D) 310
(E) 420

8. Ana is a girl and has the same number of brothers as sisters. Andrew is a boy and has twice as many sisters as brothers. Ana and Andrew are the children of Emma. How many children does Emma have?

(A) 2
(B) 3
(C) 5
(D) 7
(E) 8

9. A trainer on a Project Planning Module conducts batches of soft skill training for different companies. The trainer sets the batch size (the number of participants) of any batch such that he can make groups of equal numbers without leaving out any of the participants. For a particular batch he decides that he should be able to make teams of 3 participants each, teams of 5 participants each, and teams of 6 participants each, successfully without leaving out anyone in the batch. Which one of the following best describes the batch size (number of participants) that he chooses for the program?

(A) Exactly 30 participants.
(B) At least 30 participants.
(C) Less than 30 participants.
(D) More than 30 participants.
(E) Participants in groups of 30 or its multiples.

10. In a multi-voting system, voters can vote for more than one candidate. Two candidates A and B are contesting the election. 100 voters voted for A. Fifty out of 250 voters voted for both candidates. If each voter voted for at least one of the two candidates, then how many candidates voted only for B?

(A) 50
(B) 100
(C) 150
(D) 200
(E) 250

11. There are 750 male and female participants in a meeting. Half the female participants and one-quarter of the male participants are Democrats. One-third of all the participants are Democrats. How many of the Democrats are female?

(A) 75
(B) 100
(C) 125
(D) 175
(E) 225

12.

Column A		Column B
	In a jar, 2/5 of the marbles are red, 1/4 are green, and 1/5 are blue.	
Number of red marbles		Number of green and blue marbles

➢ Very Hard

13. A survey of n people in the town of Eros found that 50% of them prefer Brand A. Another survey of 100 people in the town of Angie found that 60% prefer Brand A. In total, 55% of all the people surveyed together prefer Brand A. What is the total number of people surveyed?

    (A) 50
    (B) 100
    (C) 150
    (D) 200
    (E) 250

Answers and Solutions to Problem Set Y

➢ Easy

1. Let j be the total number of marbles in the jar. Then $60\%j$ must be red (given), and the remaining $40\%j$ must be green (given). Now,

Column A equals 40% of the red marbles = $40\%(60\%j) = .40(.60j) = .24j$.

Column B equals 60% of the green marbles = $60\%(40\%j) = .60(.40j) = .24j$.

Since both columns equal $.24j$, the answer is (C).

➢ Medium

2. Let the number of pigeons be p and the number of rabbits be r. Since the head count together is 12,

$$p + r = 12 \qquad (1)$$

Since each pigeon has 2 legs and each rabbit has 4 legs, the total leg count is

$$2p + 4r = 32 \qquad (2)$$

Dividing equation (2) by 2 yields $p + 2r = 16$. Subtracting this equation from equation (1) yields

$$(p + r) - (p + 2r) = 12 - 16$$
$$p + r - p - 2r = -4$$
$$r = 4$$

Substituting this into equation (1) yields $p + 4 = 12$, which reduces to $p = 8$.

Hence, the number of pigeons is $p = 8$, and the number of rabbits is $r = 4$. The answer is (D).

3. Any number divisible by both 7 and 10 is a common multiple of 7 and 10. The least common multiple of 7 and 10 is 70. Hence, Column A reduces to the fraction of numbers from 0 through 1000 that are divisible by 70.

Similarly, any number divisible by both 5 and 14 is a common multiple of 5 and 14. The LCM of 5 and 14 is 70. Hence, Column B reduces to the fraction of numbers from 0 through 1000 that are divisible by 70.

Since the statement in each column is now the same, the columns are equal and the answer is (C).

➢ Hard

4. The sum of $13/n$, $18/n$, and $29/n$ is $\frac{13+18+29}{n} = \frac{60}{n}$. Now, if $60/n$ is to be an integer, n must be a factor of 60. Since the factors of 60 are 1, 2, 3, 4, 5, 6, 10, 12, 15, 20, 30, and 60, there are 12 possible values for n. The answer is (E).

5. The sum of $3n$, $9n$, and $11n$ is $23n$. Since this is to be greater than 200, we get the inequality $23n > 200$. From this, we get $n > 200/23 \approx 8.7$. Since n is an integer, $n > 8$. Now, we are given that $5 \le n \le 20$. Hence, the values for n are 9 through 20, a total of 12 numbers. The answer is (C).

6. We are given that that 59% of the employees E are workers. Since the factory consists of only workers, executives, and clerks, the remaining $100 - 59 = 41\%$ of the employees must include only executives and clerks. Since we are given that the number of executives is 460 and the number of clerks is 360, which sum to $460 + 360 = 820$, we have the equation $(41/100)E = 820$, or $E = 100/41 \times 820 = 2000$. The answer is (B).

7. Let A be the set of families having exactly one car. Then the question is how many families are there in set A.

Next, let B be the set of families having exactly two cars, and let C be the set of families having more than two cars.

Then the set of families having at least one car is the collection of the three sets A, B, and C.

The number of families in the three sets A, B, and C together is 250 (given) and the number of families in the two sets B and C together is 60 (given).

Now, since set A is the difference between a set containing the three families of A, B, and C and a set of families of B and C only, the number of families in set A equals

(the number of families in sets A, B, and C together) – (the number of families in sets B and C) =

$$250 - 60 =$$

$$190$$

The answer is (B).

8. Let the number of female children Emma has be n. Since Anna herself is one of them, she has $n - 1$ sisters. Hence, as given, she must have the same number ($= n - 1$) of brothers. Hence, the number of male children Emma has is $n - 1$. Since Andrew is one of them, Andrew has $(n - 1) - 1 = n - 2$ brothers. Now, the number of sisters Andrew has (includes Anna) is n (= the number of female children). Since Andrew has twice as many sisters as brothers, we have the equation $n = 2(n - 2)$. Solving the equation for n yields $n = 4$. Hence, Emma has 4 female children, and the number of male children she has is $n - 1 = 4 - 1 = 3$. Hence, the total number of children Emma has is $4 + 3 = 7$. The answer is (D).

9. The trainer wants to make teams of either 3 participants each or 5 participants each or 6 participants each successfully without leaving out any one of the participants in the batch. Hence, the batch size must be a multiple of all three numbers 3, 5, and 6. Hence, the batch size must be a multiple of the least common multiple of 3, 5, and 6, which is 30. The answer is (E).

10. There are three kinds of voters:

1) Voters who voted for A only. Let the count of such voters be a.
2) Voters who voted for B only. Let the count of such voters be b.
3) Voters who voted for both A and B. The count of such voters is 50 (given).

Since the total number of voters is 250, we have

$a + b + 50 = 250$

$a + b = 200$ (1) By subtracting 50 from both sides

Now, we have that 100 voters voted for A. Hence, we have

(Voters who voted for A only) + (Voters who voted for both A and B) = 100

Forming this as an equation yields

$$a + 50 = 100$$
$$a = 50$$

Substituting this in equation (1) yields $50 + b = 200$. Solving for b yields $b = 150$.

The answer is (C).

11. Let m be the number of male participants and f be the number of female participants in the meeting. The total number of participants is given as 750. Hence, we have

$$m + f = 750$$

Now, we have that half the female participants and one-quarter of the male participants are Democrats. Let d equal the number of the Democrats. Then we have the equation

$$f/2 + m/4 = d$$

Now, we have that one-third of the total participants are Democrats. Hence, we have the equation

$$d = 750/3 = 250$$

Solving the three equations yields the solution $f = 250$, $m = 500$, and $d = 250$. The number of female democratic participants equals half the female participants equals $250/2 = 125$. The answer is (C).

12. As given, 2/5 of the marbles in the jar are red, 1/4 are green, and 1/5 are blue. Hence, the green and blue marbles comprise 9/20 (= 1/4 + 1/5) of the marbles in the jar. Since 9/20 > 2/5, the combined number of green and blue marbles is greater than number of red marbles. The answer is (B).

➢ **Very Hard**

13. 50% of n people from Eros prefer brand A. 50% of n is $50/100 \times n = n/2$.

60% of 100 people from Angie prefer brand A. 60% of 100 is $60/100 \times 100 = 60$.

Of the total $n + 100$ people surveyed, $n/2 + 60$ prefer brand A. Given that this is 55%, we have

$$\frac{\frac{n}{2} + 60}{n + 100} \times 100 = 55$$

Solving the equation yields

$$\frac{\frac{n}{2} + 60}{n + 100} \times 100 = 55$$

$$\frac{n}{2} + 60 = \frac{55}{100}(n + 100)$$

$$\frac{n}{2} + 60 = \frac{11}{20}n + 55$$

$$0 = 11n/20 - n/2 + 55 - 60 \qquad \text{subtracting } n/2 \text{ and 60 from both sides}$$

$$0 = n/20 - 5$$

$$5 = n/20 \qquad \text{adding 5 to both sides}$$

$$n = 20 \times 5 = 100 \qquad \text{multiplying both sides by 20}$$

Hence, the total number of people surveyed is $n + 100 = 100 + 100 = 200$. The answer is (D).

Probability & Statistics

PROBABILITY

We know what probability means, but what is its formal definition? Let's use our intuition to define it. If there is no chance that an event will occur, then its probability of occurring should be 0. On the other extreme, if an event is certain to occur, then its probability of occurring should be 100%, or 1. Hence, our *probability* should be a number between 0 and 1, inclusive. But what kind of number? Suppose your favorite actor has a 1 in 3 chance of winning the Oscar for best actor. This can be measured by forming the fraction 1/3. Hence, a *probability* is a fraction where the top is the number of ways an event can occur and the bottom is the total number of possible events:

$$P = \frac{Number\ of\ ways\ an\ event\ can\ occur}{Number\ of\ total\ possible\ events}$$

Example: ***Flipping a coin***

What's the probability of getting heads when flipping a coin?

There is only one way to get heads in a coin toss. Hence, the top of the probability fraction is 1. There are two possible results: heads or tails. Forming the probability fraction gives 1/2.

Example: ***Tossing a die***

What's the probability of getting a 3 when tossing a die?

A die (a cube) has six faces, numbered 1 through 6. There is only one way to get a 3. Hence, the top of the fraction is 1. There are 6 possible results: 1, 2, 3, 4, 5, and 6. Forming the probability fraction gives 1/6.

Example: ***Drawing a card from a deck***

What's the probability of getting a king when drawing a card from a deck of cards?

A deck of cards has four kings, so there are 4 ways to get a king. Hence, the top of the fraction is 4. There are 52 total cards in a deck. Forming the probability fraction gives 4/52, which reduces to 1/13. Hence, there is 1 chance in 13 of getting a king.

Example: ***Drawing marbles from a bowl***

What's the probability of drawing a blue marble from a bowl containing 4 red marbles, 5 blue marbles, and 5 green marbles?

There are five ways of drawing a blue marble. Hence, the top of the fraction is 5. There are 14 (= 4 + 5 + 5) possible results. Forming the probability fraction gives 5/14.

Example: ***Drawing marbles from a bowl (second drawing)***

What's the probability of drawing a red marble from the same bowl, given that the first marble drawn was blue and was not placed back in the bowl?

There are four ways of drawing a red marble. Hence, the top of the fraction is 4. Since the blue marble from the first drawing was not replaced, there are only 4 blue marbles remaining. Hence, there are 13 (= 4 + 4 + 5) possible results. Forming the probability fraction gives 4/13.

Consecutive Probabilities

What's the probability of getting heads twice in a row when flipping a coin twice? Previously we calculated the probability for the first flip to be 1/2. Since the second flip is not affected by the first (these are called *mutually exclusive* events), its probability is also 1/2. Forming the product yields the probability of two heads in a row: $\frac{1}{2} \times \frac{1}{2} = \frac{1}{4}$.

What's the probability of drawing a blue marble and then a red marble from a bowl containing 4 red marbles, 5 blue marbles, and 5 green marbles? (Assume that the marbles are not replaced after being selected.) As calculated before, there is a 5/14 likelihood of selecting a blue marble first and a 4/13 likelihood of selecting a red marble second. Forming the product yields the probability of a red marble immediately followed by a blue marble: $\frac{5}{14} \times \frac{4}{13} = \frac{20}{182} = \frac{10}{91}$.

These two examples can be generalized into the following rule for calculating consecutive probabilities:

To calculate consecutive probabilities, multiply the individual probabilities.

This rule applies to two, three, or any number of consecutive probabilities.

Either-Or Probabilities

What's the probability of getting either heads or tails when flipping a coin once? Since the only possible outcomes are heads or tails, we expect the probability to be 100%, or 1: $\frac{1}{2} + \frac{1}{2} = 1$. Note that the events heads and tails are mutually exclusive. That is, if heads occurs, then tails cannot (and vice versa).

What's the probability of drawing a red marble or a green marble from a bowl containing 4 red marbles, 5 blue marbles, and 5 green marbles? There are 4 red marbles out of 14 total marbles. So the probability of selecting a red marble is 4/14 = 2/7. Similarly, the probability of selecting a green marble is 5/14. So the probability of selecting a red or green marble is $\frac{2}{7} + \frac{5}{14} = \frac{9}{14}$. Note again that the events are mutually exclusive. For instance, if a red marble is selected, then neither a blue marble nor a green marble is selected.

These two examples can be generalized into the following rule for calculating *either-or* probabilities:

To calculate *either-or* probabilities, add the individual probabilities (only if the events are mutually exclusive).

The probabilities in the two immediately preceding examples can be calculated more naturally by adding up the events that occur and then dividing by the total number of possible events. For the coin example, we get 2 events (heads or tails) divided by the total number of possible events, 2 (heads and tails): 2/2 = 1. For the marble example, we get 9 (= 4 + 5) ways the event can occur divided by 14 (= 4 + 5 + 5) possible events: 9/14.

If it's more natural to calculate the *either-or* probabilities above by adding up the events that occur and then dividing by the total number of possible events, why did we introduce a second way of calculating the probabilities? Because in some cases, you may have to add the individual probabilities. For example, you may be given the individual probabilities of two mutually exclusive events and be asked for the probability that either could occur. You now know to merely add their individual probabilities.

Geometric Probability

In this type of problem, you will be given two figures, with one inside the other. You'll then be asked what is the probability that a randomly selected point will be in the smaller figure. These problems are solved with the same principle we have been using: $Probability = \frac{desired\ outcome}{possible\ outcomes}$.

Example: In the figure to the right, the smaller square has sides of length 2 and the larger square has sides of length 4. If a point is chosen at random from the large square, what is the probability that it will be from the small square?

Applying the probability principle, we get $Probability = \frac{\text{area of the small square}}{\text{area of the large square}} = \frac{2^2}{4^2} = \frac{4}{16} = \frac{1}{4}$.

STATISTICS

Statistics is the study of the patterns and relationships of numbers and data. There are four main concepts that may appear on the test:

Median

When a set of numbers is arranged in order of size, the *median* is the middle number. For example, the median of the set {8, 9, 10, 11, 12} is 10 because it is the middle number. In this case, the median is also the mean (average). But this is usually not the case. For example, the median of the set {8, 9, 10, 11, 17} is 10 because it is the middle number, but the mean is $11 = \frac{8+9+10+11+17}{5}$. If a set contains an even number of elements, then the median is the average of the two middle elements. For example, the median of the set {1, 5, 8, 20} is $6.5\left(=\frac{5+8}{2}\right)$.

Example: What is the median of 0, –2, 256 , 18, $\sqrt{2}$?

Arranging the numbers from smallest to largest (we could also arrange the numbers from the largest to smallest; the answer would be the same), we get –2, 0, $\sqrt{2}$, 18, 256. The median is the middle number, $\sqrt{2}$.

Mode

The *mode* is the number or numbers that appear most frequently in a set. Note that this definition allows a set of numbers to have more than one mode.

Example: What is the mode of 3, –4, 3 , 7, 9, 7.5 ?

The number 3 is the mode because it is the only number that is listed more than once.

Example: What is the mode of 2, π, 2 , –9, π, 5 ?

Both 2 and π are modes because each occurs twice, which is the greatest number of occurrences for any number in the list.

Range

The *range* is the distance between the smallest and largest numbers in a set. To calculate the range, merely subtract the smallest number from the largest number.

Example: What is the range of 2, 8, 1 , –6, π, 1/2 ?

The largest number in this set is 8, and the smallest number is –6. Hence, the range is 8 – (–6) = 8 + 6 = 14.

Standard Deviation

On the test, you are not expected to know the definition of standard deviation. However, you may be presented with the definition of standard deviation and then be asked a question based on the definition. To make sure we cover all possible bases, we'll briefly discuss this concept.

Standard deviation measures how far the numbers in a set vary from the set's mean. If the numbers are scattered far from the set's mean, then the standard deviation is large. If the numbers are bunched up near the set's mean, then the standard deviation is small.

Example: Which of the following sets has the larger standard deviation?

A = {1, 2, 3, 4, 5}
B = {1, 4, 15, 21, 27}

All the numbers in Set A are within 2 units of the mean, 3. All the numbers in Set B are greater than 5 units from the mean, 15. Hence, the standard deviation of Set B is greater.

Problem Set Z:

➢ Easy

1. The minimum temperatures from Monday through Sunday in the first week of July in southern Iceland are observed to be –2°C, 4°C, 4°C, 5°C, 7°C, 9°C, 10°C. What is the range of the temperatures?

    (A) –10°C
    (B) –8°C
    (C) 8°C
    (D) 10°C
    (E) 12°C

➢ Medium

2. What is the probability that the product of two integers (not necessarily different integers) randomly selected from the numbers 1 through 20, inclusive, is odd?

    (A) 0
    (B) 1/4
    (C) 1/2
    (D) 2/3
    (E) 3/4

3. Two data sets S and R are defined as follows:

    Data set *S*: 28, 30, 25, 28, 27
    Data set *R*: 22, 19, 15, 17, 21, 25

    By how much is the median of data set *S* greater than the median of data set *R*?

    (A) 5
    (B) 6
    (C) 7
    (D) 8
    (E) 9

4. If x and y are two positive integers and $x + y = 5$, then what is the probability that x equals 1?

    (A) 1/2
    (B) 1/3
    (C) 1/4
    (D) 1/5
    (E) 1/6

5. The following values represent the number of cars owned by the 20 families on Pearl Street.

    1, 1, 2, 3, 2, 5, 4, 3, 2, 4, 5, 2, 6, 2, 1, 2, 4, 2, 1, 1

    What is the probability that a family randomly selected from Pearl Street has at least 3 cars?

    (A) 1/6
    (B) 2/5
    (C) 9/20
    (D) 13/20
    (E) 4/5

6. The following frequency distribution shows the number of cars owned by the 20 families on Pearl Street.

x	The number of families having x number of cars
1	2
2	2
3	a
4	4
5	5
6	2

What is the probability that a family randomly selected from the street has at least 4 cars?

(A) 1/10
(B) 1/5
(C) 3/10
(D) 9/20
(E) 11/20

7. Thirty airmail and 40 ordinary envelopes are the only envelopes in a bag. Thirty-five envelopes in the bag are unstamped, and 5/7 of the unstamped envelopes are airmail letters. What is the probability that an envelope picked randomly from the bag is an unstamped airmail envelope?

(A) 1/7
(B) 1/3
(C) 13/35
(D) 17/38
(E) 23/70

8. Set S is the set of all numbers from 1 through 100, inclusive. What is the probability that a number randomly selected from the set is divisible by 3?

(A) 1/9
(B) 33/100
(C) 34/100
(D) 1/3
(E) 66/100

9. What is the probability that the sum of two different numbers randomly picked (without replacement) from the set $S = \{1, 2, 3, 4\}$ is 5?

(A) 1/5
(B) 3/16
(C) 1/4
(D) 1/3
(E) 1/2

10. The ratio of the number of red balls, to yellow balls, to green balls in a urn is 2 : 3 : 4. What is the probability that a ball chosen at random from the urn is a red ball?

(A) 2/9
(B) 3/9
(C) 4/9
(D) 5/9
(E) 7/9

11. The frequency distribution for x is as given below. What is the range of f?

x	f
0	1
1	5
2	4
3	4

(A) 0
(B) 1
(C) 3
(D) 4
(E) 5

12. The table shows the distribution of a team of 16 engineers by gender and level.

	Junior Engineers	Senior Engineers	Lead Engineers
Male	3	4	2
Female	2	4	1

If one engineer is selected from the team, what is the probability that the engineer is a male senior engineer?

(A) 7/32
(B) 1/4
(C) 7/16
(D) 1/2
(E) 3/4

13. A prize of $200 is given to anyone who solves a hacker puzzle independently. The probability that Tom will win the prize is 0.6, and the probability that John will win the prize is 0.7. What is the probability that both will win the prize?

(A) 0.35
(B) 0.36
(C) 0.42
(D) 0.58
(E) 0.88

14. If the probability that Mike will miss at least one of the ten jobs assigned to him is 0.55, then what is the probability that he will do all ten jobs?

(A) 0.1
(B) 0.45
(C) 0.55
(D) 0.85
(E) 1

15. The probability that Tom will win the Booker prize is 0.5, and the probability that John will win the Booker prize is 0.4. There is only one Booker prize to win. What is the probability that at least one of them wins the prize?

(A) 0.2
(B) 0.4
(C) 0.7
(D) 0.8
(E) 0.9

16. The following values represent the exact number of cars owned by the 20 families on Pearl Street.

    1, 1, 2, 3, 2, 5, 4, 3, 2, 4, 5, 2, 6, 2, 1, 2, 4, 2, 1, 1

    This can be expressed in frequency distribution format as follows:

x	The number of families having x number of cars
1	5
2	7
3	a
4	3
5	b
6	1

What are the values of a and b, respectively?

(A) 1 and 1
(B) 1 and 2
(C) 2 and 1
(D) 2 and 2
(E) 2 and 3

17. In a box of 5 eggs, 2 are rotten.

Column A	Column B
The probability that one egg chosen at random from the box is rotten	The probability that two eggs chosen at random from the box are rotten

➢ **Hard**

18. A meeting is attended by 750 professionals. 450 of the attendees are females. Half the female attendees are less than thirty years old, and one-fourth of the male attendees are less than thirty years old. If one of the attendees of the meeting is selected at random to receive a prize, what is the probability that the person selected is less than thirty years old?

(A) 1/8
(B) 1/2
(C) 3/8
(D) 2/5
(E) 3/4

19. Each Employee at a certain bank is either a clerk or an agent or both. Of every three agents, one is also a clerk. Of every two clerks, one is also an agent. What is the probability that an employee randomly selected from the bank is both an agent and a clerk?

(A) 1/2
(B) 1/3
(C) 1/4
(D) 1/5
(E) 2/5

20. Every one who passes the test will be awarded a degree. The probability that Tom passes the test is 0.3, and the probability that John passes the test is 0.4. The two events are independent of each other. What is the probability that at least one of them gets the degree?

(A) 0.28
(B) 0.32
(C) 0.5
(D) 0.58
(E) 0.82

21. A national math examination has 4 statistics problems. The distribution of the number of students who answered the questions correctly is shown in the chart. If 400 students took the exam and each question was worth 25 points, then what is the average score of the students taking the exam?

Question Number	Number of students who solved the question
1	200
2	304
3	350
4	250

(A) 1 point
(B) 25 points
(C) 26 points
(D) 69 points
(E) 263.5 points

➤ **Very Hard**

22. There are 58 balls in a jar. Each ball is painted with at least one of two colors, red or green. It is observed that 2/7 of the balls that have red color also have green color, while 3/7 of the balls that have green color also have red color. What is the probability that a ball randomly picked from the jar will have both red and green colors?

(A) 6/14
(B) 2/7
(C) 6/35
(D) 6/29
(E) 6/42

Answers and Solutions to Problem Set Z

➢ Easy

1. The *range* is the greatest measurement minus the smallest measurement. The greatest of the seven temperature measurements is 10°C, and the smallest is –2°C. Hence, the required range is 10 – (–2) = 12°C. The answer is (E).

➢ Medium

2. The product of two integers is odd when both integers are themselves odd. Hence, the probability of the product being odd equals the probability of both numbers being odd. Since there is one odd number in every two numbers (there are 10 odd numbers in the 20 numbers 1 through 20, inclusive), the probability of a number being odd is 1/2. The probability of both numbers being odd (mutually exclusive case) is 1/2 × 1/2 = 1/4. The answer is (B).

3. The definition of *median* is "When a set of numbers is arranged in order of size, the *median* is the middle number. If a set contains an even number of elements, then the median is the average of the two middle elements."

Data set *S* (arranged in increasing order of size) is 25, 27, 28, 28, 30. The median of the set is the third number 28.

Data set *R* (arranged in increasing order of size) is 15, 17, 19, 21, 22, 25. The median is the average of the two middle numbers (the 3rd and 4th numbers): (19 + 21)/2 = 40/2 = 20.

The difference of 28 and 20 is 8. The answer is (D).

4. The possible positive integer solutions x and y of the equation $x + y = 5$ are $\{x, y\} = \{1, 4\}, \{2, 3\}, \{3, 2\}$, and $\{4, 1\}$. Each solution is equally probable. Exactly one of the 4 possible solutions has x equal to 1. Hence, the probability that x equals 1 is one in four ways, which equals 1/4. The answer is (C).

5. From the distribution given, the 4th, 6th, 7th, 8th, 10th, 11th, 13th, and 17th families, a total of 8, have at least 3 cars. Hence, the probability of selecting a family having at least 3 cars out of the available 20 families is 8/20, which reduces to 2/5. The answer is (B).

6. From the distribution given, there are

4 families having exactly 4 cars
5 families having exactly 5 cars
2 families having exactly 6 cars

Hence, there are 4 + 5 + 2 = 11 families with at least 4 cars. Hence, the probability of picking one such family from the 20 families is 11/20. The answer is (E).

7. We have that 30 airmail and 40 ordinary envelopes are the only envelopes in the bag. Hence, the total number of envelopes is 30 + 40 = 70. We also have that 35 envelopes in the bag are unstamped, and 5/7 of these envelopes are airmail letters. Now, 5/7 × 35 = 25. So the remaining 35 – 25 = 10 are ordinary unstamped envelopes. Hence, the probability of picking such an envelope from the bag is

(Number of unstamped ordinary envelopes) / (Total number of envelopes) =

10/70 =

1/7

The answer is (A).

8. The count of the numbers 1 through 100, inclusive, is 100.

Now, let $3n$ represent a number divisible by 3, where n is an integer.

Since we have the numbers from 1 through 100, we have $1 \le 3n \le 100$. Dividing the inequality by 3 yields $1/3 \le n \le 100/3$. The possible values of n are the integer values between 1/3 (≈ 0.33) and 100/3 (≈ 33.33). The possible numbers are 1 through 33, inclusive. The count of these numbers is 33.

Hence, the probability of randomly selecting a number divisible by 3 is 33/100. The answer is (B).

9. The first selection can be done in 4 ways (by selecting any one of the numbers 1, 2, 3, and 4 of the set S). Hence, there are 3 elements remaining in the set. The second number can be selected in 3 ways (by selecting any one of the remaining 3 numbers in the set S). Hence, the total number of ways the selection can be made is $4 \times 3 = 12$.

The selections that result in the sum 5 are 1 and 4, 4 and 1, 2 and 3, 3 and 2, a total of 4 selections. So, 4 of the 12 possible selections have a sum of 5. Hence, the probability is the fraction $4/12 = 1/3$. The answer is (D).

10. Let the number of red balls in the urn be $2k$, the number of yellow balls $3k$, and the number of green balls $4k$, where k is a common factor of the three. Now, the total number of balls in the urn is $2k + 3k + 4k = 9k$. Hence, the fraction of red balls from all the balls is $2k/9k = 2/9$. This also equals the probability that a ball chosen at random from the urn is a red ball. The answer is (A).

11. The *range* of f is the greatest value of f minus the smallest value of f: $5 - 1 = 4$. The answer is (D).

12. From the distribution table, we know that the team has exactly 4 male senior engineers out of a total of 16 engineers. Hence, the probability of selecting a male senior engineer is $4/16 = 1/4$. The answer is (B).

13. Let $P(A)$ = The probability of Tom solving the problem = 0.6, and let $P(B)$ = The probability of John solving the problem = 0.7. Now, since events A and B are independent (Tom's performance is independent of John's performance and vice versa), we have

$$P(A \text{ and } B) =$$

$$P(A) \times P(B) =$$

$$0.6 \times 0.7 =$$

$$0.42$$

The answer is (C).

14. There are only two cases:

1) Mike will miss at least one of the ten jobs.
2) Mike will not miss any of the ten jobs.

Hence, (The probability that Mike will miss at least one of the ten jobs) + (The probability that he will not miss any job) = 1. Since the probability that Mike will miss at least one of the ten jobs is 0.55, this equation becomes

$$0.55 + (\text{The probability that he will not miss any job}) = 1$$

$$(\text{The probability that he will not miss any job}) = 1 - 0.55$$

$$(\text{The probability that he will not miss any job}) = 0.45$$

The answer is (B).

15. Probability of Tom winning the prize is 0.5. Hence, probability of Tom not winning is 1 – 0.5 = 0.5.

Probability of John winning is 0.4. Hence, probability of John not winning is 1 – 0.4 = 0.6.

So, the probability of both Tom and John not winning equals

Probability of Tom not winning × Probability of John not winning =

0.5 × 0.6 =

0.3

The probability of one of them (at least) winning + The probability of neither winning = 1 (because these are the only cases.)

Hence, The probability of one of them (at least) not winning = 1 – The probability of neither winning = 1 – 0.3 = 0.7.

The answer is (C).

16. In the frequency distribution table, the first column represents the number of cars and the second column represents the number of families having the particular number of cars. Now, from the data given, the number of families having exactly 3 cars is 2, and the number of families having exactly 5 cars is 2. Hence, $a = 2$ and $b = 2$. The answer is (D).

17. Since 2 of the 5 eggs are rotten, the chance of selecting a rotten egg the first time is 2/5. For the second selection, there is only one rotten egg, out of the 4 remaining eggs. Hence, there is a 1/4 chance of selecting a rotten egg again. Hence, the probability of selecting 2 rotten eggs in a row is 2/5 × 1/4 = 1/10. Since 2/5 > 1/10, Column A is greater than Column B. The answer is (A).

➢ Hard

18. The number of attendees at the meeting is 750 of which 450 are female. Hence, the number of male attendees is 750 – 450 = 300. Half of the female attendees are less than 30 years old. One half of 450 is 450/2 = 225. Also, one-fourth of the male attendees are less than 30 years old. One-fourth of 300 is 300/4 = 75.

Now, the total number of (male and female) attendees who are less than 30 years old is 225 + 75 = 300.

So, out of the total 750 attendees 300 attendees are less than 30 years old. Hence, the probability of randomly selecting an attendee less than 30 years old (equals the fraction of all the attendees who are less than 30 years old) is 300/750 = 2/5. The answer is (D).

19. The employees of the bank can be categorized into three groups:

1) Employees who are only Clerks. Let c be the count.
2) Employees who are only Agents. Let a be the count.
3) Employees who are both Clerks and Agents. Let x be the count.

Hence, the total number of employees is $c + a + x$.
The total number of clerks is $c + x$.
The total number of agents is $a + x$.

We are given that of every three agents one is also a clerk. Hence, we have that one of every three agents is also a clerk (both agent and clerk). Forming the ratio yields $\frac{x}{a+x} = \frac{1}{3}$. Solving for a yields $a = 2x$.

We are given that of every two clerks, one is also an agent. Hence, we have that one of every two clerks is also an agent (both clerk and agent). Forming the ratio yields $\frac{x}{c+x} = \frac{1}{2}$. Solving for c yields $c = x$.

Now, the probability of selecting an employee who is both an agent and a clerk from the bank is $\dfrac{x}{c+a+x} = \dfrac{x}{x+2x+x} = \dfrac{x}{4x} = \dfrac{1}{4}$.

The answer is (C).

20. The probability that Tom passes is 0.3. Hence, the probability that Tom does not pass is 1 – 0.3 = 0.7.

The probability that John passes is 0.4. Hence, the probability that John does not pass is 1 – 0.4 = 0.6.

At least one of them gets a degree in three cases:

1) Tom passes and John does not
2) John passes and Tom does not
3) Both Tom and John pass

Hence, the probability of at least one of them passing equals

(The probability of Tom passing and John not) +
(The probability of John passing and Tom not) +
(The probability of both passing)

(The probability of Tom passing and John not) =
(The probability of Tom passing) × (The probability of John not) =
0.3 × 0.6 =
0.18

(The probability of John passing and Tom not) =
(The probability of John passing) × (The probability of Tom not) =
0.4 × 0.7 =
0.28

(The probability of both passing) =
(The probability of Tom passing) × (The probability of John passing) =
0.3 × 0.4 =
0.12

Hence, the probability of at least one passing is 0.18 + 0.28 + 0.12 = 0.58. The answer is (D).

Method II:
The probability of Tom passing is 0.3. Hence, the probability of Tom not passing is 1 – 0.3 = 0.7.

The probability of John passing is 0.4. Hence, the probability of John not passing is 1 – 0.4 = 0.6.

At least one of Tom and John passes in all the cases except when both do not pass.

Hence,

The probability of at least one passing =

1 – (the probability of neither passing) =

1 – (The probability of Tom not passing) × (The probability of John not passing) =

1 – 0.7 × 0.6 =

1 – 0.42 =

0.58

The answer is (D).

21. The average score of the students is equal to the net score of all the students divided by the number of students. The number of students is 400 (given). Now, let's calculate the net score. Each question carries 25 points, the first question is solved by 200 students, the second one by 304 students, the third one by 350 students, and the fourth one by 200 students. Hence, the net score of all the students is

$$200 \times 25 + 304 \times 25 + 350 \times 25 + 250 \times 25 =$$

$$25(200 + 304 + 350 + 250) =$$

$$25(1104)$$

Hence, the average score equals

$$25(1104)/400 =$$

$$1104/16 =$$

$$69$$

The answer is (D).

➤ **Very Hard**

22. Let T be the total number of balls, R the number of balls having red color, G the number having green color, and B the number having both colors.

Hence, the number of balls having only red is $R - B$, the number having only green is $G - B$, and the number having both is B. Now, the total number of balls is $T = (R - B) + (G - B) + B = R + G - B$.

We are given that 2/7 of the balls having red color have green also. This implies that $B = 2R/7$. Also, we are given that 3/7 of the green balls have red color. This implies that $B = 3G/7$. Solving for R and G in these two equations yields $R = 7B/2$ and $G = 7B/3$. Substituting this into the equation $T = R + G - B$ yields $T = 7B/2 + 7B/3 - B$. Solving for B yields $B = 6T/29$. Hence, the probability of selecting such a ball is the fraction $(6T/29)/T = 6/29$. The answer is (D).

Permutations & Combinations

Suppose you must seat 3 of 5 delegates in 3 chairs. And suppose you are interested in the order in which they sit. You will first select 3 of the 5 delegates, and then choose the order in which they sit. The first act is a combination, the second is a permutation. Effectively, the permutation comes after the combination. The delegates in each combination can be ordered in different ways, which can be called permutations of the combination.

Now, if you can select 3 of the 5 delegates in m ways and each selection can be ordered in n ways, then the total number of possible arrangements (permutations) is $m \cdot n$.

Now, let's count the number of permutations of 3 objects taken from a set of 4 objects {A, B, C, D}. Let's call the set {A, B, C, D} a base set.

We must first choose 3 objects from the base set, which yields the following selections:

{A, B, C}, {B, C, D}, {A, C, D}, {A, B, D}

These are combinations. We have 4 selections (combinations) here.

If {E1, E2, E3} represents one of the four combinations above, then the following are its possible permutations:

E1	E2	E3
E1	E3	E2
E2	E1	E3
E2	E3	E1
E3	E1	E2
E3	E2	E1

You can use this scheme to find the permutations of each of the 4 selections (combinations) we formed above. For example, for the selection {A, B, C}, the following are the six permutations:

A – B – C
A – C – B
B – A – C
B – C – A
C A B
C – B – A

Thus, we have 6 permutations for each selection. For practice, you may wish to list the permutations for the remaining 3 selections: {B, C, D}, {A, C, D}, and {A, B, D}.

Summary:
Here, {A, B, C, D} is the base set. We formed 4 combinations that use 3 elements each. Then we formed 6 permutations for each of the 4 combinations. Hence, the problem has in total $6 + 6 + 6 + 6 = 4 \times 6 = 24$ permutations.

Note 1: A combination might have multiple permutations. The reverse is never true.

Note 2: A permutation is an ordered combination.

Note 3: With combinations, AB = BA. With permutations, AB ≠ BA.

Combinations and their Permutations

Here is another discussion of the distinction between permutations and combinations. The concept is repeated here because it forms the basis for the rest of the chapter.

Combinations are the selections (subsets) of a base set.

For example, the possible combinations of two elements each of the set {A, B} are

A, B or B, A
(Both are the same combination)

The permutations (the combination ordered in different ways) of the combination are

A – B and B – A
(The permutations are different)

How to distinguish between a Combination and a Permutation

At the risk of redundancy, here is yet another discussion of the distinction between permutations and combinations.

As combinations, {A, B, C} and {B, A, C} are the same because each has the same number of each type of object: A, B, and C as in the base set.

But, as permutations, A – B – C and B – A – C are not the same because the ordering is different, though each has the same number of each type of object: A, B, and C as in the base set In fact, no two arrangements that are not identical are ever the same permutation.

Hence, with combinations, look for selections, while with permutations, look for arrangements.

The following definitions will help you distinguish between Combinations and Permutations

Permutations are *arrangements* (order is important) of objects formed from an original set (base set) such that each new arrangement has an order different from the original set. So, the positions of objects is important.

Combinations are sets of objects formed by *selecting* (order not important) objects from an original set (base set).

To help you remember, think "**P**ermutation … **P**osition."

Combinations with Repetitions: Permutations with Repetitions

Here, repetition of objects is allowed in selections or the arrangements.

Suppose you have the base set {A, B, C}. Allowing repetitions, the objects can repeat in the combinations (selections).

Hence, the allowed *selections* of 2 elements are {A, A}, {A, B} or {B, A}, {B, B}, {B, C} or {C, B}, {C, C}, {C, A} or {A, C} in total 6.

The corresponding *permutations* are

A – A for {A, A}
A – B and B – A for {A, B}
B – B for {B, B}
B – C and C – B for {B, C}
C – C for {C, C}
C – A and A – C for {C, A}

The total number of combinations is 6, and the total number of permutations is 9. We have 3 objects to choose for 2 positions; allowing repetitions, the calculation is $3^2 = 9$.

Note that {A, B} and {B, A} are the same combination because each has an equal number of A's and B's.

By allowing repetitions, you can chose the same object more than once and therefore can have the same object occupying different positions.

In general, *permutation* means "permutation without repetition," unless stated otherwise.

Indistinguishable Objects

Suppose we replace C in the base set {A, B, C} with A. Then, we have {A, B, A}. Now the A's in the first and third positions of the set are indistinguishable and make some of the combinations and permutations formed earlier involving C redundant (because some identical combinations and permutations will be formed). Hence, replacing distinguishable objects with indistinguishable ones reduces the number of combinations and permutations.

Combinations (repetition not allowed) with Indistinguishable Objects

Consider the set {A, B, A}. Here, for example, ABA (2 A's and 1 B as in the base set) is an allowed combination but ABB (containing 2 B's not as in base set) is not because B occurs only once in the base set.

The corresponding permutations are listed in Table IV.

Permutations (repetition not allowed) with Indistinguishable Objects

The corresponding permutations are listed in Table IV.

Observe that {A, B, C} has permutations ABC, ACB, BAC, BCA, CAB, and CBA (6 permutations); and {A, B, A} has permutations ABA, AAB, BAA, ~~BAA~~, ~~AAB~~, and ~~ABA~~ (we crossed out the last three permutations because they are identical to the first three). So, there are 3 permutations.

Combinations (repetition allowed) with Indistinguishable Objects

Again, consider the set {A, B, A}. Here, for example, ABA is an allowed combination and ABB is an allowed combination.

All the allowed permutations are listed in Table III.

Permutations (repetition allowed) with Indistinguishable Objects

The corresponding permutations are listed in Table III.

Summary:

- ➢ Repetition problems have the objects repeating in the combinations or permutations that are formed from a base set.
- ➢ Problems with indistinguishable objects, instead, have the objects repeating in the base set itself.
- ➢ Allowing repetition increases the number of selections (combinations) and therefore the number of permutations.
- ➢ Using indistinguishable objects in the base set reduces the number of selections (combinations) and the number of permutations.

Table I
The base set is {A, B, C}
Permutations with Repetitions allowed. [$n = 3$, $r = 3$]

First Position (3 ways allowed: A, B, C)	Second Position (3 ways allowed: A, B, C)	Third Position (3 ways allowed: A, B, C)	Word Formed	Count
A	A	A	AAA	1
		B	AAB	2
		C	AAC	3
	B	A	ABA	4
		B	ABB	5
		C	ABC	6
	C	A	ACA	7
		B	ACB	8
		C	ACC	9
B	A	A	BAA	10
		B	BAB	11
		C	BAC	12
	B	A	BBA	13
		B	BBB	14
		C	BBC	15
	C	A	BCA	16
		B	BCB	17
		C	BCC	18
C	A	A	CAA	19
		B	CAB	20
		C	CAC	21
	B	A	CBA	22
		B	CBB	23
		C	CBC	24
	C	A	CCA	25
		B	CCB	26
		C	CCC	27

Total number of ways: 27

Table II
The base set is {A, B, C}
The Permutations (not allowing Repetitions) are as follows [$n = 3$, $r = 3$]. Shaded entries are redundant and therefore not counted. (That is, we pick only the entries in which no object is repeated.) Shaded entries the ones having the same object repeating and therefore not counted.

First Position (3 ways allowed: A, B, C)	Second Position (3 ways allowed: A, B, C)	Third Position (3 ways allowed: A, B, C)	Word Formed	Count
A	A	A	AAA	A repeat
		B	AAB	A repeat
		C	AAC	A repeat
	B	A	ABA	A repeat
		B	ABB	B repeat
		C	ABC	1
	C	A	ACA	A repeat
		B	ACB	2
		C	ACC	C repeat
B	A	A	BAA	A repeat
		B	BAB	B repeat
		C	BAC	3
	B	A	BBA	B repeat
		B	BBB	B repeat
		C	BBC	B repeat
	C	A	BCA	4
		B	BCB	B repeat
		C	BCC	C repeat
C	A	A	CAA	A repeat
		B	CAB	5
		C	CAC	C repeat
	B	A	CBA	6
		B	CBB	B repeat
		C	CBC	C repeat
	C	A	CCA	C repeat
		B	CCB	C repeat
		C	CCC	C repeat

Total number of ways: 6

There is only 1 combination (without repetition), because any of the 6 words (ABC or ACB or BAC or BCA or CAB or CBA) formed in the above table is the same combination (is a single combination).

Permutations (repetition allowed) using Indistinguishable Objects

By replacing C with A in the base set {A, B, C}, we get {A, B, A}. Reducing the repetitive (redundant) permutations yields

Table III

The Permutations (allowing Repetitions) are as follows [$n = 3$, $r = 3$], and two of the three objects are indistinguishable. The table is derived by replacing C with A in Table I and eliminating the repeating entries. Shaded entries are redundant and therefore not counted. (That is, we pick only one of the indistinguishable permutations.)

First Position (3 ways allowed: A, B, C)	Second Position (3 ways allowed: A, B, C)	Third Position (3 ways allowed: A, B, C)	Word Formed	Count
A	A	A	AAA	1
		B	AAB	2
		A	AAA	already counted
	B	A	ABA	3
		B	ABB	4
		A	ABA	already counted
	A	A	AAA	already counted
		B	AAB	already counted
		A	AAA	already counted
B	A	A	BAA	5
		B	BAB	6
		A	BAA	already counted
	B	A	BBA	7
		B	BBB	8
		A	BBA	already counted
	A	A	BAA	already counted
		B	BAB	already counted
		A	BAA	already counted
A	A	A	AAA	already counted
		B	AAB	already counted
		A	AAA	already counted
	B	A	ABA	already counted
		B	ABB	already counted
		A	ABA	already counted
	A	A	AAA	already counted
		B	AAB	already counted
		A	AAA	already counted

Total number of ways: 8

Permutations (repetition not allowed) with Indistinguishable Objects

Indistinguishable objects are items that repeat in the original set. For example, replace C in the above set with A. Then the new base set would be {A, B, A}. Hence, if we replace C with A in the Table II, we get the repetitions in the permutations. Reducing the repetitive permutations yields

Table IV
The Permutations (not allowing Repetitions) with Indistinguishable objects are as follows [$n = 3$, $r = 3$]. The table is derived by replacing C with A in Table II and eliminating the repeating entries. Shaded entries are redundant and therefore not counted. (That is, we pick only one of the indistinguishable permutations)

First Position (3 ways allowed: A, B, C)	Second Position (3 ways allowed: A, B, C)	Third Position (3 ways allowed: A, B, C)	Word Formed	Count
A	B	A	ABA	1
	A	B	AAB	2
B	A	A	BAA	3
	A	A	BAA	already counted
A	A	B	AAB	already counted
	A	A	ABA	already counted

Total number of ways: 3

So far, we have discussed the types of the problems. When trying to solve a problem, it is very helpful to identify its type. Once this is done, we need to count the number of possibilities.

Distinction between Indistinguishable Objects Problems and Repetition Problems

Suppose you are to arrange the letters of the word SUCCESS.

The base set is {S, U, C, C, E, S, S}. There are 3 S's, which are indistinguishable objects. Hence, the letter S, can be used a maximum of 3 times in forming a new word if repetition is not allowed. So, SSSSUCE is not a possible arrangement.

If repetition is allowed, you can use S as many times as you wish, regardless of the number of S's in the base word (for example, even if there is only 1 S, you can use it up to maximum allowed times). Hence, SSSSSSS is a possible arrangement.

Counting

There are three models of counting we can use.

We already discussed that if there are m combinations possible from a base set and if there are n permutations possible for each combination, then the total number of permutations possible is $m \cdot n$.

This is also clear from the *Fundamental Principle of Counting*.

Model 1:
The Fundamental Principle of Counting:

Construct a tree diagram (we used tables above) to keep track of all possibilities. Each decision made produces a new branch. Finally, count all the allowed possibilities.

The previous tables are examples of tree diagrams. They can represent possibilities as trees. The possibilities are also counted in the tables.

Model 2:
Divide a work into mutually independent jobs and multiply the number of ways of doing each job to find the total number of ways the work can be done. For example, if you are to position three letters in 3 slots, you can divide the work into the jobs as

1) Choose one of three letters A, B, and C for the first position
2) Choose one of the remaining 2 letters for the second position
3) Choose the only remaining letter for the third position

This can be done in $3 \times 2 \times 1 = 6$ ways. The model is a result of the Fundamental Principle of Counting.

Model 3:
Models 1 and 2 are fundamental. Model 3 uses at least one of the first two models. Here, we use the following formula:

Total Number of Permutations = Number of Combinations × Number of Permutations of Each Combination

Predominantly, we use the model for calculating combinations. The total number of permutations and the number of permutations for each combination can be calculated using either or both models 1 and 2 in many cases.

Cyclic Permutations

A *cyclic permutation* is a permutation that shifts all elements of a given ordered set by a certain amount, with the elements that are shifted off the end inserted back at the beginning in the same order, i.e., cyclically. In other words, a rotation.

For example, {A, B, C, D}, {B, C, D, A}, {C, D, A, B}, and {D, A, B, C} are different linear permutations but the same cyclic permutation. The permutations when arranged in cyclic order, starting from, say, A and moving clockwise, yield the same arrangement {A, B, C, D}. The following figure helps visualize this.

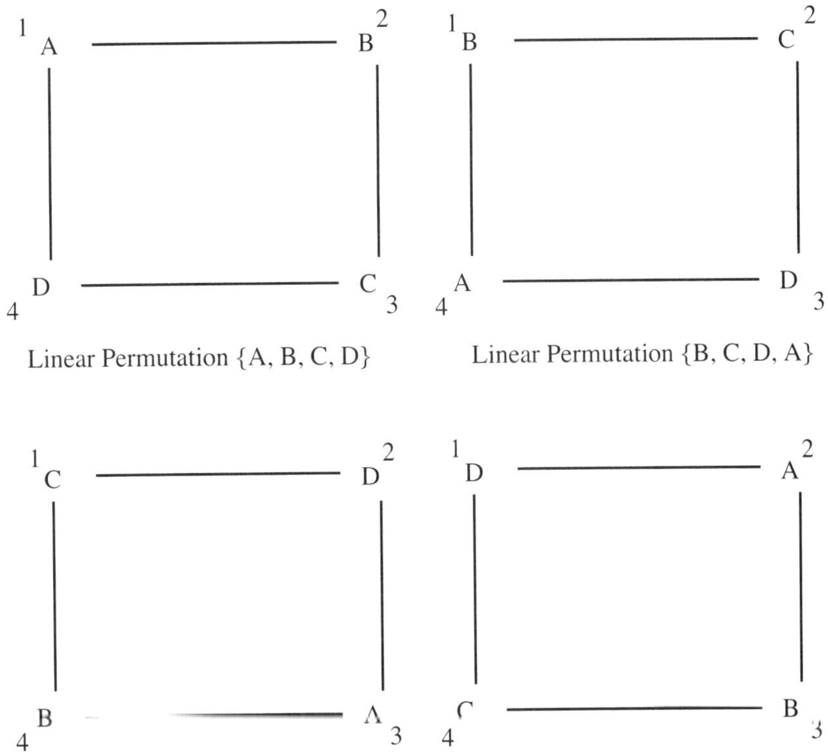

Linear Permutation {A, B, C, D}

Linear Permutation {B, C, D, A}

Linear Permutation {C, D, A, B}

Linear Permutation {D, A, B, C}

Cyclic arrangements of the cyclic permutations.

For the r placement positions (for the example in the figure, r equals 4), we get r permutations, each is an equivalent cyclic permutation. Hence, the number of cyclic permutations equals

(The number of ordinary permutations) $\div r$

Hence, for ${}_nP_r$ permutations, ${}_nP_r \div r$ cyclic permutations exist. Simply said, r linear permutations would be same cyclic permutation.

Also, {A, B, C, D} and {A, C, B, D} are different linear permutations and different cyclic permutations, because arranging them in cyclic order yields different sequences.

Factorial

The **factorial** of a non-negative integer n, denoted by n!, is the product of all positive integers less than or equal to n. That is, $n! = n(n-1)(n-2) \cdot \cdot \cdot 3 \cdot 2 \cdot 1$. For example, $4! = 4 \cdot 3 \cdot 2 \cdot 1 = 24$. Note: 0! is defined to be 1.

Formulas

Verify that the following formulas apply to the scenarios mentioned above. These formulas should be memorized.

Formula 1: If you have n items to choose from and you choose $\boldsymbol{r}$ of them, then the number of permutations *with repetitions allowed* is

$$\underbrace{n \cdot n \cdot \ldots n}_{(r \text{ times})} = n^r$$

Formula 2: The formula for permutations *with repetitions not allowed* is

$${}_nP_r = \frac{n!}{(n-r)!}$$

Formula 3: The formula for combinations *with repetitions not allowed* is

$${}_nC_r = \frac{n!}{r!(n-r)!}$$

Formula 4: We know that k *distinguishable* objects have $k!$ *different* arrangements (permutations). But a set of k *indistinguishable* objects, will have only 1 *indistinguishable* permutation. Hence, if we have P permutations for k *distinguishable* objects, we will have $\frac{P}{k!}$ permutations for k *indistinguishable* objects because we now treat the earlier $k!$ arrangements as one.

The case is similar when we have more than one set of indistinguishable objects. Suppose the word ABCDEF has ${}_nP_r$ permutations (not allowing repetitions); then the word AAABBC will have $\frac{{}_nP_r}{3! \cdot 2!}$ permutations because here we have a set of 3 indistinguishable objects A and a set of 2 indistinguishable objects B.

There are formulas for the other problem models, but they are not needed for the test. We can always use the Fundamental Principal of Counting for them.

Formula 5: For r linear positions (for the example in the figure, r equals 4), we get r permutations, each of which is an equivalent cyclic permutation. Hence, the number of cyclic permutations is

(The number of ordinary permutations) $\div r$

The formulas in this section will be referenced while we solve the problems.

Problem Solving Strategy

In permutation and combination problems, it is very important to recognize the type of problem. Many students mistakenly approach a combination problem as a permutation, and vice versa. The steps below will help you determine the problem type.

Solving a permutation or combination problem involves two steps:

1) Recognizing the problem type: permutation vs. combination.
2) Using formulas or models to count the possibilities.

We have three questions to ask ourselves in order to identify the problem type:

1) **Is it a permutation or combination?**
Check any two typical arrangements with the same combination. If the two arrangements are counted only once, it is a combination problem. Otherwise, it is a permutation.

For example, if you are asked for a lock code, then 321 and 123 could be two possibilities, and the two numbers are formed from the same combination (Same number of 1's, 2's, and 3's). So, lock codes must be permutations.

For another example, suppose you have 5 balls numbering 1 through 5. If you are asked to select 3 out of the 5 balls and you are only interested in the numbers on the balls, not the order in which they are taken, then you have a combination problem.

Problems that by definition connote ordering (though not directly stated) are permutations. For example, 3 digits form a 3-digit number. Here, the 3-digit number connotes ordering. For another example, if you are to answer 3 questions, you probably would not be asked to answer a particular question more than once. So, you would not allow repetition in the calculations. Though not often needed, such logical assumptions are allowed and sometimes expected.

If the problem itself defines slots for the arrangements, it is a permutation problem. Words like "arrange" define slots for the arrangements. We will explain this in more detail later in the problems.

Generally, "arrangements" refer to permutations, and "selections" refer to combinations. These words often flag the problem type.

Other words indicating permutations are "alteration," "shift," "transformation," and "transmutation," all of which connote ordering.

For example:
In how many ways can the letters of the word XYZ be *transformed* to form new words?
In how many ways can the letters of the word XYZ be *altered* to form new words?

Some words indicating combinations are "aggregation," "alliance," "association," "coalition," "composition," "confederation," "gang," "league," and "union," (all of which have nothing to do with arrangements but instead connote selections.)

For example:
In how many ways can a coalition of 2 countries be formed from 4 countries?
(Here, a coalition is the same whether you say country A and B are a coalition or country B and country A are a coalition.)

2) **Are repetitions allowed?**
Check whether, based on the problem description, the results of a permutation/combination can have repetitions.

For example:
If you are to list countries in a coalition, you can hardly list a country twice.
(Here, repetition automatically is not allowed unless specified otherwise.)

If you have 3 doors to a room, you could use the same door for both entering and exiting.
(Here, repetition is automatically allowed.)

3) **Are there any indistinguishable objects in the base set?**
Check the base set: the objects from which a permutation or a combination are drawn. If any indistinguishable objects (repetitions at base set level) are available, collect them. This is easy since it only requires finding identical objects in a base set, which is usually given.

For example, if the original question is to find the words formed from the word GARGUNTUNG, then, in this step, you collect the information: G exists thrice, U exists twice, and so on.

Once the problem type is recognized, use the corresponding formula or model to solve it.

Problem Set AA:

1. There are 3 doors to a lecture room. In how many ways can a lecturer enter and leave the room?

 (A) 1
 (B) 3
 (C) 6
 (D) 9
 (E) 12

2. There are 3 doors to a lecture room. In how many ways can a lecturer enter the room from one door and leave from another door?

 (A) 1
 (B) 3
 (C) 6
 (D) 9
 (E) 12

3. How many possible combinations can a 3-digit safe code have?

 (A) ${}_9C_3$
 (B) ${}_9P_3$
 (C) 3^9
 (D) 9^3
 (E) 10^3

4. Goodwin has 3 different colored pants and 2 different colored shirts. In how many ways can he choose a pair of pants and a shirt?

 (A) 2
 (B) 3
 (C) 5
 (D) 6
 (E) 12

5. In how many ways can 2 doors be selected from 3 doors?

 (A) 1
 (B) 3
 (C) 6
 (D) 9
 (E) 12

6. In how many ways can 2 doors be selected from 3 doors for entering and leaving a room?

 (A) 1
 (B) 3
 (C) 6
 (D) 9
 (E) 12

7. In how many ways can a room be entered and exited from the 3 doors to the room?

 (A) 1
 (B) 3
 (C) 6
 (D) 9
 (E) 12

8. There are 5 doors to a lecture room. Two are red and the others are green. In how many ways can a lecturer enter the room and leave the room from different colored doors?

(A) 1
(B) 3
(C) 6
(D) 9
(E) 12

9. Four pool balls—A, B, C, D—are randomly arranged in a straight line. What is the probability that the order will actually be A, B, C, D ?

(A) 1/4
(B) $\dfrac{1}{{}_4C_4}$
(C) $\dfrac{1}{{}_4P_4}$
(D) 1/2!
(E) 1/3!

10. A basketball team has 11 players on its roster. Only 5 players can be on the court at one time. How many different groups of 5 players can the team put on the floor?

(A) 5^{11}
(B) ${}_{11}C_5$
(C) ${}_{11}P_5$
(D) 11^5
(E) $11! \cdot 5!$

11. How many different 5-letter words can be formed from the word ORANGE using each letter only once?

(A) ${}_6P_6$
(B) 36
(C) ${}_6C_6$
(D) 6^6
(E) ${}_6P_5$

12. How many unequal 5-digit numbers can be formed using each digit of the number 11235 only once?

(A) 5!
(B) ${}_5P_3$
(C) $\dfrac{{}_5C_5}{2!}$
(D) $\dfrac{{}_5P_5}{2!\cdot 3!}$
(E) $\dfrac{{}_5C_5}{2!\cdot 3!}$

13. How many different six-digit numbers can be formed using all of the following digits:

3, 3, 4, 4, 4, 5

(A) 10
(B) 20
(C) 30
(D) 36
(E) 60

14. This is how Edward's Lotteries work. First, 9 different numbers are selected. Tickets with exactly 6 of the 9 numbers randomly selected are printed such that no two tickets have the same set of numbers. Finally, the winning ticket is the one containing the 6 numbers drawn from the 9 randomly. There is exactly one winning ticket in the lottery system. How many tickets can the lottery system print?

(A) ${}_9P_6$
(B) ${}_9P_3$
(C) ${}_9C_9$
(D) ${}_9C_6$
(E) 6^9

15. How many different strings of letters can be made by reordering the letters of the word SUCCESS?

(A) 20
(B) 30
(C) 40
(D) 60
(E) 420

16. A company produces 8 different types of candies, and sells the candies in gift packs. How many different gift packs containing exactly 3 different candy types can the company put on the market?

(A) ${}_8C_2$
(B) ${}_8C_3$
(C) ${}_8P_2$
(D) ${}_8P_3$
(E) $\frac{{}_8P_3}{2!}$

17. Fritz is taking an examination that consists of two parts, A and B, with the following instructions:

Part A contains three questions, and a student must answer two.
Part B contains four questions, and a student must answer two.
Part A must be completed before starting Part B.

In how many ways can the test be completed?

(A) 12
(B) 15
(C) 36
(D) 72
(E) 90

18. A menu offers 2 entrees, 3 main courses, and 3 desserts. How many different combinations of dinner can be made? (A dinner must contain an entrée, a main course, and a dessert.)

(A) 12
(B) 15
(C) 18
(D) 21
(E) 24

19. In how many ways can 3 red marbles, 2 blue marbles, and 5 yellow marbles be placed in a row?

(A) $3!\cdot2!\cdot5!$

(B) $\frac{12!}{10!}$

(C) $\frac{10!}{3!}\cdot\frac{10!}{2!}\cdot\frac{10!}{5!}$

(D) $\frac{10!}{3!\cdot2!\cdot5!}$

(E) $\frac{10!}{(3!\cdot2!\cdot5!)^2}$

20. The retirement plan for a company allows employees to invest in 10 different mutual funds. Six of the 10 funds grew by at least 10% over the last year. If Sam randomly selected 4 of the 10 funds, what is the probability that 3 of Sam's 4 funds grew by at least 10% over last year?

(A) $\frac{{}_6C_3}{{}_{10}C_4}$

(B) $\frac{{}_6C_3\cdot{}_4C_1}{{}_{10}C_4}$

(C) $\frac{{}_6C_3\cdot{}_4C_1}{{}_{10}P_4}$

(D) $\frac{{}_6P_3\cdot{}_4P_1}{{}_{10}C_4}$

(E) $\frac{{}_6P_3\cdot{}_4P_1}{{}_{10}P_4}$

21. The retirement plan for a company allows employees to invest in 10 different mutual funds. Six of the 10 funds grew by at least 10% over the last year. If Sam randomly selected 4 of the 10 funds, what is the probability that *at least* 3 of Sam's 4 funds grew by at least 10% over the last year?

(A) $\frac{{}_6C_3}{{}_{10}C_4}$

(B) $\frac{{}_6C_3\cdot{}_4C_1}{{}_{10}C_4}$

(C) $\frac{{}_6C_3\cdot{}_4C_1+{}_6C_4}{{}_{10}P_4}$

(D) $\frac{{}_6P_3\cdot{}_4P_1}{{}_{10}C_4}$

(E) $\frac{{}_6C_3\cdot{}_4C_1+{}_6C_4}{{}_{10}C_4}$

22. In how many ways can the letters of the word ACUMEN be rearranged such that the vowels always appear together?

(A) $3!\cdot3!$

(B) $\frac{6!}{2!}$

(C) $\frac{4!\cdot3!}{2!}$

(D) $4!\cdot3!$

(E) $\frac{3!\cdot3!}{2!}$

23. In how many ways can the letters of the word ACCLAIM be rearranged such that the vowels always appear together?

(A) $\dfrac{7!}{2!\cdot 2!}$

(B) $\dfrac{4!\cdot 3!}{2!\cdot 2!}$

(C) $\dfrac{4!\cdot 3!}{2!}$

(D) $\dfrac{5!}{2!\cdot 2!}$

(E) $\dfrac{5!}{2!}\cdot\dfrac{3!}{2!}$

24. In how many ways can the letters of the word GARGANTUNG be rearranged such that all the G's appear together?

(A) $\dfrac{8!}{3!\cdot 2!\cdot 2!}$

(B) $\dfrac{8!}{2!\cdot 2!}$

(C) $\dfrac{8!\cdot 3!}{2!\cdot 2!}$

(D) $\dfrac{8!}{2!\cdot 3!}$

(E) $\dfrac{10!}{3!\cdot 2!\cdot 2!}$

25. In how many ways can the letters of the word GOSSAMERE be rearranged such that all S's and M's appear in the middle?

(A) $\dfrac{9!}{2!\cdot 2!}$

(B) $\dfrac{{}_7P_6}{2!\cdot 2!}$

(C) $\dfrac{{}_7P_6}{2!}\cdot\dfrac{{}_3P_3}{2!}$

(D) $\dfrac{{}_6P_6}{2!}\cdot\dfrac{{}_3P_3}{2!}$

(E) $\dfrac{{}_{10}P_6}{2!}\cdot\dfrac{{}_3P_3}{2!}$

26. How many different four-letter words can be formed (the words need not be meaningful) using the letters of the word GREGARIOUS such that each word starts with G and ends with R?

(A) ${}_8P_2$

(B) $\dfrac{{}_8P_2}{2!\cdot 2!}$

(C) ${}_8P_4$

(D) $\dfrac{{}_8P_4}{2!\cdot 2!}$

(E) $\dfrac{{}_{10}P_2}{2!\cdot 2!}$

27. A coin is tossed five times. What is the probability that the fourth toss would turn a head?

(A) $\frac{1}{{}_5P_3}$

(B) $\frac{1}{{}_5P_9}$

(C) $\frac{1}{2}$

(D) $\frac{1}{2!}$

(E) $\frac{1}{2^5}$

28. In how many of ways can 5 balls be placed in 4 tins if any number of balls can be placed in any tin?

(A) ${}_5C_4$
(B) ${}_5P_4$
(C) 5^4
(D) 4^5
(E) 5^5

29. On average, a sharpshooter hits the target once every 3 shots. What is the probability that he will hit the target in 4 shots?

(A) 1
(B) 1/81
(C) 1/3
(D) 65/81
(E) 71/81

30. On average, a sharpshooter hits the target once every 3 shots. What is the probability that he will not hit the target until 4th shot?

(A) 1
(B) 8/81
(C) 16/81
(D) 65/81
(E) 71/81

31. A new word is to be formed by randomly rearranging the letters of the word ALGEBRA. What is the probability that the new word has consonants occupying only the positions currently occupied by consonants in the word ALGEBRA?

(A) 2/120
(B) 1/24
(C) 1/6
(D) 2/105
(E) 1/35

32. Chelsea has 5 roses and 2 jasmines. A bouquet of 3 flowers is to be formed. In how many ways can it be formed if at least one jasmine must be in the bouquet?

(A) 5
(B) 20
(C) 25
(D) 35
(E) 40

33. In how many ways can 3 boys and 2 girls be selected from a group of 6 boys and 5 girls?

    (A) 10
    (B) 20
    (C) 50
    (D) 100
    (E) 200

34. In how many ways can a committee of 5 members be formed from 4 women and 6 men such that at least 1 woman is a member of the committee?

    (A) 112
    (B) 156
    (C) 208
    (D) 246
    (E) 252

35. In how many ways can 5 boys and 4 girls be arranged in a line so that there will be a boy at the beginning and at the end?

    (A) $\frac{3!}{5!}\cdot 7!$
    (B) $\frac{5!}{6!}\cdot 7!$
    (C) $\frac{5!}{3!}\cdot 7!$
    (D) $\frac{3!}{5!}\cdot 7!$
    (E) $\frac{5!}{7!}\cdot 7!$

36. In how many ways can the letters of the word MAXIMA be arranged such that all vowels are together?

    (A) 12
    (B) 18
    (C) 30
    (D) 36
    (E) 72

37. In how many ways can the letters of the word MAXIMA be arranged such that all vowels are together and all consonants are together?

    (A) 12
    (B) 18
    (C) 30
    (D) 36
    (E) 42

38. In how many ways can 4 boys and 4 girls be arranged in a row such that no two boys and no two girls are next to each other?

    (A) 1032
    (B) 1152
    (C) 1254
    (D) 1432
    (E) 1564

39. In how many ways can 4 boys and 4 girls be arranged in a row such that boys and girls alternate their positions (that is, boy girl)?

(A) 1032
(B) 1152
(C) 1254
(D) 1432
(E) 1564

40. The University of Maryland, University of Vermont, and Emory University have each 4 soccer players. If a team of 9 is to be formed with an equal number of players from each university, how many number of ways can the selections be done?

(A) 3
(B) 4
(C) 12
(D) 16
(E) 25

41. In how many ways can 5 persons be seated around a circular table?

(A) 5
(B) 24
(C) 25
(D) 30
(E) 120

42. In how many ways can 5 people from a group of 6 people be seated around a circular table?

(A) 56
(B) 80
(C) 100
(D) 120
(E) 144

43. What is the probability that a word formed by randomly rearranging the letters of the word ALGAE is the word ALGAE itself?

(A) 1/120
(B) 1/60
(C) 2/7
(D) 2/5
(E) 1/30

Answers and Solutions to Problem Set AA

1. There are 3 doors to a lecture room. In how many ways can a lecturer enter and leave the room?

   (A) 1
   (B) 3
   (C) 6
   (D) 9
   (E) 12

Recognizing the Problem:

1) Is it a permutation or a combination problem?
Here, order is important. Suppose A, B, and C are the three doors. Entering by door A and leaving by door B is not the same way as entering by door B and leaving by door A. Hence, AB ≠ BA implies the problem is a *permutation* (order is important).

2) Are repetitions allowed?
Since the lecturer can enter and exit through the same door, *repetition* is allowed.

3) Are there any indistinguishable objects in the base set?
Doors are different. They are not indistinguishable, so *no indistinguishable objects*.

Hence, we have a permutation problem, with repetition allowed and no indistinguishable objects.

Method I (Using known formula for the scenario):
Apply Formula 1, n^r, from the Formula section. Here, $n = 3$, (three doors to choose from), $r = 2$, 2 slots (one for entry door, one for exit door).

Hence, $n^r = 3^2 = 9$, and the answer is (D).

Method II (Model 2):
The lecturer can enter the room in 3 ways and exit in 3 ways. So, in total, the lecturer can enter and leave the room in 9 (= 3 · 3) ways. The answer is (D). This problem allows repetition: the lecturer can enter by a door and exit by the same door.

Method III (Model 3):
Let the 3 doors be A, B, and C. We must choose 2 doors: one to enter and one to exit. This can be done in 6 ways: {A, A}, {A, B}, {B, B}, {B, C}, {C, C}, and {C, A}. Now, the order of the elements is important because entering by A and leaving by B is not same as entering by B and leaving by A. Let's permute the combinations, which yields

A – A
A – B and B – A
{B, B}
B – C and C – B
C – C
C – A and A – C

The total is 9, and the answer is (D).

2. There are 3 doors to a lecture room. In how many ways can a lecturer enter the room from one door and leave from another door?

(A) 1
(B) 3
(C) 6
(D) 9
(E) 12

This problem is the same as the previous one, except entering and leaving must be done by different doors (since the doors are different, repetition is not allowed).

Hence, we have a permutation (there are two slots individually defined: one naming the entering door and one naming the leaving door), without repetition, and no indistinguishable objects (doors are different).

Recognizing the Problem:

1) Is it a permutation or a combination problem?
Here, order is important. Suppose A, B, and C are the three doors. Entering by door B and leaving by door C is not same as entering by door C and leaving by door B. Hence, BC ≠ CB implies the problem is a *permutation* (order is important).

2) Are repetitions allowed?
We must count the number of possibilities in which the lecturer enters and exits by different doors, so *repetition is not allowed.*

3) Are there any indistinguishable objects in the base set?
Doors are different. They are not indistinguishable, so *no indistinguishable objects.*

Hence, we have a permutation problem, with repetition not allowed and no indistinguishable objects.

Method I (Using known formula for the scenario):
Apply Formula 2, ${}_nP_r$, from the Formula section. Here, $n = 3$ (three doors to choose), $r = 2$ slots (one for entry door, one for exit door) to place them in.

The calculation is

$$
\begin{aligned}
{}_nP_r = {}_3P_2 &= \\
\frac{3!}{(3-2)!} &= \\
\frac{3!}{1!} &= \\
\frac{3 \cdot 2 \cdot 1}{1} &= \\
&6
\end{aligned}
$$

The answer is (C).

Method II (Model 2):
The lecturer can enter the room in 3 ways and exit in 2 ways (not counting the door entered). Hence, in total, the number of ways is $3 \cdot 2 = 6$ (by Model 2) or by the Fundamental Principle of Counting $2 + 2 + 2 = 6$. The answer is (C). This is a problem with repetition not allowed.

Method III (Model 3):
Let the 3 doors be A, B, and C. Hence, the base set is {A, B, C}. We have to choose 2 doors—one to enter and one to exit. This can be done in 3 ways: {A, B}, {B, C}, {C, A} [The combinations {A, A}, {B, B}, and {C, C} were eliminated because repetition is not allowed]. Now, the order of the permutation is

important because entering by A and leaving by B is not considered same as entering by B and leaving by A. Let's permute the combinations:

{A, B}
{A, C}
{B, A}
{B, C}
{C, A}
{C, B}

The total is 6, and the answer is (C).

3. How many possible combinations can a 3-digit safe code have?

(A) ${}_9C_3$
(B) ${}_9P_3$
(C) 3^9
(D) 9^3
(E) 10^3

The safe combination could be 433 or 334; the combinations are the same, but their ordering is different. Since order is important for the safe combinations, this is a permutation problem.

A safe code can be made of any of the numbers {0, 1, 2, 3, 4, 5, 6, 7, 8, 9}. No two objects in the set are indistinguishable. Hence, the base set does not have any indistinguishable objects.

Repetitions of numbers in the safe code are possible. For example, 334 is a possible safe code.

Hence, the problem is a permutation, with repetition and no indistinguishable objects. Hence, use Formula 1, n^r [here, $n = 10$, $r = 3$]. The number of codes is $10^3 = 1000$. The answer is (E).

Safe codes allow 0 to be first digit. Here, the same arrangement rules apply to each of the 3 digits. So, this is a uniform arrangement problem. We can use any formula or model here. But there are non-uniform arrangement problems. For example, if you are to form a 3-digit number, the first digit has an additional rule: it cannot be 0 (because in this case the number would actually be 2-digit number). In such scenarios, we need to use model I or II. The number of ways the digits can be formed by model II is

$$9 \cdot 10 \cdot 10 = 900$$

4. Goodwin has 3 different colored pants and 2 different colored shirts. In how many ways can he choose a pair of pants and a shirt?

(A) 2
(B) 3
(C) 5
(D) 6
(E) 12

Model 2:
The pants can be selected in 3 ways and the shirt in 2 ways. Hence, the pair can be selected in $3 \cdot 2 = 6$ ways. The answer is (D).

5. In how many ways can 2 doors be selected from 3 doors?

(A) 1
(B) 3
(C) 6
(D) 9
(E) 12

It appears that order is not important in this problem: the doors are mentioned but not defined. Also, since we are *selecting* doors, it is a combination problem.

The base set is the 3 doors [$n = 3$]. The doors are different, so there are no indistinguishable objects in the base set.

The arranged sets are the 2 doors [$r = 2$] we select. A door cannot be selected twice because "we select 2 doors" clearly means 2 different doors.

Hence, the problem is a combination, with no indistinguishable objects and no repetitions. Hence, using Formula 3, ${}_nC_r$, yields

$$
\begin{aligned}
{}_3C_2 = \frac{n!}{r!(n-r)!} &= \\
\frac{3!}{2!(3-2)!} &= \\
\frac{3!}{2!\cdot 1!} &= \\
\frac{3\cdot 2\cdot 1}{(2\cdot 1)\cdot 1} &= \\
&3
\end{aligned}
$$

The answer is (B).

6. In how many ways can 2 doors be selected from 3 doors for entering and leaving a room?

(A) 1
(B) 3
(C) 6
(D) 9
(E) 12

The problem statement almost ended at "3 doors." The remaining part "entering and leaving" only explains the reason for the selection. Hence, this does not define the slots. So, this is a combination problem. Moreover, we are asked to *select*, not to *arrange*. Hence, the problem is a combination, with no indistinguishable objects and no repetitions allowed. Using Formula 3, the number of ways the room can be entered and left is

$$
\begin{aligned}
{}_nC_r = {}_3C_2 = \frac{n!}{r!(n-r)!} &= \\
\frac{3!}{2!(3-2)!} &= \\
\frac{3!}{2!\cdot 1!} &= \\
\frac{3\cdot 2\cdot 1}{(2\cdot 1)\cdot 1} &= \\
&3
\end{aligned}
$$

The answer is (B).

7. In how many ways can a room be entered and exited from the 3 doors to the room?

(A) 1
(B) 3
(C) 6
(D) 9
(E) 12

There is specific stress on "entered" and "exited" doors. Hence, the problem is not combinational; it is a permutation (order/positioning is important).

The problem type is "no indistinguishable objects and repetitions allowed". Hence, by Formula 1, the number of ways the room can be entered and exited is $n^r = 3^2 = 9$. The answer is (C).

8. There are 5 doors to a lecture room. Two are red and the others are green. In how many ways can a lecturer enter the room and leave the room from different colored doors?

(A) 1
(B) 3
(C) 6
(D) 9
(E) 12

There are 2 red and 3 green doors. We have two cases:

> The room can be entered from a red door (2 red doors, so 2 ways) and can be left from a green door (3 green doors, so 3 ways): $2 \cdot 3 = 6$.
>
> The room can be entered from a green door (3 green doors, so 3 ways) and can be left from a red door (2 red doors, so 2 ways): $3 \cdot 2 = 6$.

Hence, the total number of ways is

$$2 \cdot 3 + 3 \cdot 2 = 6 + 6 = 12$$

The answer is (E).

9. Four pool balls—A, B, C, D—are randomly arranged in a straight line. What is the probability that the order will actually be A, B, C, D ?

(A) 1/4
(B) $\dfrac{1}{{}_4C_4}$
(C) $\dfrac{1}{{}_4P_4}$
(D) 1/2!
(E) 1/3!

This is a permutation problem (order is important).

A ball cannot exist in two slots, so repetition is not allowed.

Each ball is given a different identity A, B, C, and D, so there are no indistinguishable objects.

Here, $n = 4$ (number of balls to arrange) in $r = 4$ (positions). We know the problem type, and the formula to use. Hence, by Formula 2, the number of arrangements possible is ${}_4P_4$, and {A, B, C, D} is just one of the arrangements. Hence, the probability is 1 in ${}_4P_4$, or $\dfrac{1}{{}_4P_4}$. The answer is (C).

10. A basketball team has 11 players on its roster. Only 5 players can be on the court at one time. How many different groups of 5 players can the team put on the floor?

(A) 5^{11}
(B) ${}_{11}C_5$
(C) ${}_{11}P_5$
(D) 11^5
(E) $11! \cdot 5!$

The task is only to select a group of 5, not to order them. Hence, this is a combination problem. There are 11 players; repetition is not possible among them (one player cannot be counted more than once); and they are not given the same identity. Hence, there are no indistinguishable objects. Using Formula 3, groups of 5 can be chosen from 11 players in ${}_{11}C_5$ ways. The answer is (B).

11. How many different 5-letter words can be formed from the word ORANGE using each letter only once?

(A) ${}_6P_3$
(B) 36
(C) ${}_6C_6$
(D) 6^6
(E) ${}_6P_5$

In the problem, order is important because ORGAN is a word formed from ORANGE and ORNAG is a word formed from ORANGE, but they are not the same word. Repetition is not allowed, since each letter in the original word is used only once.

The problem does not have indistinguishable objects because no two letters of the word ORANGE are the same. Hence, by Formula 2, ${}_nP_r$, the answer is ${}_6P_5$, which is choice (E).

12. How many unequal 5-digit numbers can be formed using each digit of the number 11235 only once?

(A) $5!$
(B) ${}_5P_3$
(C) $\dfrac{{}_5C_5}{2!}$
(D) $\dfrac{{}_5P_5}{2!\cdot 3!}$
(E) $\dfrac{{}_5C_5}{2!\cdot 3!}$

The word "unequal" indicates that this is a permutation problem, because 11532 is the same combination as 11235, but they are not equal. Hence, they are permutations, different arrangements in a combination.

The indistinguishable objects in the base set {1, 1, 2, 3, 5} are the two 1's.

Since each digit of the number 11235 (objects in the base set) is used only once, repetitions are not allowed.

Hence, by Formula 4, the number of unequal 5-digit numbers that can be formed is

$$\frac{{}_5P_5}{2!} =$$
$$\frac{5!}{0!\cdot 2!} =$$
$${}_5P_3$$

The answer is (B).

13. How many different six-digit numbers can be formed using all of the following digits:

3, 3, 4, 4, 4, 5

(A) 10
(B) 20
(C) 30
(D) 36
(E) 60

Forming a six-digit number is a permutation because the value of the number changes with the different arrangements.

Since we have indistinguishable numbers in the base set, the regular permutations generate repeating numbers. But we are asked for only different six-digit numbers. So, we count only 1 for each similar permutation.

There are two sets of indistinguishable objects in the base set: two 3's and three 4's.

No repetitions are allowed since all elements in the base set are to be used in each number.

Hence, by Formula 4, the formula for permutations with no repetitions and with distinguishable objects, the number of six-digit numbers that can be formed is

$$\frac{{}_6P_6}{2!\cdot 3!} = \frac{6!}{2!\cdot 3!} = 60$$

The answer is (E).

14. This is how Edward's Lotteries work. First, 9 different numbers are selected. Tickets with exactly 6 of the 9 numbers randomly selected are printed such that no two tickets have the same set of numbers. Finally, the winning ticket is the one containing the 6 numbers drawn from the 9 randomly. There is exactly one winning ticket in the lottery system. How many tickets can the lottery system print?

(A) ${}_9P_6$
(B) ${}_9P_3$
(C) ${}_9C_9$
(D) ${}_9C_6$
(E) 6^9

The only condition is that the winning ticket has the same set of numbers as the drawn numbers (in whatever order). Hence, order is not important. Now, count the combinations.

Since the numbers in the base set (9 numbers) are different, the base set does not have indistinguishable objects.

Six of the 9 different numbers are selected by the lottery system, so no repetitions are allowed.

By Formula 3, the formula for combinations with no repetitions and no indistinguishable objects, the number of possible selections by the lottery system is ${}_9C_6$.

Since only one winning ticket (winning combination) exists per lottery system, there is of ${}_9C_6$ tickets per lottery system. The answer is (D).

15. How many different strings of letters can be made by reordering the letters of the word SUCCESS?

(A) 20
(B) 30
(C) 40
(D) 60
(E) 420

The word SUCCESS is a different word from SUSSECC, while they are the same combination. Hence, this is a permutation problem, not a combination problem.

There are two sets of indistinguishable objects in the base set: 2 C's and 3 S's.

Each letter is used only once in each reordering (so do not allow repetition).

Hence, we have a permutation problem, with indistinguishable objects and no repetitions. Using Formula 4, the formula for permutations with no repetitions but with distinguishable objects in the Formula section, yields $n = 7$ (base word has 7 letters), and $r = 7$ (each new word will have 7 letters). The repetitions are 2 C's and 3 S's. Hence, the total number of permutations is

$$\frac{{}_7P_7}{2!\cdot 3!} =$$
$$\frac{6\cdot 5\cdot 4\cdot 3\cdot 2\cdot 1}{2\cdot 6} =$$
$$5\cdot 4\cdot 3 =$$
$$60$$

The answer is (E).

16. A company produces 8 different types of candies, and sells the candies in gift packs. How many different gift packs containing exactly 3 different candy types can the company put on the market?

(A) ${}_8C_2$
(B) ${}_8C_3$
(C) ${}_8P_2$
(D) ${}_8P_3$
(E) $\frac{{}_8P_3}{2!}$

The phrase "8 different candies" indicates the base set does not have indistinguishable objects.

Since no placement slots are defined, this is a combination problem. We need only to choose 3 of 8 candies; we do not need to order them.

Repetitions are not allowed in the sets formed.

By Formula 3, the formula for permutations with no repetitions but with distinguishable objects yields ${}_8C_3$ which is Choice (B)

17. Fritz is taking an examination that consists of two parts, A and B, with the following instructions:

    Part A contains three questions, and a student must answer two.
    Part B contains four questions, and a student must answer two.
    Part A must be completed before starting Part B.

    In how many ways can the test be completed?

    (A) 12
    (B) 15
    (C) 36
    (D) 72
    (E) 90

The problem has two parts.

Each part is a permutation problem with no indistinguishable objects (no 2 questions are the same in either part), and repetitions are not allowed (the same question is not answered twice).

Hence, the number of ways of answering the first part is ${}_3P_2$ (2 questions to answer from 3), and the number of ways of answering the second part is ${}_4P_2$ (2 questions to answer from 4).

By the Fundamental Principle of Counting, the two parts can be done in

$$ {}_3P_2 \cdot {}_4P_2 = 6 \cdot 12 = 72 \text{ ways} $$

The answer is (D).

Method II [Model 2]:

The first question in Part A can be chosen to be one of the 3 questions in Part A.

The second question in Part A can be chosen to be one of the remaining 2 questions in Part A.

The first question in Part B can be chosen to be one of the 4 questions in Part B.

The second question in Part B can be chosen to be one of the remaining 3 questions in Part B.

Hence, the number of choices is

$$ 3 \cdot 2 \cdot 4 \cdot 3 = 72 $$

The answer is (D).

Method III [Fundamental Principle of Counting combined with Model 2]:

The first question in part A can be chosen to be one of the 3 questions in Part A.

The second question in part A can be chosen to be one of the 3 questions in Part A allowing the repetitions. Hence, number of permutations is $3 \cdot 3 = 9$. There are 3 repetitions [Q1 & Q1, Q2 & Q2, Q3 & Q3]. The main question does not allow repetitions since you would not answer the same question again. Deleting them, we have $9 - 3 = 6$ ways for Part A.

The first question in Part B can be chosen to be one of the 4 questions in Part B.

The second question in Part B can be chosen to be one of 4 questions in Part B. Hence, the number of permutations is $4 \cdot 4 = 16$. There are 4 repetitions [Q1 & Q1, Q2 & Q2, Q3 & Q3, Q4 & Q4]. The main question does not allow repetitions since you would not answer the same question again. Deleting them, we have $16 - 4 = 12$ ways for Part A.

Hence, the number of choices is

$$ 6 \cdot 12 = 72 $$

The answer is (D).

18. A menu offers 2 entrees, 3 main courses, and 3 desserts. How many different combinations of dinner can be made? (A dinner must contain an entrée, a main course, and a dessert.)

    (A) 12
    (B) 15
    (C) 18
    (D) 21
    (E) 24

The problem is a mix of 3 combinational problems. The goal is to choose 1 of 2 entrees, then 1 of 3 main courses, then 1 of 3 desserts. The choices can be made in 2, 3, and 3 ways, respectively. Hence, the total number of ways of selecting the combinations is $2 \cdot 3 \cdot 3 = 18$. The answer is (C).

We can also count the combinations by the Fundamental Principle of Counting:

		Dessert 1
	Main Course 1	Dessert 2
		Dessert 3
		Dessert 1
Entrée 1	Main Course 2	Dessert 2
		Dessert 3
		Dessert 1
	Main Course 3	Dessert 2
		Dessert 3
		Dessert 1
	Main Course 1	Dessert 2
		Dessert 3
		Dessert 1
Entrée 2	Main Course 2	Dessert 2
		Dessert 3
		Dessert 1
	Main Course 3	Dessert 2
		Dessert 3
		Total 18

The Fundamental Principle of Counting states:

The total number of possible outcomes of a series of decisions, making selections from various categories, is found by multiplying the number of choices for each decision.

Counting the number of choices in the final column above yields 18.

19. In how many ways can 3 red marbles, 2 blue marbles, and 5 yellow marbles be placed in a row?

(A) $3!\cdot2!\cdot5!$

(B) $\dfrac{12!}{10!}$

(C) $\dfrac{10!}{3!}\cdot\dfrac{10!}{2!}\cdot\dfrac{10!}{5!}$

(D) $\dfrac{10!}{3!\cdot2!\cdot5!}$

(E) $\dfrac{10!}{(3!\cdot2!\cdot5!)^2}$

Since the question is asking for the number of ways the marbles can be placed adjacent to each other, this is a permutation problem.

The base set has 3 red marbles (indistinguishable objects), 2 blue marbles (indistinguishable objects) and 5 yellow marbles (indistinguishable objects). The possible arrangements are 3 + 2 + 5 = 10 positions.

The same marble cannot be used twice, so no repetitions are allowed. Formula 4, ${}_nP_r$ and the method for indistinguishable objects (that is, divide the number of permutations, ${}_nP_r$, by the factorial count of each indistinguishable object [see Formulas section]) yield the number of permutations:

$$\frac{{}_{10}P_{10}}{3!\cdot2!\cdot5!}=\frac{10!}{3!\cdot2!\cdot5!}$$

The answer is (D).

20. The retirement plan for a company allows employees to invest in 10 different mutual funds. Six of the 10 funds grew by at least 10% over the last year. If Sam randomly selected 4 of the 10 funds, what is the probability that 3 of Sam's 4 funds grew by at least 10% over last year?

(A) $\dfrac{{}_6C_3}{{}_{10}C_4}$

(B) $\dfrac{{}_6C_3\cdot{}_4C_1}{{}_{10}C_4}$

(C) $\dfrac{{}_6C_3\cdot{}_4C_1}{{}_{10}P_4}$

(D) $\dfrac{{}_6P_3\cdot{}_4P_1}{{}_{10}C_4}$

(E) $\dfrac{{}_6P_3\cdot{}_4P_1}{{}_{10}P_4}$

There are 6 winning funds that grew more than 10%, and 4 losing funds that grew less than 10%.

The problem can be split into 3 sub-problems:

We have the specific case where Sam must choose 4 funds, 3 of which are winning, so the remaining fund must be losing. Let's evaluate the number of ways this can be done. [Note: The order in which the funds are chosen is not important because whether the first 3 funds are winning and the 4th one is losing, or the first fund is losing and the last 3 are winning; only 3 of 4 funds will be winning ones. Hence, this is a combination problem.] The problem has 2 sub-problems:

1. Sam must choose 3 of the 6 winning funds. This can be done in ${}_6C_3$ ways.

2. Sam must choose one losing fund (say the 4th fund). There are 10 – 6 = 4 losing funds. Hence, the 4th fund can be any one of the 4 losing funds. The selection can be done in ${}_4C_1$ ways.

Hence, the total number of ways of choosing 3 winning funds and 1 losing one is ${}_6C_3 \cdot {}_4C_1$.

3. Sam could have chosen 4 funds in ${}_{10}C_4$ ways.

Hence, the probability that 3 of Sam's 4 funds grew by at least 10% over last year is

$$\frac{{}_6C_3 \cdot {}_4C_1}{{}_{10}C_4} = \frac{20 \cdot 4}{210} = \frac{8}{21}$$

The answer is (B).

21. The retirement plan for a company allows employees to invest in 10 different mutual funds. Six of the 10 funds grew by at least 10% over the last year. If Sam randomly selected 4 of the 10 funds, what is the probability that *at least* 3 of Sam's 4 funds grew by at least 10% over the last year?

(A) $\frac{{}_6C_3}{{}_{10}C_4}$

(B) $\frac{{}_6C_3 \cdot {}_4C_1}{{}_{10}C_4}$

(C) $\frac{{}_6C_3 \cdot {}_4C_1 + {}_6C_4}{{}_{10}P_4}$

(D) $\frac{{}_6P_3 \cdot {}_4P_1}{{}_{10}C_4}$

(E) $\frac{{}_6C_3 \cdot {}_4C_1 + {}_6C_4}{{}_{10}C_4}$

There are 6 winning funds that grew more than 10%, and 4 losing funds that grew less than 10%.

The problem can be split into 3 sub-problems:

1) Sam has to choose 3 winning funds. This can be done in ${}_6C_3$ ways.
2) Sam has to choose 1 losing fund. This can be done in ${}_4C_1$ ways.

Or

3) Sam has to choose all 4 funds to be winning funds. This can be done in ${}_6C_4$ ways.

This is how Sam chooses at least 3 winning funds.

Hence, the total number of ways of choosing *at least* 3 winning funds is ${}_6C_3 \cdot {}_4C_1 + {}_6C_4$.

If there were no restrictions (such as choosing at least 3 winning funds), Sam would have chosen funds in ${}_{10}C_4$ ways.

Hence, the probability that *at least* 3 of Sam's 4 funds grew by at least 10% over the last year is

$$\frac{{}_6C_3 \cdot {}_4C_1 + {}_6C_4}{{}_{10}C_4}$$

The answer is (E).

22. In how many ways can the letters of the word ACUMEN be rearranged such that the vowels always appear together?

    (A) $3!\cdot 3!$
    (B) $\dfrac{6!}{2!}$
    (C) $\dfrac{4!\cdot 3!}{2!}$
    (D) $4!\cdot 3!$
    (E) $\dfrac{3!\cdot 3!}{2!}$

The word "rearranged" indicates that this is a permutation problem.

The base set {A, C, U, M, E, N} has no indistinguishable objects.

Repetition is not allowed.

Since the 3 vowels must appear together, treat the three as an inseparable unit. Hence, reduce the base set to {{A, U, E}, C, M, N}. Now, there are 4 different units in the base set, and they can be arranged in ${}_4P_4 = 4!$ ways. The unit {A, U, E} can itself be internally arranged in ${}_3P_3 = 3!$ ways. Hence, by The Fundamental Principle of Counting, the total number of ways of arranging the word is $4!\cdot 3!$. The answer is (D).

23. In how many ways can the letters of the word ACCLAIM be rearranged such that the vowels always appear together?

    (A) $\dfrac{7!}{2!\cdot 2!}$
    (B) $\dfrac{4!\cdot 3!}{2!\cdot 2!}$
    (C) $\dfrac{4!\cdot 3!}{2!}$
    (D) $\dfrac{5!}{2!\cdot 2!}$
    (E) $\dfrac{5!}{2!}\cdot\dfrac{3!}{2!}$

The word "rearranged" indicates that this is a permutation problem.

Since the 3 vowels A, A, and I must appear together, treat the three as an inseparable unit. Hence, reduce the base set to {{A, A, I}, C, C, L, M}.

The set has two indistinguishable objects, C's.

Also, repetitions are not allowed since we rearrange the word.

Hence, the number of permutations that can be created with units of the set is $\dfrac{{}_5P_5}{2!} = \dfrac{5!}{2!}$.

Now, let's see how many permutations we can create with the unit {A, A, I}.

The unit {A, A, I} has two indistinguishable objects, A's.

Also, repetitions are not allowed.

Hence, by Formula 4, the number of ways of permuting it is $\dfrac{{}_3P_3}{2!} = \dfrac{3!}{2!}$.

Hence, by The Fundamental Principle of Counting, the total number of ways of rearranging the letters is

$$\frac{5!}{2!}\cdot\frac{3!}{2!}$$

The answer is (E).

24. In how many ways can the letters of the word GARGANTUNG be rearranged such that all the G's appear together?

(A) $\dfrac{8!}{3!\cdot 2!\cdot 2!}$

(B) $\dfrac{8!}{2!\cdot 2!}$

(C) $\dfrac{8!\cdot 3!}{2!\cdot 2!}$

(D) $\dfrac{8!}{2!\cdot 3!}$

(E) $\dfrac{10!}{3!\cdot 2!\cdot 2!}$

The word "rearranged" indicates that this is a permutation problem.

Since all 3 G's are together, treat them a single inseparable unit. Hence, the base set reduces to {{G, G, G}, A, R, A, N, T, U, N}. There are 8 independent units, 2 A's (indistinguishable), and two N's (indistinguishable). No unit is used twice, so there are no repetitions. Hence, by Formula 4, the number of arrangements is $\frac{{}_8P_8}{2!\cdot 2!} = \frac{8!}{2!\cdot 2!}$.

The 3 G's can be rearranged amongst themselves in $\frac{{}_3P_3}{3!} = \frac{3!}{3!} = 1$ way. Hence, the total number of ways the letters can be rearranged is

$$\frac{8!}{2!\cdot 2!}\cdot 1 = \frac{8!}{2!\cdot 2!}$$

The answer is (B).

25. In how many ways can the letters of the word GOSSAMERE be rearranged such that all S's and M's appear in the middle?

(A) $\dfrac{9!}{2!\cdot 2!}$

(B) $\dfrac{{}_7P_6}{2!\cdot 2!}$

(C) $\dfrac{{}_7P_6}{2!}\cdot\dfrac{{}_3P_3}{2!}$

(D) $\dfrac{{}_6P_6}{2!}\cdot\dfrac{{}_3P_3}{2!}$

(E) $\dfrac{{}_{10}P_6}{2!}\cdot\dfrac{{}_3P_3}{2!}$

The word "rearranged" indicates that this is a permutation problem.

Since S and M must appear in the middle, treat them as an inseparable unit and reserve the middle seat for them. Correspondingly, bracket them in the base set. The new base set becomes {{S, S, M}, G, O, A, E, R, E}. Hence, we have the following arrangement:

____ ____ ____ {S, S, M} ____ ____ ____

Now, the remaining 6 units G, O, A, E, R, and E can be arranged in the 6 blank slots; and for each arrangement, every permutation inside the unit {S, S, M} is allowed.

Hence, the blank slots can be filled in $\frac{_6P_6}{2!}$ (E repeats twice) ways.

And the unit {S, S, M} can be internally arranged in $\frac{_3P_3}{2!}$ ways.

Hence, by Model 2, the total number of ways the letters can be rearranged is

$$\frac{_6P_6}{2!} \cdot \frac{_3P_3}{2!}$$

The answer is (D).

26. How many different four-letter words can be formed (the words need not be meaningful) using the letters of the word GREGARIOUS such that each word starts with G and ends with R?

(A) $_8P_2$
(B) $\frac{_8P_2}{2! \cdot 2!}$
(C) $_8P_4$
(D) $\frac{_8P_4}{2! \cdot 2!}$
(E) $\frac{_{10}P_2}{2! \cdot 2!}$

Place one G in the first slot and one R in the last slot:

G _ _ R

The remaining letters, {G, R, E, A, I, O, U, S}, can be arranged in the remaining 2 slots in $_8P_2$ (no indistinguishable objects nor repetition). The answer is (A).

Note: Since the two G's in the base word are indistinguishable, the word G_1G_2AR is the same as G_2G_1AR. Hence, the internal arrangement of the G's or, for the same reason, the R's is not important.

27. A coin is tossed five times. What is the probability that the fourth toss would turn a head?

(A) $\frac{1}{_5P_3}$
(B) $\frac{1}{_5P_9}$
(C) $\frac{1}{2}$
(D) $\frac{1}{2!}$
(E) $\frac{1}{2^3}$

The fourth toss is independent of any other toss. The probability of a toss turning heads is 1 in 2, or simply 1/2. Hence, the probability of the fourth toss being a head is 1/2. The answer is (C).

Method II:
Each toss has 2 outcomes. Hence, 5 tosses have $2 \cdot 2 \cdot 2 \cdot \ldots 2$ (5 times) $= 2^5$ outcomes (permutation with repetition over $r = 2$ and $n = 5$ [repetitions allowed: the second and the fourth toss may both yield heads or tails]).

Reserve the third toss for a head. Now, the number of ways the remaining 4 tosses can be tossed is 2^4 (repetitions allowed). The probability is $\frac{2^4}{2^5} = \frac{1}{2}$. The answer is (C).

28. In how many of ways can 5 balls be placed in 4 tins if any number of balls can be placed in any tin?

    (A) $_5C_4$
    (B) $_5P_4$
    (C) 5^4
    (D) 4^5
    (E) 5^5

The first ball can be placed in any one of the four tins.

Similarly, the second, the third, the fourth, and the fifth balls can be placed in any one of the 4 tins.

Hence, the number of ways of placing the balls is $4 \cdot 4 \cdot 4 \cdot 4 \cdot 4 = 4^5$. The answer is (D).

Note: We used Model 2 here.

29. On average, a sharpshooter hits the target once every 3 shots. What is the probability that he will hit the target in 4 shots?

    (A) 1
    (B) 1/81
    (C) 1/3
    (D) 65/81
    (E) 71/81

The sharpshooter hits the target once in 3 shots. Hence, the probability of hitting the target is 1/3. The probability of not hitting the target is 1 – 1/3 = 2/3.

Now, (the probability of not hitting the target even once in 4 shots) + (the probability of hitting at least once in 4 shots) equals 1, because these are the only possible cases.

Hence, the probability of hitting the target at least once in 4 shots is

1 – (the probability of not hitting even once in 4 shots)

The probability of not hitting in the 4 chances is $\frac{2}{3} \cdot \frac{2}{3} \cdot \frac{2}{3} \cdot \frac{2}{3} = \frac{16}{81}$. Now, 1 – 16/81 = 65/81. The answer is (D).

This methodology is similar to Model 2. You might try analyzing why. Clue: The numerators of $\frac{2}{3} \cdot \frac{2}{3} \cdot \frac{2}{3} \cdot \frac{2}{3} = \frac{16}{81}$ are the number of ways of doing the specific jobs, and the denominators are the number of ways of doing all possible jobs.

30. On average, a sharpshooter hits the target once in every 3 shots. What is the probability that he will not hit the target until the 4th shot?

(A) 1
(B) 8/81
(C) 16/81
(D) 65/81
(E) 71/81

The sharpshooter hits the target once in every 3 shots. Hence, the probability of hitting the target is 1/3. The probability of not hitting the target is 1 – 1/3 = 2/3.

He will not hit the target on the first, second, and third shots, but he will hit it on the fourth shot. The probability of this is

$$\frac{2}{3}\cdot\frac{2}{3}\cdot\frac{2}{3}\cdot\frac{1}{3}=\frac{8}{81}$$

The answer is (B).

This methodology is similar to Model 2. You might try analyzing why. Clue: The numerators of $\frac{2}{3}\cdot\frac{2}{3}\cdot\frac{2}{3}\cdot\frac{2}{3}=\frac{16}{81}$ are the number of ways of doing the specific jobs, and the denominators are the number of ways of doing all possible jobs.

31. A new word is to be formed by randomly rearranging the letters of the word ALGEBRA. What is the probability that the new word has consonants occupying only the positions currently occupied by consonants in the word ALGEBRA?

(A) 2/120
(B) 1/24
(C) 1/6
(D) 2/105
(E) 1/35

If we do not put restrictions on the arrangements of the consonants, then by Formula 4 the number of words that can be formed from the word ALGEBRA is $\frac{7!}{2!}$ (A repeats).

If we constrain that the positions of consonants is reserved only for consonants, then the format of the new arrangement should look like this

A, L, G, E, B, R, A

V, *C*, *C*, *V*, *C*, *C*, *V*

V for vowels, *C* for consonants.

The 4 slots for consonants can be filled in ${}_4P_4 = 4!$ ways, and the 3 slots for vowels can be filled in $\frac{3!}{2!}$ (A repeats) ways. Hence, by Formula 2, the total number of arrangements in the format is $4!\left(\frac{3!}{2!}\right)$.

Hence, the probability is

$$\frac{4!\left(\frac{3!}{2!}\right)}{\frac{7!}{2!}}=\frac{1}{35}$$

The answer is (E).

32. Chelsea has 5 roses and 2 jasmines. A bouquet of 3 flowers is to be formed. In how many ways can it be formed if at least one jasmine must be in the bouquet?

    (A) 5
    (B) 20
    (C) 25
    (D) 35
    (E) 40

This is a selection problem because whether you choose a jasmine first or a rose first does not matter.

The 3 flowers in the bouquet can be either 1 jasmine and 2 roses or 2 jasmines and 1 rose.

1 of 2 jasmines can be selected in ${}_2C_1$ ways.

2 of 5 roses can be selected in ${}_5C_2$ ways.

The subtotal is ${}_2C_1 \cdot {}_5C_2 = \dfrac{2!}{1! \cdot 1!} \cdot \dfrac{5!}{3! \cdot 2!} = 2 \cdot 10 = 20$.

2 of 2 jasmines can be selected in ${}_2C_2$ ways.

1 of 5 roses can be selected in ${}_5C_1$ ways.

The subtotal is ${}_2C_2 \cdot {}_5C_1 = \dfrac{2!}{2! \cdot 0!} \cdot \dfrac{5!}{4! \cdot 1!} = 1 \cdot 5 = 5$.

The grand total is 20 + 5 = 25 ways. The answer is (C).

33. In how many ways can 3 boys and 2 girls be selected from a group of 6 boys and 5 girls?

    (A) 10
    (B) 20
    (C) 50
    (D) 100
    (E) 200

We have two independent actions to do:

1) Select 3 boys from 6 boys.
2) Select 2 girls from 5 girls.

Selection is a combination problem since selection does not include ordering. Hence, by Model 2, the number of ways is

$$({}_6C_3 \text{ ways for boys}) \cdot ({}_5C_2 \text{ ways for girls}) =$$
$$\left(\frac{6!}{3! \cdot 3!}\right) \cdot \left(\frac{5!}{2! \cdot 3!}\right) =$$
$$20 \cdot 10 =$$
$$200$$

The answer is (E).

34. In how many ways can a committee of 5 members be formed from 4 women and 6 men such that at least 1 woman is a member of the committee?

(A) 112
(B) 156
(C) 208
(D) 246
(E) 252

Forming members of committee is a selection action and therefore this is a combination problem. Whether you select A first and B next or vice versa, it will only be said that A and B are members of the committee.

The number of ways of forming the committee of 5 from 4 + 6 = 10 people is ${}_{10}C_5$. The number of ways of forming a committee with no women (5 members to choose from 6 men) is ${}_6C_5$. Hence, the number of ways of forming the combinations is

$$\begin{aligned} {}_{10}C_5 - {}_6C_5 &= \\ \frac{10!}{5! \cdot 5!} - \frac{6!}{5!} &= \\ 252 - 6 &= \\ 246 \end{aligned}$$

The answer is (D).

35. In how many ways can 5 boys and 4 girls be arranged in a line so that there will be a boy at the beginning and at the end?

(A) $\frac{3!}{5!} \cdot 7!$
(B) $\frac{5!}{6!} \cdot 7!$
(C) $\frac{5!}{3!} \cdot 7!$
(D) $\frac{3!}{5!} \cdot 7!$
(E) $\frac{5!}{7!} \cdot 7!$

The arrangement is a permutation, and there are no indistinguishable objects because no two boys or girls are identical. The first and the last slots hold two of the 5 boys, and the remaining slots are occupied by the 4 girls and the 3 remaining boys.

The first and the last slots can hold 2 of the 5 boys in ${}_5P_2$ ways, and the 3 boys and the 4 girls position themselves in the middle slots in ${}_7P_7$ ways. Hence, there are $\frac{5!}{3!} \cdot 7!$ possible arrangements. The answer is (C).

36. In how many ways can the letters of the word MAXIMA be arranged such that all vowels are together?

(A) 12
(B) 18
(C) 30
(D) 36
(E) 72

The base set can be formed as {{A, I, A}, M, X, M}. The unit {A, I, A} arranges itself in $\frac{{}_3P_3}{2!}$ ways. The 4 units in the base set can be arranged in ${}_4P_4/2!$ ways. Hence, the total number of ways of arranging the letters is

$$\frac{{}_3P_3}{2!}\cdot\frac{{}_4P_4}{2!}=\frac{3!}{2!}\cdot\frac{4!}{2!}=3\cdot 12=36$$

The answer is (D).

37. In how many ways can the letters of the word MAXIMA be arranged such that all vowels are together and all consonants are together?

(A) 12
(B) 18
(C) 30
(D) 36
(E) 42

Since vowels are together and consonants are together, arrange the base set as {{A, I, A}, {M, X, M}}. Here, {A, I, A} and {M, X, M} are two inseparable units.

The two units can be mutually arranged in ${}_2P_2$ ways.

Each unit has 3 objects, 2 of which are indistinguishable.

Hence, the number of permutations of each is $\frac{{}_3P_2}{2!}$.

Hence, the total number of arrangements possible is

$$\left({}_2P_2\right)\left(\frac{{}_3P_2}{2!}\right)\left(\frac{{}_3P_2}{2!}\right)=$$
$$(2)(3)(3)=$$
$$18$$

The answer is (B).

38. In how many ways can 4 boys and 4 girls be arranged in a row such that no two boys and no two girls are next to each other?

   (A) 1032
   (B) 1152
   (C) 1254
   (D) 1432
   (E) 1564

Form the base set as {{B1, B2, B3, B4}, {G1, G2, G3, G4}}; Looking at the problem, either B's or G's occupy the odd slots and the other one occupies the even slots. Choosing one to occupy an odd slot set can be done in $_2P_1$ ways, and the other one is automatically filled by the other group.

Now, fill B's in odd slots (the number of ways is $_4P_4$), and fill G's in even slots (the number of ways is $_4P_4$ ways). The total number of ways of doing this is $_4P_4 \cdot {}_4P_4$.

The number of ways of doing all of this is $2! \cdot {}_4P_4 \cdot {}_4P_4 = 2 \cdot 24 \cdot 24 = 1152$. The answer is (B).

39. In how many ways can 4 boys and 4 girls be arranged in a row such that boys and girls alternate their positions (that is, boy girl)?

   (A) 576
   (B) 1152
   (C) 1254
   (D) 1432
   (E) 1564

Form the base set as {{B1, B2, B3, B4}, {G1, G2, G3, G4}}; the set {B1, B2, B3, B4} occupies alternate positions, as does the set {G1, G2, G3, G4}.

Now there are odd slots and even slots. Each odd slot alternates, and each even slot alternates. Therefore, we have two major slots: *even* and *odd* and two units to occupy them: {B1, B2, B3, B4} and {G1, G2, G3, G4}. This can be done in $_2P_1$ ways.

An alternate explanation for this is: The person starting the row can be chosen in 2 ways (i.e., either boys start the first position and arrange alternately or girls start and do the same), either B starts first or G starts first.

Either way, the positions {1, 3, 5, 7} are reserved for one of the two groups B or G, and the positions {2, 4, 6, 8} are reserved for the other group. Arrangements in each position set can be done in $_4P_4$ ways.

Hence, the total number of arrangements is $2! \cdot {}_4P_4 \cdot {}_4P_4 = 2 \cdot 24 \cdot 24 = 1152$. The answer is (B).

Mathematically, this problem is the same as the previous one. Just the expression (wording) of the problem is different.

40. The University of Maryland, University of Vermont, and Emory University have each 4 soccer players. If a team of 9 is to be formed with an equal number of players from each university, how many number of ways can the selections be done?

   (A) 3
   (B) 4
   (C) 12
   (D) 16
   (E) 25

Selecting 3 of 4 players can be done in $_4C_3 = \dfrac{4!}{3! \cdot 1!} = 4$ ways.

The selection from the 3 universities can be done in $3 \cdot 4 = 12$ ways.

The answer is (C).

41. In how many ways can 5 persons be seated around a circular table?

(A) 5
(B) 24
(C) 25
(D) 30
(E) 120

For a circular table, we use cyclical permutations, not linear permutations. Hence, 1 in every r linear permutations (here $n = 5$ and $r = 5$) is a circular permutation. There are ${}_5P_5$ linear permutations. Hence,

$$\frac{{}_5P_5}{5} = \frac{5!}{5} = 4 \cdot 3 \cdot 2 \cdot 1 = 24 \text{ permutations}$$

The answer is (B).

42. In how many ways can 5 people from a group of 6 people be seated around a circular table?

(A) 56
(B) 80
(C) 100
(D) 120
(E) 144

For a circular table, we use cyclical permutations, not linear permutations. Hence, 1 in every r linear permutations (here $r = 5$) is a circular permutation. There are ${}_6P_5$ linear permutations and therefore $\frac{{}_6P_5}{5}$ circular permutations. Now,

$$\frac{{}_6P_5}{5} = \frac{6!}{5} = 1 \cdot 2 \cdot 3 \cdot 4 \cdot 6 = 144 \text{ permutations}$$

The answer is (E).

43. What is the probability that a word formed by randomly rearranging the letters of the word ALGAE is the word ALGAE itself?

(A) 1/120
(B) 1/60
(C) 2/7
(D) 2/5
(E) 1/30

The number of words that can be formed from the word ALGAE is $\frac{5!}{2!}$ (A repeats). ALGAE is just one of the words. Hence, the probability is $\frac{1}{\frac{5!}{2!}} = \frac{2!}{5!} = \frac{2}{120} = \frac{1}{60}$. The answer is (B).

Functions

DEFINITION

A function is a special relationship (correspondence) between two sets such that for each element x in its domain there is assigned one and only one element y in its range.

Notice that the correspondence has two parts:

1) For each x there is assigned *one* y. (This is the ordinary part of the definition.)
2) For each x there is assigned *only one* y. (This is the special part of the definition.)

The second part of the definition of a function creates the uniqueness of the assignment: There cannot be assigned two values of y to one x. In mathematics, uniqueness is very important. We know that $2 + 2 = 4$, but it would be confusing if $2 + 2$ could also equal something else, say 5. In this case, we could never be sure that the answer to a question was the *right* answer.

The correspondence between x and y is usually expressed with the function notation: $y = f(x)$, where y is called the dependent variable and x is called the independent variable. In other words, the value of y depends on the value of x plugged into the function. For example, the square root function can be written as $y = f(x) = \sqrt{x}$. To calculate the correspondence for $x = 4$, we get $y = f(4) = \sqrt{4} = 2$. That is, the square root function assigns the unique y value of 2 to the x value of 4. Most expressions can be turned into functions. For example, the expression $2^x - \frac{1}{x}$ becomes the function

$$f(x) = 2^x - \frac{1}{x}$$

DOMAIN AND RANGE

We usually identify a function with its correspondence, as in the example above. However, a function consists of three parts: a domain, a range, and correspondence between them.

➤ **The *domain* of a function is the set of x values for which the function is defined.**

For example, the function $f(x) = \frac{1}{x-1}$ is defined for all values of $x \neq 1$, which causes division by zero. There is an infinite variety of functions with restricted domains, but only two types of restricted domains appear on the GRE: division by zero and even roots of negative numbers. For example, the function $f(x) = \sqrt{x-2}$ is defined only if $x - 2 \geq 0$, or $x \geq 2$. The two types of restrictions can be combined. For example, $f(x) = \frac{1}{\sqrt{x-2}}$. Here, $x - 2 \geq 0$ since it's under the square root symbol. Further $x - 2 \neq 0$, or $x \neq 2$, because that would cause division by zero. Hence, the domain is all $x > 2$.

➢ **The *range* of a function is the set of *y* values that are assigned to the *x* values in the domain.**

For example, the range of the function $y = f(x) = x^2$ is $y \geq 0$ since a square is never negative. The range of the function $y = f(x) = x^2 + 1$ is $y \geq 1$ since $x^2 + 1 \geq 1$. You can always calculate the range of a function algebraically, but is usually better to graph the function and read off its range from the *y* values of the graph.

GRAPHS

The graph of a function is the set of ordered pairs $(x, f(x))$, where x is in the domain of f and $y = f(x)$.

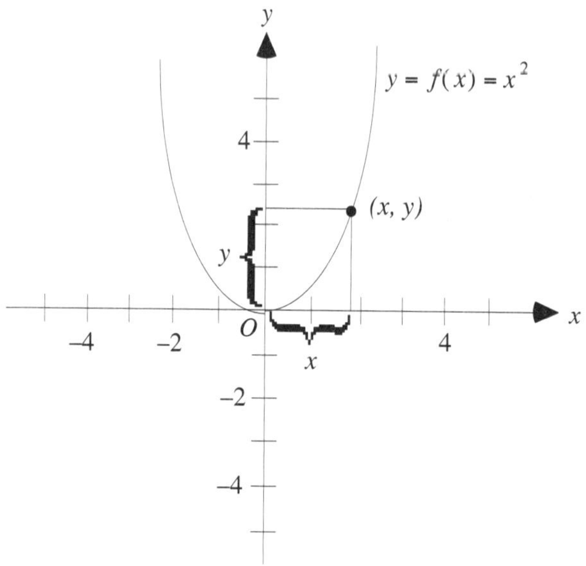

For this function, the domain is all x and the range is all $y \geq 0$ (since the graph touches the x-axis at the origin and is above the x-axis elsewhere).

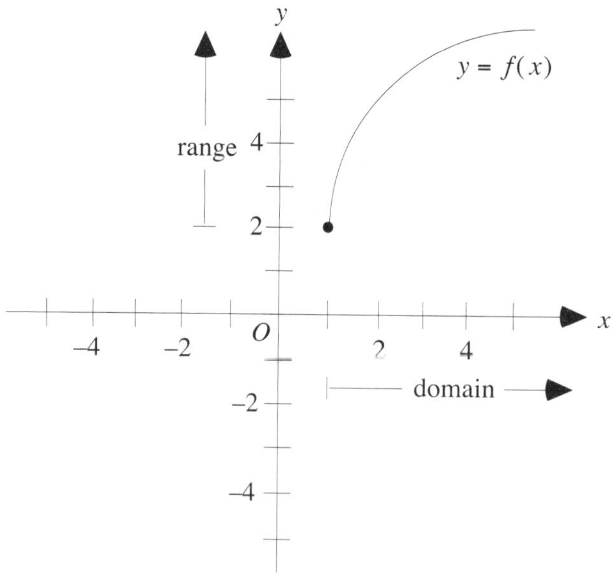

For this function, the domain is all $x \geq 1$ and the range is all $y \geq 2$.

TRANSLATIONS OF GRAPHS

Many graphs can be obtained by shifting a base graph around by adding positive or negative numbers to various places in the function. Take for example, the absolute value function $y = |x|$. Its graph is

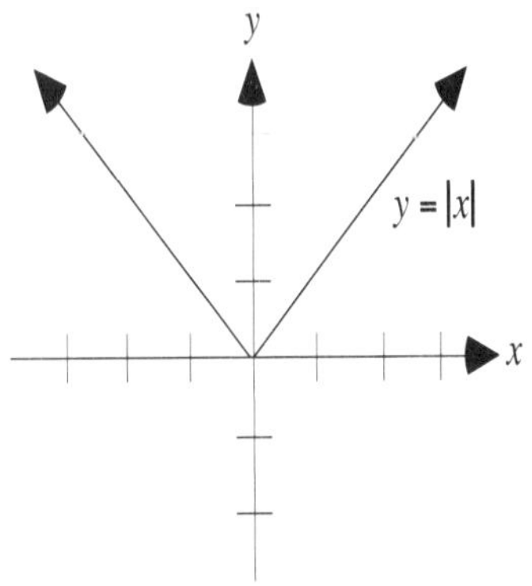

(Notice that sometimes an arrow is added to a graph to indicate the graph continues indefinitely and sometimes nothing is used. To indicate that a graph stops, a dot is added to the terminal point of the graph. Also, notice that the domain of the absolute value function is all x because you can take the absolute value of any number. The range is $y \geq 0$ because the graph touches the x-axis at the origin, is above the x-axis elsewhere, and increases indefinitely.)

To shift this base graph up one unit, we add 1 outside the absolute value symbol, $y = |x| + 1$:

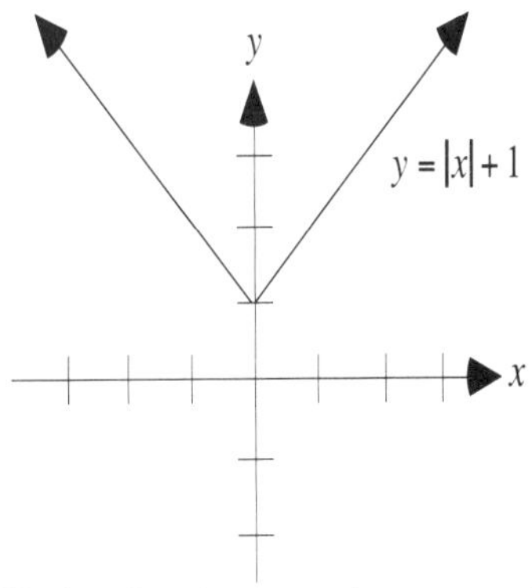

(Notice that the range is now $y \geq 1$.)

To shift the base graph down one unit, we subtract 1 outside the absolute value symbol, $y = |x| - 1$:

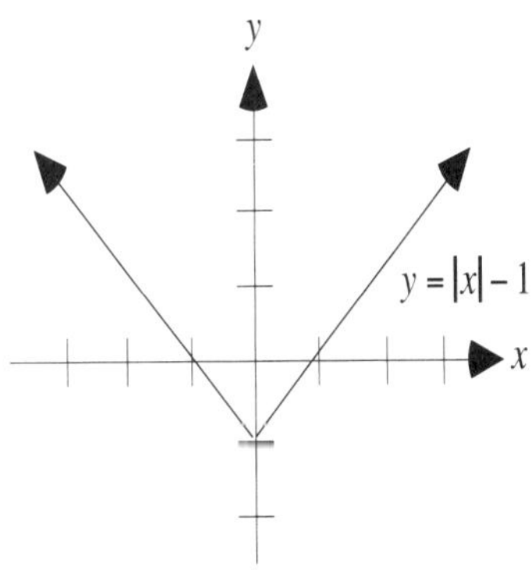

(Notice that the range is now $y \geq -1$.)

To shift the base graph to the right one unit, we subtract 1 inside the absolute value symbol, $y = |x - 1|$:

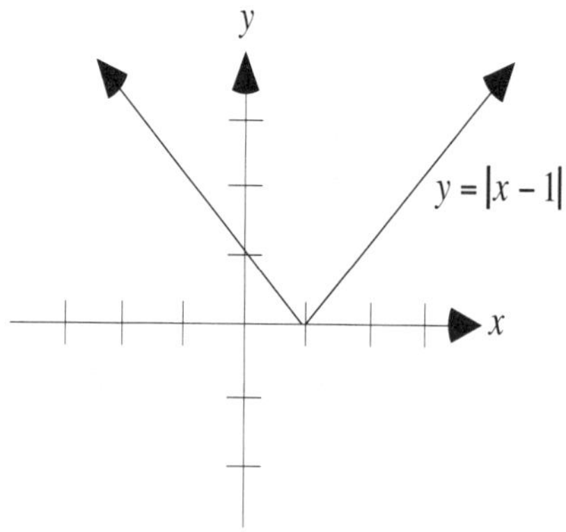

(Notice that the range did not change; it's still $y \geq 0$. Notice also that subtracting 1 moved the graph to right. Many students will mistakenly move the graph to the left because that's where the negative numbers are.)

To shift the base graph to the left one unit, we add 1 inside the absolute value symbol, $y = |x + 1|$:

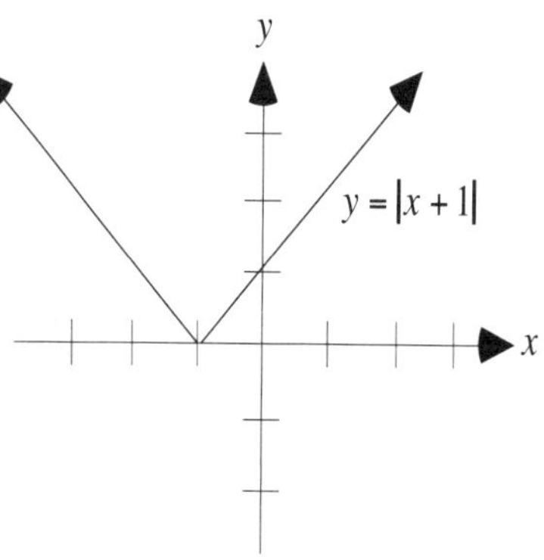

(Notice that the range did not change; it's still $y \geq 0$. Notice also that adding 1 moved the graph to left. Many students will mistakenly move the graph to the right because that's where the positive numbers are.)

The pattern of the translations above holds for all functions. So to move a function $y = f(x)$ up c units, add the positive constant c to the exterior of the function: $y = f(x) + c$. To move a function $y = f(x)$ to the right c units, subtract the constant c in interior of the function: $y = f(x - c)$. To summarize, we have

To shift up c units:	$y = f(x) + c$
To shift down c units:	$y = f(x) - c$
To shift to the right c units:	$y = f(x - c)$
To shift to the left c units:	$y = f(x + c)$

REFLECTIONS OF GRAPHS

Many graphs can be obtained by reflecting a base graph by multiplying various places in the function by negative numbers. Take for example, the square root function $y = \sqrt{x}$. Its graph is

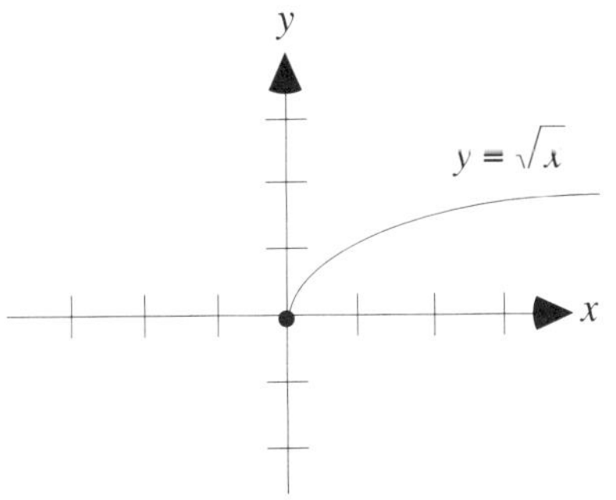

(Notice that the domain of the square root function is all $x \geq 0$ because you cannot take the square root of a negative number. The range is $y \geq 0$ because the graph touches the x-axis at the origin, is above the x-axis elsewhere, and increases indefinitely.)

To reflect this base graph about the x-axis, multiply the exterior of the square root symbol by negative one, $y = -\sqrt{x}$:

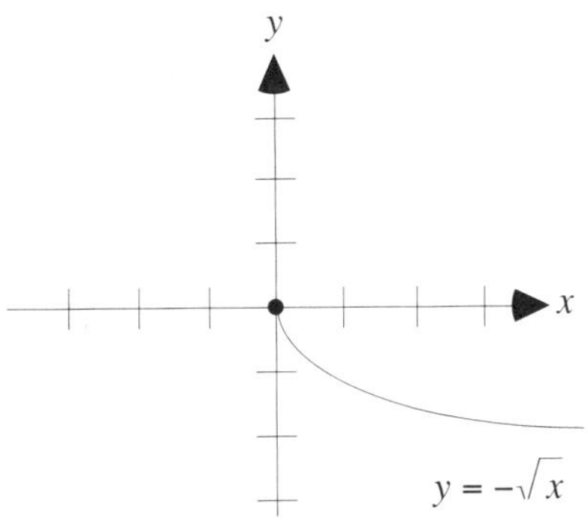

(Notice that the range is now $y \leq 0$ and the domain has not changed.)

To reflect the base graph about the y-axis, multiply the interior of the square root symbol by negative one, $y = \sqrt{-x}$:

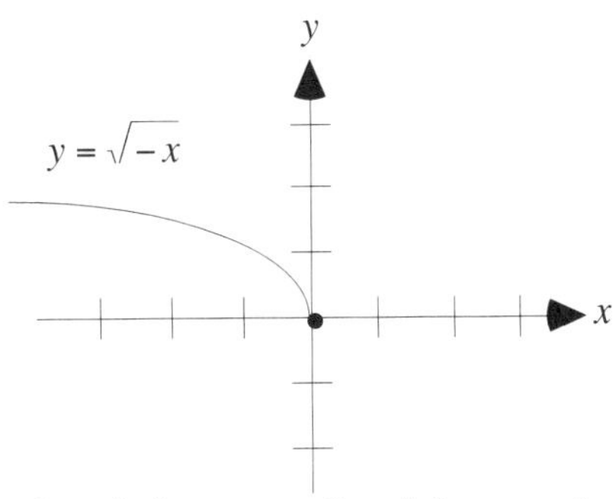

(Notice that the domain is now $x \leq 0$ and the range has not changed.)

The pattern of the reflections above holds for all functions. So to reflect a function $y = f(x)$ about the x-axis, multiply the exterior of the function by negative one: $y = -f(x)$. To reflect a function $y = f(x)$ about the y-axis, multiply the exterior of the function by negative one: $y = f(-x)$. To summarize, we have

To reflect about the x-axis:	$y = -f(x)$
To reflect about the y-axis:	$y = f(-x)$

Reflections and translations can be combined. Let's reflect the base graph of the square root function $y = \sqrt{x}$ about the x-axis, the y-axis and then shift it to the right 2 units and finally up 1 unit:

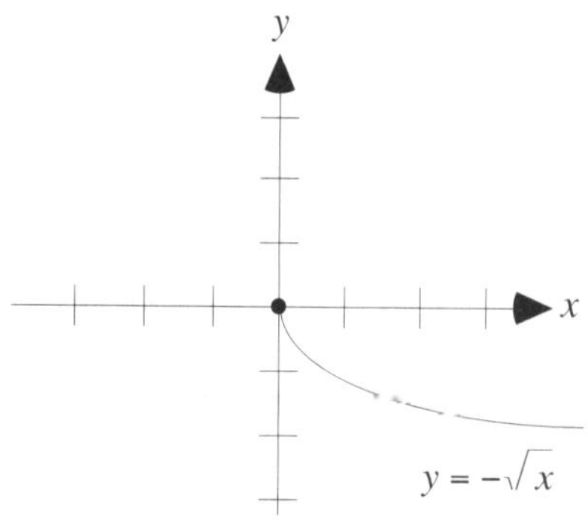

(Notice that the domain is still $x \geq 0$ and the range is now $y \leq 0$.)

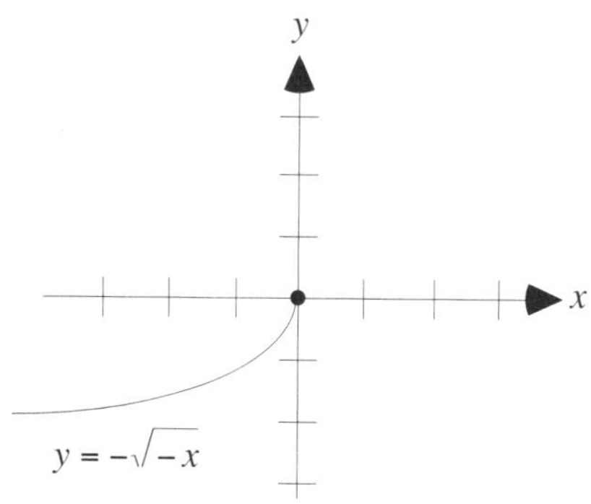

(Notice that the domain is now $x \leq 0$ and the range is still $y \leq 0$.)

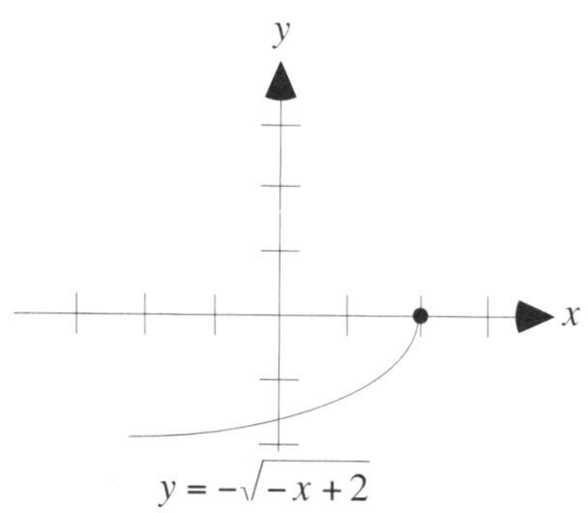

(Notice that the domain is now $x \leq 2$ and the range is still $y \leq 0$.)

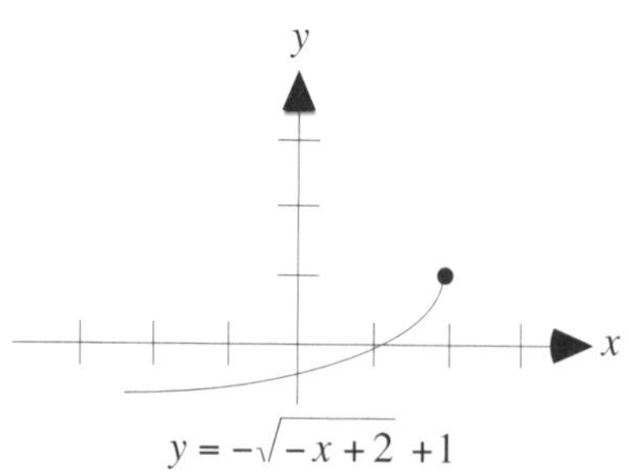

(Notice that the domain is still $x \leq 2$ and the range is now $y \leq 1$.)

EVALUATION AND COMPOSITION OF FUNCTIONS

EVALUATION

We have been using the function notation $f(x)$ intuitively; we also need to study what it actually means. You can think of the letter f in the function notation $f(x)$ as the name of the function. Instead of using the equation $y = x^3 - 1$ to describe the function, we can write $f(x) = x^3 - 1$. Here, f is the name of the function and $f(x)$ is the value of the function at x. So $f(2) = 2^3 - 1 = 8 - 1 = 7$ is the value of the function at 2. As you can see, this notation affords a convenient way of prompting the evaluation of a function for a particular value of x.

Any letter can be used as the independent variable in a function. So the above function could be written $f(p) = p^3 - 1$. This indicates that the independent variable in a function is just a "placeholder." The function could be written without a variable as follows:

$$f(\ ) = (\ )^3 - 1$$

In this form, the function can be viewed as an input/output operation. If 2 is put into the function $f(2)$, then $2^3 - 1$ is returned.

In addition to plugging numbers into functions, we can plug expressions into functions. Plugging $y + 1$ into the function $f(x) = x^2 - x$ yields

$$f(y+1) = (y+1)^2 - (y+1)$$

You can also plug other expressions in terms of x into a function. Plugging $2x$ into the function $f(x) = x^2 - x$ yields

$$f(2x) = (2x)^2 - 2x$$

This evaluation can be troubling to students because the variable x in the function is being replaced by the same variable. But the x in function is just a placeholder. If the placeholder were removed from the function, the substitution would appear more natural. In $f(\ ) = (\ )^2 - (\ )$, we plug $2x$ into the left side $f(2x)$ and it returns the right side $(2x)^2 - 2x$.

COMPOSITION

We have plugged numbers into functions and expressions into functions; now let's plug in other functions. Since a function is identified with its expression, we have actually already done this. In the example above with $f(x) = x^2 - x$ and $2x$, let's call $2x$ by the name $g(x)$. In other words, $g(x) = 2x$. Then the composition of f with g (that is plugging g into f) is

$$f(g(x)) = f(2x) = (2x)^2 - 2x$$

You probably won't see the notation $f(g(x))$ on the test. But you probably will see one or more problems that ask you perform the substitution. For another example, let $f(x) = \frac{1}{x+1}$ and let $g(x) = x^2$. Then $f(g(x)) = \frac{1}{x^2+1}$ and $g(f(x)) = \left(\frac{1}{x+1}\right)^2$. Once you see that the composition of functions merely substitutes one function into another, these problems can become routine. Notice that the composition operation $f(g(x))$ is performed from the inner parentheses out, not from left to right. In the operation $f(g(2))$, the number 2 is first plugged into the function g and then that result is plugged in the function f.

A function can also be composed with itself. That is, substituted into itself. Let $f(x) = \sqrt{x} - 2$. Then $f(f(x)) = \sqrt{\sqrt{x} - 2} - 2$.

Example: The graph of $y = f(x)$ is shown to the right. If $f(-1) = v$, then which one of the following could be the value of $f(v)$?

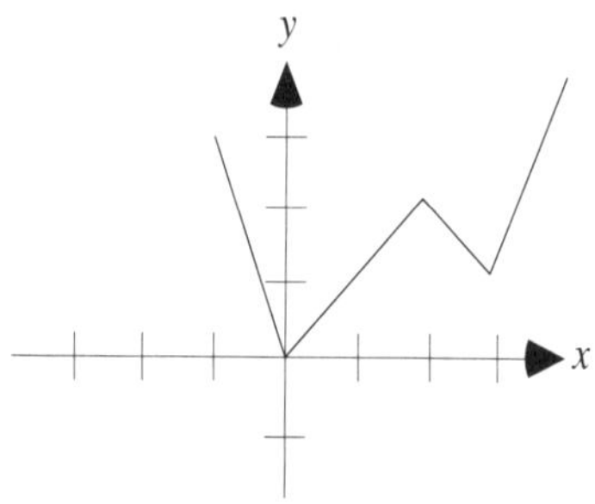

(A) 0
(B) 1
(C) 2
(D) 2.5
(E) 3

Since we are being asked to evaluate $f(v)$ and we are told that $v = f(-1)$, we are just being asked to compose $f(x)$ with itself. That is, we need to calculate $f(f(-1))$. From the graph, $f(-1) = 3$. So $f(f(-1)) = f(3)$. Again, from the graph, $f(3) = 1$. So $f(f(-1)) = f(3) = 1$. The answer is (B).

QUADRATIC FUNCTIONS

Quadratic functions (parabolas) have the following form:

$$y = f(x) = ax^2 + bx + c$$

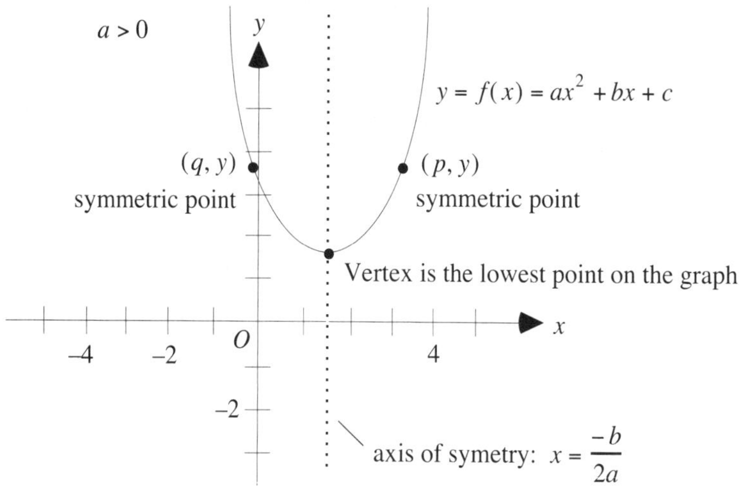

The lowest or highest point on a quadratic graph is called the vertex. The x–coordinate of the vertex occurs at $x = \dfrac{-b}{2a}$. This vertical line also forms the axis of symmetry of the graph, which means that if the graph were folded along its axis, the left and right sides of the graph would coincide.

In graphs of the form $y = f(x) = ax^2 + bx + c$ if $a > 0$, then the graph opens up.

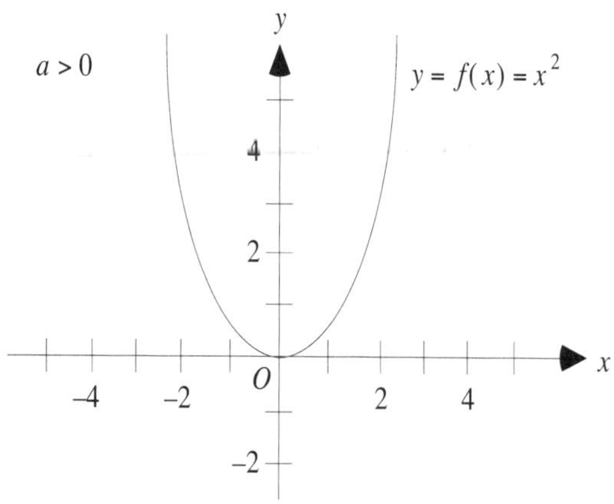

If $a < 0$, then the graph opens down.

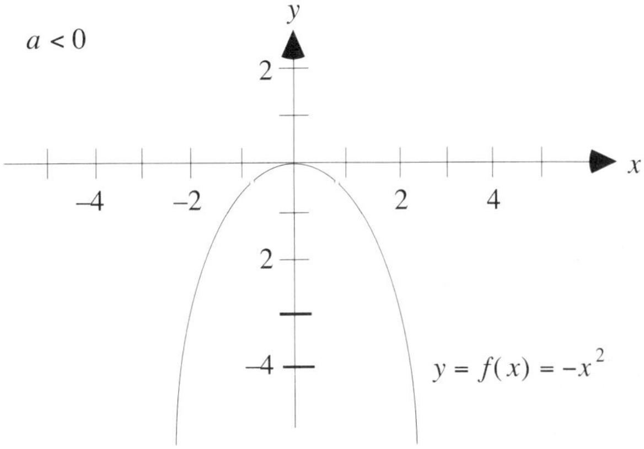

By completing the square, the form $y = ax^2 + bx + c$ can be written as $y = a(x-h)^2 + k$. You are not expected to know this form on the test. But it is a convenient form since the vertex occurs at the point (h, k) and the axis of symmetry is the line $x = h$.

We have been analyzing quadratic functions that are vertically symmetric. Though not as common, quadratic functions can also be horizontally symmetric. They have the following form:

$$x = g(y) = ay^2 + by + c$$

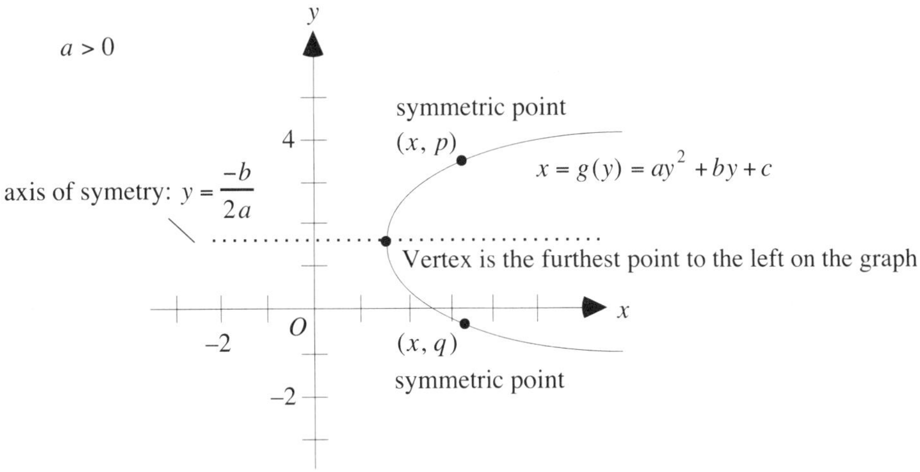

The furthest point to the left on this graph is called the vertex. The y-coordinate of the vertex occurs at $y = \frac{-b}{2a}$. This horizontal line also forms the axis of symmetry of the graph, which means that if the graph were folded along its axis, the top and bottom parts of the graph would coincide.

In graphs of the form $x = ay^2 + by + c$ if $a > 0$, then the graph opens to the right and if $a < 0$ then the graph opens to the left.

Example: The graph of $x = -y^2 + 2$ and the graph of the line k intersect at $(0, p)$ and $(1, q)$. Which one of the following is the smallest possible slope of line k?

(A) $-\sqrt{2} - 1$
(B) $-\sqrt{2} + 1$
(C) $\sqrt{2} - 1$
(D) $\sqrt{2} + 1$
(E) $\sqrt{2} + 2$

Let's make a rough sketch of the graphs. Expressing $x = -y^2 + 2$ in standard form yields $x = -1y^2 + 0 \cdot y + 2$. Since $a = -1$, $b = 0$, and $c = 2$, the graph opens to the left and its vertex is at (2, 0).

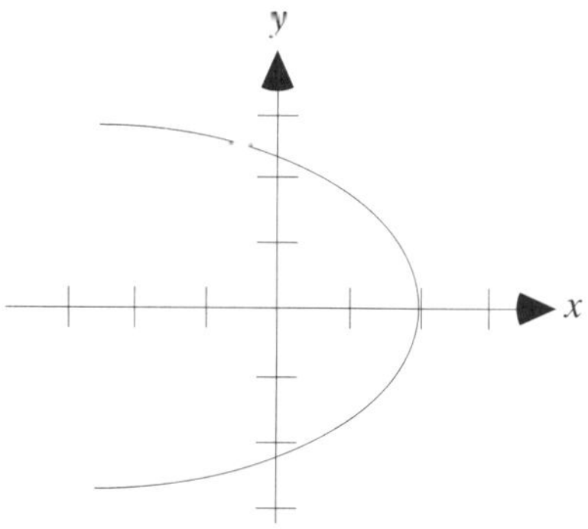

Since p and q can be positive or negative, there are four possible positions for line k (the y-coordinates in the graphs below can be calculated by plugging $x = 0$ and $x = 1$ into the function $x = -y^2 + 2$):

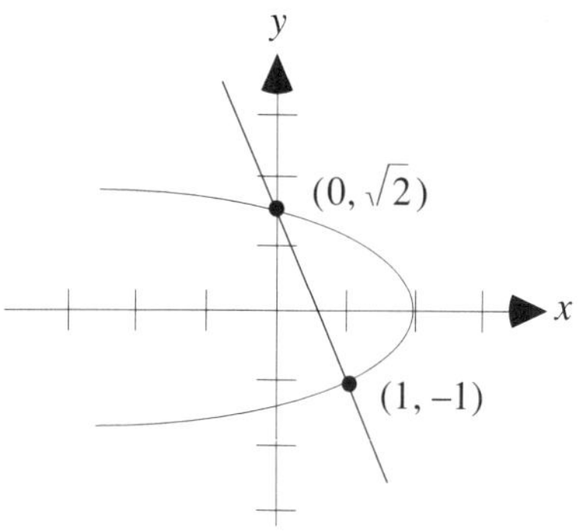

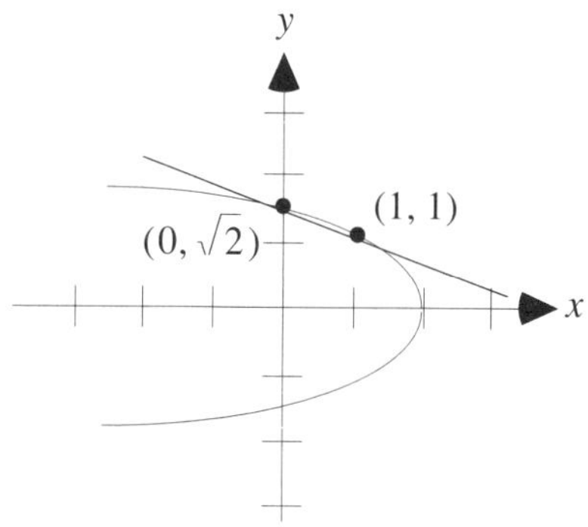

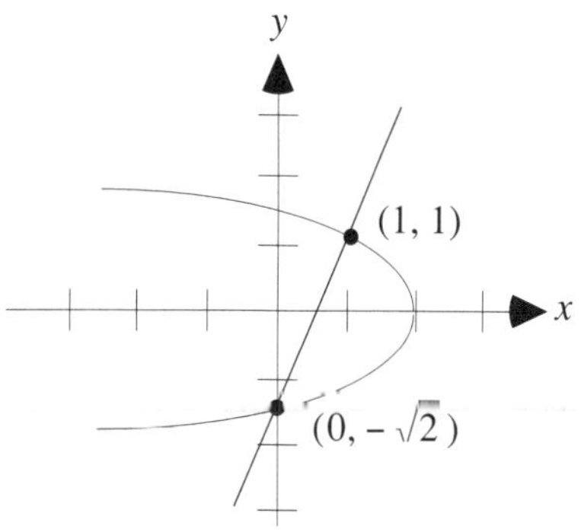

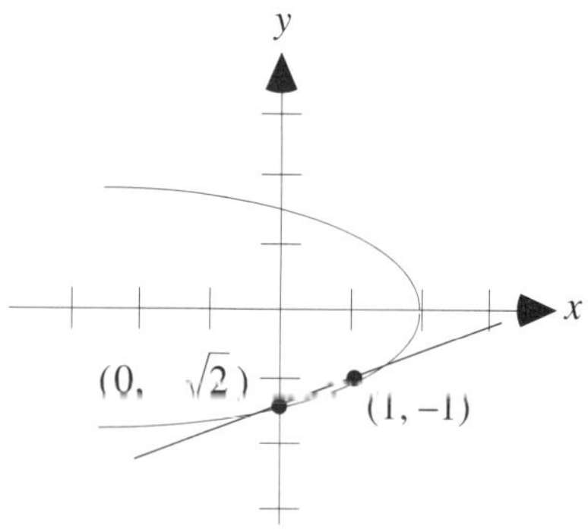

Since the line in the first graph has the steepest negative slope, it is the smallest possible slope. Calculating the slope yields

$$m = \frac{\sqrt{2} - (-1)}{0 - 1} = \frac{\sqrt{2} + 1}{-1} = -\left(\sqrt{2} + 1\right) = -\sqrt{2} - 1$$

The answer is (A).

QUALITATIVE BEHAVIOR OF GRAPHS AND FUNCTIONS

In this rather vague category, you will be asked how a function and its graph are related. You may be asked to identify the zeros of a function based on its graph. The zeros, or roots, of a function are the x-coordinates of where it crosses the x-axis. Or you may be given two graphs and asked for what x values are their functions equal. The functions will be equal where they intersect.

Example: The graphs of $y = f(x)$ and $y = 1$ are shown to the right. For how many x values does $f(x)$ equal 1?

(A) 0
(B) 1
(C) 2
(D) 3
(E) 4

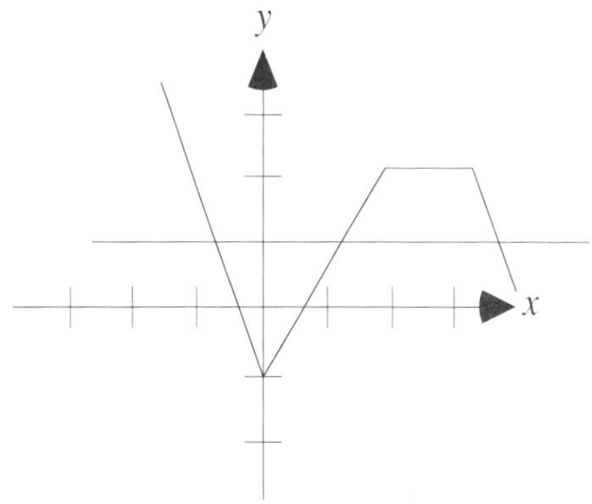

The figure shows that the graphs intersect at three points. At each of these points, both graphs have a height, or y-coordinate, of 1. The points are approximately (–.8, 1), (1.2, 1), and (4, 1). Hence, $f(x) = 1$ for three x values. The answer is (D).

FUNCTIONS AS MODELS OF REAL-LIFE SITUATIONS

Functions can be used to predict the outcomes of certain physical events or real-life situations. For example, a function can predict the maximum height a projectile will reach when fired with an initial velocity, or the number of movie tickets that will be sold at a given price.

Example: The graph to the right shows the number of music CDs sold at various prices. At what price should the CDs be marked to sell the maximum number of CDs?

(A) 0
(B) 5
(C) 10
(D) 15
(E) 20

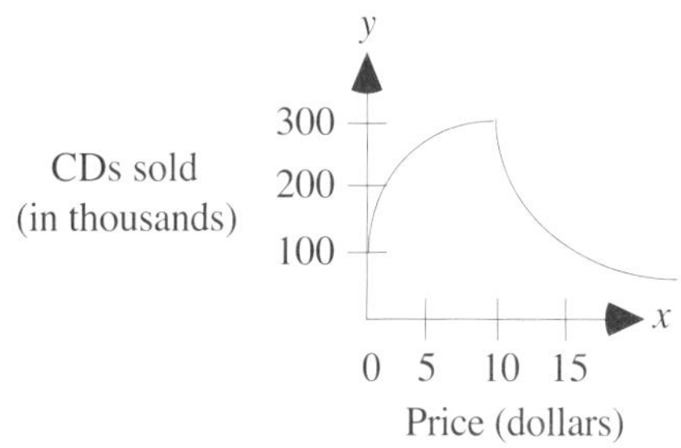

As you read the graph from left to right, it shows that sales initially increase rapidly and then slow to a maximum of about 300,000. From there, sales drop precipitously and then slowly approach zero as the price continues to increase. From the graph, sales of 300,000 units on the y-axis correspond to a price of about \$10 on the x-axis. The answer is (C).

Problem Set BB:

➢ Medium

1. The functions f and g are defined as $f(x, y) = 2x + y$ and $g(x, y) = x + 2y$. What is the value of $f(3, 4)$?

(A) $f(4, 3)$
(B) $f(3, 7)$
(C) $f(7, 4)$
(D) $g(3, 4)$
(E) $g(4, 3)$

2. A function $f(x)$ is defined for all real numbers by the expression $(x - 1.5)(x - 2.5)(x - 3.5)(x - 4.5)$. For which one of the following values of x, represented on the number line, is $f(x)$ negative?

(A) Point A
(B) Point B
(C) Point C
(D) Point D
(E) Point E

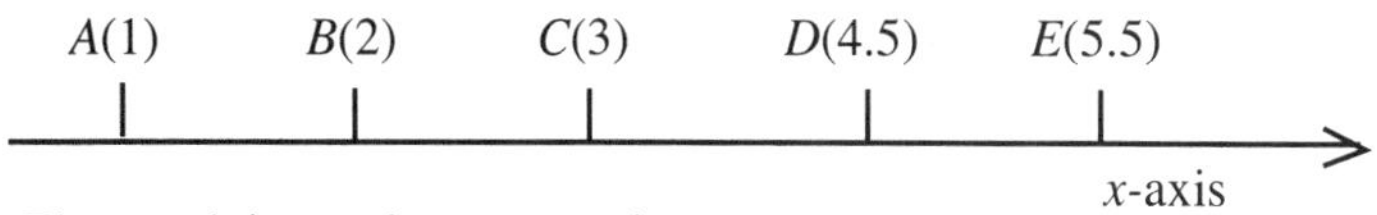

The graph is not drawn to scale.

3.

Column A		Column B
	A function $f(x)$ is defined for all real numbers as $f(x) = (x - 1)(x - 2)(x - 3)(x - 4)$	
$f(2.5)$		$f(3.5)$

4. The functions f and g are defined as $f(x, y)$ equals average of x and y and $g(x, y)$ equals the greater of the numbers x and y. Then $f(3, 4) + g(3, 4) =$

(A) 6
(B) 6.5
(C) 7
(D) 7.5
(E) 8.5

5.

Column A		Column B
	The functions f and g are defined as $f(x, y) = 2x + y$ and $g(x, y) = x + 2y$.	
$f(3, 4) + g(3, 4)$		$f(4, 3) + g(4, 3)$

6.

Column A		Column B
	The function f is defined for all positive integers n as $f(n) = \frac{n}{n+1}$	
$f(1) \times f(2)$		$f(2) \times f(3)$

$$g(x) = (2x - 3)^{1/4} + 1$$

7. In the function above, for what values of x is $g(x)$ a real number?

(A) $x \geq 0$
(B) $x \geq 1/2$
(C) $x \geq 3/2$
(D) $x \geq 2$
(E) $x \geq 3$

x	–1	0	1	2
$f(x)$	1	3	1	–5

8. The table above shows the values of the quadratic function f for several values of x. Which one of the following best represents f?

(A) $f(x) = -2x^2$
(B) $f(x) = x^2 + 3$
(C) $f(x) = -x^2 + 3$
(D) $f(x) = -2x^2 - 3$
(E) $f(x) = -2x^2 + 3$

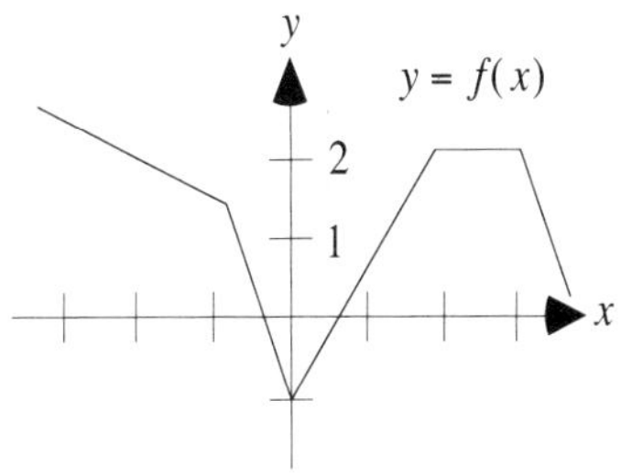

9. In the function above, if $f(k) = 2$, then which one of the following could be a value of k?

(A) –1
(B) 0
(C) 0.5
(D) 2.5
(E) 4

10. Let the function h be defined by $h(x) = \sqrt{x} + 2$. If $3h(v) = 18$, then which one of the following is the value of $h\left(\frac{v}{4}\right)$?

(A) –4
(B) –1
(C) 0
(D) 2
(E) 4

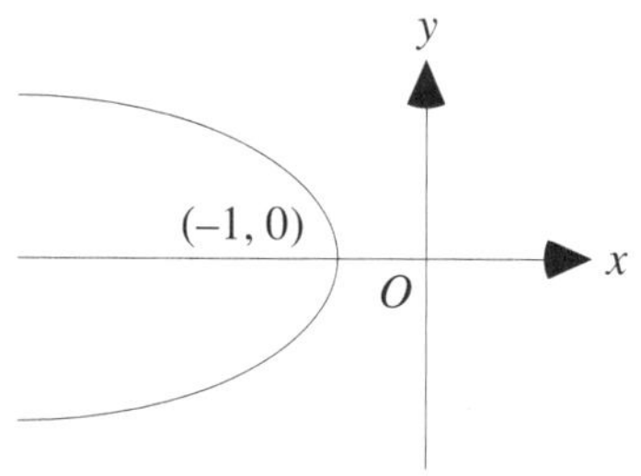

11. The graph above shows a parabola that is symmetric about the x-axis. Which one of the following could be the equation of the graph?

(A) $x = -y^2 - 1$
(B) $x = -y^2$
(C) $x = -y^2 + 1$
(D) $x = y^2 - 1$
(E) $x = (y + 1)^2$

12. A pottery store owner determines that the revenue for sales of a particular item can be modeled by the function $r(x) = 50\sqrt{x} - 40$, where x is the number of the items sold. How many of the items must be sold to generate \$110 in revenue?

(A) 5
(B) 6
(C) 7
(D) 8
(E) 9

➤ **Hard**

13. The functions f and g are defined as $f(x) = 2x - 3$ and $g(x) = x + 3/2$. For what value of y is $f(y) = g(y - 3)$?

(A) 1
(B) 3/2
(C) 2
(D) 3
(E) 5

14. A function is defined for all positive numbers x as $f(x) = a\sqrt{x} + b$. What is the value of $f(3)$, if $f(4) - f(1) = 2$ and $f(4) + f(1) = 10$?

(A) 1
(B) 2
(C) $2\sqrt{3}$
(D) $2\sqrt{3} + 2$
(E) $2\sqrt{3} - 2$

15.

Column A		Column B
	The function $f(x, y)$ is defined as the geometric mean of x and y (geometric mean of x and y equals $\sqrt{xy}$), and the function $g(x, y)$ is defined as the least common multiple of x and y. a and b are two different prime numbers.	
$f(a, b)$		$g(a, b)$

16. At time $t = 0$, a projectile was fired upward from an initial height of 10 feet. Its height after t seconds is given by the function $h(t) = p - 10(q - t)^2$, where p and q are positive constants. If the projectile reached a maximum height of 100 feet when $t = 3$, then what was the height, in feet, of the projectile when $t = 4$?
(A) 62
(B) 70
(C) 85
(D) 89
(E) 90

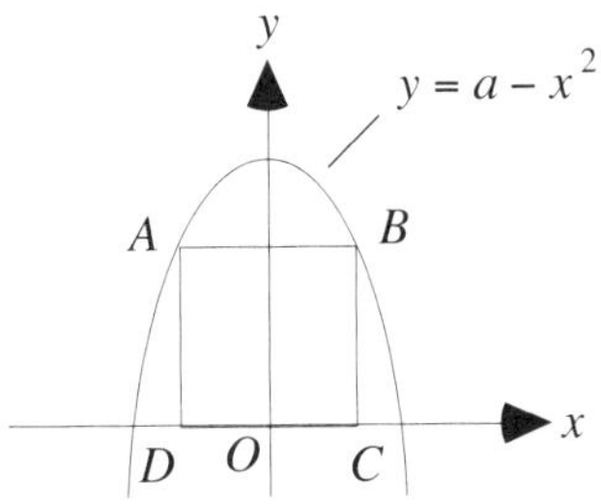

17. The figure above shows the graph of $y = a - x^2$ for some constant a. If the square $ABCD$ intersects the graph at points A and B and the area of the square is 16, what is the value of a ?
(A) 2
(B) 4
(C) 6
(D) 8
(E) 10

➢ **Very Hard**

18. If the function $f(x)$ is defined for all real numbers x as the maximum value of $2x + 4$ and $12 + 3x$, then for which one of the following values of x will $f(x)$ actually equal $2x + 4$?
(A) –4
(B) –5
(C) –6
(D) –7
(E) –9

Answers and Solutions to Problem Set BB

➢ Medium

1. We are given the function rules $f(x, y) = 2x + y$ and $g(x, y) = x + 2y$. Swapping arguments in g yields $g(y, x) = y + 2x = f(x, y)$.

Hence, $f(3, 4) = g(4, 3)$. The answer is (E).

2. Choice A: The point A represents $x = 1$. Now, $f(1) = (1 - 1.5)(1 - 2.5)(1 - 3.5)(1 - 4.5) = (-0.5)(-1.5)(-2.5)(-3.5)$ = product of four (an even number of) negative numbers. The result is positive. Reject.

Choice B. The point B represents $x = 2$. Now, $f(2) = (2 - 1.5)(2 - 2.5)(2 - 3.5)(2 - 4.5) = (0.5)(-0.5)(-1.5)(-2.5)$ = product of a positive number and three (an odd number of) negative numbers. The result is negative. Hence, correct.

Choice C: The point C represents $x = 3$. Now, $f(3) = (3 - 1.5)(3 - 2.5)(3 - 3.5)(3 - 4.5) = (1.5)(0.5)(-0.5)(-1.5)$ = Product of two positive numbers and two (an even number of) negative numbers. The result is positive. Reject.

Choice D: The point D represents $x = 4.5$. Now, $f(4.5) = (4.5 - 1.5)(4.5 - 2.5)(4.5 - 3.5)(4.5 - 4.5) = 3 \times 2 \times 1 \times 0 = 0$, not a negative number. Reject.

Choice E: The point E represents $x = 5.5$. Now, $f(5.5) = (5.5 - 1.5)(5.5 - 2.5)(5.5 - 3.5)(5.5 - 4.5) = 4 \times 3 \times 2 \times 1$ = product of positive numbers. The result is positive. Reject.

The answer is (B).

3. By the given rule, $f(x) = (x - 1)(x - 2)(x - 3)(x - 4)$. Hence, Column A equals

$$f(2.5) = (2.5 - 1)(2.5 - 2)(2.5 - 3)(2.5 - 4) = (1.5)(0.5)(-0.5)(-1.5)$$

The product of two positive numbers (1.5 and 0.5) and two negative numbers (–0.5 and –1.5) is positive.

Column B equals

$$f(3.5) = (3.5 - 1)(3.5 - 2)(3.5 - 3)(3.5 - 4) = (2.5)(1.5)(0.5)(-0.5)$$

The product of three positive numbers (2.5, 1.5, and 0.5) and a negative number (–0.5) is negative.

Since a positive number is greater than a negative number, Column A > Column B. The answer is (A).

4. $f(3, 4)$ = the average of 3 and 4 = $\dfrac{3+4}{2} = \dfrac{7}{2} = 3.5$.

$g(3, 4)$ = the greater number of 3 and 4, which is 4.

Hence, $f(3, 4) + g(3, 4) = 3.5 + 4 = 7.5$.

The answer is (D).

5. We are given the function rules $f(x, y) = 2x + y$ and $g(x, y) = x + 2y$. Swapping the arguments of g yields $g(y, x) = y + 2x = f(x, y)$. Hence, we have that $f(x, y) = g(y, x)$. Hence, $f(3, 4) = g(4, 3)$, and $f(4, 3) = g(3, 4)$.

Subtracting the two equations yields $f(3, 4) - f(4, 3) = g(4, 3) - g(3, 4)$.

Adding $f(4, 3)$ and $g(3, 4)$ to both sides yields

$$f(3, 4) + g(3, 4) = f(4, 3) + g(4, 3)$$

Hence, Column A equals Column B, and the answer is (C).

6. The function f is defined as $f(n)=\dfrac{n}{n+1}$. Substituting

$$n=1 \text{ yields } f(1)=\frac{1}{1+1}=\frac{1}{2}.$$
$$n=2 \text{ yields } f(2)=\frac{2}{2+1}=\frac{2}{3}.$$
$$n=3 \text{ yields } f(3)=\frac{3}{3+1}=\frac{3}{4}.$$

Column A = $f(1)\times f(2)=\dfrac{1}{2}\times\dfrac{2}{3}=\dfrac{1}{3}$.

Column B = $f(2)\times f(3)=\dfrac{2}{3}\times\dfrac{3}{4}=\dfrac{2}{4}=\dfrac{1}{2}$.

Since 1/2 is greater than 1/3, Column B is greater than Column A. The answer is (B).

7. Let's change the fractional notation to radical notation: $g(x)=\sqrt[4]{2x-3}+1$. Since we have an even root, the expression under the radical must be greater than or equal to zero. Hence, $2x-3\ge 0$. Adding 3 to both sides of this inequality yields $2x\ge 3$. Dividing both sides by 2 yields $x\ge 3/2$. The answer is (C).

8. We need to plug the x table values into each given function to find the one that returns the function values in the bottom row of the table. Let's start with $x=0$ since zero is the easiest number to calculate with. According to the table $f(0)=3$. This eliminates Choice (A) since $f(0)=-2(0)^2=-2(0)=0$; and it eliminates Choice (D) since $f(0)=-2(0)^2-3=-2\cdot 0-3=0-3=-3$. Now, choose $x=1$. The next easiest number to calculate with. According to the table $f(1)=1$. This eliminates Choice (B) since $f(1)=1^2+3=1+3=4$; and it eliminates Choice (C) since $f(1)=-(1)^2+3=-1+3=2$. Hence, by process of elimination, the answer is (E).

9. The graph has a height of 2 for every value of x between 2 and 3; it also has a height of 2 at about $x=-2$. The only number offered in this interval is 2.5. This is illustrated by the dot and the thick line in the following graph:

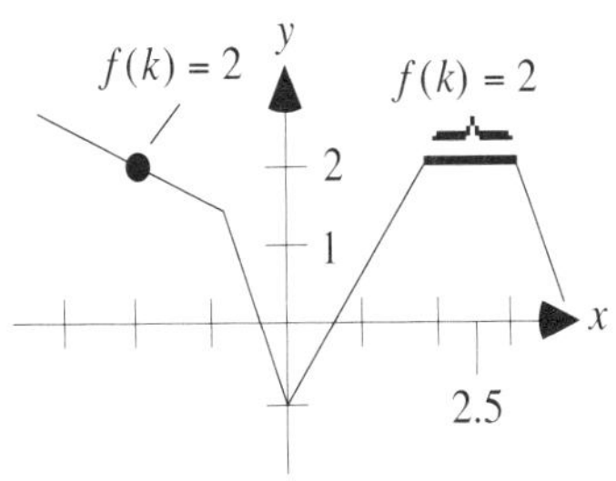

The answer is (D).

10. Evaluating the function $h(x)=\sqrt{x}+2$ at v yields $h(v)=\sqrt{v}+2$. Plugging this into the equation $3h(v) = 18$ yields

$$3\left(\sqrt{v}+2\right)=18$$

$$\sqrt{v}+2=6 \qquad \text{by dividing both sides by 3}$$

$$\sqrt{v}=4 \qquad \text{by subtracting 2 from both sides}$$

$$\left(\sqrt{v}\right)^2=4^2 \qquad \text{by squaring both sides}$$

$$v=16 \qquad \text{since } \left(\sqrt{v}\right)^2=v$$

Plugging $v = 16$ into $h\left(\frac{v}{4}\right)$ yields

$$h\left(\frac{v}{4}\right)=h\left(\frac{16}{4}\right)=h(4)=\sqrt{4}+2=2+2=4$$

The answer is (E).

11. Since the graph is symmetric about the x-axis, its base graph is $x=y^2$. Since the graph opens to the left, we know that the exterior of the base function is multiplied by negative one: $-y^2$. Since the graph is shifted one unit to the left, we know that one is subtracted from the exterior of the function: $x=-y^2-1$. The answer is (A).

12. We are asked to find the value of x for which revenue is \$110. In mathematical terms, we need to solve the equation $r(x) = 110$. Since $r(x)=50\sqrt{x}-40$, we get

$$50\sqrt{x}-40=110$$

$$50\sqrt{x}=150$$

$$\sqrt{x}=3$$

$$\left(\sqrt{x}\right)^2=3^2$$

$$|x|=9$$

$$x=9 \quad \text{or} \quad x=-9$$

Since $x = -9$ has no physical interpretation for this problem, we know that $x = 9$. The answer is (E).

➤ **Hard**

13. The given function definitions are $f(x) = 2x - 3$ and $g(x) = x + 3/2$. Putting $x = y$ in the definition of f yields $f(y) = 2y - 3$.

Putting $x = y - 3$ in the definition of g yields $g(y - 3) = (y - 3) + 3/2 = y - 3 + 3/2 = y - 3/2$.

Now, $f(y)$ equals $g(y - 3)$ when $2y - 3 = y - 3/2$. Solving for y yields $y = 3/2$. The answer is (B).

14. The rule for the function f on positive integers x is $f(x) = a\sqrt{x} + b$. Putting $x = 1$ in the rule yields $f(1) = a\sqrt{1} + b = a + b$. Putting $x = 4$ in the rule yields $f(4) = a\sqrt{4} + b = 2a + b$. Now, we are given that $f(4) - f(1) = 2$ and $f(4) + f(1) = 10$. Substituting the known results in the two equations yields

$$f(4) - f(1) = 2$$
$$(2a + b) - (a + b) = 2$$
$$a = 2$$

$$f(4) + f(1) = 10$$
$$(2a + b) + (a + b) = 10$$
$$3a + 2b = 10$$
$$3(2) + 2b = 10 \qquad \text{by putting } a = 2 \text{ in the equation}$$
$$b = 2 \qquad \text{by solving the equation for } b$$

Hence, the rule can be rephrased as $f(x) = 2\sqrt{x} + 2$. Putting $x = 3$ in the rule yields $f(3) = 2\sqrt{3} + 2$. Hence, the answer is (D).

15. Function f is defined to be the geometric mean of its arguments, and the geometric mean of a and b is $\sqrt{ab}$. Hence, Column A equals $f(a, b) = \sqrt{ab}$. Function g with arguments a and b is defined to be the LCM of a and b, and we are given that a and b are different prime numbers. Hence, the least common multiple is ab. For example, the LCM of 5 and 7 is 35 (= 5 · 7). Thus, Column B equals $g(a, b) = ab$.

The square root of a number (here ab) greater than 1 (since a and b are primes, their product is greater than 1) is always less than the number itself. Hence, $\sqrt{ab} < ab$ and therefore Column A < Column B. The answer is (B).

16. **Method I:**
Recall that when a quadratic function is written in the form $y = a(x - h)^2 + k$, its vertex (in this case, the maximum height of the projectile) occurs at the point (h, k). So let's rewrite the function $h(t) = p - 10(q - t)^2$ in the form $h(t) = a(t - h)^2 + k$. Notice that we changed y to $h(t)$ and x to t.

$$\begin{aligned} h(t) &= p - 10(q - t)^2 \\ &= -10(q - t)^2 + p \\ &= -10(-[-q + t])^2 + p \\ &= -10(-1)^2(t - q)^2 + p \\ &= -10(+1)(t - q)^2 + p \\ &= -10(t - q)^2 + p \end{aligned}$$

In this form, we can see that the vertex (maximum) occurs at the point (q, p). We are given that the maximum height of 100 occurs when t is 3. Hence, $q = 3$ and $p = 100$. Plugging this into our function yields

$$h(t) = -10(t - q)^2 + p = -10(t - 3)^2 + 100$$

We are asked to find the height of the projectile when $t = 4$. Evaluating our function at 4 yields

$$\begin{aligned} h(4) &= -10(4 - 3)^2 + 100 \\ &= -10(1)^2 + 100 \\ &= -10 \cdot 1 + 100 \\ &= -10 + 100 \\ &= 90 \end{aligned}$$

The answer is (E).

Method II:
In this method, we are going to solve a system of two equations in two unknowns in order to determine the values of p and q in the function $h(t) = p - 10(q - t)^2$. At time $t = 0$, the projectile had a height of 10 feet. In other words, $h(0) = 10$. At time $t = 3$, the projectile had a height of 100 feet. In other words, $h(3) = 100$. Plugging this information into the function $h(t) = p - 10(q - t)^2$ yields

$$h(0) = 10 \Rightarrow 10 = p - 10(q - 0)^2$$
$$h(3) = 100 \Rightarrow 100 = p - 10(q - 3)^2$$

Now, we solve this system of equations by subtracting the bottom equation from the top equation:

$$\begin{array}{r} 10 = p - 10q^2 \\ (-)\ \ 100 = p - 10(q - 3)^2 \\ \hline -90 = -10q^2 + 10(q - 3)^2 \end{array}$$

Solving this equation for q yields

$$-90 = -10q^2 + 10(q - 3)^2$$
$$-90 = -10q^2 + 10(q^2 - 6q + 9)$$
$$-90 = -10q^2 + 10q^2 - 60q + 90$$
$$-90 = -60q + 90$$
$$-180 = -60q$$
$$3 = q$$

Plugging $q = 3$ into the equation $10 = p - 10q^2$ yields

$$10 = p - 10 \cdot 3^2$$
$$10 = p - 10 \cdot 9$$
$$10 = p - 90$$
$$100 = p$$

Hence, the function $h(t) = p - 10(q - t)^2$ becomes $h(t) = 100 - 10(3 - t)^2$. We are asked to find the height of the projectile when $t = 4$. Evaluating this function at 4 yields

$$h(4) = 100 - 10(3 - 4)^2$$
$$= 100 - 10(-1)^2$$
$$= 100 - 10 \cdot 1$$
$$= 100 - 10$$
$$= 90$$

The answer is (E).

17. Let s denote the length of a side of square $ABCD$. Since the area of the square is 16, we get $s^2 = 16$. Taking the square root of both sides of this equation yields $s = 4$. Hence, line segment AB has length 4. Since the parabola is symmetric about the y-axis, Point B is 2 units from the y-axis (as is Point A). That is, the x-coordinate of Point B is 2. Since line segment BC has length 4, the coordinates of Point B are (2, 4). Since the square and the parabola intersect at Point B, the point (2, 4) must satisfy the equation $y = a - x^2$:

$$4 = a - 2^2$$
$$4 = a - 4$$
$$8 = a$$

The answer is (D).

➢ Very Hard

18. $f(x)$ equals the maximum of $2x + 4$ and $12 + 3x$. Hence, the question asks for which value of x is $2x + 4$ is greater than or equal to $12 + 3x$.* Expressing the inequality yields

$$2x + 4 \geq 12 + 3x$$

Subtracting $12 + 2x$ from both sides yields

$$x \leq -8$$

The answer is (E) since it is the only answer-choice that satisfies the inequality.

* Note: By symmetry, we could also ask for which value of x is $12 + 3x$ is greater than or equal to $2x + 4$.

Miscellaneous Problems

Example 1: The language Q has the following properties:

(1) ABC is the base word.
(2) If C immediately follows B, then C can be moved to the front of the code word to generate another word.

Which one of the following is a code word in language Q?

(A) CAB (B) BCA (C) AAA (D) ABA (E) CCC

From (1), ABC is a code word.

From (2), the C in the code word ABC can be moved to the front of the word: CAB.

Hence, CAB is a code word and the answer is (A).

Example 2: Bowl S contains only marbles. If 1/4 of the marbles were removed, the bowl would be filled to 1/2 of its capacity. If 100 marbles were added, the bowl would be full. How many marbles are in bowl S?

(A) 100 (B) 200 (C) 250 (D) 300 (E) 400

Let n be the number of marbles in the bowl, and let c be the capacity of the bowl. Then translating *"if 1/4 of the marbles were removed, the bowl would be filled to 1/2 of its capacity"* into an equation yields

$$n-\frac{1}{4}n=\frac{1}{2}c\text{, or }\frac{3}{2}n=c$$

Next, translating *"if 100 marbles were added, the bowl would be full"* into an equation yields

$$100+n=c$$

Hence, we have the system:

$$\frac{3}{2}n=c$$

$$100+n=c$$

Combining the two above equations yields

$$\frac{3}{2}n=100+n$$

$$3n=200+2n$$

$$n=200$$

The answer is (B).

Method II (Plugging in):
Suppose there are 100 marbles in the bowl—choice (A). Removing 1/4 of them would leave 75 marbles in the bowl. Since this is 1/2 the capacity of the bowl, the capacity of the bowl is 150. But if we add 100 marbles to the original 100, we get 200 marbles, not 150. This eliminates (A).

Next, suppose there are 200 marbles in the bowl—choice (B). Removing 1/4 of them would leave 150 marbles in the bowl. Since this is 1/2 the capacity of the bowl, the capacity of the bowl is 300. Now, if we add 100 marbles to the original 200, we get 300 marbles—the capacity of the bowl. The answer is (B).

Problem Set CC:

➢ **Easy**

1. A stockholder holds one share each of two different companies A and B. Last month, the value of a share of Company A increased by 13 dollars and that of Company B decreased by 8 dollars. How much did the net value of the two shares increase last month?

   (A) 2
   (B) 3
   (C) 4
   (D) 5
   (E) 6

The following two questions below refer to the statement below:

Neel is a distributor and Nick is a retailer. Neel purchased electronic shavers of a particular type at \$4 a piece and sold all of them to Nick at \$6 a piece. Nick sold all the shavers at \$8 a piece to consumers.

2.

Column A	Column B
Profit made by Neel in the deal	Profit made by Nick in the deal

➢ **Medium**

3. The percentage of profit is defined as $\frac{\text{profit}}{\text{cost}} \cdot 100$.

Column A	Column B
The percentage of profit that Neel made in the deal	The percentage of profit that Nick made in the deal

4. Which one of the following products has the greatest value?

   (A) 6.00×0.20
   (B) 6.01×0.19
   (C) 6.02×0.18
   (D) 6.03×0.17
   (E) 6.04×0.16

5. A fund was invested 25 years ago. Its value is approximately \$300,000 now. If the value of the fund doubled each year for the last 10 years, how long ago was the value of the fund exactly half of the current value?

   (A) Half a year ago.
   (B) 1 year ago.
   (C) 2 years ago.
   (D) 5 years ago.
   (E) 10 years ago.

6. Column A: $15^2 + 17^2 + 19^2$ | Column B: $(15 + 17 + 19)^2$

7. The average temperature in New Orland from January through August is 36°C. The minimum and the maximum temperatures between September and December are 26°C and 36°C, respectively.

Column A: The average temperature for the year | Column B: 36°C

8. John was born on February 28, 1999. It was a Sunday. February of that year had only 28 days, and the year had exactly 365 days. His brother Jack was born on the same day of the year 2000. February of the year 2000 had 29 days. On which day was Jack born?

(A) Monday
(B) Tuesday
(C) Friday
(D) Saturday
(E) Sunday

➢ Hard

9. Park, Jack, and Galvin distributed prize money of 120 dollars among themselves. Park received 3/10 of what Jack and Galvin together received. Jack received 3/11 of what Park and Galvin together received.

Column A: The amount received by Park | Column B: The amount received by Jack

10. Jack, Karl, Marc, and Kate are friends. They collected just enough money to buy a car. Jack contributed 1/3 of what his three friends contributed together. Karl contributed 1/4 of what his three friends contributed together. Marc contributed 2/5 of what his three friends contributed together.

Column A: The amount paid by Jack | Column B: The amount paid by Marc

Questions 11 and 12 refer to the discussion below:

A manufacturer sells goods at \$4 per unit to stockists after a 10% profit. The stockists then sell the goods to distributors at 25% profit. The distributor adds a 20% profit on it and sells it to a retailer.

11. At what price (per unit) did the retailer purchase the goods?

(A) 0.6
(B) 6
(C) 6.3
(D) 6.6
(E) 7

12. If the retailer sells the goods to the end customer at 10% profit, then what does each unit of the goods cost to a customer?

(A) 4
(B) 5
(C) 6
(D) 6.6
(E) 7

13. The letters of the word JOHNY can be jumbled in 120 ways. In how many of them does the letter 'H' appear in the middle?

(A) 1
(B) 20
(C) 24
(D) 26
(E) 30

14. Craig invited four friends to watch a TV show. He arranged 5 seats in a row. The number of ways he and his four friends can sit in the row is n. In how many of these ways can Craig sit in the middle?

(A) n
(B) $n/2$
(C) $n/3$
(D) $n/4$
(E) $n/5$

15. Eric and Ortega and their teammates watch a movie. They all sit in a row, and they can sit in n different ways. In how many of the ways can Eric sit to the right of Ortega?

(A) n
(B) $n/2$
(C) $n/3$
(D) $n/4$
(E) $n/5$

16. If distinct numbers x, y, z, and p are chosen from the numbers –2, 2, 1/2, –1/3, what is the largest possible value of the expression $\frac{x^2 y}{z - p}$?

(A) 26/4
(B) 34/5
(C) 38/5
(D) 48/5
(E) 52/5

17. A bank pays interest to its customers on the last day of the year. The interest paid to a customer is calculated as 10% of the average monthly balance maintained by the customer. John is a customer at the bank. On the last day, when the interest was accumulated into his account, his bank balance doubled to $5680. What is the average monthly balance maintained by John in his account during the year?

(A) 2840
(B) 5680
(C) 6840
(D) 7540
(E) 28400

Use the table shown to answer the next two questions.

	% (by weight) of Protein	% (by weight) of Fat	Cost of 10 grams of the food
Food A	10	30	$2
Food B	20	10	$3
Food C	30	35	$7

18. Which one of the following diets or the combination of diets supplies more protein for every dollar?

(A) Food A
(B) Food B
(C) Food C
(D) 60% food B and 40% food C
(E) 60% food A and 40% food C

19. Which diet supplies more protein than fat in each gram?

(A) Food A
(B) Food B
(C) Foods A and B in the ratio 60 : 40
(D) Foods B and C in the ratio 60 : 40
(E) Foods A and C in the ratio 60 : 40

20. Forty tiles of dimensions 1 foot × 2 foot each are required to completely cover a floor. How many tiles of dimensions 2 foot × 4 foot each would be required to completely cover the same floor?

(A) 10
(B) 20
(C) 80
(D) 160
(E) 320

21. If s and t are positive integers and $s/t = 39.12$, then which one of the following could t equal?

(A) 8
(B) 13
(C) 15
(D) 60
(E) 75

22. What is the minimum number of tiles of size 16 by 24 required to form a square by placing the tiles adjacent to one another other?

(A) 6
(B) 8
(C) 11
(D) 16
(E) 24

The following two questions refer to the statement below:

Nancy purchased a consignment of c red roses at a cost of d dollars. She made bouquets with the roses and each bouquet contained d roses. She sold each bouquet at a price of c dollars. Overall, she made a profit.

23.

Column A	Column B
c	d

24.

Column A	Column B
Cost (in dollars) of each flower in the consignment	1 dollar

25.

720 cents can buy 72 bananas.

If the price of each banana is reduced by 1 cent, 720 cents can buy 72 + x bananas.

If the price of each banana is reduced by 2 cents, 720 cents can buy 72 + y bananas.

Column A	Column B
$2x$	y

26.

The money John has is just enough to buy him either 100 apples and 150 oranges, or 50 apples and 225 oranges.

Column A	Column B
Cost of 100 apples	Cost of 150 oranges

27.

Set $A = \{-2, -1, 0, 1, 2\}$
Set $B = \{-4, -2, 0, 2, 4,\}$

Column A	Column B
The sum of the product of each element in A with each element in B	0

➢ Very Hard

28. In 2003, there are 28 days in February and there are 365 days in the year. In 2004, there are 29 days in February and there are 366 days in the year. If the date March 11, 2003 is a Tuesday, then which one of the following would the date March 11, 2004 be?

    (A) Monday
    (B) Tuesday
    (C) Wednesday
    (D) Thursday
    (E) Sunday

29. There are 5 packers A, B, C, D, and E in a company. The five packers A, B, C, D and E charge \$66, \$52, \$46, \$32, and \$28, respectively, to pack each item. The time taken by the packers to pack one item is 20 minutes, 24 minutes, 30 minutes, 40 minutes and 48 minutes, respectively. All the items are sold at the end of the day. Each item earns a profit of 100 dollars, and the packers are paid from this profit. If each packer works 8 hours a day, which packer contributes the most to the net profit of the company?

    (A) Packer A
    (B) Packer B
    (C) Packer C
    (D) Packer D
    (E) Packer E

Answers and Solutions to Problem Set CC

➢ Easy

1. We have that the value of a share of Company A increased by 13 dollars and that of Company B decreased by 8 dollars. Hence, the net increase in the combined value of the two shares is 13 – 8 = 5. The answer is (D).

2. Neel purchased the shavers at \$4 a piece and sold them at \$6 a piece. Hence, he gained 6 – 4 = 2 dollars on each piece. On n such pieces, he gained (n items)(2 dollars on each item) = $2n$ dollars. Hence, Column A equals $2n$.

Nick purchased the same items at \$6 each and sold them at \$8 each. Hence, he gained 8 – 6 = 2 dollars on each piece. On the n pieces, he gained (n items)(2 dollars on each item) = $2n$ dollars. Hence, Column B also equals $2n$.

Since Column A equals Column B, the answer is (C).

➢ Medium

3. Neel purchased the shavers at \$4 a piece. Hence, n pieces cost him $4n$ dollars.

Nick purchased the shavers from Neel at \$6 a piece. Hence, n pieces cost him $6n$ dollars.

Column A: The percentage of profit made by Neel equals $\frac{\text{profit}}{\text{cost}} \cdot 100 = \frac{2n}{4n} \cdot 100 = \frac{1}{2} \cdot 100 = 50\%$.

Column B: The percentage of profit made by Nick equals $\frac{\text{profit}}{\text{cost}} \cdot 100 = \frac{2n}{6n} \cdot 100 = \frac{1}{3} \cdot 100 \approx 33.3\%$.

Hence, Column A is greater than Column B, and the answer is (A).

4. Each answer-choice has two factors. The first factor of each answer-choice varies from 6.00 to 6.04, and the second factor varies from 0.16 to 0.20. The percentage change in the first factor is very small (0.67%) compared to the large (almost 25%) change in the second factor. Hence, we can approximate the first factor with 6.00, and the answer-choice that has the greatest second factor [choice (A)] is the biggest. Hence, the answer is (A).

Method II:
All the answer-choices are positive. Hence, we can use the ratios of the answer-choices to find which choice is the greatest:

$$\frac{\text{Choice (A)}}{\text{Choice (B)}} = \frac{6.00 \times 0.2}{6.02 \times 0.19} = \frac{600 \times 20}{601 \times 19} = \frac{600 \times 20}{(600+1) \times (20-1)} = \frac{600 \times 20}{600 \times 20 + 20 - 600 - 1} = \frac{600 \times 20}{600 \times 20 - 581} =$$

Positive Numerator ÷ Lesser Positive Denominator > 1
Hence, Choice (A) > Choice (B). Reject Choice (B).

$$\frac{\text{Choice (A)}}{\text{Choice (C)}} = \frac{6.00 \times 0.2}{6.02 \times 0.18} = \frac{600 \times 20}{602 \times 18} = \frac{600 \times 20}{(600+2) \times (20-2)} = \frac{600 \times 20}{600 \times 20 + 40 - 1200 - 4} = \frac{600 \times 20}{600 \times 20 - 1164} =$$

Positive Numerator ÷ Lesser Positive Denominator > 1
Hence, Choice (A) > Choice (C). Reject Choice (C).

$$\frac{\text{Choice (A)}}{\text{Choice (D)}} = \frac{6.00 \times 0.2}{6.03 \times 0.17} = \frac{600 \times 20}{603 \times 17} = \frac{600 \times 20}{(600+3) \times (20-3)} = \frac{600 \times 20}{600 \times 20 + 60 - 1800 - 9} = \frac{600 \times 20}{600 \times 20 - 1749} =$$

Positive Numerator ÷ Lesser Positive Denominator > 1
Hence, Choice (A) > Choice (D). Reject Choice (D).

$$\frac{\text{Choice (A)}}{\text{Choice (E)}} = \frac{6.00 \times 0.2}{6.04 \times 0.16} = \frac{600 \times 20}{604 \times 16} = \frac{600 \times 20}{(600+4) \times (20-4)} = \frac{600 \times 20}{600 \times 20 + 80 - 2400 - 16} = \frac{600 \times 20}{600 \times 20 - 2336} =$$

Positive Numerator ÷ Lesser Positive Denominator > 1
Hence, Choice (A) > Choice (E). Reject Choice (E).

The answer is (A).

5. Since the value of the fund doubled each year for the last 10 years, its value would halve *each* year going back for the period. Hence, the answer is (B).

6. We know that the square of a sum of n positive numbers is always greater than the sum of the squares of the n numbers. For example, $25 = (2 + 3)^2 > 2^2 + 3^2 = 13$. So, $(15 + 17 + 19)^2$ is greater than $15^2 + 17^2 + 19^2$. Hence, Column B is greater than Column A, and the answer is (B).

7. We are given that the average temperature from January through August is 36°C.

The average of a set of numbers always lies between the smallest number and the greatest number in the set. The minimum and the maximum temperatures between September and December are 26°C and 36°C, respectively. Hence, the average temperature of the period lies between 26°C and 36°C. So, the average temperature for the period September through December is less than 36°C.

So, the overall temperature for the year is less than 36°C. Hence, Column A is less than Column B, and the answer is (B).

8. The period February 28, 1999 through February 28, 2000 (not including the former date) does not include the complete month of February 2000 (which actually had 29 days). Hence, the length of the period is exactly 365 days (equal to the length of a normal year). Now, dividing 365 by 7 (the number of days in a week) yields a quotient of 52 and a remainder of 1. Hence, the exact length of the period is 52 weeks and one day. Hence, the day February 28 of the year 2000 would advance by one day over the date February 28 of the year 1999. Hence, since February 28, 1999 is a Sunday, the date February 28, 2000 is a Monday. The answer is (A).

➤ Hard

9. Let the amounts received by Park, Jack, and Galvin be P, J, and G, respectively.

Since the prize money of \$120 was distributed to Park, Jack, and Galvin, the amount that Jack and Galvin together received equals 120 – (the amount received by Park) = $120 - P$.

Since we are given that Park received 3/10 of what Jack and Galvin together received, we have the equation

$$P = (3/10)(120 - P)$$
$$P = 3/10 \cdot 120 - 3/10 \cdot P$$
$$P + 3/10 \cdot P = 3/10 \cdot 120$$
$$13/10 \cdot P = 3/10 \cdot 120$$
$$P = 3/13 \cdot 120 = \text{Column A}$$

Similarly, since we are given that Jack received 3/11 of what Park and Galvin together received $(120 - J)$, we have the equation

$$J = (3/11)(120 - J)$$
$$J = 3/11 \cdot 120 - 3/11 \cdot J$$
$$J + 3/11 \cdot J = 3/11 \cdot 120$$
$$14/11 \cdot J = 3/11 \cdot 120$$
$$J = 3/14 \cdot 120 = \text{Column B}$$

Since $3/13 \cdot 120$ is greater than $3/14 \cdot 120$, Column A > Column B and the answer is (A).

10. Marc contributed 2/5 of what his three friends contributed together, while Jack contributed only 1/3 of what his three friends contributed together. Clearly, Marc must have contributed more than Jack. Let's see this in detail:

Let the total contribution of the four friends be T, the contribution by Jack be J, and the contribution by Marc be M.

Now, the contribution by the three friends other than Jack is $T - J$. Jack contributed 1/3 of this. Hence, we have $J = (1/3)(T - J)$, or $J = T/4$ (by solving for J).

Also, the contribution by the three friends other than Marc is $T - M$. As given, Marc contributed 2/5 of this. Hence, we have $M = (2/5)(T - M)$, or $M = 2T/7$ (by solving for M).

Now, $2T/7$ is greater than $T/4$ and therefore Marc contributed more. The answer is (B).

11. The supply chain can be visually mapped as shown:

The manufacturer sells the goods to stockists at $4 per unit.

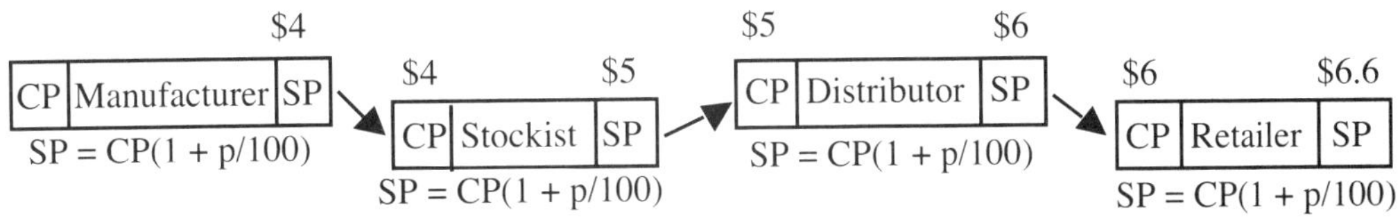

The stockist now sells the goods to distributors at a profit of 25%. By the formula, Selling price = (Cost price)(1 + profit percent/100), the selling price of the stockist (which also equals the cost price to the distributors) is $4(1 + 25/100) = 4(1 + 1/4) = 4(5/4) = \$5$.

Then the distributor sells the goods to retailers after a profit of 20%. Again since the Selling price = (Cost price)(1 + profit percent/100), the selling price of the distributor is $5(1 + 20/100) = 5(1 + 1/5) = 5(6/5) = \$6$. This is also the cost price of the retailer. Hence, the answer is (B).

12. The cost price to the end customer is the selling price of the retailer. The goods cost $6 per unit to the retailer (from the solution of the previous question). He sells it to the end customer after 10% profit. Hence, by the formula, selling price = (cost price)(1 + profit percent/100) = $\$6(1 + 10/100) = \$6.6$. The answer is (D).

13. To form a word by jumbling the five-lettered word JOHNY, the letter 'H' can be placed in any one of the five relative positions with equal probability. Hence, one fifth of the jumbled words will have the letter 'H' in the middle. We know that a total of 120 words can be formed by jumbling the word JOHNY. Hence, one-fifth of them ($1/5 \times 120 = 24$ words) have the letter 'H' in the middle. The answer is (C).

Method II
Let the letters of the words be represented by 5 compartments:

Placing H in the middle compartment gives

		H		

Now, there are 4 letters remaining for the first position (J, O, N, Y):

4		H		

Since one of these letters will be used for the first position, there are 3 letters available for the second position:

4	3	H		

Similarly, there are 2 letters available for the fourth position and 1 letter for the fifth position:

4	3	H	2	1

Multiplying the options gives

$$4 \cdot 3 \cdot 2 \cdot 1 = 24$$

The answer is (C).

14. In all the arrangements that can be formed, Craig is equally likely to sit in any one of the 5 seats. Hence, in one-fifth of all the arrangements, Craig sits in the middle seat. Since there are n possible arrangements, he sits in the middle seat $n/5$ times. The answer is (E).

15. In any arrangement, it is equally likely that Eric sits to the right of Ortega, or not. Hence, in exactly half of the n possible arrangements, Eric sits to the right of Ortega. Now, half of n is $n/2$. The answer is (B).

16. The numerator is x^2y. To maximize the value of the expression $\frac{x^2y}{z-p}$, this has to be as big as possible and should be positive if possible. Since x^2 is positive (whether x is negative or positive) use up a negative number for x. Since x^2 is in the numerator, choose a big number (in absolute value) for x, choose –2. Also, choose a big positive number for y in order to maximize the value of the given expression. So, choose 2 for y.

The denominator is $z - p$. Make this a positive value by choosing a positive value for the minimum value of z and a maximum value (in absolute value) for p. So, choose $z = 1/2$ and $p = -1/3$. So $z - p = 1/2 - (-1/3) = 1/2 + 1/3 = 5/6$.

Hence, $\frac{x^2y}{z-p} = \frac{(-2)^2 2}{\frac{5}{6}} = \frac{8}{\frac{5}{6}} = 8 \times \frac{6}{5} = \frac{48}{5}$. The answer is (D).

17. Since the balance in John's account doubled on the last day to \$5680, from the accumulation of the interest, the interest added into his account must equal 5680/2 = 2840 dollars.

Since the interest is calculated as 10% of average monthly balance maintained, we have the equation

10% of Average Monthly Balance = 2840
10/100 × Average Monthly Balance = 2840
1/10 × Average Monthly Balance = 2840
Average Monthly Balance = 10 × 2840 = 28400

The answer is (E).

18. The protein per dollar equals the protein (in grams)/weight (in grams) × weight (in grams)/money(dollars) = 1/10 × (protein (in grams)/(weight (in grams)) × weight (in grams)/money (in dollars) 10 gram cost of the diet. Now, calculating the value for each diet we have

	% (by weight) of protein	Cost of 10 grams of the food	Proteins per dollar
Food A	10% or 10/100 × 10 =1 gram	\$2	1/2 = 0.5
Food B	20% or 20/100 × 10 =2 grams	\$3	2/3 = 0.667
Food C	30% or 30/100 × 10 = 3 grams	\$7	3/7 = 0.43

Among the three foods, Food B has the greatest value (0.667 grams of protein per dollar). Any combination of this food with either of the two other foods A and C would only decrease the value of proteins per dollar. Hence, the answer is (B).

19. The protein-fat ratio for the foods here equals the ratio of "% of protein" to "% of fat." The ratios for foods A, B and C are, respectively, 10/30 = 1/3, 20/20 = 1/1, 30/35 = 6/7. In all of these, protein is less than fat.

Choice (C): Now, considering the combination in choice (C), food A to B ratio is 60 : 40.
Hence, choose 6 gm of food A and 4 gm of food B, so that foods are in this ratio.
Food A has 10% protein and 30% fat. 10% of 6 gm is 0.6 and 30% of 6 is 1.8.
Food B has 20% protein and 10% fat. 20% of 4 gm is 0.8 and 10% of 4 is 0.4.
The net sum: Protein = 0.6 (from A) + 0.8 (from B) = 1.4.
The net sum: Fat = 1.8 (from A) + 0.4 (from B) = 2.2.
The food has less protein than Fat. Reject the choice.

Choice (D): Now, considering the combination in choice (D), food B to C ratio is 60 : 40. Hence, choose 6 gm of food B and 4 gm of food C, so that foods are in this ratio.

Food B has 20% protein and 10% fat. 20% of 6 gm is 1.2 and 10% of 6 is 0.6.
Food C has 30% protein and 35% fat. 30% of 4 gm is 1.2 and 35% of 4 is 1.4.

The net sum: Protein = 1.2 (from B) + 1.2 (from C) = 2.4.
The net sum: Fat = 0.6 (from B) + 1.4 (from C) = 2.0.
The food has more protein than Fat. This is a correct choice.

Choice (E): Both the foods A and C have higher fat percentage than protein percentage. So, any combination of the food will not result in a higher protein than fat. Reject the choice.

The answer is (D).

20. The area of a 2 × 4 tile is 4 times as large as the area of a 1 × 2 tile. Hence, we need only 1/4 as many large tiles to cover the same area as the 40 small tiles. Hence, 40/4 = 10. The answer is (A).

21. We have that $s/t = 39.12$. Solving for s yields $s =$

$39.12t =$
$(39 + 0.12)t =$
$39t + 0.12t =$
$39 \times (\text{a positive integer}) + 0.12t =$
$(\text{a positive integer}) + 0.12t$

s is a positive integer only when $0.12t$ is also a positive integer. Now, $0.12t$ equals $12/100 \times t = 3/25 \times t$ and would result in an integer only when the denominator of the fraction (i.e., 25) is canceled out by t. This happens when t is a multiple of 25. The answer is (E), the only answer-choice that is a multiple of 25.

22. Let m columns and n rows be formed from the tiles of size 16 × 24. Let the columns be formed by the 16 inch sides and the rows be formed by the 24 inch sides. Then the total length of all the rows is $16m$, and the total length of all the columns is $24n$.

Since the result is a square, $16m = 24n$ or $m/n = 24/16 = 3/2$. The minimum possible values of m and n are 3 and 2, respectively. Hence, the total number of tiles required is $mn = 3 \times 2 = 6$. The answer is (A).

Method II
Since the tiles form a square, a side of the square formed must be a multiple of both 16 and 24. The least such number is 48. Since 48/16 = 3, there are three columns of length 16. Since 48/2 = 24, there are two rows of length 24. Hence, the total number of tiles required is 3 × 2 = 6. The answer is (A).

23. The consignment of c flowers cost Nancy d dollars. Hence, effectively, each flower cost her d/c dollars. Now, she made bouquets of d flowers each and sold them at the price of c dollars. Hence, effectively, the selling price of each flower is c/d dollars. Since, overall she made a profit, the selling price per flower must be greater than the cost per flower. Hence, we have $c/d > d/c$. Multiplying both sides of the inequality by cd yields $c^2 > d^2$. Square rooting both sides yields $c > d$. Hence, Column A is greater than Column B, and the answer is (A).

24. The consignment of c flowers cost Nancy d dollars. Hence, effectively, each flower cost her d/c dollars. Hence, Column A equals d/c. Now, she made bouquets of d flowers each, and sold them at the price of c dollars. Hence, effectively, the selling price of each flower is c/d dollars. Since, overall she made a profit, the selling price per flower must be greater than the cost per flower. Hence, we have $c/d > d/c$. Multiplying both sides of the inequality by cd yields $c^2 > d^2$. Square rooting both sides yields $c > d$. Dividing both sides of the inequality $c > d$ by c yields $1 > d/c$. Hence, Column A is less than Column B, and the answer is (B).

25. Since 72 bananas cost 720 cents, each banana costs 720/72 = 10 cents.

When price is reduced by 1, the new price is 10 – 1 = 9 cents. Hence, 720 cents can buy 80 (= 720/9) bananas. Equating this to $72 + x$ yields $72 + x = 80$, or $x = 80 - 72 = 8$.

When price is reduced by 2, the new price is 10 – 2 = 8 cents. Hence, 720 cents can buy 90 (= 720/8) bananas. Equating this to $72 + y$ yields $72 + y = 90$, or $y = 90 - 72 = 18$.

Now, Column A equals $2x = 2(8) = 16$, and this is less than 18 (= Column B). The answer is (B).

26. Let the cost of each apple be a, and the cost of each orange be b. Since 100 apples and 150 oranges cost the same as 50 apples and 225 oranges, we have the equation

$$100a + 150b = 50a + 225b$$
$$50a = 75b$$
$$a = (75/50)b = 3b/2$$
$$2a/3 = b$$

Hence, the apples cost 3/2 times as much as the oranges. Now, 100 apples cost $100a$, and 150 oranges cost $150b = 150(2a/3)$ [Using the known equation $2a/3 = b$] $= 100a$. Hence, both columns equal $100a$, and the answer is (C).

27. Column A = The sum of the product of each element in A with each element in B

$$= (-2) \times -4 + (-2) \times -2 + (-2) \times 0 + (-2) \times 2 + (-2) \times 4$$
$$+ (-1) \times -4 + (-1) \times -2 + (-1) \times 0 + (-1) \times 2 + (-1) \times 4$$
$$+ 0 \times -4 + 0 \times -2 + 0 \times 0 + 0 \times 2 + 0 \times 4$$
$$+ 1 \times -4 + 1 \times -2 + 1 \times 0 + 1 \times 2 + 1 \times 4$$
$$+ 2 \times -4 + 2 \times -2 + 2 \times 0 + 2 \times 2 + 2 \times 4$$

$$= (-2)\{-4 - 2 + 0 + 2 + 4\}$$
$$+ (-1)\{-4 - 2 + 0 + 2 + 4\}$$
$$+ 0\{-4 - 2 + 0 + 2 + 4\}$$
$$+ 1\{-4 - 2 + 0 + 2 + 4\}$$
$$+ 2\{-4 - 2 + 0 + 2 + 4\}$$

$$= -2 \times 0 + (-1) \times 0 + 0 \times 0 + 2 \times 0 + 4 \times 0$$

$$= 0 = \text{Column B}$$

The answer is (C).

> **Very Hard**

28. The period March 11, 2003 through March 11, 2004, not including the former date, includes the complete month of February 2004. So, the length of the period is 366 days (equal to the length of a leap year). The number 366 has a quotient of 52 and a remainder of 2 when divided by 7. Hence, the length of the period is 52 weeks and 2 days. So, the date March 11, 2004 is 2 days advanced over the date March 11, 2003, which is given to be a Tuesday. The second day after a Tuesday is a Thursday. Hence, March 11, 2004 is a Thursday. The answer is (D).

29. Each item earns a profit of 100 dollars for the company. If a packer charges x dollars for each item, the profit that the company would effectively get on each item packed would be $100 - x$.

Each worker works for 8 hours (= 480 minutes). Hence, if a worker takes t minutes to pack an item, then he would pack $480/t$ items each day. So, the net profit on the $480/t$ items is $\frac{480}{t}(100-x)$. The expression $\frac{480}{t}(100-x)$ is a maximum when $\frac{1}{t}(100-x)$ is maximum.

Hence, select the answer-choice that yields the maximum value of the expression $\frac{1}{t}(100-x)$ or $\frac{100-x}{t}$:

Choice (A): Packer A: $t = 20$ minutes, $x = 66$ dollars. Hence, $\frac{100-x}{t} = \frac{100-66}{20} = \frac{34}{20} = \frac{17}{10}$.

Choice (B): Packer B: $t = 24$ minutes, $x = 52$ dollars. Hence, $\frac{100-x}{t} = \frac{100-52}{24} = \frac{48}{24} = 2$, which is greater than 17/10. Hence, eliminate choice (A).

Choice (C): Packer C: $t = 30$ minutes, $x = 46$ dollars. Hence, $\frac{100-x}{t} = \frac{100-46}{30} = \frac{54}{30} = \frac{9}{5}$, which is less than 2. Hence, eliminate choice (C).

Choice (D): Packer D: $t = 40$ minutes, $x = 32$ dollars. Hence, $\frac{100-x}{t} = \frac{100-32}{40} = \frac{68}{40} = \frac{17}{10}$, which is less than 2. Hence, eliminate choice (D).

Choice (E): Packer E: $t = 48$ minutes, $x = 28$ dollars. Hence, $\frac{100-x}{t} = \frac{100-28}{48} = \frac{72}{48} = \frac{3}{2}$, which is less than 2. Hence, eliminate choice (E).

Hence, the correct choice is (B).

Part Two

Summary of Math Properties

Arithmetic

1. A *prime number* is an integer that is divisible only by itself and 1.
2. An even number is divisible by 2, and can be written as $2x$.
3. An odd number is not divisible by 2, and can be written as $2x + 1$.
4. Division by zero is undefined.
5. Perfect squares: 1, 4, 9, 16, 25, 36, 49, 64, 81 . . .
6. Perfect cubes: 1, 8, 27, 64, 125 . . .
7. If the last digit of a integer is 0, 2, 4, 6, or 8, then it is divisible by 2.
8. An integer is divisible by 3 if the sum of its digits is divisible by 3.
9. If the last digit of a integer is 0 or 5, then it is divisible by 5.
10. Miscellaneous Properties of Positive and Negative Numbers:

    A. The product (quotient) of positive numbers is positive.
    B. The product (quotient) of a positive number and a negative number is negative.
    C. The product (quotient) of an even number of negative numbers is positive.
    D. The product (quotient) of an odd number of negative numbers is negative.
    E. The sum of negative numbers is negative.
    F. A number raised to an even exponent is greater than or equal to zero.

$$\begin{aligned} &even \times even = even \\ &odd \times odd = odd \\ &even \times odd = even \end{aligned}$$

$$\begin{aligned} &even + even = even \\ &odd + odd = even \\ &even + odd = odd \end{aligned}$$

11. Consecutive integers are written as $x, x + 1, x + 2, \ldots$
12. Consecutive even or odd integers are written as $x, x + 2, x + 4, \ldots$
13. The integer zero is neither positive nor negative, but it is even: $0 = 2 \cdot 0$.
14. Commutative property: $x + y = y + x$. Example: $5 + 4 = 4 + 5$.
15. Associative property: $(x + y) + z = x + (y + z)$. Example: $(1 + 2) + 3 = 1 + (2 + 3)$.
16. Order of operations: Parentheses, Exponents, Multiplication, Division, Addition, Subtraction.
17. $-\frac{x}{y} = \frac{-x}{y} = \frac{x}{-y}$. Example: $-\frac{2}{3} = \frac{-2}{3} = \frac{2}{-3}$
18.

$$33\frac{1}{3}\% = \frac{1}{3} \qquad 20\% = \frac{1}{5}$$

$$66\frac{2}{3}\% = \frac{2}{3} \qquad 40\% = \frac{2}{5}$$

$$25\% = \frac{1}{4} \qquad 60\% = \frac{3}{5}$$

$$50\% = \frac{1}{2} \qquad 80\% = \frac{4}{5}$$

19.

$\frac{1}{100} = .01$	$\frac{1}{10} = .1$	$\frac{2}{5} = .4$
$\frac{1}{50} = .02$	$\frac{1}{5} = .2$	$\frac{1}{2} = .5$
$\frac{1}{25} = .04$	$\frac{1}{4} = .25$	$\frac{2}{3} = .666\ldots$
$\frac{1}{20} = .05$	$\frac{1}{3} = .333\ldots$	$\frac{3}{4} = .75$

20. Common measurements:
1 foot = 12 inches
1 yard = 3 feet
1 quart = 2 pints
1 gallon = 4 quarts
1 pound = 16 ounces

21. Important approximations: $\sqrt{2} \approx 1.4$ $\quad \sqrt{3} \approx 1.7 \quad \pi \approx 3.14$

22. *"The remainder is r when p is divided by q"* means $p = qz + r$; the integer z is called the quotient. For instance, *"The remainder is 1 when 7 is divided by 3"* means $7 = 3 \cdot 2 + 1$.

23. $$Probability = \frac{number\ of\ outcomes}{total\ number\ of\ possible\ outcomes}$$

Algebra

24. Multiplying or dividing both sides of an inequality by a negative number reverses the inequality. That is, if $x > y$ and $c < 0$, then $cx < cy$.

25. Transitive Property: If $x < y$ and $y < z$, then $x < z$.

26. Like Inequalities Can Be Added: If $x < y$ and $w < z$, then $x + w < y + z$.

27. Rules for exponents:

$$x^a \cdot x^b = x^{a+b} \qquad \text{Caution, } x^a + x^b \neq x^{a+b}$$
$$\left(x^a\right)^b = x^{ab}$$
$$(xy)^a = x^a \cdot y^a$$
$$\left(\frac{x}{y}\right)^a = \frac{x^a}{y^a}$$
$$\frac{x^a}{x^b} = x^{a-b}\text{, if } a > b. \qquad \frac{x^a}{x^b} = \frac{1}{x^{b-a}}\text{, if } b > a.$$
$$x^0 = 1$$

28. There are only two rules for roots that you need to know for the GRE:

$$\sqrt[n]{xy} = \sqrt[n]{x}\sqrt[n]{y} \qquad \text{For example, } \sqrt{3x} = \sqrt{3}\sqrt{x}.$$

$$\sqrt[n]{\frac{x}{y}} = \frac{\sqrt[n]{x}}{\sqrt[n]{y}} \qquad \text{For example, } \sqrt[3]{\frac{x}{8}} = \frac{\sqrt[3]{x}}{\sqrt[3]{8}} = \frac{\sqrt[3]{x}}{2}.$$

Caution: $\sqrt[n]{x+y} \neq \sqrt[n]{x} + \sqrt[n]{y}$.

29. Factoring formulas:

$$x(y + z) = xy + xz$$
$$x^2 - y^2 = (x + y)(x - y)$$
$$(x - y)^2 = x^2 - 2xy + y^2$$
$$(x + y)^2 = x^2 + 2xy + y^2$$
$$-(x - y) = y - x$$

30. Adding, multiplying, and dividing fractions:

$\frac{x}{y} + \frac{z}{y} = \frac{x + z}{y}$ and $\frac{x}{y} - \frac{z}{y} = \frac{x - z}{y}$ Example: $\frac{2}{4} + \frac{3}{4} = \frac{2 + 3}{4} = \frac{5}{4}$.

$\frac{w}{x} \cdot \frac{y}{z} = \frac{wy}{xz}$ Example: $\frac{1}{2} \cdot \frac{3}{4} = \frac{1 \cdot 3}{2 \cdot 4} = \frac{3}{8}$.

$\frac{w}{x} \div \frac{y}{z} = \frac{w}{x} \cdot \frac{z}{y}$ Example: $\frac{1}{2} \div \frac{3}{4} = \frac{1}{2} \cdot \frac{4}{3} = \frac{4}{6} = \frac{2}{3}$.

31. $x\% = \frac{x}{100}$

32. Quadratic Formula: $x = \frac{-b \pm \sqrt{b^2 - 4ac}}{2a}$ are the solutions of the equation $ax^2 + bx + c = 0$.

Geometry

33. There are four major types of angle measures:

An **acute angle** has measure less than 90°:

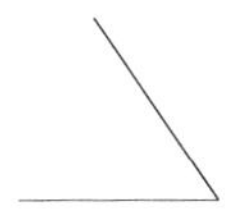

A **right angle** has measure 90°:

An **obtuse angle** has measure greater than 90°:

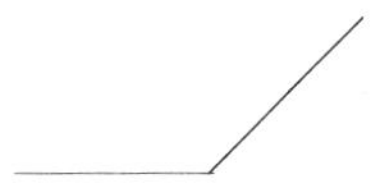

A **straight angle** has measure 180°:

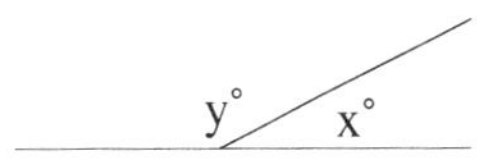

x + y = 180°

34. Two angles are supplementary if their angle sum is 180°:

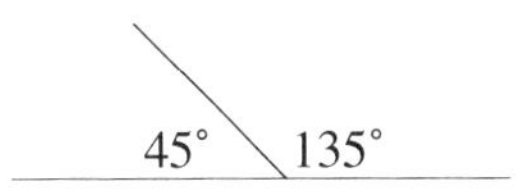

45 + 135 = 180

35. Two angles are complementary if their angle sum is 90°:

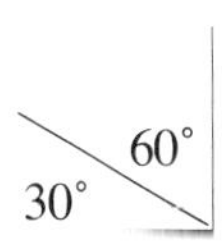

30 + 60 = 90

36. Perpendicular lines meet at right angles:

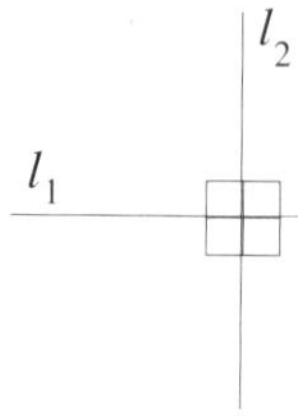

$l_1 \perp l_2$

37. When two straight lines meet at a point, they form four angles. The angles opposite each other are called vertical angles, and they are congruent (equal). In the figure to the right, $a = b$, and $c = d$.

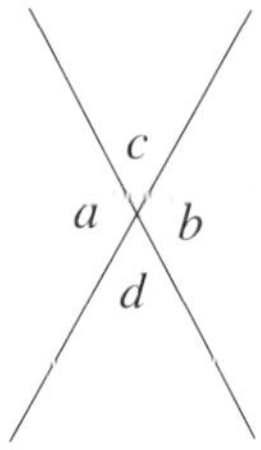

$a = b$ and $c = d$

38. When parallel lines are cut by a transversal, three important angle relationships exist:

Alternate interior angles are equal.

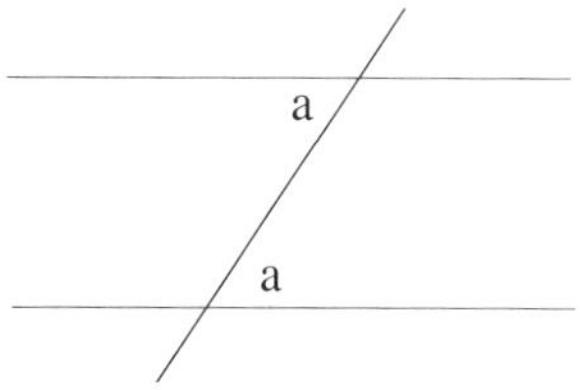

Corresponding angles are equal.

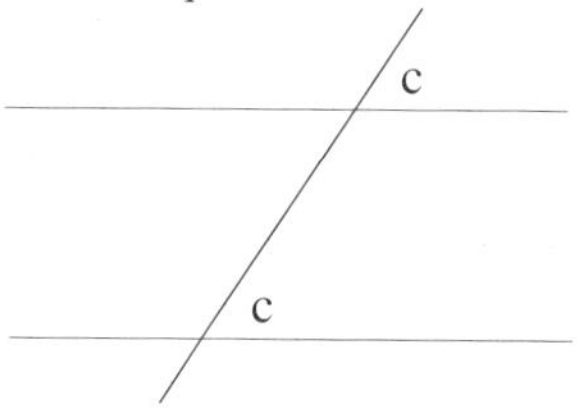

Interior angles on the same side of the transversal are supplementary.

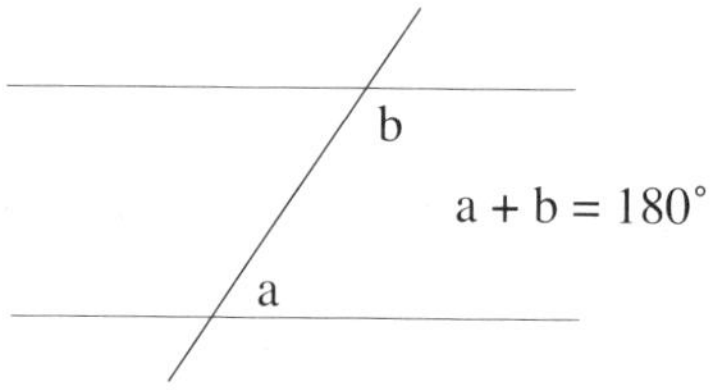

39. The shortest distance from a point not on a line to the line is along a perpendicular line.

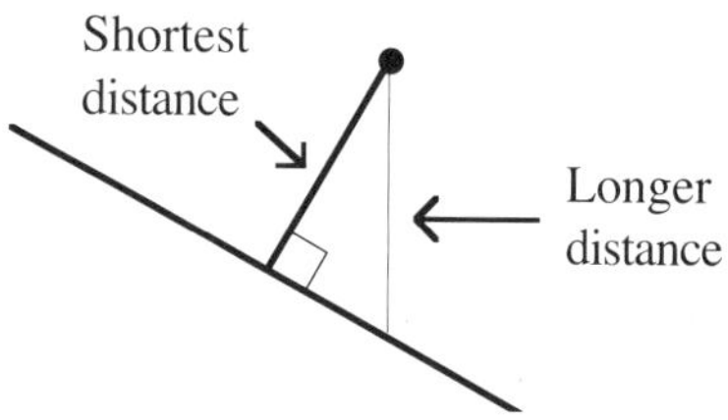

40. A triangle containing a right angle is called a *right triangle*. The right angle is denoted by a small square:

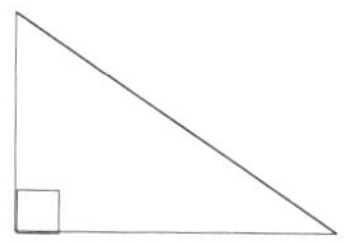

41. A triangle with two equal sides is called isosceles. The angles opposite the equal sides are called the base angles:

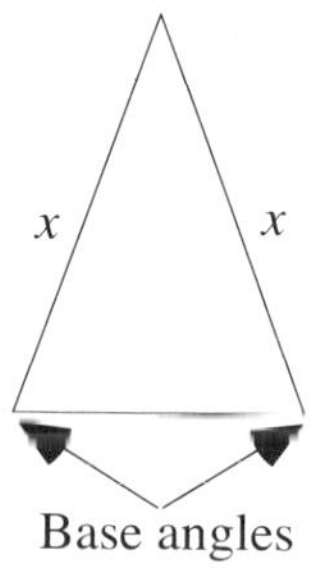

42. In an equilateral triangle, all three sides are equal and each angle is 60°:

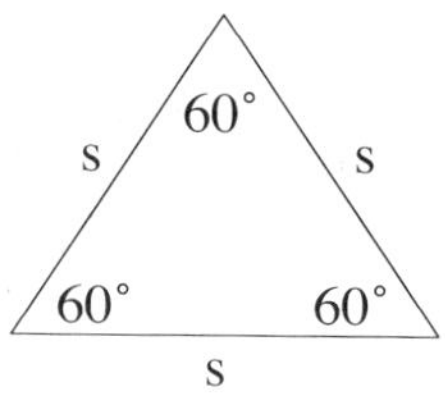

43. The altitude to the base of an isosceles or equilateral triangle bisects the base and bisects the vertex angle:

Isosceles:
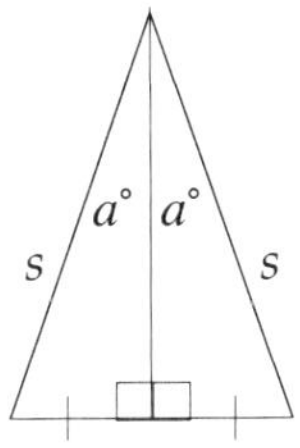

Equilateral:
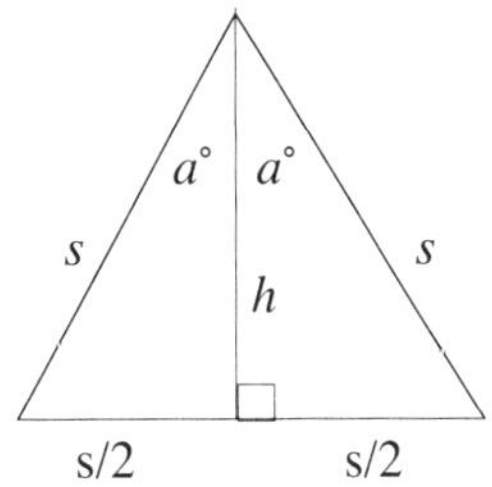

$h = \frac{s\sqrt{3}}{2}$

44. The angle sum of a triangle is 180°:

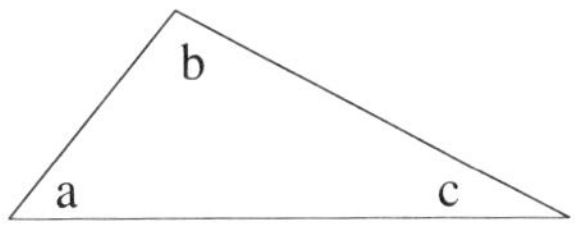

a + b + c = 180°

45. The area of a triangle is $\frac{1}{2}bh$, where b is the base and h is the height.

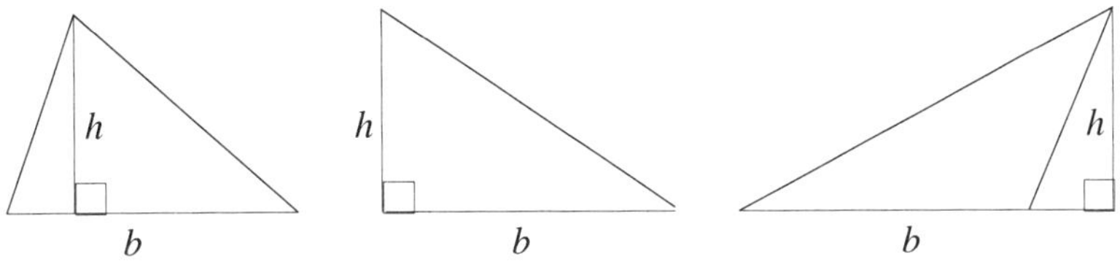

$A = \frac{1}{2}bh$

46. In a triangle, the longer side is opposite the larger angle, and vice versa:

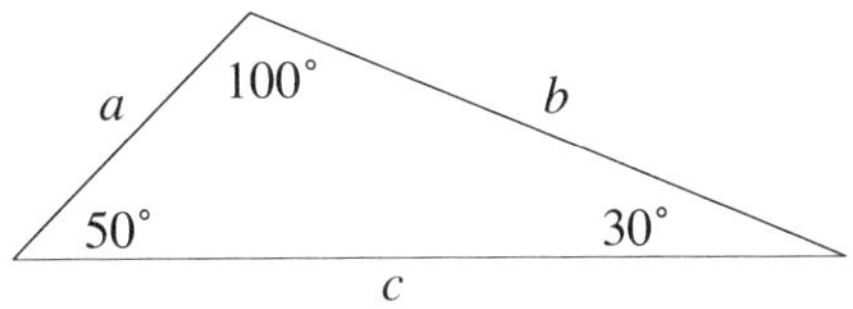

50° is larger than 30°, so side b is longer than side a.

47. Pythagorean Theorem (right triangles only): The square of the hypotenuse is equal to the sum of the squares of the legs.

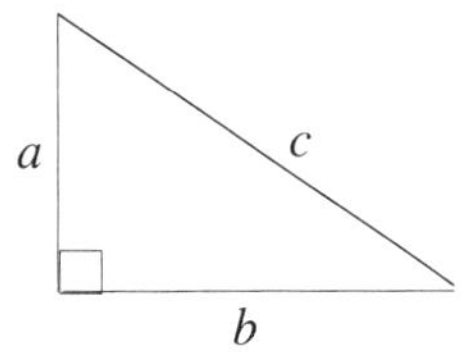

$c^2 = a^2 + b^2$

48. A Pythagorean triple: the numbers 3, 4, and 5 can always represent the sides of a right triangle and they appear very often: $5^2 = 3^2 + 4^2$.

49. Two triangles are similar (same shape and usually different size) if their corresponding angles are equal. If two triangles are similar, their corresponding sides are proportional:

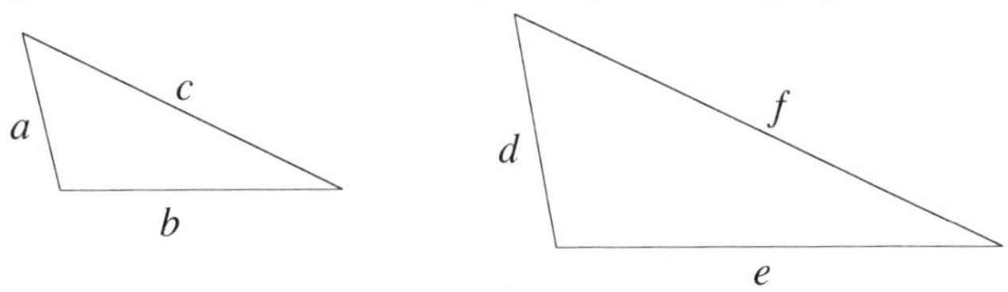

$$\frac{a}{d} = \frac{b}{e} = \frac{c}{f}$$

50. If two angles of a triangle are congruent to two angles of another triangle, the triangles are similar.

In the figure to the right, the large and small triangles are similar because both contain a right angle and they share $\angle A$.

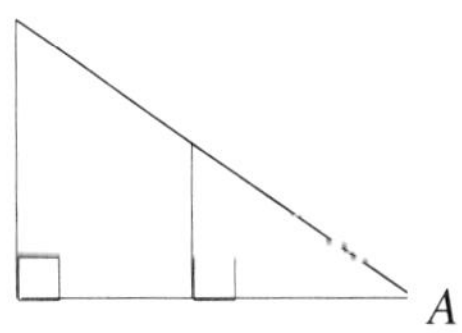

51. Two triangles are congruent (identical) if they have the same size and shape.

52. In a triangle, an exterior angle is equal to the sum of its remote interior angles and is therefore greater than either of them:

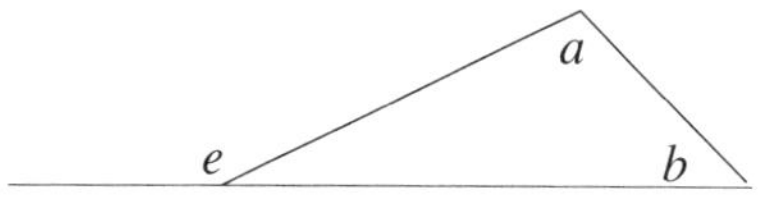

$e = a + b$ and $e > a$ and $e > b$

53. In a triangle, the sum of the lengths of any two sides is greater than the length of the remaining side:

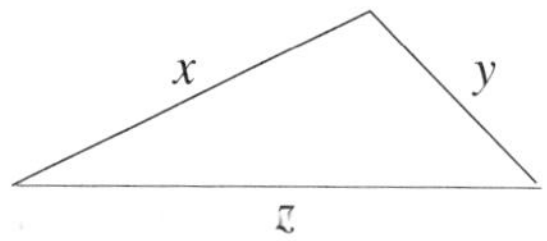

$x + y > z$

$y + z > x$

$x + z > y$

54. In a 30°–60°–90° triangle, the sides have the following relationships:

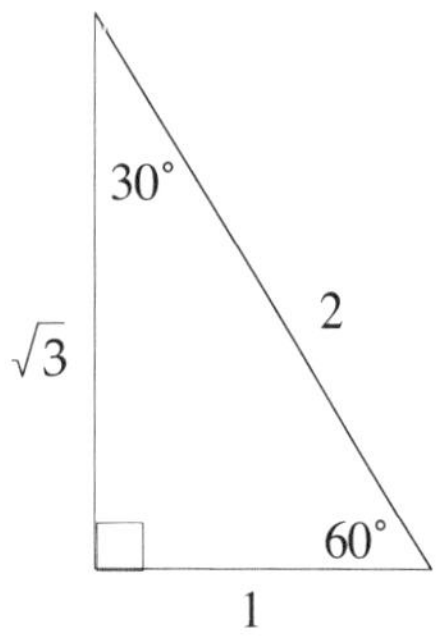

In general —>

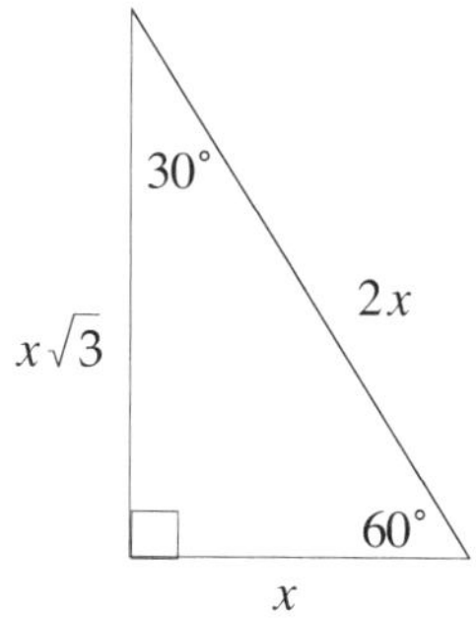

55. In a 45°–45°–90° triangle, the sides have the following relationships:

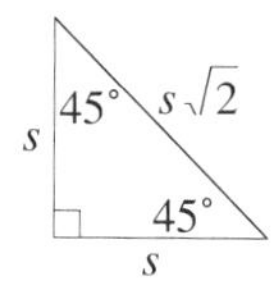

56. Opposite sides of a parallelogram are both parallel and congruent:

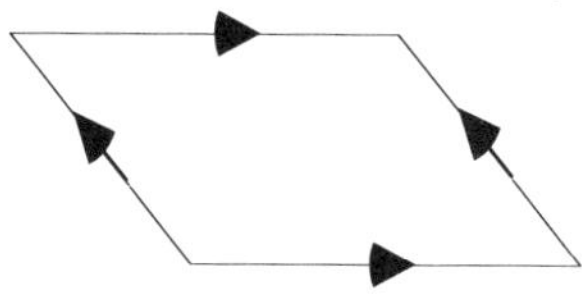

57. The diagonals of a parallelogram bisect each other:

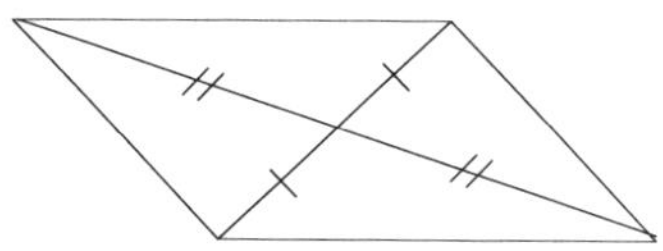

58. A parallelogram with four right angles is a *rectangle*. If w is the width and l is the length of a rectangle, then its area is $A = lw$ and its perimeter is $P = 2w + 2l$:

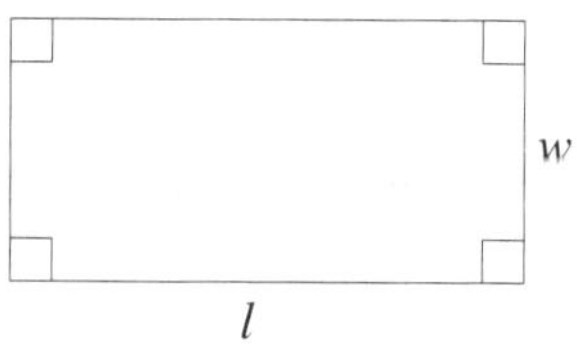

$A = l \cdot w$

$P = 2w + 2l$

59. If the opposite sides of a rectangle are equal, it is a square and its area is $A = s^2$ and its perimeter is $P = 4s$, where s is the length of a side:

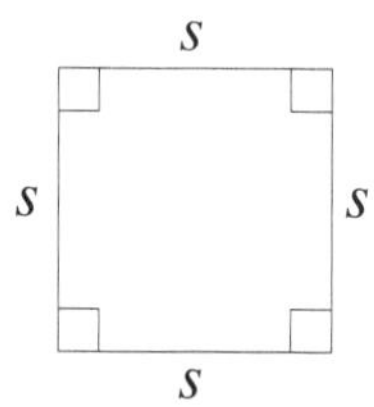

$A = s^2$

$P = 4s$

60. The diagonals of a square bisect each other and are perpendicular to each other:

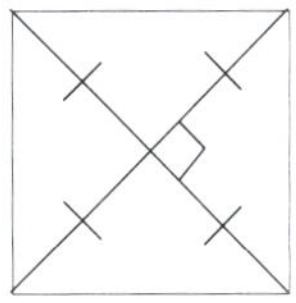

61. A quadrilateral with only one pair of parallel sides is a *trapezoid*. The parallel sides are called *bases*, and the non-parallel sides are called *legs*:

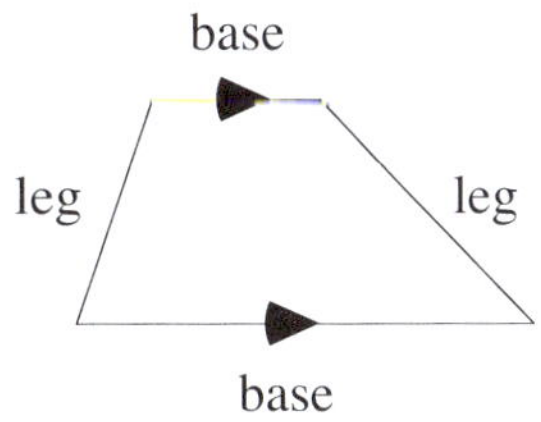

62. The area of a trapezoid is the average of the bases times the height:

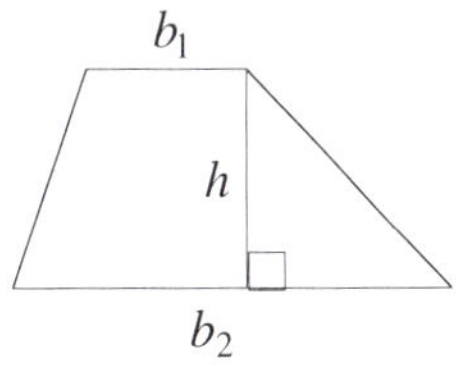

$$A = \left(\frac{b_1 + b_2}{2}\right)h$$

63. The volume of a rectangular solid (a box) is the product of the length, width, and height. The surface area is the sum of the area of the six faces:

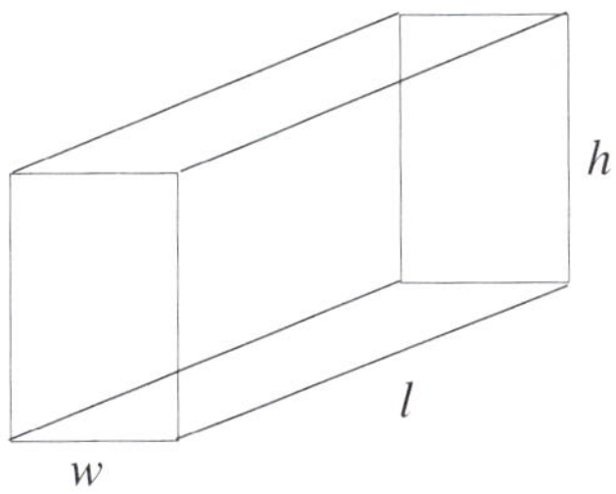

$$V = l \cdot w \cdot h$$
$$S = 2wl + 2hl + 2wh$$

64. If the length, width, and height of a rectangular solid (a box) are the same, it is a cube. Its volume is the cube of one of its sides, and its surface area is the sum of the areas of the six faces:

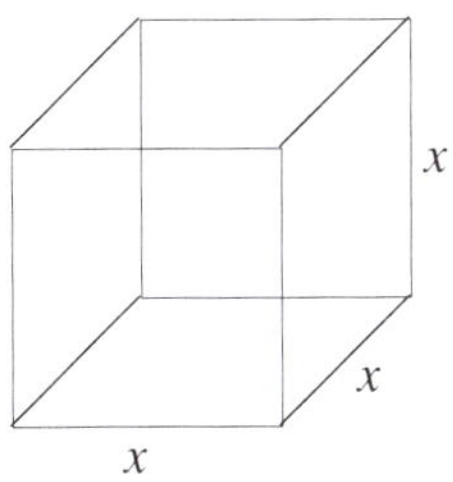

$$V = x^3$$
$$S = 6x^2$$

65. The volume of a cylinder is $V = \pi r^2 h$, and the lateral surface (excluding the top and bottom) is $S = 2\pi rh$, where r is the radius and h is the height:

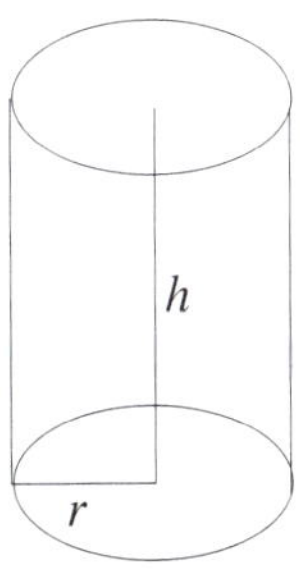

$$V = \pi r^2 h$$
$$S = 2\pi rh + 2\pi r^2$$

66. A line segment form the circle to its center is a *radius*.
A line segment with both end points on a circle is a *chord*.
A chord passing though the center of a circle is a *diameter*.
A diameter can be viewed as two radii, and hence a diameter's length is twice that of a radius.
A line passing through two points on a circle is a *secant*.
A piece of the circumference is an *arc*.
The area bounded by the circumference and an angle with vertex at the center of the circle is a *sector*.

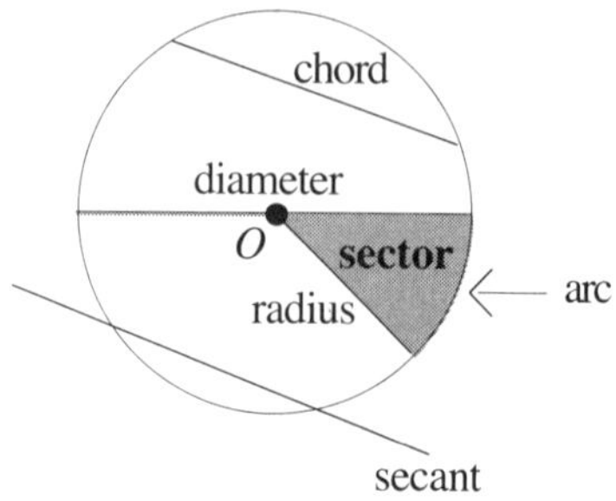

67. A tangent line to a circle intersects the circle at only one point. The radius of the circle is perpendicular to the tangent line at the point of tangency:

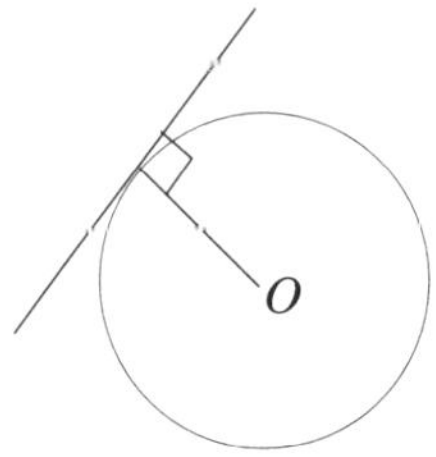

68. Two tangents to a circle from a common exterior point of the circle are congruent:

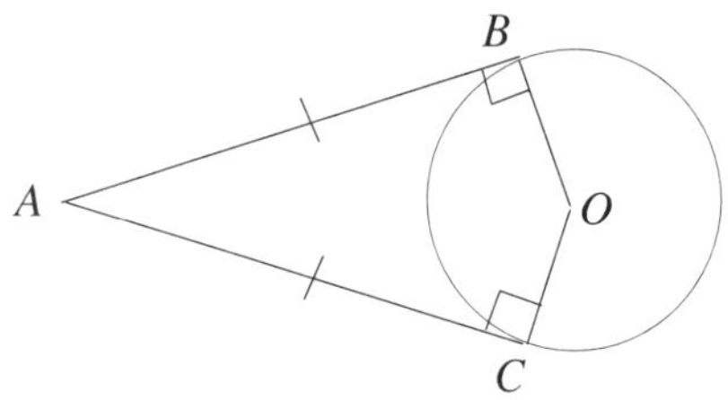

$AB \cong AC$

69. An angle inscribed in a semicircle is a right angle:

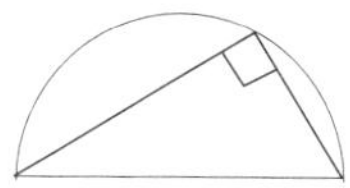

70. A central angle has by definition the same measure as its intercepted arc.

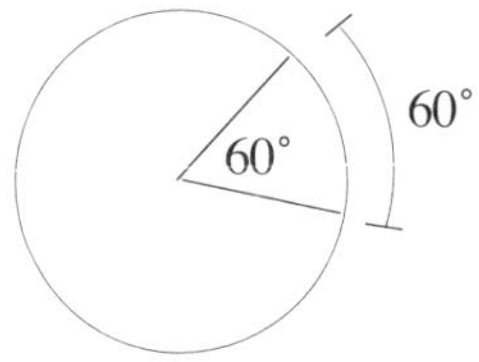

71. An inscribed angle has one-half the measure of its intercepted arc.

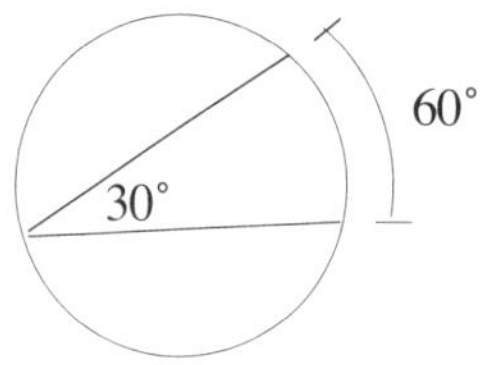

72. The area of a circle is πr^2, and its circumference (perimeter) is $2\pi r$, where r is the radius:

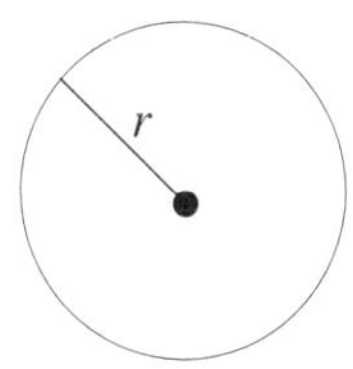

$A = \pi r^2$

$C = 2\pi r$

73. To find the area of the shaded region of a figure, subtract the area of the unshaded region from the area of the entire figure.

74. When drawing geometric figures, don't forget extreme cases.

Miscellaneous

75. To compare two fractions, cross-multiply. The larger product will be on the same side as the larger fraction.

76. Taking the square root of a fraction between 0 and 1 makes it larger.

    Caution: This is not true for fractions greater than 1. For example, $\sqrt{\frac{9}{4}} = \frac{3}{2}$. But $\frac{3}{2} < \frac{9}{4}$.

77. Squaring a fraction between 0 and 1 makes it smaller.

78. $ax^2 \neq (ax)^2$. In fact, $a^2x^2 = (ax)^2$.

79. $\frac{1/a}{b} \neq \frac{1}{a/b}$. In fact, $\frac{1/a}{b} = \frac{1}{ab}$ and $\frac{1}{a/b} = \frac{b}{a}$.

80. $-(a + b) \neq -a + b$. In fact, $-(a + b) = -a - b$.

81. $\textit{percentage increase} = \frac{\textit{increase}}{\textit{original amount}}$

82. Systems of simultaneous equations can most often be solved by merely adding or subtracting the equations.

83. When counting elements that are in overlapping sets, the total number will equal the number in one group plus the number in the other group minus the number common to both groups.

84. The number of integers between two integers inclusive is one more than their difference.

85. Elimination strategies:
    A. On hard problems, if you are asked to find the least (or greatest) number, then eliminate the least (or greatest) answer-choice.
    B. On hard problems, eliminate the answer-choice "not enough information."
    C. On hard problems, eliminate answer-choices that merely repeat numbers from the problem.
    D. On hard problems, eliminate answer-choices that can be derived from elementary operations.
    E. After you have eliminated as many answer-choices as you can, choose from the more complicated or more unusual answer-choices remaining.

86. To solve a fractional equation, multiply both sides by the LCD (lowest common denominator) to clear fractions.

87. You can cancel only over multiplication, not over addition or subtraction. For example, the c's in the expression $\frac{c + x}{c}$ cannot be canceled.

88. The average of N numbers is their sum divided by N, that is, $\textit{average} = \frac{\textit{sum}}{N}$.

89. *Weighted average:* The average between two sets of numbers is closer to the set with more numbers.

90. $\textit{Average Speed} = \frac{\textit{Total Distance}}{\textit{Total Time}}$

91. $\textit{Distance} = \textit{Rate} \times \textit{Time}$

92. *Work* = *Rate* × *Time*, or $W = R \times T$. The amount of work done is usually 1 unit. Hence, the formula becomes $1 = R \times T$. Solving this for R gives $R = \frac{1}{T}$.

93. *Interest* = *Amount* × *Time* × *Rate*

94. Principles for solving quantitative comparisons

    A. You can add or subtract the same term (number) from both sides of a quantitative comparison problem.

    B. You can multiply or divide both sides of a quantitative comparison problem by the same positive term (number). (Caution: this cannot be done if the term can ever be negative or zero.)

    C. When using substitution on quantitative comparison problems, you must plug in all five major types of numbers: positives, negatives, fractions, 0, and 1. Test 0, 1, 2, –2, and 1/2, in that order.

    D. If there are only numbers (i.e., no variables) in a quantitative comparison problem, then "not-enough-information" cannot be the answer.

95. Substitution (Special Cases):

    A. In a problem with two variables, say, x and y, you must check the case in which $x = y$. (This often gives a double case.)

    B. When you are given that $x < 0$, you must plug in negative whole numbers, negative fractions, and –1. (Choose the numbers –1, –2, and –1/2, in that order.)

    C. Sometimes you have to plug in the first three numbers (but never more than three) from a class of numbers.

Part Three

Diagnostic/ Review Test

This diagnostic test appears at the end of the book because it is probably best for you to use it as a review test. Unless your math skills are very strong, you should thoroughly study every chapter. Afterwards, you can use this diagnostic/review test to determine which chapters you need to work on more. If you do not have much time to study, this test can also be used to concentrate your studies on your weakest areas.

1. If $3x + 9 = 15$, then $x + 2 =$

(A) 2
(B) 3
(C) 4
(D) 5
(E) 6

2. If $a = 3b$, $b^2 = 2c$, $9c = d$, then $\frac{a^2}{d} =$

(A) 1/2
(B) 2
(C) 10/3
(D) 5
(E) 6

3.

$$a + b + c/2 = 60$$
$$-a - b + c/2 = -10$$

Column A	Column B
b	c

4. $3 - (2^3 - 2[3 - 16 \div 2]) =$

(A) –15
(B) –5
(C) 1
(D) 2
(E) 30

5. $(x - 2)(x + 4) - (x - 3)(x - 1) = 0$

(A) –5
(B) –1
(C) 0
(D) 1/2
(E) 11/6

6. $-2^4 - \left(x^2 - 1\right)^2 =$

(A) $-x^4 + 2x^2 + 15$
(B) $-x^4 - 2x^2 + 17$
(C) $-x^4 + 2x^2 - 17$
(D) $-x^4 + 2x^2 - 15$
(E) $-x^4 + 2x^2 + 17$

7. The smallest prime number greater than 48 is

(A) 49
(B) 50
(C) 51
(D) 52
(E) 53

8. If a, b, and c are consecutive integers and $a < b < c$, which of the following must be true?

(A) b^2 is a prime number
(B) $\frac{a+c}{2} = b$
(C) $a + b$ is even
(D) $\frac{ab}{3}$ is an integer
(E) $c - a = b$

9. $\sqrt{(42-6)(20+16)} =$

(A) 2
(B) 20
(C) 28
(D) 30
(E) 36

10. $\left(4^x\right)^2 =$

(A) 2^{4x}
(B) 4^{x+2}
(C) 2^{2x+2}
(D) 4^{x^2}
(E) 2^{2x^2}

11. If $8^{13} = 2^z$, then $z =$

(A) 10
(B) 13
(C) 19
(D) 26
(E) 39

12. 1/2 of 0.2 percent equals

(A) 1
(B) 0.1
(C) 0.01
(D) 0.001
(E) 0.0001

13. $\dfrac{4}{\frac{1}{3}+1} =$

(A) 1
(B) 1/2
(C) 2
(D) 3
(E) 4

14. If $x + y = k$, then $3x^2 + 6xy + 3y^2 =$

(A) k
(B) $3k$
(C) $6k$
(D) k^2
(E) $3k^2$

15. $8x^2 - 18 =$

(A) $8(x^2 - 2)$
(B) $2(2x + 3)(2x - 3)$
(C) $2(4x + 3)(4x - 3)$
(D) $2(2x + 9)(2x - 9)$
(E) $2(4x + 3)(x - 3)$

16. For which values of x is the following inequality true: $x^2 < 2x$.

(A) $x < 0$
(B) $0 < x < 2$
(C) $-2 < x < 2$
(D) $x < 2$
(E) $x > 2$

17. If x is an integer and $y = -3x + 7$, what is the least value of x for which y is less than 1?

(A) 1
(B) 2
(C) 3
(D) 4
(E) 5

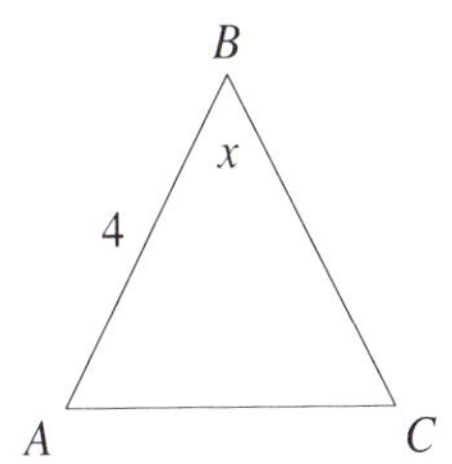

Note, figure not drawn to scale

18. In the figure above, triangle ABC is isosceles with base AC. If $x = 60°$, then $AC =$

(A) 2
(B) 3
(C) 4
(D) 14/3
(E) $\sqrt{30}$

19. A unit square is circumscribed about a circle. If the circumference of the circle is $q\pi$, what is the value of q?

(A) 1
(B) 2
(C) π
(D) 2π
(E) 5π

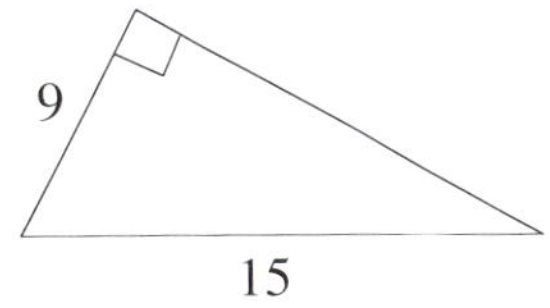

20. What is the area of the triangle above?

(A) 20
(B) 24
(C) 30
(D) 54
(E) 64

21. If the average of $2x$ and $4x$ is 12, then $x =$

(A) 1
(B) 2
(C) 3
(D) 4
(E) 24

22. The average of x, y, and z is 8 and the average of y and z is 4. What is the value of x?

(A) 4
(B) 9
(C) 16
(D) 20
(E) 24

23. If the ratio of two numbers is 6 and their sum is 21, what is the value of the larger number?

(A) 1
(B) 5
(C) 12
(D) 17
(E) 18

24. What percent of $3x$ is $6y$ if $x = 4y$?

(A) 50%
(B) 40%
(C) 30%
(D) 20%
(E) 18%

25. $y = 3x$ and $x > 2$

Column A	Column B
10% of y	40% of x

26. How many ounces of water must be added to a 30-ounce solution that is 40 percent alcohol to dilute the solution to 25 percent alcohol?

(A) 9
(B) 10
(C) 15
(D) 16
(E) 18

27. What is the value of the 201st term of a sequence if the first term of the sequence is 2 and each successive term is 4 more than the term immediately preceding it?

(A) 798
(B) 800
(C) 802
(D) 804
(E) 806

28. A particular carmaker sells four models of cars, and each model comes with 5 options. How many different types of cars does the carmaker sell?

(A) 15
(B) 16
(C) 17
(D) 18
(E) 20

29. Define a @ b to be $a^3 - 1$.

Column A	Column B
x @ 1	x @ 10

30. Define the symbol * by the following equation: $x^* = 1 - x$, for all non-negative x. If $((1-x)^*)^* = (1-x)^*$, then $x =$

(A) 1/2
(B) 3/4
(C) 1
(D) 2
(E) 3

1. Dividing both sides of the equation by 3 yields

$$x + 3 = 5$$

Subtracting 1 from both sides of this equation (because we are looking for $x + 2$) yields

$$x + 2 = 4$$

The answer is (C).

2.

$$\frac{a^2}{d} =$$

$$\frac{(3b)^2}{9c} = \quad \text{since } a = 3b \text{ and } 9c = d$$

$$\frac{9b^2}{9c} =$$

$$\frac{b^2}{c} =$$

$$\frac{2c}{c} = \quad \text{since } b^2 = 2c$$

$$2$$

The answer is (B).

3. Merely adding the two equations yields

$$c = 50$$

Next, multiplying the bottom equation by –1 and then adding the equations yields

$$\begin{array}{rr} & a + b + c/2 = 60 \\ (+) & \underline{a + b - c/2 = 10} \\ & 2a + 2b = 70 \end{array}$$

Dividing this equation by 2 yields

$$a + b = 35$$

This equation does not allow us to determine whether the value of b is larger, smaller, or equal to 50. For example, if $a = 0$, then $b = 35$. In this case, Column B is larger. Now suppose, is $a = -15$, then $b = 50$. In this case, the columns are equal. This is a double case and therefore the answer is (D), not enough information to decide.

4.

$$3 - (2^3 - 2[3 - 16 \div 2]) =$$ Within the innermost parentheses, division is performed before subtraction:

$$3 - (2^3 - 2[3 - 8]) =$$

$$3 - (2^3 - 2[-5]) =$$

$$3 - (8 - 2[-5]) =$$

$$3 - (8 + 10) =$$

$$3 - 18 =$$

$$-15$$

The answer is (A).

5. Multiplying (using foil multiplication) both terms in the expression yields

$$x^2 + 4x - 2x - 8 - (x^2 - x - 3x + 3) = 0$$

(Notice that parentheses are used in the second expansion but not in the first. Parentheses must be used in the second expansion because the negative sign must be distributed to *every* term within the parentheses.)

Combining like terms yields

$$x^2 + 2x - 8 - (x^2 - 4x + 3) = 0$$

Distributing the negative sign to every term within the parentheses yields

$$x^2 + 2x - 8 - x^2 + 4x - 3 = 0$$

(Note, although distributing the negative sign over the parentheses is an elementary operation, many, if not most, students will apply the negative sign to only the first term:

$$-x^2 - 4x + 3$$

The writers of the test are aware of this common mistake and structure the test so that there are many opportunities to make this mistake.)

Grouping like terms together yields

$$(x^2 - x^2) + (2x + 4x) + (-8 - 3) = 0$$

Combining the like terms yields

$$6x - 11 = 0$$
$$6x = 11$$
$$x = 11/6$$

The answer is (E).

6.

$$-2^4 - (x^2 - 1)^2 =$$
$$-16 - [(x^2)^2 - 2x^2 + 1] =$$
$$-16 - [x^4 - 2x^2 + 1] =$$
$$-16 - x^4 + 2x^2 - 1 =$$
$$-x^4 + 2x^2 - 17$$

The answer is (C).

Notice that $-2^4 = -16$, not 16. This is one of the most common mistakes on the test. To see why $-2^4 = -16$ more clearly, rewrite -2^4 as follows:

$$-2^4 = (-1)2^4$$

In this form, it is clearer that the exponent, 4, applies only to the number 2, not to the number –1. So $-2^4 = (-1)2^4 = (-1)16 = -16$.

To make the answer positive 16, the –2 could be placed in parentheses:

$$(-2)^4 = [(-1)2]^4 = (-1)^4\, 2^4 = (+1)16 = 16$$

7. Since the question asks for the *smallest* prime greater then 48, we start with the smallest answer-choice. Now, 49 is not prime since $49 = 7 \cdot 7$. Next, 50 is not prime since $50 = 5 \cdot 10$. Next, 51 is not prime since $51 = 3 \cdot 17$. Next, 52 is not prime since $52 = 2 \cdot 26$. Finally, 53 *is* prime since it is divisible by only itself and 1. The answer is (E).

Note, an integer is prime if it greater than 1 and divisible by only itself and 1. The number 2 is the smallest prime (and the only even prime) because the only integers that divide into it evenly are 1 and 2. The number 3 is the next larger prime. The number 4 is not prime because $4 = 2 \cdot 2$. Following is a partial list of the prime numbers. You should memorize it.

2, 3, 5, 7, 11, 13, 17, 19, 23, 29, 31, . . .

8. Recall that an integer is prime if it is divisible by only itself and 1. In other words, an integer is prime if it cannot be written as a product of two other integers, other than itself and 1. Now, $b^2 = bb$. Since b^2 can be written as a product of b and b, it is not prime. Statement (A) is false.

Turning to Choice (B), since a, b, and c are consecutive integers, in that order, b is one unit larger than a: $b = a + 1$, and c is one unit larger than b: $c = b + 1 = (a + 1) + 1 = a + 2$. Now, plugging this information into the expression $\frac{a+c}{2}$ yields

$$\frac{a+c}{2} =$$
$$\frac{a+(a+2)}{2} =$$
$$\frac{2a+2}{2} =$$
$$\frac{2a}{2} + \frac{2}{2} =$$
$$a + 1 =$$
$$b$$

The answer is (B).

Regarding the other answer-choices, Choice (C) is true in some cases and false in others. To show that it can be false, let's plug in some numbers satisfying the given conditions. How about $a = 1$ and $b = 2$. In this case, $a + b = 1 + 2 = 3$, which is odd, not even. This eliminates Choice (C). Notice that to show a statement is false, we need only find one exception. However, to show a statement is true by plugging in numbers, you usually have to plug in more than one set of numbers because the statement may be true for one set of numbers but not for another set. We'll discuss in detail later the conditions under which you can say that a statement is true by plugging in numbers.

Choice (D) is not necessarily true. For instance, let $a = 1$ and $b = 2$. Then $\frac{ab}{3} = \frac{1 \cdot 2}{3} = \frac{2}{3}$, which is not an integer. This eliminates Choice (D).

Finally, $c - a = b$ is not necessarily true. For instance, let $a = 2$, $b = 3$, and $c = 4$. Then $c - a = 4 - 2 = 2 \neq 3$. This eliminates Choice (E).

9.

$$\sqrt{(42-6)(20+16)} =$$
$$\sqrt{(36)(36)} =$$
$$\sqrt{36}\sqrt{36} = \qquad \text{from the rule } \sqrt{xy} = \sqrt{x}\sqrt{y}$$
$$6 \cdot 6 =$$
$$36$$

The answer is (E).

10.

$$\left(4^x\right)^2 =$$
$$4^{2x} = \qquad \text{by the rule } \left(x^a\right)^b = x^{ab}$$
$$\left(2^2\right)^{2x} = \qquad \text{by replacing 4 with } 2^2$$
$$(2)^{4x} \qquad \text{by the rule } \left(x^a\right)^b = x^{ab}$$

The answer is (A). Note, this is considered to be a hard problem.

As to the other answer-choices, Choice (B) wrongly adds the exponents x and 2. The exponents are added when the same bases are multiplied:

$$a^x a^y = a^{x+y}$$

For example: $2^3 2^2 = 2^{3+2} = 2^5 = 32$. Be careful not to multiply unlike bases. For example, do not add exponents in the following expression: $2^3 4^2$. The exponents cannot be added here because the bases, 2 and 4, are not the same.

Choice (C), first changes 4 into 2^2, and then correctly multiplies 2 and x: $\left(2^2\right)^x = 2^{2x}$. However, it then errs in adding $2x$ and 2: $\left(2^{2x}\right)^2 \neq 2^{2x+2}$.

Choice (D) wrongly squares the x. When a power is raised to another power, the powers are multiplied:

$$\left(x^a\right)^b = x^{ab}$$

So $\left(4^x\right)^2 = 4^{2x}$.

Choice (E) makes the same mistake as in Choice (D).

11. The number 8 can be written as 2^3. Plugging this into the equation $8^{13} = 2^z$ yields

$$\left(2^3\right)^{13} = 2^z$$

Applying the rule $\left(x^a\right)^b = x^{ab}$ yields

$$2^{39} = 2^z$$

Since the bases are the same, the exponents must be the same. Hence, $z = 39$, and the answer is (E).

12. Recall that percent means to divide by 100. So .2 percent equals .2/100 = .002. (Recall that the decimal point is moved to the left one space for each zero in the denominator.) Now, as a decimal 1/2 = .5.

In percent problems, "of" means multiplication. So multiplying .5 and .002 yields

$$\begin{array}{r} .002 \\ \times \quad .5 \\ \hline .001 \end{array}$$

Hence, the answer is (D).

13.

$$\frac{4}{\frac{1}{3}+1} =$$

$$\frac{4}{\frac{1}{3}+\frac{3}{3}} = \quad \text{by creating a common denominator of 3}$$

$$\frac{4}{\frac{1+3}{3}} =$$

$$\frac{4}{\frac{4}{3}} =$$

$$4 \cdot \frac{3}{4} = \quad \text{Recall: "to divide" means to invert and multiply}$$

$$3 \quad \text{by canceling the 4's}$$

Hence, the answer is (D).

14.

$3x^2 + 6xy + 3y^2 =$

$3(x^2 + 2xy + y^2) =$ by factoring out the common factor 3

$3(x + y)^2 =$ by the perfect square trinomial formula $x^2 + 2xy + y^2 = (x + y)^2$

$3k^2$

Hence, the answer is (E).

15. $8x^2 - 18 =$
$2(4x^2 - 9) =$ by the distributive property $ax + ay = a(x + y)$
$2(2^2x^2 - 3^2) =$
$2([2x]^2 - 3^2) =$
$2(2x + 3)(2x - 3)$ by the difference of squares formula $x^2 - y^2 = (x + y)(x - y)$

The answer is (B).

It is common for students to wrongly apply the difference of squares formula to a perfect square:

$$(x - y)^2 \neq (x + y)(x - y)$$

The correct formulas follow. Notice that the first formula is the square of a difference, and the second formula is the difference of two squares.

Perfect square trinomial: $(x - y)^2 = x^2 - 2xy + y^2$

Difference of squares: $x^2 - y^2 = (x + y)(x - y)$

It is also common for students to wrongly distribute the 2 in a perfect square:

$$(x - y)^2 \neq x^2 - y^2$$

Note, there is no factoring formula for a sum of squares: $x^2 + y^2$. It cannot be factored.

16. First, replace the inequality symbol with an equal symbol: $x^2 = 2x$

Subtracting $2x$ from both sides yields $x^2 - 2x = 0$

Factoring by the distributive rule yields $x(x - 2) = 0$

Setting each factor to 0 yields $x = 0$ and $x - 2 = 0$

Or $x = 0$ and $x = 2$

Now, the only numbers at which the expression can change sign are 0 and 2. So 0 and 2 divide the number line into three intervals. Let's set up a number line and choose test points in each interval:

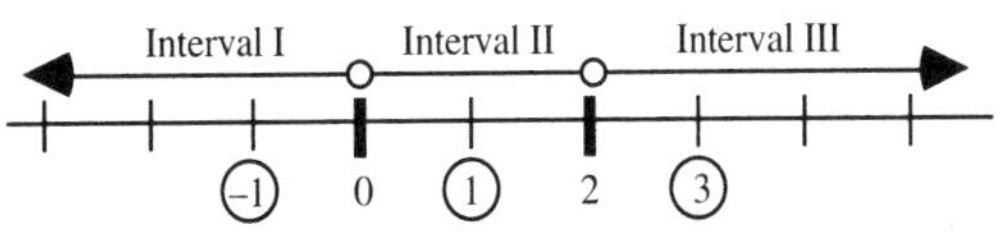

When $x = -1$, $x^2 < 2x$ becomes $1 < -2$. This is false. Hence, no numbers in Interval I satisfy the inequality. When $x = 1$, $x^2 < 2x$ becomes $1 < 2$. This is true. Hence, all numbers in Interval II satisfy the inequality. That is, $0 < x < 2$. When $x = 3$, $x^2 < 2x$ becomes $9 < 6$. This is false. Hence, no numbers in Interval III satisfy the inequality. The answer is (B). The graph of the solution follows:

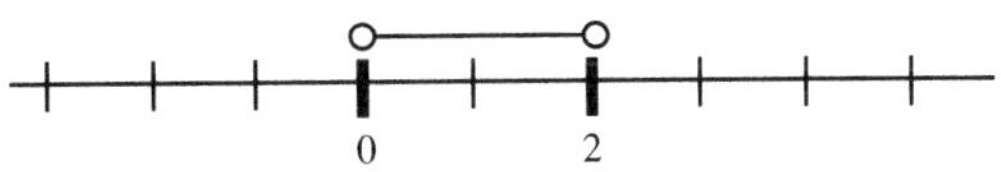

17. Since y is to be less than 1 and $y = -3x + 7$, we get

$-3x + 7 < 1$
$-3x < -6$ by subtracting 7 from both sides of the inequality
$x > 2$ by dividing both sides of the inequality by -3

(Note that the inequality changes direction when we divide both sides by a negative number. This is also the case if you multiply both sides of an inequality by a negative number.)

Since x is an integer and is to be as small as possible, $x = 3$. The answer is (C).

18. Since the triangle is isosceles, with base AC, the base angles are congruent (equal). That is, $A = C$. Since the angle sum of a triangle is 180, we get

$$A + C + x = 180$$

Replacing C with A and x with 60 gives

$$\begin{gathered} A + A + 60 = 180 \\ A + A + 60 = 180 \\ 2A + 60 = 180 \\ 2A = 120 \\ A = 60 \end{gathered}$$

Hence, the triangle is equilateral (all three sides are congruent). Since we are given that side AB has length 4, side AC also has length 4. The answer is (C).

19. Since the unit square is circumscribed about the circle, the diameter of the circle is 1 and the radius of the circle is $r = d/2 = 1/2$. This is illustrated in the following figure:

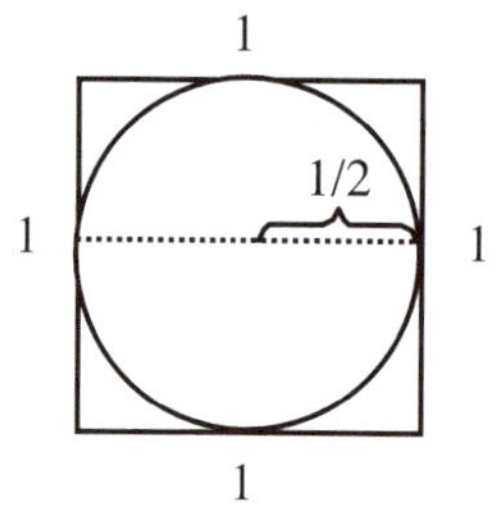

Now, the circumference of a circle is given by the formula $2\pi r$. For this circle the formula becomes $2\pi r = 2\pi(1/2) = \pi$. We are told that the circumference of the circle is $q\pi$. Setting these two expressions equal yields

$$\pi = q\pi$$

Dividing both sides of this equation by π yields

$$1 = q$$

The answer is (A).

20. Let x be the unknown side of the triangle. Applying the Pythagorean Theorem yields

$9^2 + x^2 = 15^2$	
$81 + x^2 = 225$	by squaring the terms
$x^2 = 144$	by subtracting 81 from both sides of the equation
$x = \pm\sqrt{144}$	by taking the square root of both sides of the equation
$x = 12$	since we are looking for a length, we take the positive root

In a right triangle, the legs are the base and the height of the triangle. Hence, $A = \frac{1}{2}bh = \frac{1}{2} \cdot 9 \cdot 12 = 54$. The answer is (D).

21. Since the average of $2x$ and $4x$ is 12, we get

$$\begin{gathered} \frac{2x + 4x}{2} = 12 \\ \frac{6x}{2} = 12 \\ 3x = 12 \\ x = 4 \end{gathered}$$

The answer is (D).

22. Recall that the average of N numbers is their sum divided by N. That is, average = sum/N. Since the average of x, y, and z is 8 and the average of y and z is 4, this formula yields

$$\frac{x+y+z}{3}=8$$
$$\frac{y+z}{2}=4$$

Solving the bottom equation for $y + z$ yields $y + z = 8$. Plugging this into the top equation gives

$$\frac{x+8}{3}=8$$
$$x+8=24$$
$$x=16$$

The answer is (C).

23. Let the two numbers be x and y. Now, a ratio is simply a fraction. Forming the fraction yields $x/y = 6$, and forming the sum yields $x + y = 21$. Solving the first equation for x yields $x = 6y$. Plugging this into the second equation yields

$$6y+y=21$$
$$7y=21$$
$$y=3$$

Plugging this into the equation $x = 6y$ yields

$$x=6(3)=18$$

The answer is (E).

24. Let $z\%$ represent the unknown percent. Now, when solving percent problems, "of" means times. Translating the statement "What percent of $3x$ is $6y$" into an equation yields

$$z\%(3x)=6y$$

Substituting $x = 4y$ into this equation yields

$$z\%(3\cdot 4y)=6y$$
$$z\%(12y)=6y$$
$$z\%=\frac{6y}{12y}$$
$$z\%=1/2=.50=50\%$$

The answer is (A).

25. Translating Column A into a mathematical expression yields $.10y$. Translating Column B into a mathematical expression yields $.40x$. Since $y = 3x$, Column A becomes $.10y = .10(3x) = .30x$. Since .40 is larger than .30 and x is positive, Column B is larger. The answer is (B).

26. Let x be the amount of water added. Since there is no alcohol in the water, the percent of alcohol in the water is $0\%x$. The amount of alcohol in the original solution is $40\%(30)$, and the amount of alcohol in the final solution will be $25\%(30 + x)$. Now, the concentration of alcohol in the original solution plus the concentration of alcohol in the added solution (water) must equal the concentration of alcohol in the resulting solution:

$$40\%(30)+0\%x=25\%(30+x)$$

Multiplying this equation by 100 to clear the percent symbol yields

$$40(30)+0=25(30+x)$$
$$1200=750+25x$$
$$450=25x$$
$$18=x$$

The answer is (E).

27. Except for the first term, each term of the sequence is found by adding 4 to the term immediately preceding it. In other words, we are simply adding 4 to the sequence 200 times. This yields

$$4 \cdot 200 = 800$$

Adding the 2 in the first term gives 800 + 2 = 802. The answer is (C).

We can also solve this problem formally. The first term of the sequence is 2, and since each successive term is 4 more than the term immediately preceding it, the second term is 2 + 4, and the third term is (2 + 4) + 4, and the fourth term is [(2 + 4) + 4] + 4, etc. Regrouping yields (note that we rewrite the first term as 2 + 4(0). You'll see why in a moment.)

$$2 + 4(0), 2 + 4(1), 2 + 4(2), 2 + 4(3), \dots$$

Notice that the number within each pair of parentheses is 1 less than the numerical order of the term. For instance, the *first* term has a 0 within the parentheses, the *second* term has a 1 within the parentheses, etc. Hence, the nth term of the sequence is

$$2 + 4(n - 1)$$

Using this formula, the 201st term is $2 + 4(201 - 1) = 2 + 4(200) = 2 + 800 = 802$.

28. For the first model, there are 5 options. So there are 5 different types of cars in this model. For the second model, there are the same number of different types of cars. Likewise, for the other two types of models. Hence, there are 5 + 5 + 5 + 5 = 20 different types of cars. The answer is (E).

This problem illustrates the *Fundamental Principle of Counting*:

> If an event occurs m times, and each of the m events is followed by a second event which occurs k times, then the first event follows the second event $m \cdot k$ times.

29. This is considered to be a hard problem. However, it is actually quite easy. By the definition given, the function @ merely cubes the term on the left and then subtracts 1 from it (the value of the term on the right is irrelevant). In each column, the term on the left is x. Hence, in each case, the result is $x^3 - 1$. This shows the two expressions are equal, and the answer is (C).

30.

$$((1-x)*)* = (1-x)*$$
$$(1-(1-x))* = (1-x)*$$
$$(1-1+x)* = (1-x)*$$
$$(x)* = (1-x)*$$
$$1-x = 1-(1-x)$$
$$1-x = 1-1+x$$
$$1-x = x$$
$$1 = 2x$$
$$1/2 = x$$

The answer is (A).

Study Plan

Use the list below to review the appropriate chapters for any questions you missed.

Equations: Page 259
Questions: 1, 2, 3
Algebraic Expressions: Page 322
Questions: 4, 5, 6
Number Theory: Page 37
Questions: 7, 8
Exponents & Roots: Page 304
Questions: 9, 10, 11
Fractions & Decimals: Page 243
Questions: 12, 13

Factoring: Page 316
Questions: 14, 15
Inequalities: Page 230
Questions: 16, 17
Geometry: Page 78
Questions: 18, 19, 20
Averages: Page 274
Questions: 21, 22
Ratio & Proportion: Page 285
Question: 23

Percents: Page 330
Questions: 24, 25
Word Problems: Page 376
Question: 26
Sequences & Series: Page 397
Question: 27
Counting: Page 405
Question: 28
Defined Functions: Page 28
Questions: 29, 30